Specific Intermolecular Interactions of Organic Compounds

Alexei K. Baev

Specific Intermolecular Interactions of Organic Compounds

Springer

Prof. Alexei K. Baev
International Sakharov
Environmental Uni
Dept. of Chemistry
Dolgobrodskaya Street 23
220009 Minsk
Belarus
alexeibaev@mail.ru

ISBN 978-3-642-21621-3 e-ISBN 978-3-642-21622-0
DOI 10.1007/978-3-642-21622-0
Springer Heidelberg Dordrecht London New York

Library of Congress Control Number: 2011941194

Printed on acid-free paper

Springer is part of Springer Science+Business Media (www.springer.com)

Preface

Since the time of discovery of noncovalency bonds the energy of the hydrogen bond in water remains the reliably measured value. Yet quite a number of versatile organic compounds have been synthesized; certain success is attained in the field of structural studies, in measuring intramolecular bonds energies, in carbohydrates, nucleic acids, and proteins chemistry, and in the development of the chemistry of organoelemental compounds. The lack in the scientific publications of the information on the energy of hydrogen bonds and specific intermolecular interactions between the simplest organic compounds (methane, ethane, ethylene, methyl ether, and formic acid) provides special difficulties in their investigation. This impedes the progress in the development of the theoretical concepts of the solution chemistry, in the establishment of correspondence between the structure and energy in various classes of organic compounds, the understanding of the solvates structure, and the specific features of interspecies interaction. Even more difficulties exist in the development in the field of solution chemistry of an independent supramolecular chemistry, the chemistry of intermolecular bonds corresponding to the association between two or more chemical species of large size. A vigorous development is observed in the physical chemistry of organoelemental compounds serving as a certain bridge connecting the inorganic and organic chemistry where frequently appear concepts contradicting the current notions existing in these fields, but which are key points for creating hypotheses, new ideas, and developing of theories inadequate to the existing ones. Let us mention here the statement of Sidgewick that the molecules with the electronic configuration of a noble gas are incapable of association, whereas our investigation established the existence in the vapor of dimers of carbonyl compounds of the elements from the subgroup of chromium, iron, and nickel.

In the physical chemistry of organic compounds and their solutions is firmly established and successfully developed the concept of the correspondence between the structure and the energy, determined by the investigation of molecular structures, of the condensed states, and of the distribution of electrons and charges on the atoms in the molecule, and by measuring the thermodynamic properties. One of the advantages of the thermodynamic methods consists in the successful application of the notion of the bond vacancies even at the lack of the sufficient information on the molecular structure and the distribution of the charge on the functional groups of the molecule. The discovery of the nature of the hydrogen bonds and specific intermolecular interactions, establishment of new previously unknown types, and their energies makes it possible to solve practical problems; in particular, it opens further opportunities in the development of the chemistry of biologically important substances and in the understanding of the empirical correlations between the thermodynamical properties and molecular parameters measured by various methods.

The key problem elaborated in the monograph is the concept that a carbon atom in the four-valency state not only is capable but actually forms the fifth coordination in the intermolecular interactions with an energy comparable with the energy of the hydrogen bond in water. The unique feature of the behavior of the carbon atom with an essentially unshared $2s^2$(c)-electron pair is its ability to act as a donor or an acceptor depending on the properties of the element participating in the interaction, to take part in the formation of solvate structure, and in the formation of specific interaction. These qualities are also inherent to molecules of amino acids, carbohydrates, and proteins.

We do not plan to consider comprehensively in this monograph all the problems of hydrogen bonds and the specificity of intermolecular interaction involving the pentacoordinate carbon atom, in all classes of organic and organoelemental compounds. We limit the scope of the discussion to the principal organic compounds: ethers, ketones, alcohols, carboxylic acids, and hydrocarbons, which always stimulated the appearance of new hypotheses and development of theories, essentially necessary for all classes of organic, organoelemental compounds, hydrocarbons, amino acids, and peptides. The information on fundamental aspects of this monograph may be obtained from its table of contents. At the same time the special importance of the theory of the hydrogen bonds and of specific intermolecular interaction, in substantiation of their types and energy determination, should be stressed. The refining of the current concepts in the field of hydrogen bonds and specific interactions in liquid and crystalline organic compounds, solutions of nonelectrolytes, of the theoretical regularities are necessary for the understanding of phenomena in organic and complex biologic systems and for further purposeful development of the key problems of the specific interactions and hydrogen bonds in the biologically important substances.

The critical analysis of the total scope of thermochemical and thermodynamic properties of alkyl derivatives of elements of the II–VI main groups established in the investigations of the author inevitably infringes on the views regarding the classical model of the sp^3-hybridized valence orbitals of the carbon atom whose attraction prevented the doubts in its imperfection (Chap. 1). As a result of this analysis the phenomenon of the reverse dative bond between the carbon atoms of the alkyl chain and the contacting central atom of the molecule was substantiated and also the capability of the terminal methyl groups to participate in the specific intermolecular interactions revealing the pentacoordinated state. These fundamental statements led to the founded rejection of the model of the sp^3-hybridized carbon atom based on the quantum-chemical calculations of the methyl derivatives of the elements of subgroups of boron and tin, and methyl compounds of nitrogen and oxygen, additionally confirming the capability of a pentacoordinate carbon to take part in the formation of the specific interaction. The implementation of these concepts into the theoretical footing of organic and inorganic chemists may be somewhat inertial due to the unconventional ideas and the approach to the theory of the hydrogen bonds and specific interaction. But the fact that here are compiled the self-consistent energies of the hydrogen bonds of various types and specific intermolecular interactions adequately reflecting their nature, the data, which are highly necessary for the specialists, gives a hope that they will be gratefully acknowledged.

The scientific idea of the reverse dative bond in organoelemental compounds with a hydrocarbon group confirmed with the quantum-chemical calculations in its development underlies the new concepts on the nature of the bonds in the alkyls of nontransition elements and in organic compounds with a heteroatom. This idea ensures the understanding of the chemistry of liquid solvents and liquid and solid nonelectrolytes (Chap. 2). The results of thermodynamic calculation are attractive for they provide an unambiguous solution of the problem of measuring the energy of the hydrogen bonds and the specific intermolecular interactions whose quantitative value remained unknown and even unestimated approximately. The growing number of carbon atoms in the alkyl chain results in more stable specific interactions formed by the terminal methyl group of the propyl fragment (Chap. 3). This fact is due to the weakening of the influence of the reverse dative bond on the terminal methyl group with the growing number of carbon atoms in the alkyl chain. From the large number of the established rules of the energy changes in the specific interactions of the saturated and unsaturated cyclic ethers and organic cyclic oxides, we report one among them regarding the regular decrease in the energy of low-stability hydrogen bonds between the hydrogen atoms of the CH_2 groups and the oxygen atom of the oxides with the growing number of the methylene groups in the molecular ring.

The alternative character of the theory and the necessity of its further development prompted the elaboration of the principles of proving various types of hydrogen bonds and specific intermolecular interactions and the procedures of thermodynamic calculation of their energies based on the enthalpies of vaporization that were suitable for all classes of organic compounds. These principles are reported in Chap. 3 and are elaborated in the successive chapters.

In Chap. 4, thermodynamic analysis is performed of the structural and thermodynamic characteristics of liquid and solid symmetric and unsymmetrical ketones with saturated and unsaturated alkyl chains, cyclic ketones, diketones, and oxyketones. As a result the self-consistent energy values were obtained for the specific intermolecular interactions and the H-bonds of low stability formed by the oxyketones. An important attention is paid to the new conception in the theory of the structure and stabilization of the molecules of ketones, alcohols, to the extrastabilizing effect of the isostructural methyl groups ensuring their reduced values of the enthalpy characteristics in the vaporization processes and the equal contribution into the enthalpy characteristic of the vaporization of four CH_2 groups with contiguous methyl groups. This principle is valid for various classes of organic compounds and reflects the analogy of the structural fragments at the number of CH_2 groups exceeding eight in ketones, alcohols, and obviously in more complex organic substances with the heteroatoms in the molecules and in the complex biological systems. A special attention is paid to the thermodynamic investigation of saturated and unsaturated ketones with a large number of carbon atoms and isostructural methyl groups.

In Chap. 5, various types of hydrogen bonds and specific intermolecular interaction were substantiated and their nature was revealed in liquid and solid monohydric and polyhydric alcohols with saturated and unsaturated chains, with the open and cyclic structure and isostructural alkyl groups. Theoretical problems are discussed and the regularities in the stability series of bonds are substantiated. The results obtained impelled to carry out thermodynamic investigation of isostructural and acetylene alcohols, and their solutions with polyatomic saturated and unsaturated ketones. The findings obtained confirmed the developed concept of the stabilizing effect of the isostructural methyl group.

In Chaps. 6 and 7, the theory of the hydrogen bond finds further development and are extended the theoretical notions and rules to liquid and solid aldehydes, esters, and carboxylic acids. The performed thermodynamic analysis of the enthalpy and structural entropy characteristics of vaporization of carboxylic acids associated in vapor resulted in valid energy values of various types of hydrogen bonds and specific intermolecular interactions of saturated and unsaturated mono- and dibasic acids, acids with an isostructural methyl group, hydroxycarboxylic acids, methoxy- and ethoxycarboxylic acids, peroxycarboxylic acids, and cyclic carboxylic acids. The established array of valid self-consistent energies of various types of hydrogen bonds and specific intermolecular interactions is illustrated by the regular series of their stabilization.

Chapter 8 illustrates the application of the developed theory of specific interactions to alkanes and alkenes and the established regularities of the specific interactions existing in their condensed state.

In writing the book the author tried to reveal the depth of the analyzed problem and to provide its clear description remembering that "formulas are not the essence of the theory, and a real theoretician uses them sparingly expressing with words all that is possible to express in words" (L. Boltzmann). The author is deeply grateful to Professor Josef Barthel (Regensburg University, Editor-in-Chief of the Journal of Molecular Liquids) for his suggestion and demand of the author to write a monograph basing on the theme of the article submitted to the *Journal of Molecular Liquids*. The ideas reported in this article underlie the development of the thermodynamic theory of the hydrogen bond and the specific interactions originating from the pentacoordinated carbon atom.

Minsk Belarus

Alexei K. Baev

Synopsis

The book presents the development of the thermodynamic theory of specific intermolecular interactions, nonstereotypic ideas, and approaches to the theory of H-bonding and specific interactions involving motivated concepts on a five-coordinated carbon atom based on the discussion of a wide spectrum of compounds (ethers, ketones, alcohols, carboxylic acids, and hydrocarbons). New types of hydrogen bonds and specific interactions are substantiated and on the basis of elaborated methodology their energies are determined and the system of interconnected quantitative characteristics of a stability of specific intermolecular interactions, reflecting both their nature and law-governed changes, is created. A new conception of the extra stabilizing effect of isostructural methyl group is considered in the theory of a structure and stability of molecules of organic compounds and in the energy of interactions.

Contents

Abbreviation

DA	Donor–acceptor bond
DE-Me	Dissociation energies of bond
Ddis.dim	Dissociation energies of dimmer molecule
$\overset{\curvearrowleft}{M \longrightarrow CO}$ -	Reverse dative bond on metal carbonyl
$\overset{\curvearrowleft}{Al \longrightarrow C}$	Reverse dative bond AlR_3
(η^E)	Dynamic viscosity
Pa (kPa)	Vapor pressure at Pascal
$T_{\text{m.p.}}$	Melting point
$T_{\text{b.p.}}$	Boiling point
$\Delta_{\text{vap}}H^0(298)$	Vaporization enthalpy
$\Delta_{\text{vap}}H^0(298)$ alc	Vaporization enthalpy of alcohols
$\Delta_{\text{vap}}H^0(298)$ cyc.c	Vaporization enthalpy of cyclic hydrocarbon
$\Delta_{\text{vap}}H^0(298)$ bz	Vaporization enthalpy of benzene
$\Delta_{\text{vap}}H^0(298)$ der.bz	Vaporization enthalpy of derivative of benzene
$\Delta_{\text{vap}}H^0(298)$ к. iso	Vaporization enthalpy of compounds with the number (k) isostructural fragments
$\Delta_{\text{vap}}H^0(298)$ c n	Vaporization enthalpy of compounds with normal structure
$\Delta_{\text{vap}}H^0(T)$ cyc.pr.	Vaporization enthalpy of cyclic propyl
$\Delta_{\text{vap}}H^0(T)$ pr	Vaporization enthalpy of propyl
$\Delta_{\text{vap}}H^0(T)$ c.bu.	Vaporization enthalpy of cyclic butane
$\Delta_{\text{vap}}H^0(298)$ frag	Contribution's energy of fragment at enthalpy
$\Delta_{\text{vap}}H^0(T)$ hc.a	Vaporization enthalpy of hydroxy carboxylic acids
$\Delta_{\text{vap}}H^0(T)$ po.a	Vaporization enthalpy of peroxy carboxylic acid
$\Delta_{\text{vap}}H^0(T)$mtc.	Methoxypropionic acid
$\Delta_{\text{evop}}H^0(T)$	Evaporation enthalpy
$\Delta_{\text{evap}}H^0(T)$ c.a.	Evaporation enthalpy of cyclic carboxylic acid
$\Delta_{\text{evap}}H^0(T)$ bz.a	Evaporation enthalpy of benzoic acid
$\Delta_{\text{melt}}H^0(T)$	Melting (heat) enthalpy

$\Delta_{pol}H^0(T)$	Enthalpy of polymorphic transformation
$\Delta_{sub}H^0(298)$	Sublimation enthalpy
$\Delta_{sub}H^0(T)$mbz.a.	Sublimation enthalpy of methylbenzoic acid
$\Delta_{sub}H^0(T)$ mbz.a	Sublimation enthalpy of methoxybenzoic acid
$\Delta_{sub}H^0(T)$ etbz.a	Sublimation enthalpy of ethylbenzoic acid
$\Delta_{sub}H^0(T)$ hbz.a	Sublimation enthalpy of hydroxylbenzoic acid
$\Delta_{sub}H^0(T)$ *iso*	Sublimation enthalpy of isostructural compound
$\Delta_{sub}H^0(298)$ d.i.p.	Sublimation enthalpy of diisopropyl
$\Delta_{sub}H^0(298)$ d.e.	Sublimation enthalpy of diethyl ether
$\Delta_{sub}H^0(T)$bz.a.	Vaporization enthalpy of benzoic acid
$\Delta_{sub}H^0(T)$mbz.a.	Sublimation enthalpy of methylbenzoic acid
$\Delta_{sub}H^0(T)$ebz.a	Sublimation enthalpy of ethylbenzoic acid
$\Delta_{sub}H^0(T)$ipbz.a	Sublimation enthalpy of isopropylbenzoic acid
$\Delta_{sub}H^0(T)$hbz.a	Sublimation enthalpy of hydroxybenzoic acid
$\Delta_{sub}H^0(T)$mtbz.a	Sublimation enthalpy of methoxybenzoic acid
C_n	Number of carbon atoms at alkyl chain
E.E.C.	Endoeffect coordination
Ddest. E	Destabilized effect
AN	Acetonitrile
FA	Formamide
AC	Acetone
NMF	*N*-methylformamide
DMF	*N,N*-dimethylformamide
DMSO	Dimethyl sulfoxide
Me	Methyl
Et	Ethyl
Pr	Propyl
Bu	Butyl
iso-p	Isopropanol
Et. al.	Ethanol
iso.m. bu.	*iso*-Methylbutanol
1-Bu	1-Butanol
1-al	1-Alcohol
iso-CH_3	Isostructural methyl group

Contribution's Energy at Enthalpy

DCH_3	*iso*-CH_3
DCH_2	CH_2-group
DCH	CH-group

Energies of Specific Intermolecular Interaction of Pentacoordinate Carbon Atom

$N–CH_3 \leftarrow N–$	$N(CH_3)_3$
$N–CH_2–CH_3 \leftarrow N$	$N(C_2H_5)_3$

Energy of Specific Interaction of Saturated Alkyl Series (Fragment)

$D–O \rightarrow CH_3$	Methyl
$D–O \rightarrow CH_3–CH_2–$	Ethyl
$D–O \rightarrow CH_2–CH_2–$	Ethylene
$D–O \rightarrow CH_3–CH_2–CH_2–$	Propyl
$D–O \rightarrow CH_3–(CH_3)–CH$	Ligand with a *iso*-CH_3-group
$D–O \rightarrow CH_2–(CH_2)_n–$	Fragment of alkyl chain
$D–O \rightarrow CH_3–(CH_2)nC–$	Alkyl chain of ketone
$D–O \rightarrow C{=}O$	Carbonyl group
$D–CH_2–CH_3 \rightarrow CH_2–CH_3$	Hydrocarbon
$DH_3C \rightarrow H–CH_2$	*iso*-CH_3
$DH_2C \rightarrow H–CH$	Methylene group
$DH–C \rightarrow H–C$	CH group

Energy of Specific Interaction of the Alkene Series Unsaturated Ligand (Fragment)

$D–O \rightarrow CH_2{=}CH–$	$CH_2{=}CH–$ fragment
$D–O \rightarrow CH_3–CH{=}CH–$	Propylene
$D–O \rightarrow CH_2{=}CH–CH_2–$	Propylene
$D–O \rightarrow CH_2{=}CH–CH_2–$	Propylene
$D–O \rightarrow CH_2{=}(CH_2)_n–$	Alkene chain
$DH–O \rightarrow CH{\equiv}C–CH_2$	Chain with the triple bond – $C{\equiv}C$
$D{=}HC–H_3C \rightarrow CH_2{=}CH–$	Unsaturated hydrocarbon (fragment)

Energies of Specific Intermolecular Interaction of Alkyl Series with Carbonyl Oxygen

$D{=}O \rightarrow CH_3–$	CH_3
$D{=}O \rightarrow CH_3–CH_2–$	C_2H_5
$D{=}O \rightarrow CH_3–CH_2–CH_2–$	C_3H_7
$D{=}O \rightarrow CH_3–(CH_2)_n–$	$CH_3–(CH_2)_n–$ alkyl series

Energies of Specific Intermolecular Interaction of Alkene Series with Carbonyl Oxygen

$D{=}O \rightarrow CH_2{=}CH–CH_2$	Propylene
$D{=}O \rightarrow CH_3–CH{=}CH$	Propylene

Energies of Specific Intermolecular Interaction of Ketone Series

$D{=}O \rightarrow CH_3–C–$	Acetonic form
$D{=}O \rightarrow CH_2–CH_2–C–$	Saturated fragment ketone
$D{=}O \rightarrow CH_3–CH(CH_3)–C–$	Fragment with isostructural CH_3 group
$D{=}O \rightarrow CH_3–(CH_2)_n–C–$	Alkyl chain of ketone

Energies of Specific Intermolecular Interaction of Ketene Series

D=O→CH_2=CH–C–	Propene
D=O→CH_3–CH=CH–C–	Butene

Energy of Hydrogen Bond of Etheric Oxygen

DH–O···H–O–	Alcohol, acid
D–O···H–C–	CH group
D–O···H–C–H	CH_2
D–O···H–CH_2	CH_3
D–O···H–N	Amino group
DN–H···N–H–	NH_3
D–O···H–C=	H–C= with double bond

Energy of Hydrogen Bond with Carbonyl Oxygen

D=O···H–C–	CH
D=O···H–CH–	CH_2
D=O···H–CH_2	CH_3; *iso*-CH_3
D=O···H–N–	Amino group

Chapter 1
Reverse Dative Bond in Organic Compounds, Molecular Complexes and Inconsistency of the sp^3-Hybridization Model with Respect to Carbon Atom

1.1 Introduction

Extensive investigation of organometallic and organoelemental compounds in various aggregate states resulted in the establishment of the thermodynamic properties of the vaporization process and enthalpy of formation, and simultaneously of their dissociation energies. The results of performed calorimetric measurements were accompanied with deep critical analysis of the methodological approach leading to reliability of the data of the thermodynamic studies. The rich experience of the calorimetric investigations shows that the final thermodynamic findings for the same compounds obtained by various research teams [1–3] are well consistent.

In many events, the established thermochemical properties and dissociation energies of organic complexes remain the only physicochemical characteristics of molecules providing the information on the features of intramolecular and intermolecular interactions. However, unlike the thermochemical properties of inorganic compounds the analogous experimental data of organoelemental compounds, in particular, of alkyl derivatives of nontransition elements, were not subjected to the critical analysis and the reasons remained unclear, and the reasons leading to inadequate character of the variation in the average cleavage energy of the chemical bonds at growing number of the carbon atoms in the alkyl chain. Analogous inconsistencies with the current concepts originating from the results of the study of inorganic and organic compounds are not isolated instances. However, the most important inconsistency corresponds to the fulfillment of the critical analysis of the total array of thermochemical and thermodynamic properties of alkyl compounds necessarily related to the classical conception of the model of the sp^3-hybridization of the valency orbitals of a carbon atom whose attractiveness prevented the doubts regarding its imperfection.

Here should be mentioned a fundamental paper of V.L. Talroze and A.L. Lubimova where for the first time cation CH_5^+ was described with a hypercoordinated structure [4]. However, these results remained unclaimed for over

A.K. Baev, *Specific Intermolecular Interactions of Organic Compounds*,
DOI 10.1007/978-3-642-21622-0_1, © Springer-Verlag Berlin Heidelberg 2012

40 years. In recent decades, fundamental works appeared on the developing nontraditional chemistry of hypercoordinated carbon [5] and nonclassical structures of organic compounds [6]. In the latter publications, the prediction approach was analyzed of isolobal relations by R. Hoffmann for considering nonclassical organic and organometallic structures.

In this chapter, the reliably established thermodynamic and thermochemical properties of organoelemental compounds are analyzed and the contradiction of these data is revealed with the forecasts based on the model of the sp^3-hybridization. As a result of the analysis, the phenomenon of a reverse dative bond is substantiated between the carbon atoms of the alkyl group and the contacting central atom, the inconsistency of the sp^3-hybridization model is proved, and the capability of the carbon atoms of terminal methyl groups to take part in the specific intermolecular interaction is revealed. The scientific conception of the phenomenon of the reverse dative bond in organoelemental compounds with the saturated and unsaturated hydrocarbon substituent in its development should underlie new notions on the bond nature in alkyls of nontransition elements and organic compounds containing a heteroatom and provide the understanding of the chemistry of liquid nonelectrolytes.

1.2 Some Problems of Liquid Molecular Complexes and Solutions

The contemporary trends in the investigation of molecular liquids are restricted to the study of organic compounds, characterized by the variety of hydrogen bonding. It is alcohols, organic acids, functional solvents, amines, and amino acids. The availability of the spatial structure in the liquid state with a three-dimensional network structure containing the chain and ring fragments [7–10] with the common donor–acceptor nature of the intermolecular hydrogen bonding is regarded as a very important special feature of these compounds. Different compounds of the homological series with the hydroxyl group are characterized by the chain intermolecular hydrogen bonding extended to long-range environment. As result of this longer range of the solvation at methyl alcohol and formic acid it is higher considerably than at ethyl, *n*-butyl alcohols and acetic acid with the characteristic for their ring structure of dimers [11]. However, an examination of the nature of the intermolecular interaction in liquid ketones, ethers rouses great difficulties. The understanding of the structure of the solvents and their solutions with the complicated molecules are becoming more problematical. That is why a liquid state of many classes of organic compounds is liable to oblivion. The nature of intermolecular interactions of these compounds is not examined on the following reasons. Firstly, historically an opinion existed that liquid organic compounds participated in the universal (nonspecific) interactions and consequently the investigation of these was complicated. Secondly, simplicity of perception and attractiveness of the comprehensive

sp^3-hybridization model of electron configuration of carbon atom at saturated hydrocarbons, alkyl compounds could not be questioned and discussed severely. However, it is not in conformity with the notion about saturated hydrocarbons as the compounds with universal intermolecular interactions are characterized by the values of the vaporization enthalpy more than 20 kJ mol^{-1}. It shows a necessity to develop new idea which will explain the nature of intermolecular interactions of liquid organic compounds identically. Demand for the new ideas is of the general interest for the examination of alkyl compounds, ketones, ethers, alcohols, and other more complicated compounds with functional groups. The reliably established experimental thermodynamic properties of the alkyl compounds are not corresponding to foretold values by sp^3-hybridization model of an electron configuration of carbon atom. This model cannot explain the reasons causing the increase in the vaporization enthalpy in the homological series, for example, of gallium alkyls compounds [12].

Compounds	$GaMe_3$	$GaEt_3$	Ga(n-Pr_3)	Ga(n-Bu_3)
$\Delta_{vap}H^0$ (298) kJ mol^{-1}	32.6 ± 0.1	43.1 ± 0.5	46.6 ± 0.5	51.6 ± 1.6

The regularities in the variation of the vaporization enthalpy are often ascribed to the growth in the molecular mass [10]. However, this suggestion is only a superficial description of the phenomenon and prevents revealing the nature of the intermolecular interaction. These regularities are observed in the homologous series of saturated and unsaturated hydrocarbons. The inconsistency of this common view is illustrated by the data on the vaporization enthalpy of various organic compounds (Table 1.1) demonstrating the lack of correlation between the molecular mass and the enthalpy characteristics, energies of the donor–acceptor bonds. Note that water possessing stable hydrogen bonds corresponds to the middle position among the compounds mentioned in Table 1.1.

Therefore, the new concepts should provide a possibility to describe from the general viewpoint the nature of the intermolecular interactions in molecular liquids and solutions applying the thermodynamical properties of the processes of vaporization and sublimation and employing them to the estimation of the energy of the hydrogen bonds and of the specific interactions of organic compounds.

Table 1.1 Vaporization enthalpy $\Delta_{vap}H^0$ (T) (kJ mol^{-1}) and energy of donor–acceptor bonds (kJ mol^{-1}) [12, 13]

Compounds	$\Delta_{vap}H^0$ (T)	DA	Compounds	$\Delta_{vap}H^0$ (T)	DA
n-Propane	18.77	–	2,2,2-Tri-fluoroethanol	43.97	13.5
n-Butane	22.39	–	Water	43.99	20.7
n-Pentane	25.77	–	*N,N*-Dimethylformamide	47.51	17.8
Acetone	29.0	7.0	n-Decanone	51.08	–
Tetramethyltin	32.0	–	Tetraethyllead	51.39	–
Acetonitrile	32.9	9.0	Dimethyl sulfoxide	52.88	24.1
Tetramethyllead	37.1	–	Formamide	64.98	24.1
Methanol	37.43	19.7	–	–	–

1.3 Contradiction of Thermodynamic Properties of Organoelemental Compounds to the sp^3-Hybridization Model

Organoelemental compounds forming a certain connecting bridge between inorganic and organic chemistry can be described both from the viewpoint of the utmost general conceptions and in the framework of the specially developed notions.

The synthesis of sandwich cene and then dibenzene, monocyclic, and polycyclic organoelemental compounds considerably shifted the direction of the studies of structures and the bonding therein. The performed simultaneously with the theoretical considerations of the intramolecular interactions experimental studies of physicochemical properties and thermodynamic characteristics made necessary an analysis of the nature of the intermolecular interactions present in various aggregate states and in the solutions of organoelemental compounds. Evidently, first in this respect were the studies on the metal carbonyls [14–16]. The discovery of strong intermolecular interactions in the solid and the liquid state and the proof of the presence of dimeric molecules in the vapor of nickel, iron, molybdenum, and tungsten carbonyls disproved the Sidgewick statement that the compounds with the 18-electron configuration of metal atoms were incapable of association. The latter statement contradicted the actual values of vaporization enthalpy of alkyl compounds of all nontransition elements. For quite a number of inorganic compounds, the nature is already established of the interaction in various aggregate states, of specific and nonspecific interactions in water and nonaqueous solutions. However, analogous investigations of organoelemental compounds were only at the beginning of development, and the obtained experimental findings were not analyzed or were treated superficially without establishing the causes and consequences, for instance, stating that heavier molecules were less volatile.

The fact that the models of *sp*-, sp^2-, and sp^3-hybridization explained the structure of molecules is incontestable. However, the sp^3-hybridization model for carbon atom in alkyl compounds prevents the possibility to understand a trivial fact of high volatility of methane, methyl compounds of nontransition elements as compared, e.g., with the ethyl analogs [13, 16]. The existing now wide range of experimental data on the thermodynamic properties of nonelectrolytes solutions corresponding to the mixing enthalpy [17, 18] was not discussed at all. At the same time, the extremal point on the diagrams of mixing heats with any combination of alkyl compounds of nontransition elements is used as a proof for the formation of a stable complex in the system based on the fact of the existence of a stable and volatile complex of the composition R_3MAsR_3 in alkyl compounds of the elements of the third and fifth groups. In solutions where the solvent and the solute are organic components with established types of specific intermolecular interactions still are conserved the hydrogen bonds and heterobonds in the molecular complexes [19–21].

Yet an attempt was made to explain the association in such systems by the formation of the carbenium ion [22]. Evidently, the key to the understanding of the features of the intermolecular interactions in alkyl groups and alkyl compound lies in the features of the electron density distribution in the molecules of these compounds. The calculations of the electron density distribution in alkyls B $(CH_3)_3$ and $N(CH_3)_3$ [23] performed using extended Hueckel method [24] showed within the limits of assumed approximations the different direction of the shift of the electron density. Its distribution in the methyl compounds of boron and nitrogen corresponds to the following charges on atoms [19, 25] $B(CH_3)_3$ – B(0.704), C (−0.525), H(0.097), $N(CH_3)_3$ – N(0.024), C(−0.113), H(0.035) and on methyl groups −0.234e and −0.008e, respectively. In the calculations of Gallais et al. [25] for the electron density distribution in a dialkyl ether, a negative charge is located on the oxygen (−0.228) and a positive charge on the methyl group (+0.114e).

The mentioned data for these compounds show that the methyl groups at boron and oxygen possess the charges of opposite signs, and in the series B–N–O the positive charge on the boron changes to a negative charge on the oxygen atom. This is governed by the fact that the acceptor properties of the central atom in the alkyls of elements in subgroups of zinc, boron, and silicon change to donor qualities in the elements of subgroups of nitrogen and oxygen in their alkyl compounds.

If it is valid that the increased energy of the M–C bond dissociation in going from $M(C_2H_5)_n$ to $M(CH_3)_n$ is caused by the chemical bond stabilization at increased shift of the electron density from the metal to carbon atom, then the presence of a strong bond in a methyl compound, for instance, of zinc (Zn–CH_3) should also involve the increased shift of the electron density to the carbon atom from the hydrogen and zinc resulting in an increased negative charge on the methyl group compared to an ethyl analog.

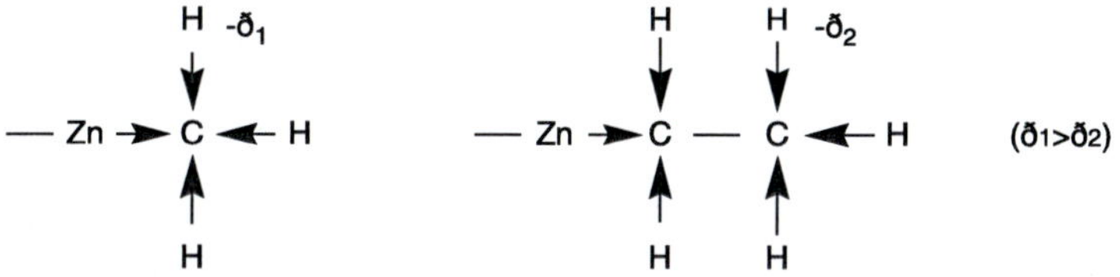

Consequently, the intermolecular interaction in the liquid dimethylzinc should be very strong and therefore the vaporization enthalpy of its methyl derivative being the integral characteristic of all existing intermolecular interactions [26] also should be essentially larger than the vaporization enthalpy of the ethyl and propyl analogs. This is a necessary conclusion of the consideration of the compounds of the elements of all main groups. However, these logical considerations based on the sp^3-hybridization model contradict the experimental values of the vaporization enthalpy, namely (Table 1.2):

$$\Delta_{vap}H^0(T)_,(ZnMe_2)<\Delta_{vap}H^0(T),(ZnEt_2)<\Delta_{vap}H^0(T),(Zn\,n-Pr_2).$$

Table 1.2 Dissociation energies (kJ mol^{-1}) of metal–carbon bond [1, 2] and vaporization enthalpy (kJ mol^{-1}) of alkyl compounds [12, 16]

Compounds	DE–Me	DE–Et	$\Delta_{vap}H^0$ (*T*)	
			EMe_n	EEt_n
ZnR_2	186.4	145	30.4 ± 0.2	38.8 ± 0.4
CdR_2	148.5	110.6	37.1 ± 0.07	46.0 ± 0.4
HgR_2	130.0	102.7	36.7 ± 0.1	–
AsR_3	238.4	185.9	27.7 ± 0.2	38.5 ± 0.7

This means that the dissociation energy of the bond Zn–C (E–C), the charge on the carbon atom of the methyl and ethyl group, and the evaporation enthalpy are not governed exclusively by the shift of the electron density from the zinc atom to carbon or from carbon atom to oxygen. Let us pay attention also to the fact that the visible correlation between the shift of the electron density in the molecule and the bond dissociation energy, for instance, Zn–C (E–C) in alkyls does not result in the correlation of each of these properties with the vaporization enthalpy in the homologous series of the element (Table 1.2). This makes it possible to conclude that the charge on the carbon atom of these alkyl groups and the charge on the zinc atom or the atom of the element of II–IV and V–VI groups is governed not only by the shift of the electron density, but by one more important property of the carbon atom that is not taken into consideration. This property is its capability to take part in the formation of the reverse dative bond at the expense of the essentially unshared $2s^2$-electron pair that significantly changes its charge and the charge of the atom of the metal or element contacting with this carbon in the molecule.

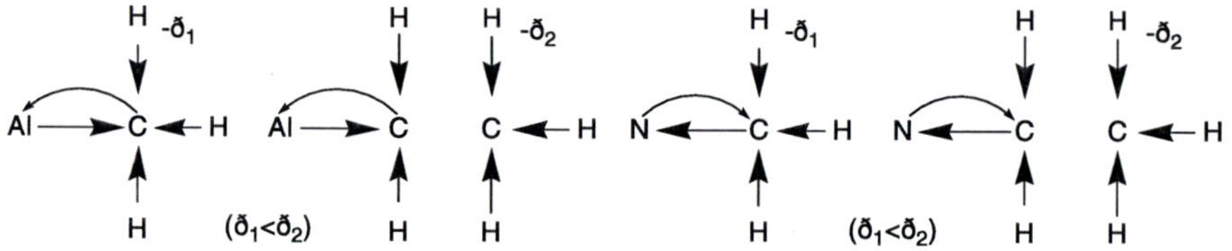

This means that the shift of the electron density from an atom, e.g., of aluminum to carbon on its 2*s*-orbital is accompanied with the subsequent transition of the part of the electron density on the $3p_z$-orbital of the aluminum atom. As a result the dissociation energy of the Al → CH_3 bond is stabilized. In the ethyl compound, the second carbon atom of the ethyl group impedes the shift of the electron density from the carbon atom contiguous to the aluminum, therefore the dissociation energy Al → CH_2–CH_3 is stabilized to a lesser degree than in the trimethylaluminum. It follows that the difference in the transmittance of the electron density from the carbon atoms of methyl and ethyl groups provides different stabilizing effect on the Al → C bond and increased negative charge ($-\delta_1 < -\delta_2$) on the methyl carbon atom of the ethyl group. Consequently, the latter methyl group forms a more stable specific intermolecular interaction, and the integral enthalpy characteristic of the ethyl compound is always larger than that of the methyl compound of the same element.

In the alkyl compounds of the elements from nitrogen and oxygen subgroups in connection with its electronegativity the shift of the electron density occurs along the alkyl chain occurs from the 2*s*-orbital of the carbon atoms of the alkyl group contacting with these elements to the orbitals of the atoms of these elements. The interacting carbon atoms of the chain in the ethyl, propyl, and other alkyl groups hamper the shift of the electron density from the carbon atom of the terminal methyl group. Consequently, the reverse transmission of the electron density occurs mainly on the contiguous carbon atom, and the carbon atom of the terminal methyl group in the ethyl and propyl substituent possesses a decreased negative charge and therefore has an increased acceptor power. As a result the bonds N ← C, O ← C are destabilized, and the specific intermolecular interactions $N–CH_3 \leftarrow N$, $N–CH_2–CH_3 \leftarrow N$ are stabilized, and the enthalpy of vaporization and sublimation of ethyl and propyl compounds grows.

Just the increased integral value of vaporization enthalpy indicates the increased donor properties of the carbon atom of the terminal methyl in the ethyl, propyl, and butyl groups ($ZnMe_2$ – 30.4, $ZnEt_2$ – 38.8, $Zn(n\text{-}Pr)_2$ – 42.1, $Zn(n\text{-}Bu)_2$ – 50.7 kJ mol^{-1}) in the alkyls of the elements of the II–IV groups and the increased acceptor properties in the alkyls of the elements of the V–VI groups [27].

The understanding of this problem requires the rejection of the sp^3-hybridization and adoption of the formation of the dative bond between the carbon atom of the alkyl group and the donor-metals or elements-acceptors contacting with it and belonging, respectively, to II–IV and V–VI groups.

The essence of this hypothesis consists in the notion that the dative bond is formed by hydrocarbon substituents with the atoms of all nontransition elements; it stabilizes complexes AR_k and affects their physicochemical properties.

Actually, the sp^3-hybridization model is inconsistent already because due to its universalism it cannot explain the intramolecular bonds in organoelemental compounds, regularities in the changes of the thermodynamic parameters, and the specificity of the intermolecular interactions.

1.4 Reverse Dative Bond in Organometallic and Organoelemental Compounds

The versatile structures of organoelemental compounds are manifested in diversity of the intramolecular interactions. In the founder of the sandwich compounds of ferrocene with equal the interatomic distances Fe–C on conditions that of the ring C_5H_5 into the formation of the $Fe–C_5H_5$ bond is restricted by the π-shell three types of bonds are presumed to be involved: covalent, donor–acceptor, and dative [28]. This originates from the fact of the presence in the C_5H_5 of an unpaired π-electron, two vacant π-orbitals, and two unshared pairs of π-electrons. Semiempirical and ab initio calculations showed that the effective charge was distributed as $Fe^{+\delta}(C_5H_5)_2^{-\delta}$. Although the charge value depends on the calculation procedure and varies in the range $\delta = 0.6 - 1.3$ [28] no experimental data contradict to this scheme.

In the dibenzenechromium $Cr(C_6H_6)_2$, covalent bonds are lacking and only donor–acceptor and dative bonds are present. According to quantum-chemical calculations, the dominant role belongs to the dative bonds and the charge distribution is like $Cr^{+\delta}(C_6H_6)_2^{-\delta}$. The study of bonds Cr–C polarity in symmetrical sandwiches $Cr(C_6H_6)_2$ and $Cr(C_5H_5)_2$ confirmed the low polarity in the mixed sandwich $Cr(C_5H_5)(C_6H_6)$. This is caused by the close location of the energies of interacting electronic levels at the dative interaction Cr–C_6H_6, therefore the considerable charge transfer along this dative bond is facilitated. An analogous close location of the energy levels is observed in $Cr(CO)_6$, where the acceptor character of the $(CO)_3$ fragment is more pronounced than that of C_6H_6. The strong acceptor properties of the CO group are manifested in the cyclopentadienyl carbonyl complexes of manganese and iron. The positive charge on the metal atom grows in going from $Mn(C_5H_5)_2$ to $Mn(C_5H_5)(CO)_3$ simultaneously with the growing charge on the carbon in the C_5H_5 ring. The π-acceptor nature of the carbonyl ligand is the most pronounced. The main bond energy values in gaseous carbonyls and carbon monoxide [28] showed that at the coordination of the CO molecules, the energy both of electrons C(1*s*) and O(1*s*) decreased because the CO ligand acted as a strong electron acceptor and at the formation of the dative π-bond M → CO significantly more electron density was transferred on the antibonding π-orbital of the ligand than was shifted to the metal atom at the formation of the donor–acceptor σ-bond M ← CO.

The M–CO bonds in the carbonyl compounds are fairly strong and belong to the strongest bonds in organometallic compounds. This is seen from the values of the average dissociation energy (DM–CO) and the energy of the cleavage of one CO group (D_1) [29] (Table 1.3).

At the same time the formation of dative bonds M → CO due to the transfer of the *d*-electron density from the metal atom to the carbonyl group does not exclude the strong association of metal carbonyls in the condensed state and the stability of gaseous dimers $M_2(CO)_{2n}$ owing to the formation of metal–metal bonds [29]

Table 1.3 Dissociation energies (kJ mol^{-1}) of M–CO bond and of dimer molecules of metal carbonyls [29]

Metal carbonyls					
Dissociation energy	$Cr(CO)_6$	$Mo(CO)_6$	$W(CO)_6$	$Fe(CO)_5$	$Ni(CO)_4$
DM–CO	105	149	176	116	144
$D_1(CO)_{n-1}$ M–CO	152	167	190	171	102
Ddis. dimer $M_2(CO)_{2n}$	25.9	29.7	33.9	32.3	23.9

if non-two-electron type. The mentioned concise list of the well-based examples of organoelemental compounds with a dative bond does not include the compounds containing saturated hydrocarbon groups.

In the series of ligands with the π-acceptor properties CO > C_2H_2 > C_2H_4 > PPh_3 the latter belongs to weaker π-acceptor. It is obvious that the appearance of the π-dative bond in the presence of a σ-bond occurs in the organometallic compounds in the case of ligands containing double bonds. With respect to energy the strength of the dative bond decreases at the replacement of the triple bond in the molecule C_2H_2 by the double bond in the molecule C_2H_4. In event of a dative bond involving a methyl group, its contribution into the stabilization of M–C bonds would be less significant.

The present idea that is not consistent with the current conceptions has not been formulated before not on account of the absence of experimental data. These data were abundant, but they contradicted the theory of the hybridization of valence atomic orbitals. However, it was suggested [28, 30] to extend both the hybridization theory and the theory of hypervalence bonds to the molecules where the coordination number of the central atom was ≤4, therewith for the hybridization only of the sp^3 type and for the nontransition elements of the first half of the period. The quantum-chemical calculations show [28] that the energy of the 2*s*-orbital of the carbon atom in methane is located considerably lower than the 1*s*-orbitals of hydrogen, and consequently in the CH_4 molecule two electrons $2s^2$(c) form an essentially unshared electron pair, and the C–H bonds arise mainly at the expense of the interactions 2*p*(C)–1*s*(H) [28]. Therefore, the C–H bonds in CH_4 belong to the type of orbital-deficient hypervalence and not to common covalence ones and the tetrahedral structure of the methane molecule is not due to sp^3-hybridization [28].

Therefore, the essentially unshared $2s^2$-electron pair of carbon atom is capable of formation of additional bonds, but these are less strong bonds than those formed by ethylene and especially acetylene with metal atoms. The general difference between acetylene, ethylene, and methane consists in the presence in the former of occupied (π) and vacant (π*) levels twice as much as in the ethylene, and the single carbon atom in methane (group CH_3 or R) possesses an essentially unshared electron pair. As a consequence, the acetylene and the ethylene form dative bonds with the atoms of transition metals having sufficiently large number of electrons (Pt, Ag). Inasmuch as the CO molecule possesses even more pronounced ability to form the dative bond M⟵C, its interaction with the atoms of transition elements extends to the elements with d^4-electrons (Cr, Mo, W). The presence of the given two types of bonds stabilizes the energy of the intramolecular interaction M⇄CO. However, the elements of the zinc subgroup (d^{10}-electrons) do not form such bonds. The difference consists in that the latter metals belong to the nontransition elements whose energy levels of *nd*-orbitals are situated by 10–15 eV higher, and the energies of p^3-orbitals by 10–20 eV lower as compared to the energies of valence *nd*-orbitals, the most important for binding the ligands. In other words, in the atoms of nontransition elements the energies *nd–np* and *np–ns* are widely separated, often by over 10 eV (in elements of V–VIII groups), and the difference in energies grows with the group number [30]. In typical ligands, the energies of the valence orbitals are as a rule somewhat lower than the energies of the valence *np*-orbitals of the elements of the main V–VIII groups. The vacant *nd*-orbitals of the central atom in the molecules of the coordination compounds of nontransition elements proved to be located at the orbitals of ligands. In the case of compounds of the fourth group elements, germanium, tin, and lead exist reliable data that their *nd*-orbitals take part in binding the ligands [28].

CH_4	4.3	NH_3	4.07	OH_2	4.79
CF_4	5.1	NF_3	2.9	OF_2	2.17

However, it corresponds to more fine contributions of the *nd*-orbitals. The same reasons sometimes impede the involvement of the valence s^2-electrons of the central atom into the bond formation: their energy frequently proves to be too low as compared to the energy of the valence orbitals of ligands [30]. However, another pattern is observed in compounds of carbon with fluorine. This is due to the fact that the energy level of the 2*ps*-orbital of fluorine is closer to the energy of 2*s*-orbitals of carbon, whereas the level of the 1*s*(H) is significantly higher than the energy level 2*s* of the carbon atom. Therefore, in the CF_4 molecule also $2s^2$-valence electrons of the carbon take part in the formation of bonds with fluorine, and thus the CF_4 molecule contains four covalent C–F bonds. Based on the growth of the mean energy (eV) of the two-center bond in going from CH_4 to CF_4, Korol'kov [30] concluded on transition from the hypervalent molecular structure in CH_4 to CF_4 structure with four covalent and respectively stronger bonds. The important point here is that the energy levels of 2*p*-orbitals of the nontransition elements are located far nearer to the carbon 2*s*-orbital than the energy levels of 1*s*(H) and 2*s*(C) orbitals. Inasmuch as in the carbon atom of the methyl group, $2s^2$-electron pair remains essentially unshared, depending on the direction of the electron density shift in the bonds M–C and E–C the carbon atom is capable to act as a donor or an acceptor due to the relative deficit or excess of the electron density. For instance, in methyl compounds of the elements of subgroups of zinc, aluminum, or germanium a significant shift of the electron density to the carbon atom should occur. In the formation of these compounds, the carbon atom of the methyl group obtaining an excess of electron density by the bond Al → C transfers a part from $2s^2$-electron pair to the free p_z-orbitals of the atoms of the elements $Al \rightleftarrows C$ forming a reverse dative bond. To the distribution of the electron density in the molecule of $B(CH_3)_3$ corresponds a negative charge on the carbon (−0.525e) and positive charge on the boron (0.704e), whereas in the $O(CH_3)_2$ molecule the negative charge is on the oxygen atom (−0.228), and positive one on each methyl group (0.114) [19, 23]. Similar situation should occur in the methyl compounds of all elements of the zinc, boron, and silicon subgroups, and also in compounds of oxygen and nitrogen with the distinction that the definite involvement of the essentially unshared $2s^2$-electron pair of carbon should be different.

As mentioned earlier, in going to the elements of the fifth and sixth groups the electronegativity of the element atoms increased, and therefore the essentially unshared $2s^2$(c)-electron pair of the methyl group acts as an acceptor in the formation of the reverse dative bond.

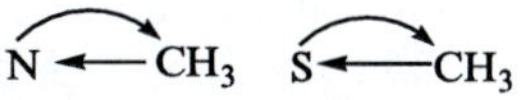

However, based on the value of the charge on the nitrogen atom (+0.024e) [23] or on oxygen (−0.228e) [24] it is not possible to estimate the contribution of the reverse dative bond into the strengthening of the N–CH_3, O–CH_3 bonds since the number of the transferred electrons is unknown. Yet the fact of the positive charge on the nitrogen atom indicates the considerable contribution of the reverse dative bond into the value of the dissociation energy of the N–CH_3 bond.

Hence, the reverse dative bond in the alkyls of the elements of the main groups is essentially different from the existing bond of analogous nature in the acetylene and ethylene compounds with the transition elements.

Limiting ourselves to the above description concerning the studied electronic structure of certain organometallic and organoelemental compounds we shall in the next section turn to a more detailed analysis of the results of quantum-chemical calculations of a number of methyl compounds of the elements from the aluminum and silicon subgroups.

1.5 Electronic Configurations of Alkyl Compounds and Inconsistency of the sp^3-Hybridization Model

The first reliable quantum-chemical calculations of the electron density of the $2s$, $2p_x$, $2p_y$, $2p_z$, and $3s$, $3p_x$, $3p_y$, $3p_z$-orbitals of boron and aluminum in the molecules AR_k were performed by Hückel method [31] using the modification developed by Hoffmann [32, 33]. In this method, the orbital is regarded as a linear combination of atomic orbitals of all valence electrons of the atoms in the molecule. Obviously, this version of the Hückel method does not utilize the hybridization and the orientation of the orbitals. The results of these calculations have a fundamental importance for they contain the necessary information on the electronic population of the p_z-orbital of the boron and aluminum atoms (Table 1.4).

Table 1.4 Electronic populations of atomic orbitals of central atom in molecules AR_k [23]

Orbital	BH_3	AlH_3	$B(CH_3)_3$	$Al(CH_3)_3$
s	0.912	0.812	0.812	0.505
p_x, p_y	0.825	0.497	0.552	0.206
p_z	0.0	0.0	0.140	0.111

The zero electronic population of the p_z-orbitals of the boron and aluminum in the hydrides AH_κ reveals that the hydrogen atom is incapable or nearly incapable to transfer the electron density to this orbital. Consequently, in the molecule BH_3 and AlH_3 the $2p_z$ and $3p_z$-orbitals of the boron and the aluminum remain unpopulated [34], and the reverse dative bond does not operate apparently due to the considerable energy difference between these levels and the 1*s*-level of the hydrogen atom. However, the electronic population of the p_z-orbital of the boron and aluminum in the AR_k molecule is 0.149 and 0.111 electron, respectively. This result shows the ability of the carbon atom of the methyl group to take part in the formation of the reverse dative bond, and this ability is incomparable with that of the hydrogen atom. Later quantum-chemical calculations [35] taking into account all electrons of the boron atom in the BH_3 molecule found some electronic population of the p_z-orbital of the boron atom (0.029), although lower than in the $B(CH_3)_3$ molecule. The high electron-acceptor property of boron and aluminum atoms is usually ascribed to the nearly unpopulated p_z-orbital atoms of these elements and the presence there of a high positive charge. The occupancy of the p_z-orbital of the boron (0.140) and aluminum (0.111) atoms in the molecules of $B(CH_3)_3$ and $Al(CH_3)_3$ shows the donor property of the carbon atom of the methyl group and the lower acceptor quality of aluminum atom compared to boron. These results raise doubts of adequacy of description of the nature by the model of the sp^3-hybridization of the carbon atom.

Hence the reverse dative bond of the carbon atom of the methyl group is substantiated by the calculations of the electron density distribution in the molecules of $B(CH_3)_3$ [23] and $Al(CH_3)_3$ [36]. In the $B(CH_3)_3$ molecule the boron atom operates as a strong π-acceptor, the boron p_z-orbital in $B(CH_3)_3$ acquires 0.3 electron, and the source of this charge is predominantly the carbon atoms of the methyl groups, namely, a covalent [19] and a reverse dative bond form.

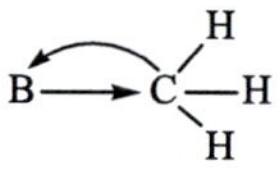

As a result the bond in $B(CH_3)_3$ is stabilized, and the excess charge on the methyl group is reduced. Consequently, the carbon atom of the methyl group possesses a weaker dative property (and the boron, a weaker acceptor property) in the formation of intermolecular bonds $B–CH_3 \rightarrow B$ in the liquid trimethylboron. In the trimethylaluminum, the p_z-orbital of the aluminum atom obtained 0.111 electrons from the carbon atoms of the methyl groups, and the reverse dative and covalent bonds are also formed.

The important feature of the performed quantum-chemical calculations [35] consists in the statement that the electron density distribution in the BH_3CO molecule is accompanied by the transfer of about 0.23 electrons from the carbon atom of the carbon monoxide to the proper p_z-orbital of the boron atom in the BH_3 molecule decreasing thus its positive charge. The electron transfer occurs mainly through an essentially unshared 2*s*-electron pair of the carbon atom to the p_z-orbital

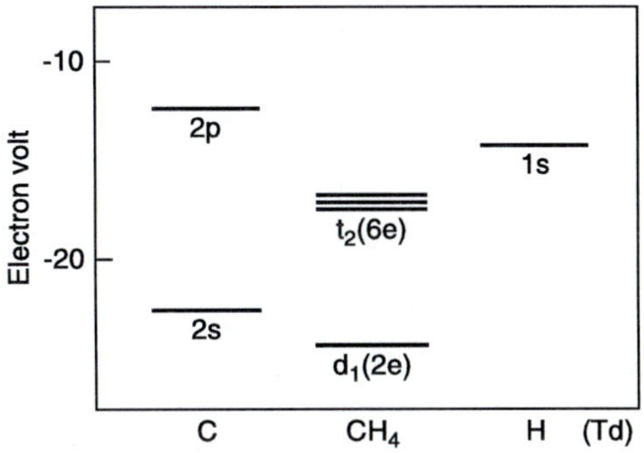

Fig. 1.1 The arrangement of 2s-, 2p-levels of carbon atom and 1s-level of hydrogen atom

of boron. Due to the electron transfer the carbon atom acquires a positive charge, and the boron atom, a large negative charge ensuring the reverse dative bond and a stable B–C bond.

The quantum-chemical calculations of the electronic structure of methane [28] showed that the 2*s*-level of the carbon atom was localized significantly lower than the 1*s*-level of hydrogen, therefore the $2s^2$-electron pair of the carbon atom remained essentially unshared (Fig. 1.1). Therefore, the bonds in the methane molecule formed as a result of the interaction $2p_C$–$1s_H$ and consequently the C–H bond belongs rather to the orbitally deficient hypervalent than to the common covalent bonds [28]. Since at the interaction of σ-orbitals of equivalent hydrogen atoms in the CH_4 molecule with three *p*-orbitals (Cartesian) the tetrahedron configuration is more feasible, the existing geometry of the CH_4 molecule is not necessarily connected with the sp^3-hybridization. Therefore, the essentially unshared 2*s*-electron pair of the carbon atom is capable of additional interaction alongside the four valence bonds providing the five-coordinated state of the carbon atom.

The lack of more informative quantum-chemical calculations of the electronic structures of AR_k molecules required more extensive research with the use of the modern approaches [37–44]. Nonempirical quantum-chemical calculations of the electronic structures of the trimethyl derivatives of nontransition metals $Al(CH_3)_3$, $Ga(CH_3)_3$, $In(CH_3)_3$, of $Al_2(CH_3)_6$ dimer, and $[Tl(CH_3)_3]_4$ tetramer [37] were performed with the use of two methods: restricted Hartree–Fock (RHF) and Becke–Lee–Yang–Parr (B3LYP) that was a version of the density functional theory (DFT) based on three-parameter functional including exchange and correlation contributions (local and nonlocal). Ab initio calculations by RHF method in the bases MINI and 6-31G* were carried out by GAMESS program [45, 46], and the ab initio calculations by B3LYP method were performed in the bases 6-31G* and LanL2DZ using GAUSSIAN-94 software [47].

The estimation of the accuracy of the results of the quantum-chemical calculations was done by evaluation of the internuclear distances between the metal and the carbon and the bond angles in $A(CH_3)_3$, $A(CH_3)_4$. The values of *R* (A–C) and bond angles calculated in all the three atomic bases were fairly consistent with the experimental data for all investigated methyl derivatives. Therewith obviously the theoretically established geometrical structure of each compound was the same as found experimentally, for instance, symmetry point group C_{3v} for $Al(CH_3)_3$. The performed quantum-chemical calculations of $Al(CH_3)_3$, $Ga(CH_3)_3$, $In(CH_3)_3$, dimer $Al_2(CH_3)_6$, tetramer $[Tl(CH_3)_3]_4$, and methane CH_4 provided the most important characteristics of the electronic structure: total energy (E), one-electron energies of the valence MO(ε), efficient atomic charges (q) corresponding to the overall electronic population of the valence orbitals of the atoms (V), multiplicity indices of two-center bonds (W), spatial distribution of the valence electron density. The values calculated by different methods are in good agreement; in total, RHF data and B3LYP data for three different bases (MINI, 6-31G*, LanL2DZ) are similar. The data obtained by the RHF method in different atomic bases are very close, those obtained by B3LYP method in bases LanL2DZ and 6-31G* are lesser and greater than the RHF data respectively, therefore we further use in the discussion the RHF data.

The directly calculated characteristics of the valence electronic structure of the trimethyl derivatives $A(CH_3)_3$ show the presence of the contribution from the configuration originating from the hypervalent interactions in the A–CH_3 fragment alongside the configuration of the covalent interactions corresponding to the sp^3-hybridization scheme of the valence orbitals of the carbon atom in CH_3. Consequently, in the valence electronic structure of $A(CH_3)_k$ one of the electron pairs of each fragment A–CH_3 is localized predominantly on the C atom and thus the possibility exists of the formation in the subsequent intermolecular interactions of the dative bond C(2e) → A(0e). In the valence electronic structure of compounds $[A(CH_3)_k]_n$ with bridging CH_3 groups, e.g., in the dimer $Al_2(CH_3)_6$ and tetramer $[Tl(CH_3)_3]_4$, the multiplicity indices of the two-center bonds A–C_b between the atoms A and bridging carbon atoms C_b are nearly the same as the multiplicity indices of the two-center bonds A–C_t with the terminal carbon atoms C_t (in particular, $W_{Al\text{–}C(b)} = 0.87$, $W_{Al\text{–}C(t)} = 0.89$). It means that in the three-center fragment A–C_b–A one two-center bond is formed in the same way as that with the terminal CH_3-group (A–C_t) (Fig. 1.2), and another one is the mentioned dative bond C(2e) → A(0e).

Therefore, the results obtained show that in the electronic structure of dimeric and tetrameric derivatives $[A(CH_3)_k]_n$ including bridging CH_3 groups the dative bonds under discussion are really existing.

Dative bond between the carbon atom C of the alkyl group R and the central atom A in compounds AR_k C(2e) → A(0e) may form involving, on the one hand, the valence electron pair of the carbon atom (in particular, $2s^2$(c)-electron pair), more properly, by the electron pair of the alkyl group R localized prevailingly on the carbon atom, and on the other hand, involving the vacant orbital localized

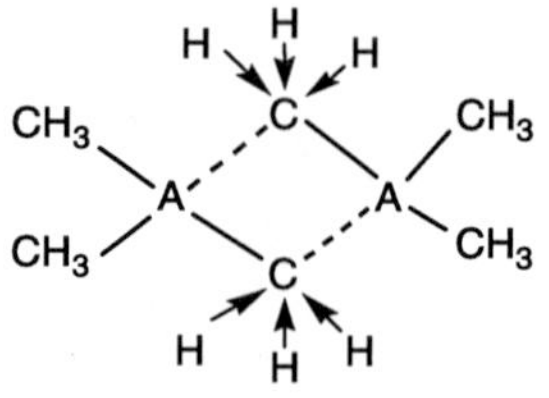

Fig. 1.2 Dimer molecule of trimethylaluminum

predominantly on atom A. In the form of the latter may appear, firstly, unoccupied one-center valence orbital of atom A or, secondly, the free multicenter orbital of compound AR_k, for instance, the free valence MO nonbonding or weakly antibonding with respect to the bonds A–R with the sufficiently low energy. These assumptions are based on the following statements. Firstly, even in the CH_4 molecule the valence $2s^2$-electron pair remains essentially localized on the C atom, and only two electrons (of the four valence electrons) of the C atom on its three valence AO $2p_x$, $2p_y$, $2p_z$ effectively take part in the hypervalent (electron-deficient) C–H bonds with all four hydrogen atoms (for the tetrahedron symmetry point group it is fully consistent with the group theory requirements). Therefore, it is probable that the same feature (the conservation of the pseudoatomic $2s^2$(c)-electron pair) is also characteristic of the electronic structure of the methyl group attached to atom A (A–CH_3) where the carbon atom forms four bonds (one C–A and three C–H) also employing two electrons on its three valence AO $2p_x$, $2p_y$, $2p_z$, moreover, due to the fact that the orbital negativity of the A atom at least does not exceed that of the hydrogen atom. Secondly, the electronic structure of carbonium cation $[CH_5]^+$ may be the most easily described by the combination of the hypervalent bonds like in the methane CH_4 between the carbon atom and four H atoms with the fifth dative bond of carbon with a hydrogen cation (formally) formed by the valence electron pair of the C atom and the vacant orbital of the hydrogen cation. Analogous structures belong to the other complexes of the pentacoordinate carbon discovered in the recent decade. Thirdly, the proper existence of methyl derivatives with the bridging (μ_2) CH_3 groups in the classic example of the $Al_2(CH_3)_6$ requires the presence in their structure, same as in the carbonium cation, of the pentacoordinate carbon atom. In the electronic structure of the bridging μ_2CH_3 group linked simultaneously to two A atoms, the carbon atom formed four bonds (with its three H atoms and one A atom) in the same way as does the terminal CH_3 group bonded to a single A atom, namely, using only two of its electrons on its valence AO $2p_x$, $2p_y$, $2p_z$. The fifth bond with the second A atom the carbon atom of the bridging μ_2CH_3 group formed apparently by the dative mechanism at the expense of its valence $2s^2$-electron pair and the vacant orbital of this A atom. Inasmuch as the tetrahedron configuration is the most feasible for the interaction of the σ-orbitals of the four equivalent substituents with the three Cartesian p-orbitals of the central atom, the considered geometry of CH_4 is not at all connected with the sp^3-hybridization [33]. Finally, fourthly, the constantly revealed experimentally efficient intermolecular interactions in alkyl (in particular, in methyl) compounds of the nontransition elements are fundamentally underlain by the same phenomena [38–44].

The energy of the interaction involving the pentacoordinate carbon atom in the $Al_2(CH_3)_6$ dimer proved to be high. We measured it from the dissociation enthalpy of the process $Al_2(CH_3)_6 = 2Al(CH_3)_3$ where two bridging bonds Al–CH_3–Al were broken (Fig. 1.3). Their energy of 44.2 ± 0.9 kJ mol^{-1} [48, 49] is four times larger than the energy of the hydrogen bond in liquid water. The revealed effective intermolecular interactions in the liquid alkyls of the elements of the main groups and their energies measured by thermodynamic properties are underlain by the operation of the essentially unshared $2s^2$-electron pair of carbon atom. The energy

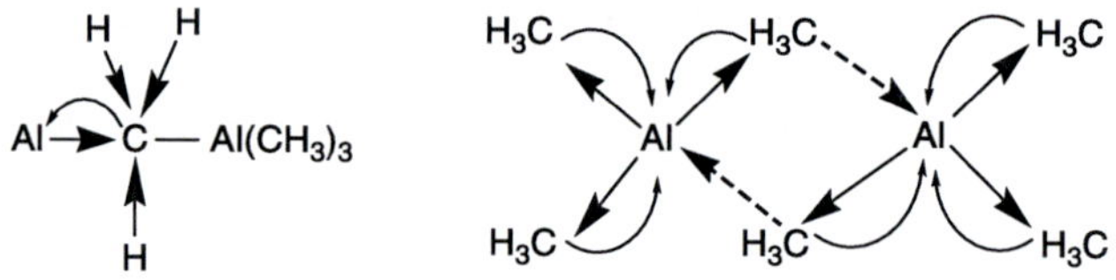

Fig. 1.3 Reverse dative bonds in dimer molecule of trimethylaluminum

of the intermolecular interactions significantly depends on the reverse dative bond existing in the molecules of the alkyl compounds (Fig. 1.3).

The divergence of the regular variations in the experimentally measured thermodynamic characteristics of the main groups elements from the properties predicted by the model of the sp^3-hybridization, the involvement of the essentially unshared $2s^2$-electron pair of the carbon atom in the dative bond and in the formation of stable specific interactions with the participation of the pentacoordinate carbon atom, the existence of dimer molecules and of the reverse dative bond in the alkyl molecules demonstrate the inconsistency of the above mentioned model [12].

Proceeding from the theoretical investigations C.A. Coulson regarded the model of the sp^3-hybridization not as a real phenomenon in the same way as the resonance structures. The advantage of the hybridization concept consisted in the conservation of the notion of a localized bond and of the coupling of two atoms involved into the formation of the bond [50]. The rejection of the sp^3-hybridization model is the route to the theory of the specific interactions in liquid and solid state, of molecular solutions; it leads to the development of the theory of the dative bond in organic and organoelemental compounds, in functional solvents. The thermodynamic aspects of the reverse dative bond were developed in [51–58].

1.6 Reverse Dative Bond and Steric Effect in Alkyls and Molecular Complexes

The notion of the dative bond in alkyl compounds [52, 59–61] confirmed by the quantum-chemical calculations of the electron density distribution finds its support and development in the mean dissociation energies of the element–carbon bond in the alkyl compounds obtained from the experimentally measured enthalpies of formation (Table 1.5). The maximum capability to the formation of a reverse dative bond should belong to the alkyl compounds with a methyl group, therefore they possess the highest mean dissociation values (Table 1.5). The replacement of the methyl by an ethyl group with a C–C bond affects the capacity of the carbon atom in contact with the atom of the nontransition element for the electron transfer from the $2s^2$ orbital of this carbon to the p_z-orbital of the atoms of the elements of the third and fourth groups and from the p_z-orbital of the atoms of the elements of the fifth

Table 1.5 Dissociation energies (kJ mol^{-1}) of element–carbon bond in alkyl compounds of II–VI groups [3, 16]

Compounds	DE–CH_3	DE–C_2H_5	Δ = DE–CH_3 − DE–C_2H_5
CdR_2	146 ± 4	113 ± 4	33.0
BR_3	379.9	344.5	29.4
AlR_3	283.3	272.6	10.7
GaR_3	256.4	224.9	31.5
GeR_4	258.2	242.9	15.3
SnR_4	226.4	194.7	31.7
PbR_4	161.4	129.6	31.7
NR_3	311.8	296.7	15.1
PR_3	285.5	230.2	55.3
AsR_3	238.4	185.9	52.5
SbR_3	223.5	179.8	43.5
OR_2	362.9	358.7	4.2
SR_2	303.6	293.5	10.1

and sixth groups to the $2s^2$ orbital of the carbon. Therefore, in these molecules the mean dissociation energies of the bond element–ethyl are always less than the mean dissociation energies of the bond DE–CH_3 (Table 1.5). Consequently, the difference Δ = DE–CH_3 − DE–C_2H_5 may be taken as a measure of the energy contribution of the reverse dative bond stabilizing the dissociation energy E–CH_3 in $E(CH_3)_n$ and E–C_2H_5 in $E(C_2H_5)_n$. This value of the stabilization energy of the E–CH_3 bond may be regarded as the relative value of the contribution of the reverse dative bond comparable for all alkyl compounds.

It is difficult to draw a clear distinction between the distribution of the electron density in the molecules of alkyl derivatives from the atoms of the elements of the II and IV groups and the reverse dative bond from the carbon atom, and also

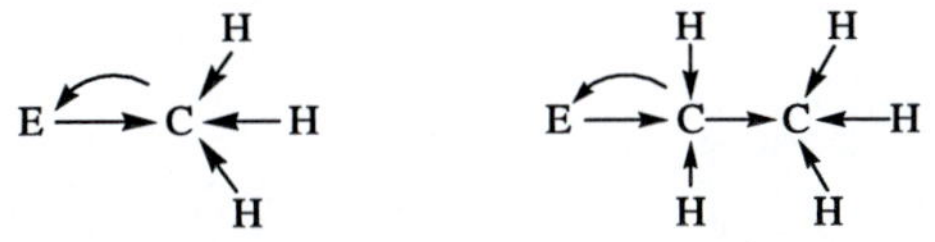

from the atoms of element of the V and VI groups to the carbon atom since these effects supplement each other.

Nonetheless, the differences in the dissociation energies of the bonds E–CH_3 and E–C_2H_5 reflect the influence of the second carbon atom of the alkyl group on the participation in the dative bond of the essentially unshared $2s^2$ electron pair of the carbon atom contacting with the atom of the corresponding element.

The lesser difference in the dissociation energies DE–Me – DE–Et in the nitrogen alkyls compared with the alkyls of phosphorus and arsenic (Table 1.5) is due to the weaker reverse dative bonds in the nitrogen compounds because of its lower donor ability. The low positive charge instead of negative one on the nitrogen atom (+0.024) and negative charge on each methyl group (−0.008) suggests the insignificant difference in the electronic occupancy in the triethylamine molecule. Even lesser difference in the dissociation energies of oxygen alkyls (4.2 kJ mol^{-1}) suggests the conclusion that the electron density distribution in dimethyl ether and diethyl ether differs insignificantly. Just this fact prevented obtaining reliable information in the framework of the approximations of the Gallais method [32] for the charges on oxygen, carbon, and hydrogen in the molecules of these compounds. On the contrary, significant difference in the electron density distribution is expectable for the methyl and ethyl compounds of phosphorus, arsenic, and also of antimony, tin, lead, boron, and gallium. The growth of the donor ability in the pair oxygen–sulfur is manifested in the difference in the energies of dissociation of bonds DS–Me – DS–Et = 10.1 kJ mol^{-1} compared to 4.2 kJ mol^{-1} in analogous oxygen compounds. Therefore, the large values of the dissociation energy of O–R bonds are caused by the shift of the electron density from the carbon atom to the oxygen. Inasmuch as the energy difference of the $2s^2$ level of the carbon atom and the *p*-orbitals of the elements grows in the series P–As–Sb, O–S–Se–Te, consequently the role of the reverse dative bond diminishes in going from arsenic to antimony.

Hence the difference in the dissociation energies of DS–Me and DS–Et is subjected to the influence of opposite effects. Its growth in going from NR_3 to PR_3 indicates the prevailing role of the reverse dative bonds, whereas the decrease in this value shows the essential diminishing influence of the dative bond due to the

Table 1.6 Dissociation energies (kJ mol^{-1}) of element–carbon bond in alkyl compounds of zinc, boron, and germanium subgroup elements

Compounds	DE–R		
	Zn–R	B–R	Ge–R
$M(CH_3)_n$	182 ± 1	371 ± 4	258.2 ± 3
$M(C_2H_5)_n$	151 ± 4	353 ± 2	242.9 ± 4
$M(C_3H_7)_n$	165 ± 12	360 ± 4	242 ± 2
$M(C_4H_9)_n$	162 ± 12	356 ± 2	247 ± 8
$M(C_6H_{13})_n$	–	343 ± 8	–
$M(C_7H_{15})_n$	–	343 ± 8	–
$M(C_8H_{17})_n$	–	343 ± 8	–
	Hg–R	Te–R	Sn–R
$M(CH_3)_n$	125 ± 1	217 ± 6	222 ± 4
$M(C_2H_5)_n$	109 ± 1	195 ± 6	192 ± 4
$M(C_3H_7)_n$	106 ± 1	199 ± 5	197 ± 8
$M(C_4H_9)_n$	–	195 ± 5	197 ± 4
$M(C_5H_{11})_n$	–	198 ± 5	–

increased difference between the energies of the *p*-orbitals of P > As > Sb and the energy of the $2s^2$-orbital of carbon.

The next important statement demonstrates that in going from methyl to ethyl and further to *n*-propyl, *n*-butyl compounds the mean dissociation energy M–R (E–R) decreases to the values negligibly varying after a distinct "*n*" index value (number of the carbon atoms in the alkyl group) (Table 1.6).

This fact indicates directly that, firstly, the growth of the number of carbon atoms in the alkyl group after a certain number (no more than three) virtually does not affect the electron density distribution between the central atom and carbon atom (and hydrogen) contacting with it, and secondly, that the growth of the number of carbon atoms in the alkyl group after a certain number (no more than three) virtually does not affect the change in the dissociation energy and the contribution of the reverse dative bond into the vaporization enthalpy and consequently into the energy of the intermolecular interaction. It should be here taken into consideration that the displacement of the electron density in an alkyl chain depends on the presence in the middle of the chain of an even and odd carbon atom [29]. This factor should affect the dissociation energy of the A–R bond as seen to a certain extent from the data in Table 1.6. However, in this case we mean finer effects masked by the accuracy in the determination of the dissociation energies of the A–R bonds. We shall consider this problem further by the example of the intermolecular interactions in the liquid alkyl compounds. The comparison of mean bond dissociation energies in the series of alkyls of the same element (Table 1.6) leads to a conclusion that their values decrease in going from the methyl to the ethyl compound and further to propyl, butyl, and up to octyl derivatives these values remain virtually constant.

The dissociation energies in propyl, butyl, and more complex compounds are not different to such extent that we would not be able to assume the equal contribution of the reverse dative bond in all these compounds into the dissociation energy. Therefore, it becomes clear that the length of the alkyl chain with more than three carbons does not affect the dissociation energy of the A–C bond. Consequently, the growing length of the carbon chain of the alkyl group does not produce a steric effect and does not affect the stability of the compound. Since the decrease in the dissociation energy in going from the methyl to the ethyl compound is accompanied with the weakening of the reverse dative bond, the steric effect may exist only between the carbon atoms contacting with the central atom of the compound AR_k. In other words, the notion of the influence of the steric effect at the growing number of the carbon atoms in the alkyl chain on the decrease in the dissociation energy in the series $Zn(CH_3)$–$Zn(n\text{-}C_4H_9)$, $Sn(CH_3)_4$–$S(n\text{-}C_4H_9)$, $Te(CH_3)_2$–$Te(n\text{-}C_4H_9)$ is incorrect. We illustrate these statements by the examples reflecting the fundamental difference in the mean values of the dissociation energy of A–C bond, selecting a molecule with a smaller radius of the central atom.

Figure 1.4 demonstrates the endoeffect of coordination at the successive increase in the number of hydrogen atoms in BH_3 whose covalent radii according to Pauling [62] amount to 0.34 and 0.30 Å. The dashed line drawn from the origin through the dissociation energy of the first hydrogen atom located in the flat structure of

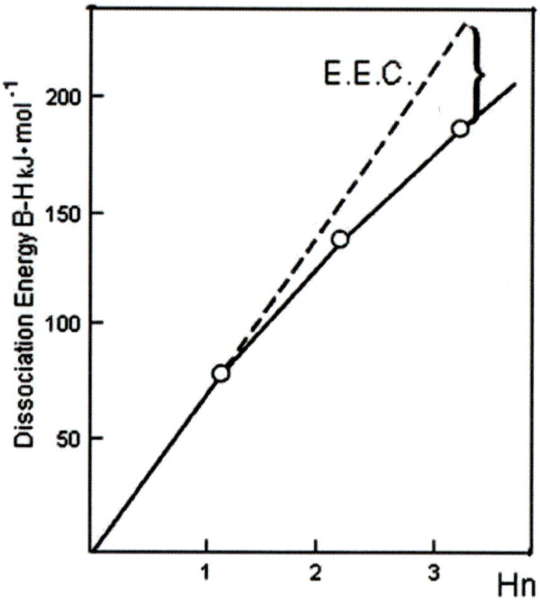

Fig. 1.4 Dependence of the dissociation energy B–H from the number of hydrogen atom in molecule BH_3

an equilateral triangle illustrates the changes in the dissociation energy in the series $BH > BH_2 > BH_3 > BH_4$ at the absence of the steric effect. Therefore, the difference between the extrapolated values (dashed line) and the experimentally obtained dissociation energies of B–H bonds (continuous line) corresponds to the magnitude of the coordination endoeffect in this case of hydrogen atoms.

The endoeffect value at the successive addition of two and three hydrogen to boron atom resulting in a flat triangular structure amounts to 8 and 22 kJ mol^{-1}, respectively, and at the addition of the fourth hydrogen atom, 35 kJ mol^{-1}.

In a flat molecule of trimethylboron at the same covalent radius of boron and the radius of the carbon atom of 0.77 Å, the coordination effect increased twice compared with analogous effect in the BH_3 molecule and amounts to 18 and 48 kJ mol^{-1} in $B(CH_3)_2$ and $B(CH_3)_3$, respectively. It means that the increased radius of the coordinated carbon atom directly interacting with boron results in growing of the Coulomb interaction between the carbon atoms of the methyl groups. Consequently, the distance between the coordinated atoms and the central boron grows from $r_{B–H} = 1.16$ Å to $r_{B–C} = 1.578$ Å [63]. The successive addition of the second, the third, and the fourth methyl group to the nitrogen atom with a covalent radius 0.70 Å resulted in the growth of the coordination endoeffect from 19 to 50 and 115 kJ mol^{-1}, respectively. Thus, the double increase in the radius of nitrogen compared to boron and reduced interatomic distances in the trimethylamine molecule $r_{N–C} = 1.451$ Å [63] practically does not lead to any difference in the coordination endoeffect of the methyl groups in the trimethylboron and trimethylamine.

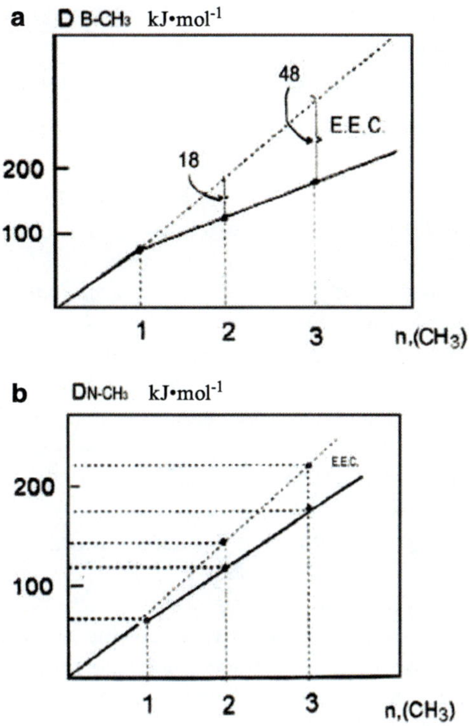

Fig. 1.5 Dependence of the dissociation energy B–CH_3 (**a**) and N–CH_3, (**b**) in molecules $B(CH_3)_3$, $N(CH_3)_3$ on the number of methyl groups, and the coordination endoeffect

An important fact consists in the equal value of the endoeffect in a flat molecule of $B(CH_3)_3$ and a trigonal pyramid of $N(CH_3)_3$ with the location of the methyl groups in the base forming a equilateral triangle. In these molecules are also different interatomic distances $r_{B\text{–}C} = 1.578$ Å and $r_{N\text{–}C} = 1.451$ Å. These facts lead to a conclusion that in the steric effect are involved only the atoms directly linked to the boron or nitrogen, hydrogen atoms in the BH_3, and only carbon atoms of methyl groups in the trimethylboron and trimethylamine (Fig. 1.5).

Hence the mean values of the dissociation energies of alkyl compounds listed in Table 1.6 include the steric effect contributed by the carbon atom contacting with the central atom of the compound AR_k.

The decrease in the donor properties of the carbon in the ethyl group of triethylamine originates from the effect of the second carbon atom of this group, therefore the cleavage energy of the ethyl group should be lower than the dissociation energy of the methyl group in the trimethylamine. In its turn the coordination endoeffect of the ethyl groups situated in the equilateral trigonal base of a pyramid structure additionally contributes to the value of the dissociation energy. Consequently, the donor properties of nitrogen atoms should decrease in the series $N(CH_3)_3$, $N(C_2H_5)_3$, and $N(C_3H_7)_3$. It follows from the analysis of the dissociation energy of the N–R bond (Table 1.6) that starting with the propyl compound these values vary insignificantly and therefore the propyl, butyl, and more complex

Table 1.7 Dissociation energy of bond B–N (kJ mol^{-1}) in alkyl complexes

Bond	Complexes				
	$H_3BN(CH_3)_3$	$(CH_3)_3BNH_3$	$(CH_3)_3BNH_2(CH_3)$	$(CH_3)_3BNH(CH_3)_2$	$(CH_3)_3BN(CH_3)_3$
DB–N	136.8	57.8	73.7	80.8	73.7

alkyl groups are characterized by hardly differing donor properties and ensure the equivalent electron density distribution in the NR_3 molecules. By the same reasons the steric effect observed in the base of the trigonal pyramidal structure of these molecules originating from the carbon atoms contacting with nitrogen should not interfere with the interaction of NR_3 molecules with the acceptor molecules BR_3. The high energy of the donor–acceptor bond in the complex $H_3BN(CH_3)_3$ should be attributed also to the high positive charge on the boron atom in BH_3. The energy of the donor–acceptor interaction between BR_3 and NR_3 is also due to the donor properties of nitrogen in the triamines which are partially consumed in overcoming the endoeffect of the fourth ligand coordination, in this case, of the nitrogen atom of the triamine, in the forming complex (Table 1.7).

The comparison of the energy levels of the core and valence electrons of the H_3BAR_k complexes showed that in the reaction of BH_3 with NH_3 and $N(CH_3)_3$ occurred a strong σ-interaction. It was demonstrated by the X-ray electron spectroscopy that the replacement of the hydrogen atom in NH_3 by the methyl group resulted in an insignificant decrease in the electron density on the nitrogen atom [64]. Nonetheless, the change of a single hydrogen for a methyl group in the NH_3 molecule sharply stabilized the complex, and that should be attributed to the shift of the electron density on the p_z-orbital of the boron atom governing the B–N in this complex. The replacement of the second hydrogen atom in the NH_3 molecule by a methyl group affects to a lesser extent the complex stabilization, and the substitution of the last hydrogen results in a slight destabilization. Thus, the destabilization of the B–N bond is due to the increased coordination endoeffect of carbon atoms in the fragment $N(CH_3)_3$ of the $(CH_3)_3BN(CH_3)_3$ complex in the base of the pyramidal structure of the trimethylamine and to the lower ability of the methyl group of this fragment to transmit the electron density as compared to the hydrogen atom. This resulted in a certain decrease in the charge on the nitrogen atom. At the replacement of three hydrogen atoms in the ammonia molecule by methyl groups, the positive charge on the nitrogen in $N(CH_3)_3$ amounts to +0.024e [25, 65, 66] (Fig. 1.6).

Due to the formation of the complex $(CH_3)_3BN(CH_3)_3$ the positive charge on the nitrogen grows by 0.792e. The charge on the boron atom decreases by 0.441e, and 0.293e is located on its p_z-orbital [19]. The electron density distribution in the complex is characterized by the location of 0.25e on the p_z-orbital of boron.

A special attention attracts the complexes with NH_3 where one hydrogen is replaced with alkyl groups with the normal or branched structures. The insignificant difference in the dissociation energies of the trimethylboron complexes with primary amines NH_2R containing methyl, ethyl, isopropyl, and *sec*-butyl groups illustrate the trend in their variation. This trend in its turn reveals the stabilization of the complex at the replacement of the methyl group by ethyl in the molecule of the

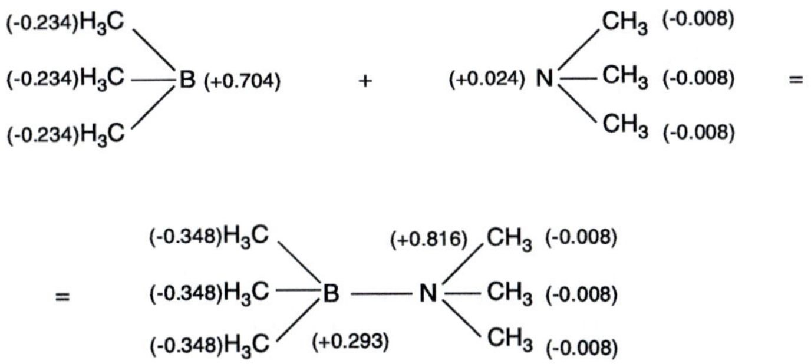

Fig. 1.6 Charge distribution in complexes $(CH_3)_3BN(CH_3)_3$, $N(CH_3)_3$, and $B(CH_3)_3$

Table 1.8 Dissociation energy of bond B–N (kJ mol^{-1}) at isostructural alkyl complexes

Complex	Bond DB–N	Complex	Bond DB–N
$(CH_3)_3BNH_2(CH_3)$	73.7	$(CH_3)_3BNH_2(i\text{-}C_4H_9)$	72.4
$(CH_3)_3BNH_2(C_2H_5)$	75.3	$(CH_3)_3BNH_2(tert\text{-}C_4H_9)$	52.4
$(CH_3)_3BNH_2(i\text{-}C_3H_7)$	72.8	–	–

primary amine due to the shift of relatively larger electron density from the ethyl group to the nitrogen atom as compared to the effect of the methyl group in the $(CH_3)_3BNH_2(CH_3)$. However, the replacement of the ethyl group by the isopropyl in the complex $(CH_3)_3BNH_2(i\text{-}C_3H_7)$ containing also same as ethyl two-carbon chain results in the decrease in the dissociation energy. Further replacement of the isopropyl group by the isobutyl moiety with the increase of the chain length to three carbon atoms is not accompanied with the larger shift of the electron density in the three-carbon chain, and the dissociation energy of these complexes remains unchanged. The location of two methyl groups on the carbon atom in contact with the nitrogen in the *tert*-butyl moiety of the complex $(CH_3)_3BNH_2(tert\text{-}C_4H_9)$ results in the reduction of the donor properties of the two-carbon chain to a great extent leading to a decrease in the dissociation energy by 22.9 kJ mol^{-1}. Thus, the isostructural methyl groups of the isopropyl, isobutyl, and *tert*-butyl moieties hamper the shift of the electron density to the nitrogen along two-carbon and three-carbon alkyl chains resulting in the decrease in the negative charge and donor properties (Table 1.8).

The comparative analysis of the electronic structure of trimethylaluminum and trimethylboron [64] by the method Xά-PB showed insignificant contribution of the Al 3p_z-orbital into the π-MO and a large charge on the aluminum atom [64]. This contribution originates from the participation of the reverse dative bond in the aluminum alkyls AlH_2R, and its decrease in the series $CH_3 > C_2H_5 > C_3H_7 > C_4H_9$ originates from the reduction in the energy of the dissociation complexes of alkylaluminum with alkyls of nitrogen and oxygen (Fig. 1.7) up to the propyl derivative. The virtually constant value of the energy of dissociation of complexes from tri-*n*-propylaluminum till tri-*n*-octylaluminum can be governed only by the constant charge of the aluminum atom and the electron density on its p_z-orbital.

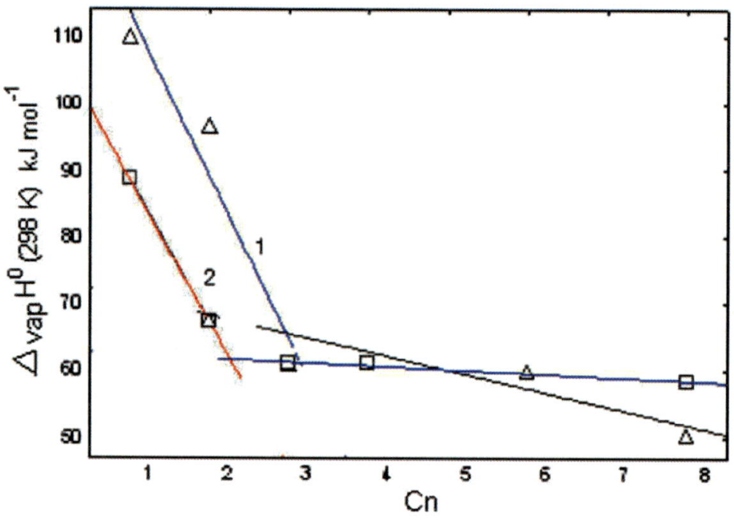

Fig. 1.7 Dependence of the dissociation energy of the donor–acceptor bond Al–O (1) and Al–N (2) in the molecular complexes $R_3AlO(C_4H_9)_2$ and $R_3AlN(C_2H_5)_3$ on the number of carbon atoms in the alkyl chain of AlR_3

Table 1.9 Dissociation energy of bond Al–N (kJ mol^{-1}) at alkyl complexes

Complex	Bond DAl–N	Complex	Bond DAl–N
$(C_3H_7)_3AlN(C_2H_5)_3$	61.1	$(C_3H_7)_3AlO(C_2H_5)_2$	65.0
$(i\text{-}C_3H_7)_3AlN(C_2H_5)_3$	75.3	$(i\text{-}C_3H_7)_3AlO(C_2H_5)_2$	71.2
–	–	$(i\text{-}C_4H_9)_3AlN(C_2H_5)_2$	93.6

The dependences shown in Fig. 1.7 permit reliable interpolation for the estimation of dissociation energy of the complexes $(C_5H_{11})_3AlN(C_2H_5)_3$, $(C_6H_{13})_3AlN(C_2H_5)_3$, and $(C_7H_{15})_3AlN(C_2H_5)_3$ equal, respectively, 60.5, 59.5, and 59.0 kJ mol^{-1}. These data show that independent of the flexible carbon chain up till the octyl group in the trigonal flat structure of the AlR_3 molecule the steric effect exists only between the carbon atoms contacting with the aluminum. The characteristic feature of the alkylaluminum complexes is their high stability in the case of branched alkyl groups (Table 1.9).

This fact in its turn indicates that the isostructural position of the methyl group governs the shift of the electron density in the carbon alkyl chain. The presence of an isostructural methyl group in the isobutyl moiety provides the stability of complex close to that of trimethylaluminum. Thus, the aluminum atoms in $Al(CH_3)_3$ and $Al(i\text{-}C_4H_9)_3$ have similar positive charges, and their $3p_z$-orbitals contribute similarly into the π-MO. This is confirmed by the effective charges of mercury in $Hg(Alk)_2$ varying from +0.023e in $Hg(CH_3)_2$ to −0.070e in $Hg(i\text{-}C_4H_9)_2$ [64]. The insignificant positive charge on the mercury atom in $Hg(CH_3)_2$ is also confirmed in [67]. Note that the majority of published results of the experimental investigations on the dissociation energies of the donor–acceptor complexes compiled in [19] are reported without indicated accuracy and are obtained in different solvents and by different methods, therefore in our analysis we relied on the most trustworthy data obtained by one procedure.

Thus, the thermodynamic analysis of a series of similar alkyl complexes of boron and aluminum with alkyl compounds of the elements from the fifth and sixth groups involving the data of quantum chemistry made it possible:

- To reveal the role of the reverse dative bond in the compounds AR_k, its part in the formation of the charges on the atoms of elements of subgroups of boron, nitrogen, and oxygen, to explain the fact of the presence electrons on the p_z-orbitals at the expense of the reverse dative bond
- To substantiate the appearance of the coordination endoeffect in alkyl compounds AR_k due to the carbon atoms of alkyl groups contacting with the central atom
- To reveal the special role of the isostructural methyl group in the stabilization of complexes, the shift of the electron density in the fragment of the molecule
- To understand the effect of the reverse dative bond on complex stabilization

The reported trends in the changes of the dissociation energies of alkyl compounds with the normal and branched structure of the alkyl groups refute the sterical effects obtained in [68] based on linear correlations for C_3H_7(1.38), *i*-C_3H_7(1.59), C_4H_9(1.59), and *i*-C_4H_9(1.69).

The significant result of the performed thermodynamic analysis is the fact that the compounds with the branched alkyl groups are insensitive to the steric effect. On the contrary, they provide more stable complexes. This fact suggests a skeptical position with respect to the correlation analysis in the chemistry of organoelemental and organic compounds considering the steric effects [68]. The true accounting for the steric effect may be based on a detailed analysis of the dissociation energy, thermodynamic properties, and the data on the electron density distribution in compounds under consideration.

The thermodynamic properties of compounds and the quantum-chemical calculations discussed in this chapter showed that the considered reverse dative bond is a fundamental property of organometallic, organoelemental compounds of the elements of main groups and of organic compounds. The reverse dative bond in alkyls stabilizes the molecules of alkyl compounds AR_k and affects all their physicochemical properties, the complexing in solutions and in vapor, their dissociation energies, and the energies of specific interactions in condensed state, in solutions of nonelectrolytes, and the solvation processes.

References

1. Skinner HA, Connor IA (1987) In: Liebman IF, Creenberg A (eds) Molecular structure and energetic. New York
2. Pilcher G (1982) In: Ribeiro da silva MAV (eds) Thermochemistry and its applications to chemical and biochemical systems. Dordrecht, Boston, Lancaster

3. Rabinovich IB, Nesterov VN, Telnov VI, Sheiman MS (1996) Thermodynamics of metalorganic compounds. University of N. Novgorod, N. Novgorod
4. Talrose VL, Lubimova AL (1952) Secondary process in the ion source of the mass spectrograph. Doclady A.S. SSSR, Russia, p 86, 99
5. Olan GA, Surya Prakash GK, Wlliams RE, Field LD, Wade K (1987) Hypercarbon chemistry. Wiley-Interscience/Wiley, New York/Toronto, ON/Singapore
6. Minkin VI, Miniaev PM (1985) No classical structure organic compounds. Univercity of Rostov on Done, Rostov on Done
7. Pullman A, Bertho H, Gissner-Prettre C, Hinton JF, Harpool D (1978) A ab initio study of a nuclear side: Iridine. J Am Chem Soc 100:3991
8. Ohtaki H, Funaki A, Role BM, Reibnegger GJ (1983) The structural of liquid formamide studied by means of X-ray diffraction and initio LCGO-MO-SCF calculation. Bull Chem Soc Jpn 56:2115
9. Ohtaki H, Itoh S (1983) Structure of liquid *N,N*-dimethylformamide studied by means of X-ray diffraction. Z Naturforsch 40A:1315
10. Ohtaki H, Ishiguro SI (1994) In: Mamontov G, Popov AJ (eds) Chemistry of nonaqueous solution. Current progress. VCH, New York, pp 178–226
11. Krestov GA, Vinogradov VI, Kesle IyM (1986) Modern problem of solution chemistry. Nauka, Moscow
12. Baev AK (1969) Problems of chemical nature of phase transformation. General and applied chemistry, vol 1. Vaisheshika Educating, Minsk, pp 197–206
13. Baev AK (1969) Phase condition and complex formation ability of halogenide metals. General and applied chemistry, vol 1. Vaisheshika Educating, Minsk, pp 207–218
14. Baev AK (2001) Russ J Izvestiya Vischich Ucheb Zaved Chem Chemo Technol 44:3–13
15. Rid R, Prausnic Dzh, Shervud T (1977) The properties of gas and liquids. New York
16. Baev AK (1987) Chemistry of gas-heterogenic systems of elementorganic compounds. Science and Technology, Minsk
17. Belousov VP, Morachevakei AG (1970) Heat mixing of liquids. Chemistry, Leningrad
18. Belousov VP, Morachevakei AG, Panov MYu (1981) Heat properties of non-electrolyte solutions. Chemistry, Leningrad, p 264
19. Gurianova EN, Goldshtain IG, Romm IP (1973) Donor-acceptor bonds. Chemistry, Moscow
20. Nikiforov MYy, Alper GA, Durov DA, Korolev VP et al (1989) Nonelectrolyte solutions in liquids. Nauca, Moscow
21. Korolkov DV (1992) Electron structure and properties of compounds no-transition elements. Chemistry, St. Petersburg
22. Mathieu J, Panico E (1975) Mechanisms reactionnels a chimie organique. Hermann, Paris
23. Ehrenson S (1968) Theory of electron effect at formation of complex of trimethylamine with trimethylborium. Theor Chim Acta 10:193
24. Boer FP, Newton MD (1966) Molecular orbital's for boron hydrides parameterized from SEF-model calculations. J Am Chem Soc 88:2361–2366
25. Gallais F, Voigt D, Laba JF (1965) Generalization comprehension "electro negativity" at calculation of effect electro negativity of atom in saturated molecule. J Chem Phys 62:761
26. Baev AK (1972) Complex formation ability halogenides of second – six groups periodical system. General and applied chemistry, vol 5. Vaisheshika Educating, Minsk, pp 35–51
27. Baev AK (1992) Structural-energetic aspects of association interactions of alkyls elements of I–VI groups. Russ J Chem Thermodyn Thermo Chem 1:143
28. Schustorovich EM (1978) Chemical bonds in coordination compounds. Knowledge, Moscow
29. Baev AK (1986) Structure and energy of metals carbonyls. Vischaia Schkola, Minsk
30. Korolkov DV (1977) Development of theory valence. Chemistry, Leningrad, p 194
31. Kato H, Yamaguche K et al (1965) Electron structure some galogenides and alkyl borium and aluminum. Bull Chem Soc Jpn 36:2144
32. Hoffmann RK (1963) Distribution of electron levels in alkanes. Chem Phys 39:1397

33. Hoffmann RJ (1964) Theory of ecsictones at the notion of secondary quantization. Chem Phys 40:2474
34. Armstrong DR, Perkins PG (1969) Calculation σ–π and π–σ* – transfer at vinyl borium. J Chem Soc A: 1044
35. Moirean MC, Veillard AH (1968) Quant mechanic calculation of the barrier of inner rotation. II Molecule borazine BH_3-NH_3. Theor Chim Acta 11:344
36. Levinson KA, Perkins PG (1970) Methylaluminium compounds. II. The electronic structure of the trimethylaluminium. Theory Chim Acta 17:15–17
37. Sheldrick GM, Sheldrick WS (1970) Crystal structure of trimethylthallium. J Chem Soc Sec A Inorg Phys Theor Chem: 28–30
38. Baev AK (2001) Pentacoordinated carbon atom in intermolecular interaction in solutions of alkyl compounds of non-transition elements. In: Book of abstract. VIth international symposium on carbonion chemistry. Marburg, Germany, p 75
39. Baev AK (2001) Thermodynamics and specific intermolecular interactions in solutions of non-electrolytes with pentacoordinated carbon atom. In: Book of abstract. 27th international conference on solution chemistry. Vaals, Netherlands, p 36
40. Baev AK (2001) Specific interactions in solvates solutions of non-electrolytes with iso-structural molecules components and pentacoordinated carbon atom. In: Book of abstract. VIIIth international conference. The problems of solution and complex formation in solution. Ivanovo, Russia, p 246
41. Baev AK, Korolkov DV (2003) Phenomenon of dative bond in elementorganic compounds. In: Book of abstract. XVth FECHEM conference on organometallic chemistry. University of Zurich, Switzerland, p 350
42. Baev AK (2003) Novel approaches to the structure of functional solvents and elementorganic compounds with saturated hydrocarbon ligands and their solvates in binary systems. In: Book of abstract. 27th international conference on solution chemistry. Debrecen, Hungary, p 79
43. Baev AK (2004) Insolvency of sp^3-gibrydixation model of carbon atom and chemistry of non-electrolytes solutions. In: Book of abstract. IXth international conference. The problems of solution and complex formation in solution. Ivanovo, Russia, pp 62–64
44. Baev AK (2004) Lack of co-ordination of thermodynamic properties of elementorganic compounds with sp^3-gibrydixation model of carbon atom and its insolvency. In: Book of abstract. Russian research symposium on thermochemistry and kalorimetry. Nizhny Novgorod, Russia, pp 40–41
45. Becke AD (1993) Density-functional thermochemistry. II. New mixing Hartri-Fock and local density-functional theory. J Chem Phys 98:5648
46. Bartolotti LJ, Parr RG, Tal Y (1998) Thomas–Fermi theory from a perturbative treatment of the hydrogenic limit energy density functional a possible link to local density functional theory. Int J Quant Chem 34:509–515
47. Frisch MA, Trucks GW, Schigell HR, Gill PM, Johnson BG, Robbs MA Gaussian 94, Revision G.3
48. Baev AK, Vitalev SE (1989) Thermodynamic vaporization and association of triethylaluminium. Vesti Akademii Navuk Belarusii 2:58–63
49. Vitalev SE, Baev AK (1993) Thermodynamic vaporization and association of trimethylaluminum. Russ J Izvestiya Vischich Ucheb Zaved Chem Chemo Technol 36:36–40
50. Coulson CA (1961) Valence. Oxford University Press, Oxford
51. Baev AK (1997) Thermodynamic analyses of structural changes in systems $Zn(CH_3)_2$-M $(CH_3)_2$ (Te, S, Se). Russ J Inorg Chem 42:662–672
52. Baev AK (1997) Insolvency sp^3-hybridization and reverse dative bond of alkyl compound and molecular complexes. In: Book of abstract. International symposium "the research of Georg wilting relevance to chemistry today". Heidelberg, Germany, p 21
53. Baev AK (1998) Solvation with pentocoordinate carbon and complication in solution of alkyls the third group elements. In: Book of abstract. VIIth international conference. The problems of solution and complex formation in solution. Ivanovo, Russia, p 2

54. Baev AK (1998) Pentacoordinated carbon in liquid state and solutions of the alkyls main group element. In: Book of abstract. XVIIth international conference on organometallic chemistry. Munich, Germany, p 18
55. Baev AR (1998) Thermodynamic study of structural changes in system dimethylcadmium-dimethylmercury. Russ J Inorg Chem 43:1325–1331
56. Baev AK, Cherniak IN (2000) Equilibrium liquid-vapor in systems trimethylaluminium-trimethylarsin and trimethylindium-trimethylphosphin. Russ J Phys Chem 74:980–985
57. Baev AK (2001) Thermodynamic parameters of solutions of binary system dimethylzinc-trimethylgallium. Russ J Phys Chem 75:1180–1182
58. Baev AR (2001) Thermodynamics of solutions trimethylaluminum and trimethylgallium. Russ J Coord Chem 27:637–640
59. Baev AK (1993) Unknown particular interacting and hypothesis of sigma-reverse bond in II–VI groups elements alkyls. In: 1st journal of organometallic chemistry conference on applied organometallic chemistry. Munchen, Germany, p 115
60. Baev AK (2009) Phenomenological significance of nonidentical sp^3-hybridization model and nature of interaction in chemistry of liquid non-electrolytes and chemical thermodynamics. In: Book of abstract. Xth international conference on physical and coordination chemistry of porphyrines and their analogous. Ivanovo, Russia, July 2009, p 86
61. Baev AK (1996) Pentacoordinated carbon atom at intermolecular interactions. Russ J Coord Chem 22:399–402
62. Pauling L (1970) General chemistry. W.N. Freeman and Company, San Francisco, CA
63. Vilkov LV, Mastrukov VS, Sadova NI (1978) Determination of geometrical structure of free molecules. Chemistry, Leningrad
64. Nefedov VI, Vovna VI (1989) Electron structure of organic and elementorganic compounds. Nauka, Moscow, p 175
65. Gallas F, Loth P, Labaire JF (1966) Study "concrete condition" σ-bond of some molecules tricoordinated phosphorus. J Chem Phys 63:413
66. Gallas F, Loth P, Labaire JF (1967) About some products adding type of 1:1 bromidealuminum and triethylaluminium with aliphatic sulphides and phosphines. Compt Rend C_{265}: 363
67. Jolly WL, Bomben KD, Eyermann CJ (1984) Core-electron binding energies for gaseous atoms and molecules. Atom Data Nucl Data Tables 31:433–493
68. Galkin VI, Salchov PD, Cherkasov PA (1995) New possibility for application of the model of frontal sterical. I. Theoretical calculations of the sterical factor at the theory of collisions and structure of activated complexes. Russ J Org Chem 29:1921–1926

Chapter 2
Nontraditional Approaches to Thermodynamic Analysis of Liquid Organic Compounds

2.1 Consequences of the Essentially Unshared $2s^2$-Electron Pair of Carbon Atom

The structure of the liquid state of one among the simplest compounds, methanol, is regarded as formed by hydrogen bonds involving the hydroxy group. This concept is preserved although a lot of publications discuss this compound [1]. The methyl group of the CH_3OH molecule possessing an essentially unshared $2s^2$-electron pair and a pronounced donor or acceptor properties depending on the interacting functional group in the contacting molecule of the nearest surrounding has valuable features necessary for the understanding of the nature of specific interactions in liquid alcohols, functional solvents, ketones, ethers, and carboxylic acids. The differences in the structures of the liquid organic compounds and their solutions govern directly the nature of the specific interactions which depend on the definite bond vacancy of the molecule, functional group, and evidently on the CH_3-group possessing the mode of the specific interaction inherent to it. The latter originates from the pseudoatomic essentially unshared $2s^2$-electron pair. Therefore the functional solvents, in particular, *N*-methylformamide, whose molecule contains various functional groups of different character, should form in the liquid state versatile types of specific interactions. Hence it is necessary to develop new approach to the investigation of liquid structures and to the analysis of their thermodynamic and other physicochemical properties. Apparently, none of the existing investigation methods provide unambiguous information on the structure of the solution and the solvent. For instance, in the case of close energy values of two specific interactions in solution of dissimilar character it is impossible to determine their structure based only on the spectral data or quantum-chemical calculations [2–4]. The processing of the X-ray pattern for liquids gives reliable information only on the mean number of neighbors in the first coordination shell. However, the X-ray patterns also cannot be solved in case of close energy values of different types of the specific interactions. The investigation of finer details of the structure meets more difficult problems of the ambiguous character of a method [2–4]. The radial distribution function (RDF)

A.K. Baev, *Specific Intermolecular Interactions of Organic Compounds*,
DOI 10.1007/978-3-642-21622-0_2,

converts the spatial distribution of atoms and molecules into a smoothed homogeneous function. This approach restricts the possibility to use the concept of molecule for understanding the structure of the short-range ordering in the liquid, e.g., at the formation in the alcohols of the hydrogen bond H-O•••H-O and the specific interactions $H–O\rightarrow CH_3$ formed by the essentially unshared $2s^2$-electron pair of the carbon atom from the methyl group in the case of small difference in their energies. At the same time two different configurations, in particular, with the larger energy difference between the hydrogen bond H-O•••H-O and the specific interaction $H–O\rightarrow CH_3$, may be described by virtually similar RDFs. This fact restricts the use of the X-ray diffraction for the evaluation of the short-range ordering in liquids. The X-ray diffraction studies extended, for example, the understanding of the structure of the liquid dimethylformamide [5]. Yet the suggested structure did not take into consideration the existence of the specific interactions $O\rightarrow CH_3$, $N–H_3C\leftarrow N$ [6] that should fundamentally correct the assumed structure and also should improve the quantum-chemical calculations of the hydrogen bond energy in liquid formamide amounting according to the calculations to 67.3 kJ mol^{-1} for the chain structure and to 63.2 kJ mol^{-1} for the ring structure [1, 7, 8]. Consequently in the probabilistic structure of liquid, the sp^3-hybridization model of the carbon atom should be rejected and the presence of the essentially unshared $2s^2$-electron pair forming the specific interaction and supplying the carbon atom with the pentacoordinate state should be taken into account. Because of this ability of the carbon atom the methyl group acquires the property of a functional group with a free bond vacancy; this fact is confirmed by the quantum-chemical calculations. It is seen by an example of the dimeric molecule of trimethylaluminum that this functional group enters into coupling chemical interaction between the carbon atom of the methyl group and the aluminum atom of another molecule, and this interaction is strictly localized on the definite orbitals of the reacting species [9]. The pseudoatomic essentially unshared $2s^2$(c)-electron pair of the carbon of the $Al(CH_3)_3$ involved into the interaction with the aluminum atom of a molecule in the nearest surrounding is localized on the p_z-orbital of the aluminum atom. The interaction is of the donor–acceptor character and belongs to the new type of specific interactions $Al\leftarrow CH_3–Al$. This type of donor–acceptor interactions extends the concept of the intermolecular interactions. Note that the weak hydrogen bond in the liquid ammonia is also a specific interaction. The same donor–acceptor bond forms the carbon atoms of methyl groups with the metal in the metal alkyls AR_k. The interactions $A–R\rightarrow A$ are formed between the molecules of the nearest surrounding in the liquid alkylmetals. The presence of the essentially unshared $2s^2$(c)-electron pair in the alkyl derivatives AR_k provides many possibilities of the formation of specific interactions of the same nature, of different types varying by energies and charges on the atoms of elements and carbon. The negative charge on the carbon in the methyl group in trimethylboron $B–CH_3$ (−0.234) considerably exceeds the charge on the similar group in the trimethylamine $N–CH_3$ (−0.023) and results in the high energy of the specific interactions $B–H_3C\rightarrow B$ and $N\rightarrow CH_3–N$ [10]. Thus, the pseudoatomic $2s^2$(c)-electron pair of the methyl group of trimethylboron is capable of donor qualities, and the methyl group of trimethylamine, of acceptor properties forming the donor–acceptor coupling interactions. Under certain conditions, the methyl groups form less stable interactions $M{-}CH3^{-\delta}{}_1 \rightarrow CH3^{-}\delta2{-}E\,(-\delta_1 > -\,\delta_2)$ with existing inequality of the negative charges $-\delta^1 > -\delta^2$.

This type of specific interactions of low stability may arise in the binary systems of metal alkyls [11, 12]. The treatment of the structures of liquid organic compounds becomes complicated with the growing number in the molecule of functional groups whose bond vacancies form the specific interactions. For the multitude of the nonelectrolyte solutions, the concentration concept of Krestov et al. [1] is the most attractive. The concept is based on the dominating structure of the solution components. In the total range of solutions, certain compositions are marked where the structure of the solvent is governed by the structure of the pure components or the structure of their mixtures, solvated molecules of the solvate. The fundamental value of this concept is the admission of the formation of solvated species. The advantage of this concept consists in regarding the system as the product of the interaction of molecules of similar compounds; therefore the solvent molecule with all its bond vacancies dictates the solvate structure. Hence the properties of such solutions originate from the changes in the energy of the specific interactions in the liquid component and the specific interactions newly formed in the solvate. This characteristic feature of the nonelectrolyte solutions restricts the free choice of structures in the interpretation of the observed solution properties as a function of its composition. Inasmuch as the solvation is understood as the sum of energy and structural changes in the process of the solvent molecule transition into the liquid phase of the solution [1], the establishment of the thermodynamic properties of the reverse process makes it possible to evaluate the energies of the specific interactions existing in the solvate using the information on the vapor composition [13, 14]. Obviously, in this case, the energy of specific interactions does not take into account the long-range solvation and the nonspecific interactions. At the analogy of the structure of the liquid component and the solvate, the entropy as a structural characteristic corresponds only to the energy difference in the specific intermolecular interactions arising in the structures of the initial liquid compounds [15].

Nontraditional approaches to the thermodynamic and other physicochemical properties of liquid and solid organic compounds of saturated and unsaturated series, organoelemental compounds, and large compounds with varying number of functional groups, bond vacancies, of nonelectrolyte solutions aimed at the evaluation of the energy of specific interactions, the hydrogen bonds included, are based on the following principles [16–18]:

1. Well-grounded rejection of the model of sp^3-hybridized carbon atom in organic and organoelemental compounds
2. The existence of an essentially unshared $2s^2$(c)-electron pair of carbon atom, the existence of a pentacoordinate carbon atom taking part in the formation of specific interactions
3. All bond vacancies of the molecules of organic compounds participate in the specific interactions

4. The enthalpy (entropy) of sublimation, vaporization, melting, and polymorphous transitions are connected with the number of the broken specific interactions and their energies
5. The dominating influence of the structure of the solution components. Several regions of composition are separated where the solvent structure is defined by the structure of pure components or the solvated species of the mixture [13–15]

Whereas the knowledge of the structure of liquid and solid organic compounds provides the understanding of the thermodynamic properties [7, 8], the knowledge of the thermodynamic characteristics of vaporization makes it possible to evaluate the energies of the specific intermolecular interactions. The fundamentally important point consists in the possibility by well-founded estimation of the bond vacancies of the molecule to get to the evaluation of the energy of the specific interactions employing the enthalpy characteristics of vaporization. In event of liquid water having a tetrahedral structure in the crystal and in the liquid state at the temperature close to the melting point, the molecule has four free bond vacancies forming four hydrogen bonds whose cleavage at the vaporization allows the evaluation from the enthalpy characteristic of vaporization (43.96 kJ mol^{-1}) [5] the energy of the hydrogen bond at 10.99 kJ mol^{-1} [$\Delta_{vap}H^0$ (298 K)/4]. At the higher temperature, tetrameric and other forms of water molecules with a reduced number in the ring form in the liquid state of water, therefore the hydrogen bond energy is an average value. Taking into account that monomer molecules are going over into the vapor the temperature increase is accompanied by some insignificant stabilization of the hydrogen bond due to the decrease in the number of molecules in the cycle. Still its different energies measured at the nonstandard temperature originate from the difference in the thermal motion. The nonspecific interactions are ignored in the thermodynamic calculations of the energies of the specific interactions and the hydrogen bonds for the energies of nonspecific interactions are as low as the error limits of the experimental measurements of the enthalpy characteristics.

2.2 Specific Interactions of Liquid Alkyls and Their Energies

Regarding the structure of a liquid as a unique system whose elementary volume is meaningful when it contains the minimum number of species obeying the statistical laws [1], the knowledge of the number and types of the bond vacancies, of interacting species provides a possibility to estimate the energy of the specific interactions. This interpretation of the structure of the molecular liquid requires definite model concepts incorporating the characteristic features of the initial species. For instance, the evaluation of the energy of the hydrogen bond in liquid water (cited above) required the knowledge of the number of its bond vacancies, of the vaporization enthalpy and the vapor composition. This approach results in a system of self-consistent energies of the specific interactions in molecular liquids.

Here the most important point is the accounting for the structure of the molecule and its bond vacancies. This approach is already applied fundamentally to functional solvents, alcohols, and carboxylic acids. However, for all these compounds only the number of hydroxy or carboxy groups was taken into consideration. The quantum-chemical calculations performed for the evaluation of the energy of the hydrogen bonds also took into account only the OH groups [7, 8]. Therefore, for estimation of the energy of previously unknown types of specific interactions in the molecular liquids of organoelemental compounds we use the above named principles. In this connection let us consider the structure of the molecules of organoelemental compounds, the vapor composition, and the structure of the crystal and the liquid.

Alkylmetals of the zinc subgroup have a linear form [19]. The alkylmetals of the boron subgroup have a structure of planar triangle. The investigations by X-ray and electron diffraction and by IR spectroscopy showed that alkyls of gallium and indium existed in vapor as monomer molecules. The trimethylaluminum is present in vapor in dimers [20, 21], triethylaluminum is less dimerized in vapor [21], and the partial pressure of $Al_2(n\text{-}C_3H_7)_6$ in vapor is incomparably lower [22]. In vapor trimethyl- and triethylaluminum was investigated by X-ray and electron diffraction and by IR spectroscopy [23–26]. The alkylaluminum exists in dimers both in the liquid and in the solid state [27]. The independent electron diffraction studies [11, 12, 19] showed that trimethylindium is a tetramer in the crystal and a monomer in vapor and in liquid. The study of the crystalline trimethylindium by the X-ray diffraction resulted in a conclusion on the existence of a group of four indium atoms and trigonal units $In(CH_3)_3$ located nearly in a plane normal to the plane of In_4. The tetrameric unit is held by the methyl groups placed between the indium atoms. As a result the carbon atom is in the pentacoordinate state [28].

The study of the structure of the trimethylindium crystal [29] furnished the structural parameters and interatomic distances in the distorted trigonal bipyramid (Fig. 2.1) with an indium atom in the center that formed two specific interactions In-$CH_3 \rightarrow In$ with the molecules of the nearest surrounding and three proper methyl groups with the indium atoms of the same contacting molecules. Consequently, the trigonal bipyramidal structure of the crystal of $In(CH_3)_3$ includes five pentacoordinate carbon atoms. Trimethylthallium also forms analogous crystal structure. The study of the vibration spectrum of trimethylgallium [28, 30] also indicated the presence in the liquid state of dimeric molecules like those of $Al_2(CH_3)_6$. However, the investigation of this compound by X-ray diffraction and IR spectroscopy [30] did not found dimers in the liquid trimethylgallium. The existence of the trimethylaluminum in the dimeric form allows a conclusion that this molecule has four bond vacancies (Fig. 2.2), and the coordination number of the aluminum atom is also four. In the series of alkyl compounds of the aluminum subgroup elements, the number of coordinated molecules grows, and in the liquid and crystal state of alkyl compounds in the series gallium–thallium round the central bipyramidal structure five molecules are coordinated (Fig. 2.2). The metal atom in alkyls of the elements of zinc subgroup MR_2 of the linear structure is capable of acquiring additionally two electron pairs at the contact with the methyl groups of the other molecules

Fig. 2.1 The structure of trimethylindium in solid state

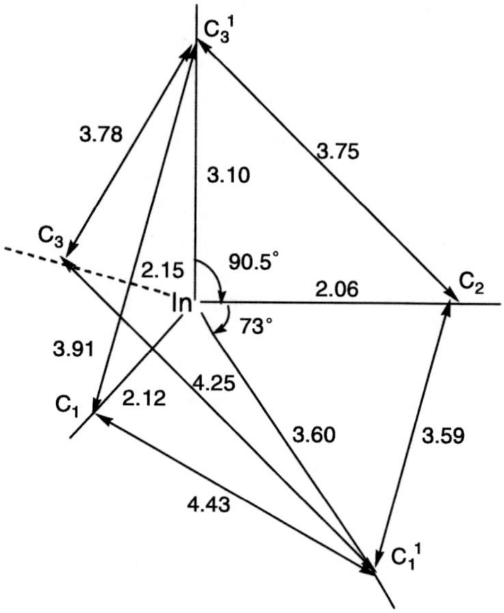

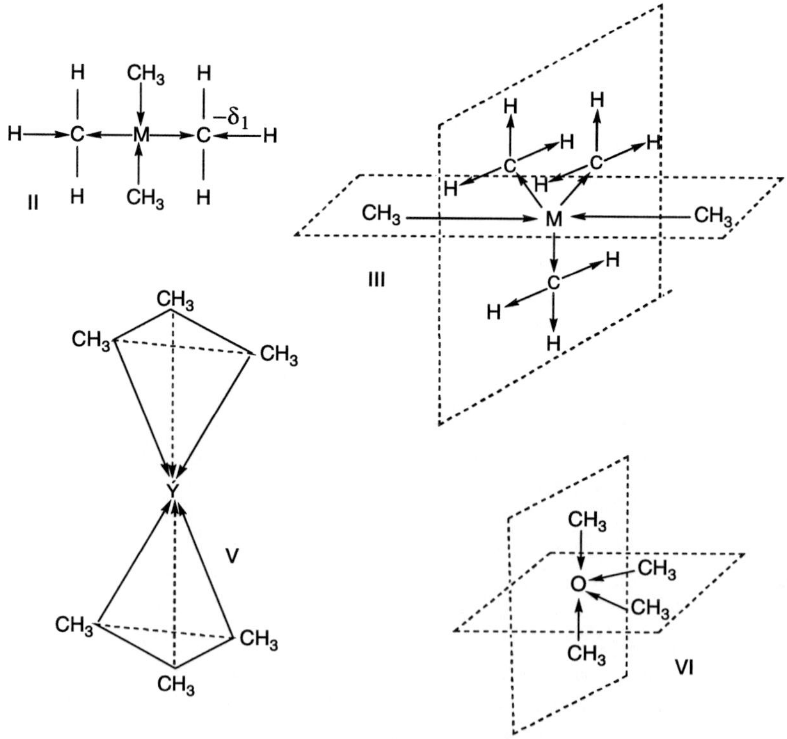

Fig. 2.2 Coordination possibilities of alkyls of the elements of main groups

(Fig. 2.2). The molecules of these alkyl derivatives interact through two CH_3 groups with the metal atoms (M = Zn, Cd, Hg), and two specific interactions M–CH_3→M are formed by the proper methyl groups.

The information on the liquid and solid state of the alkyl derivatives of elements from the fourth to sixth groups of the periodic table belonging to subgroups of carbon, nitrogen, and sulfur are scanty. From the viewpoint of bond vacancies, the alkyls of the elements from the sulfur subgroup having an angle structure and two unshared electron pairs on the atom of the central element ER_2 form the structure of the liquid state similar to that of water (Fig. 2.2), utilizing four bond vacancies. A similar structure has the alkyls of the nitrogen subgroup. All hydrogen atoms of ammonia NH_3 and carbon atoms of the CH_3 groups of the alkyl derivative ER_3 take part in the formation, respectively, of the hydrogen bonds and the specific interactions, and around the central atom are located six coordinated hydrogen atoms or methyl groups (Fig. 2.2). This is not something extraordinary since all bond vacancies should be occupied, the atoms of the nitrogen subgroup are frequently pentacoordinate, and the formation of complexes with octahedral structure is characteristic of these elements. Note that the central atoms in the alkyls $A(CH_3)_k$ of the elements from the second to fourth main groups bear positive charges and show acceptor properties in the formation of the specific interactions A–CH_3→A whereas the atoms of the elements of the fifth to sixth main groups $E(CH_3)_k$ possess negative charges and exhibit donor properties in the formation of the specific interactions E–CH_3←E (Fig. 2.3).

All alkyl compounds of the main group elements form specific interactions involving the pentacoordinate carbon atom (Fig. 2.4). The established bond vacancies of the alkyl compounds of the main group were used in the calculations of the energy of the specific interactions.

The performed calculations of the energy of the specific intermolecular interactions using the vaporization enthalpy [13] show (Table 2.1) that in some events this energy exceeds that of the hydrogen bond in the liquid water (10.99 kJ mol^{-1}) and essentially exceeds the energy of the same type interaction in the liquid ammonia (3.89 kJ mol^{-1}). Note that in the performed calculations the nonspecific interactions of low stability were disregarded for their values do not exceed the error in the experimentally measured thermodynamic characteristics.

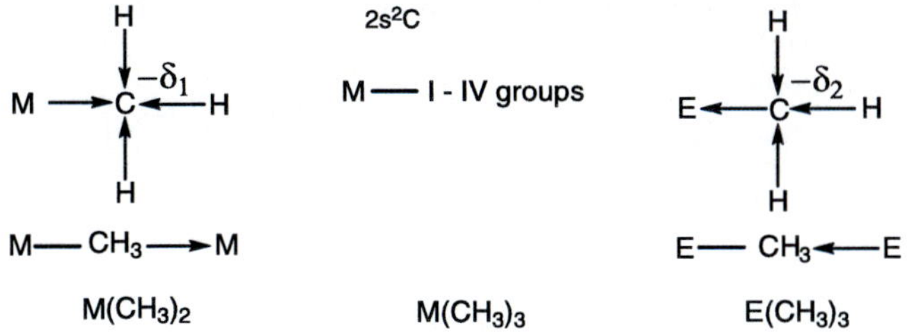

Fig. 2.3 Donor and acceptor properties of carbon atom in methyl group of alkyl compounds

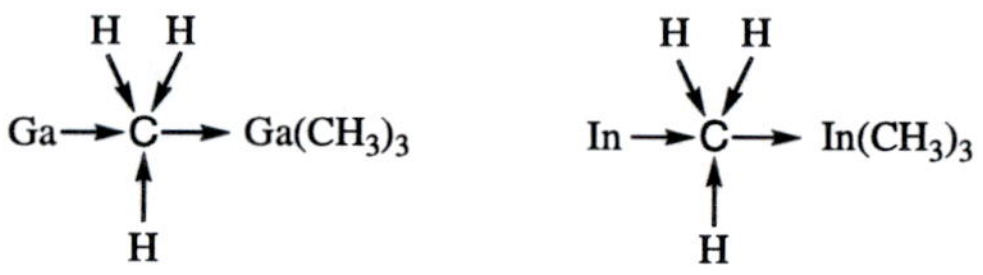

Fig. 2.4 Pentacoordinate carbon atoms in a specific intermolecular interaction

Table 2.1 Energies (kJ mol^{-1}) of specific interactions in liquid alkyls of elements of main subgroups

Compound	Bonding	Energy DM–CH_3→M	Compound	Bonding	Energy DE–CH_3←E
$Zn(CH_3)_2$	Zn–CH_3→Zn	7.60	$C(CH_3)_4$	C–CH_3→C	2.36
$Cd(CH_3)_2$	Cd–CH_3→Cd	5.22	$Sn(CH_3)_4$	Sn–CH_3→Sn	4.08
$Hg(CH_3)_2$	Hg–CH_3→Hg	8.32	$N(CH_3)_3$	N–CH_3←N	4.25
$Zn(C_2H_5)_2$	Zn–C_2H_5→Zn[a]	9.70	$P(CH_3)_3$	P–CH_3←P	4.80
$Cd(C_2H_5)_2$	Cd–C_2H_5→Cd[a]	9.48	$N(C_2H_5)_3$	N–C_2H_5←N[b]	5.86
$Al(CH_3)_3$	Al–CH_3→Al	10.30	$O(CH_3)_2$	O–CH_3←O	5.75
$Ga(CH_3)_3$	Ga–CH_3→Ga	8.42	$S(CH_3)_2$	S–CH_3←S	7.09
$In(CH_3)_3$	In–CH_3→In	8.42	$Se(CH_3)_2$	Se–CH_3←Se	7.80
$Al(C_2H_5)_3$	Al–C_2H_5→Al[a]	12.70	$O(C_2H_5)_2$	O–C_2H_5←O[b]	8.13
$Ga(C_2H_5)_3$	Ga–C_2H_5→Ga[a]	8.78	$S(C_2H_5)_2$	S–C_2H_5←S[b]	8.74

[a]M–CH_2–CH_3→M
[b]E–CH_2–CH_3←E

The regularity of energy variation of the considered specific interactions in the series of the main group elements is conserved in the methyl and methyl–ethyl derivatives. The sufficiently low errors in the experimentally measured vaporization enthalpies of alkyls of the elements from the second, third, and fifth groups [13] suggest that the deviation from the regular character in the energy variation of the specific interactions in alkylcadmium, alkylgallium, and alkylphosphorus has a profound sense. We believe that the cause of this fact consists in the electronic population of p_z-orbital of the elements of the main groups and the specific capability of the carbon atom to exhibit donor or acceptor properties depending on the properties of the other element involved in the intermolecular interaction.

The location of various numbers of electrons on dissimilar energy levels results in different donor properties of cadmium and gallium atoms. The presence of five electrons on the third energy level of the phosphorus atom provides it with acceptor properties in the interaction with the carbon atom. Thus, the carbon atom with an essentially unshared $2s^2$-electron pair exhibits the acceptor properties with the zinc and gallium atoms and, on the other hand, donor properties with the phosphorus atom. It is symptomatic that the comparative analysis of the electronic structures of the molecules of $B(CH_3)_3$ and $Al(CH_3)_3$ (Chap. 1), obtained from the X-ray data, shows the smaller contribution of the Al3p_z-orbital into the π-molecular orbital and an enhanced positive charge on the aluminum atom [31]. In dimethylmercury of the zinc subgroup, a fundamentally different succession of the 3d-levels is observed. Therefore, the dimethylmercury deviates from the regular series in the energy of the specific interactions having a low value of Hg–CH_3→Hg (8.32 kJ mol^{-1}). The alkyl compounds of all elements from the main groups obey the general rule. The energy of the specific interactions grows with the growing number of carbon atoms in the alkyl chain (Table 2.1). However, this growth cannot go on forever, and in going to the compound with larger number of atoms in the chain the energy difference of the intermolecular interactions diminishes in each subsequent step. It is proved on the alkyls of zinc and gallium.

$$\begin{aligned} & DZn(CH_3)_2(7.60) < DZn(C_2H_5)_2(9.70) < DZn(n-C_3H_7)_2(10.53) \\ & \quad < DZn(n-C_4H_9)_2(12.67) < DZn(n-C_5H_{11})_2\left(12.33\,\text{kJ mol}^{-1}\right) \end{aligned}$$

$$\Delta D\ M{-}CH_3 \rightarrow M \quad 2.10 \quad 0.83 \quad 2.14 \quad -0.34\,\text{kJ mol}^{-1}.$$

$$\begin{aligned} & DGa(CH_3)_3(6.60) < DGa(C_2H_5)_3(8.78) < DGa(C_3H_7)_3(9.32) \\ & \quad < DGa(C_4H_9)_3\left(10.32\,\text{kJ mol}^{-1}\right) \end{aligned}$$

$$\Delta D\ Ga{-}CH_3 \rightarrow Ga \quad 2.18 \quad 0.54 \quad 1.0\,\text{kJ mol}^{-1}.$$

Yet the difference in the energies of the specific interactions is always smaller between compounds with even and odd number of carbon atoms in the alkyl chain than between compounds with the odd and even number of carbon atoms in the alkyl chain. Similar sequence is observed for the dissociation energy of the M–C bond in the series of the alkyl derivatives of an element. This phenomenon is due to the different influence of the reverse dative bond on the energy of the specific interactions at the growing number of carbon atoms in the chain. This means that the shift of the electron density along the chain from the central atom to the terminal methyl group in compounds AR_k of the elements from the subgroups of zinc and boron and along the carbon–carbon chain to the central atom of the fifth and sixth groups in the complex ER_k depends essentially on the even or odd number of carbon atoms in the chain. Consequently, the action of the reverse dative bond between the central atom of the alkyl compound and the contiguous carbon atom essentially depends on the number of atoms in the chain and on whether this number is even or odd (Fig. 2.5).

The sharp decrease in the difference in the vaporization enthalpies at three carbon atoms in the chain shows that the influence of the reverse dative bond is finished and, on the other hand, the trend in the increasing energy of the specific interactions. Although the information on the charges on the central atoms of the compounds in question and on the carbon atoms of the alkyl chains is very scanty, nonetheless, it is illustrative that the charge on the mercury atom in the dimethylmercury (+0.023) turns to negative in $Hg(i\text{-}C_4H_9)_2$ (−0.070) [31]. In this connection should be noted the change in the sign of the charge of the phosphorus atom in going from PH_3 (−0.04) to $P(CH_3)_3$ (+0.070) [31], corresponding to the effect of the reverse dative bond at replacing the hydrogen by the methyl group in these compounds. These data clearly show the opposite direction of the reverse

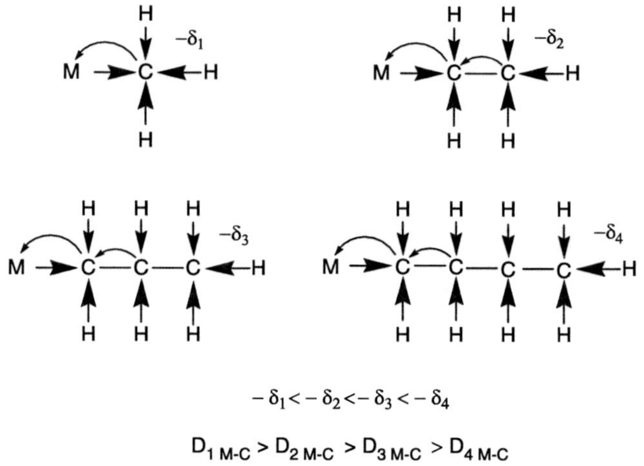

Fig. 2.5 Decrease in the influence of dative bond with growing number of carbon atoms in alkyl chain on the dissociation energy DM–C and the energy of specific intermolecular interactions

dative bond in the alkyls of the elements of second to fourth groups compared to the elements of the fifth and sixth groups. Note also the decrease in the influence of the reverse dative bond at replacement of the methyl group by ethyl and further propyl group. However, the participation of the pentacoordinate carbon atom in the specific interactions is common for all alkyl compounds.

2.3 Novel Approaches to the Thermodynamic Analysis of Liquid Functional Solvents and Energies of Specific Interaction

It was shown in the preceding paragraph by an example of the alkyl derivatives of the elements of the main groups that the knowledge of the structure of a molecule and of a number of the free bond vacancies is required for understanding the structure of the liquid state of organic compound and the types of the specific interactions. In turn the knowledge of the thermodynamics of the vaporization provides a possibility to establish the energy of the latter for the enthalpy (entropy) of various phase transitions are connected with the energy and the number of the specific interactions [15]. Therefore, the thermodynamic characteristics of the phase transitions cannot be governed by the molecular mass and even in the homologous series the specific features in the intermolecular interactions should be taken into consideration. However even in alkanes appear some complications. For instance, their melting points as a function of the number of carbon atoms in the carbon–carbon chain are described by two plots $T_{m.p.} = f(n)$ (Fig. 2.6).

As seen from the figure, the melting point grows with the chain length, and the difference of these values in the neighboring members of the homologous series is

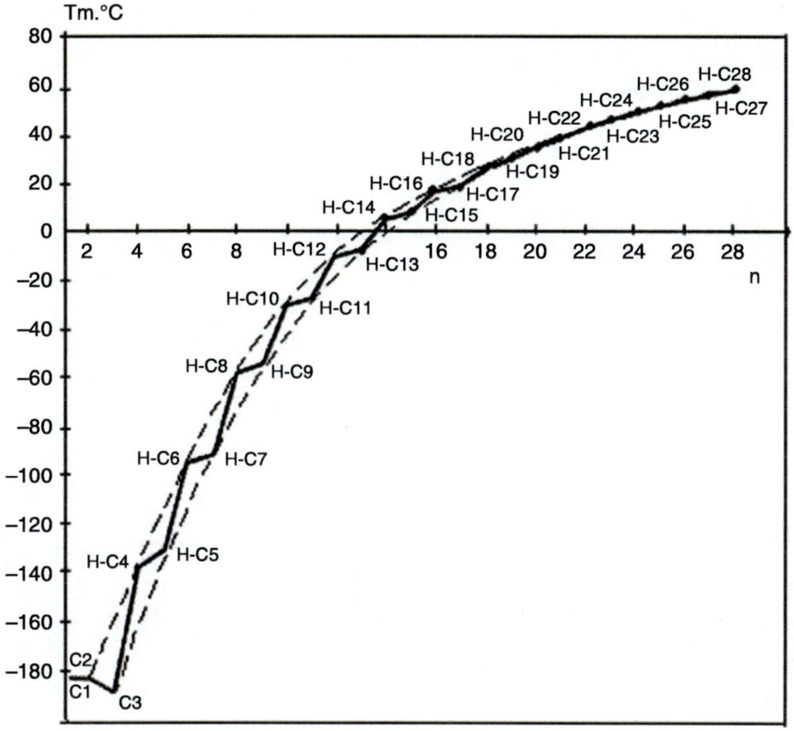

Fig. 2.6 Dependence of the melting points of alkanes on the number of carbon atoms in the molecular chain

not very large. The characteristic crystal lattice of hydrocarbons with even number of atoms n-C_6–n-C_{20} is triclinic. The compounds with a large number of carbon atoms in the chain undergo polymorphous transitions. Their low-temperature phase in alkanes of the series n-C_{20}–n-C_{26} is characterized by the triclinic crystals, and for compounds n-C_{26}–n-C_{36}, the monoclinic modification. The high-temperature phase is stable till the melting point and it possesses a hexagonal structure. The alkanes with the odd number of carbon atoms from n-C_9 till n-C_{35} undergo polymorphous transitions. They have the rhombic structure at low temperatures and the hexagonal one at high temperature. The vaporization enthalpies of the saturated alcohols of the normal structure also should be described by two converging curves corresponding to compounds with even (1) and odd (2) number of carbon atoms in the chain. This feature of alkanes properties is governed by the essentially unshared $2s^2$(c)-electron pair of the carbon atom involved into the distribution of the electron density. Figure 2.7 illustrates two analogous dependences of zinc alkyls where the alkyl substituents contain either even carbon atoms, $Zn(C_2H_5)_2$ and $Zn(C_4H_9)_2$ [33], or odd carbon atoms, $Zn(CH_3)_2$, $Zn(C_3H_7)_2$, $Zn(C_5H_{11})_2$.

Analogous but less pronounces character have the dependences of the vaporization enthalpies of the saturated hydrocarbons (Fig. 2.8). The vaporization entropy

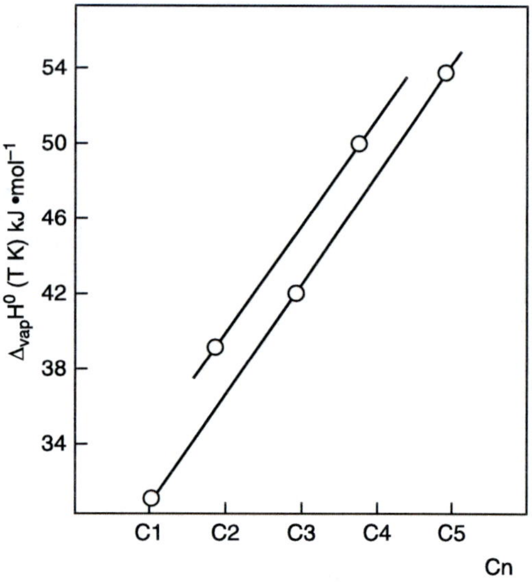

Fig. 2.7 Dependence of the vaporization enthalpy on the number of carbon atoms in molecules of zinc alkyls [32, 33]

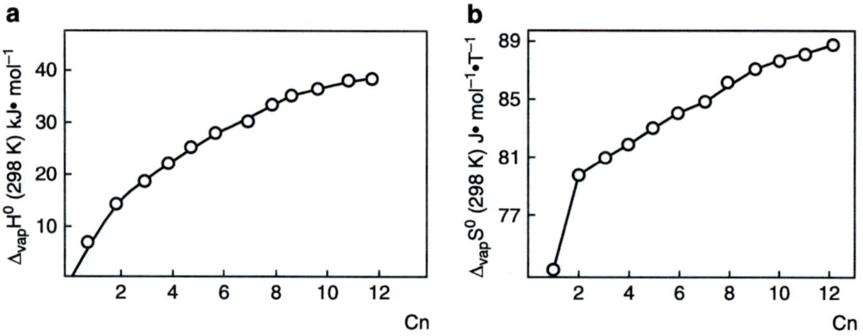

Fig. 2.8 The dependence of the vaporization enthalpy (**a**) and entropy (**b**) on the number of carbon atoms in molecules of saturated hydrocarbons $_{vap}S^0$(298 K) J mol^{-1} T^{-1}

being the characteristic of the structural transformations has a monotonic regularly growing value in compounds C_2–C_{10} illustrating the growing number of broken interactions with the increase of the amount of the CH_2 groups in the molecules.

The dependence of vaporization enthalpy of the alcohols of the normal structure (Fig. 2.9) should be divided in two parts. The first members of this homologous series, methanol, ethanol, and *n*-propanol possess an expressed capability of dimerization in the condensed and vapor phase, and *n*-butanol–*n*-decanol whose properties fit to two converging plots corresponding to the compounds with even and odd number of carbon atoms in the chain. The entropy, structural characteristic with a maximum at the *n*-propanol also indicates that the pronounced ability to dimerize in solid, liquid, and vapor phase is not characteristic of the alcohols of the

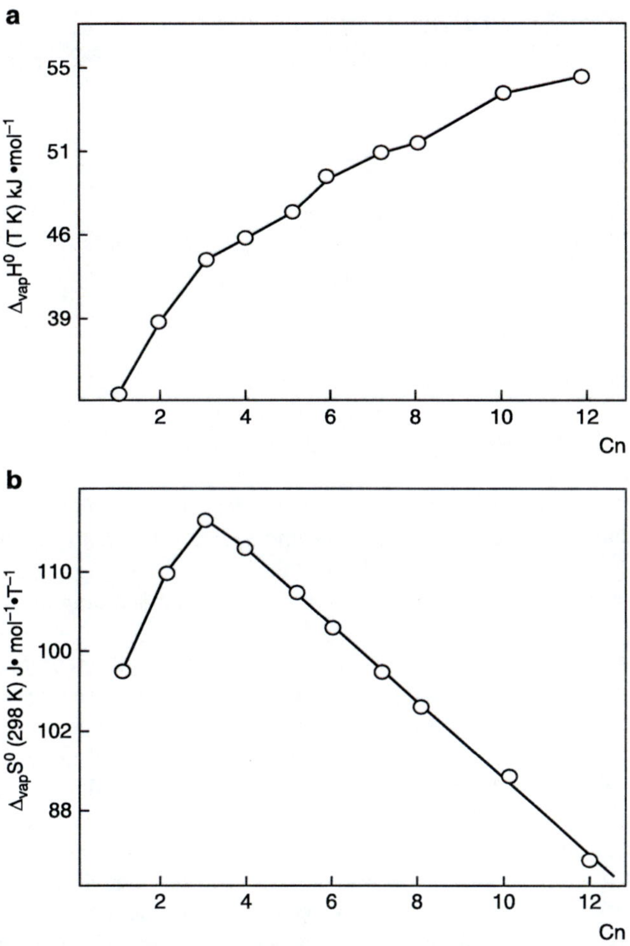

Fig. 2.9 Dependence of the vaporization enthalpy (**a**) and entropy (**b**) on the number of carbon atoms in molecules of alcohols

normal structure with over four carbon atoms in the alkyl chain (Fig. 2.9). If only the hydrogen bonds analogous to those for the homologous series of *n*-alkanes. Yet these two classes of organic compounds are distinguished by the opposite shift of the electron density in the molecules, and by the influence of the reverse dative bond on the formed hydrogen bonds and specific interactions.

Even more drastic inconsistency between the properties of compounds and the value of the molecular mass appear in formamide and its methyl derivatives as show the data of Table 2.2. The growth of the molecular mass in the series formamide < *N*-methylformamide < *N,N*-dimethylformamide is accompanied with the decreasing energy of the hydrogen bond as has been revealed by the method of the electron diffraction [8]. In this connection, the influence of the

Table 2.2 Physicochemical properties of compounds [5]

Compound	Molar mass	$T_{m.p.}$(°C)	$T_{b.p.}$(°C)	$\Delta_{vap}H^0$ (298 K)(kJ mol^{-1})
H_2O	18	0.0	100	43.94
CH_3OH	32.04	−97.68	64.35	37.43
AN	41.05	−43.84	81.60	32.9
FA	45.04	2.55	210.5	64.98
C_2H_5OH	46.07	−114.49	78.29	42.31
AC	58.09	−94.7	56.07	29.0
NMF	59.07	−3.8	180	56.25
DMF	73.09	−60.43	153.0	47.51
DMSO	78.13	−43.84	189.0	52.88

methyl group should be analyzed when it replaces a hydrogen atom in the amino group, or, more rightly, at the replacement of the hydrogen bond formed by the hydrogen of the amino group by the intermolecular interaction formed by the methyl group.

Proceeding from the curve of the radial distribution, Ohtaki and Iton [8] concluded that in liquid and crystals the molecular surrounding includes chain, ring, and ring–chain structures [5, 7, 8]. Just the simultaneous presence of chain and ring structures in the liquid formamide can provide understanding of discrepancies in the relaxation and vibration studies (Fig. 2.10).

According to quantum-chemical calculations [5], the hydrogen bond energy in the chain structure (67.8 kJ mol^{-1}) is somewhat higher than that in the ring structure (63.9 kJ mol^{-1}). At the same time it was noted that the presence of two hydrogen bonds in the ring structure provides it with a higher stability. Thus, the chain structure allows the isolation of a dimer with the above hydrogen bond energy D=O$\cdots$H–N– (Fig. 2.10).

In keeping with the concepts of Ohtaki and Iton [8], the considered hydrogen bonds in the liquid formamide and *N*-methylformamide form between the oxygen and hydrogen atoms of amino groups belonging to the contacting molecules =O$\cdots$H–N–, which are insignificantly weaker in the NMF with a mobile chain structure. Whereas the curve of the radial distribution established for liquid formamide showed the presence of large wide peaks at 350–550 and 700–900 pm corresponding to strong intermolecular interactions and a considerable molecular order, in the *N*-methylformamide the analogous curves of the radial distribution indicated the absence of a significant range of the intermolecular interactions despite the formation of the strong hydrogen bonds between the molecules. Based on the values of the energy of the hydrogen bonds calculated by the method of molecular orbitals, Ohtaki et al. concluded that the *trans*-position of the hydrogen atom with respect to the methyl group is more favorable than the *cis*-position, and the bond angle –C=O$\cdots$H–N– equal 120° is also more favorable for the formation of hydrogen bonds between *trans*-positions of the molecules. At the small difference in the energy of molecules and various dimers, the hydrogen bonds of the *N*-methylformamide molecules can form various configurations of the type "head-to-tail" of the linear polymers.

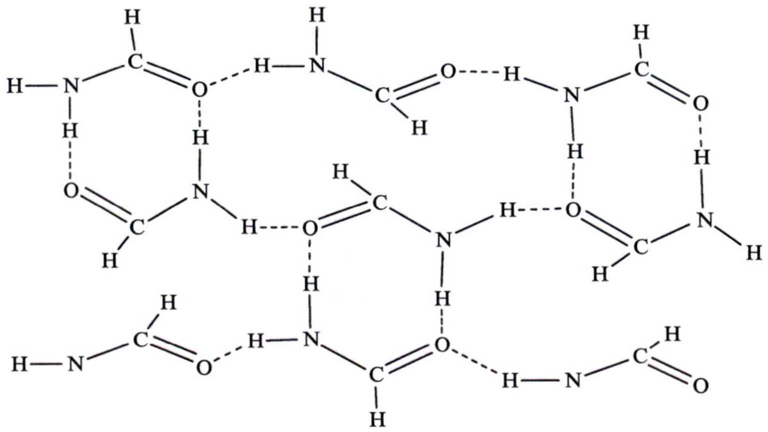

Fig. 2.10 Hydrogen-bonded network structure of liquid formamide [5]

The curves of the radial distribution obtained for *N,N*-dimethylformamide [5] showed the absence of any significant peak at $r = 300$ pm. Therefore, Ohtaki et al. concluded that in *N,N*-dimethylformamide existed an intermolecular interaction with an interatomic distance around 300 pm and that the interactions H–C=O ···H–C=O did not exist in DMF. Hence they suggested the existence of an intermolecular interaction H–C=O··· H–C=O in liquid DMF weaker than the common hydrogen bond. This suggestion was based on a small positive charge on the hydrogen atom (that was capable to have even a small negative charge), the interatomic distance *r*C–H···O=315 − 335 pm, and the angle −C=O···H = + 30−(−30) [34]. In this connection, the molecules of the *N,N*-dimethylformamide can form a chain or network structure with two coordinated molecules. It was noted that the interaction −COH··· H_3CNH in the liquid *N,N*-dimethylformamide has an extremely low stability. The interpretation of the curves of the radial distribution of formamide, *N*-methylformamide, and *N,N*-dimethylformamide [5, 35] resulted in a conclusion that in the ring structure of the liquid state of FA six hydrogen bonds −C=O···H–N− of high stability were formed, in the chain structure of NMF two hydrogen bonds of the same type but with reduced stability were formed, and in the chain-network structure of DMF two hydrogen bonds arose [5] H−C=O···H−C=O of low stability. Regarding the vaporization enthalpy as connected to the number and energy of the hydrogen bonds we calculated the energies of these interactions (Table 2.3). The obtained hydrogen bond energies with a maximum value in NMF are in conflict with the conclusions based on the analysis of radial distribution curves. The obtained sequence is conserved also when the contribution of the weak interactions −COH··· H_3CNH− and −COH··· H_3CN− into the vaporization enthalpy of NMF and DMF, respectively, is taken into account. Thus, each assumption gets in conflict with the main conclusions based on the analysis of radial distribution curves of liquid FA, NMF, and DMF.

Table 2.3 Energy of assumed hydrogen bonds obtained from the vaporization enthalpy (kJ mol^{-1})

Compounds	$\Delta_{vap}H^0$ (298 K)	Hydrogen bond	*D*
FA	64.98	$-C=O\cdots HN- = \Delta_{vap}H^0(298\,K)/6$	10.83
NMF	56.25	$-C=O\cdots HN- = \Delta_{vap}H^0(298\,K)/2$	28.13
DMF	46.89	$-HC=O\cdots HC=O = \Delta_{vap}H^0(298\,K)/2$	23.45

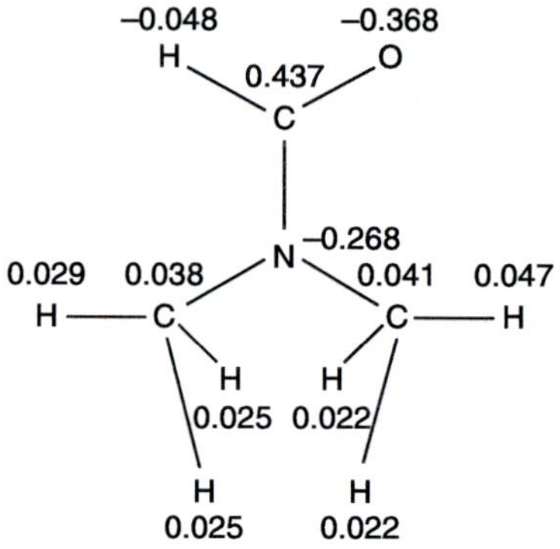

Fig. 2.11 Charge distributions in dimethylformamide

It is not unexpected for Ohtaki et al. have not possessed the additional information, and the conclusion on a small positive charge and even a small negative charge on the hydrogen atom is not right.

This discrepancy between the hydrogen bond energy and the analysis of the curves of the radial distribution is removed at the use of the nontraditional approach to the thermodynamic analysis of the interactions in liquid organic compounds formulated in Sect. 2.1. This approach is based on the rejection of the sp^3-hybridization model and on involvement of the carbon atom of the methyl group into the intermolecular interactions.

First of all note the low values of the positive charges on the hydrogen atoms of the methyl groups in the dimethylformamide (0.025, 0.025, 0.029 and 0.022, 0.028, 0.041) (Fig. 2.11). On this basis a conclusion was made [5] on the formation of very weak intermolecular interactions $-COH\cdots H_3CN-$ and $-COH\cdots H_3CNH-$ in the liquid *N*-methylformamide and *N,N*-dimethylformamide. However, this suggestion is dubious. In this connection, the attention is attracted by the high negative charge on the oxygen of the carbonyl group commeasurable with the charge on the oxygen in alcohol or ketone, and by the positive charges on the carbon atoms of the methyl groups.

The possible existence of weak hydrogen interactions $-C{-}H\cdots O-$ was suggested for hydroxy-containing organic compounds [35] between the hydrogen of a methyl group and the oxygen of a hydroxy group. From the analysis of the experimentally observed bands of the stretching vibrations of different bonds in gaseous and liquid methanol [36], it was concluded that in CH_3OH a mean content of such bonds was one per molecule. It was also indicated that these type bonds should be ten times weaker than the ordinary hydrogen bonds of methanol. Hobza and Sponer [37] suggested that the formation of this type of hydrogen bond occurred through nontraditional mechanism of the electron density redistribution and was accompanied with somewhat unusual spectral effects. These so-called nontraditional hydrogen bonds led to the blue shift in the spectra [38]. The performed quantum-chemical calculations of the methanol dimers linked by the classic hydrogen bonds H–O···O–H led Vandach et al. to the assumption that the methanol molecule is capable to form the hydrogen bond through the proton of the hydroxy group and that this type of a bond resulted in the blue shift in the corresponding IR spectra [39]. The quantum-chemical calculations using the density functional theory [40] indicate that this bond should decrease the cooperative effect in the liquid methanol. The comparison of the results of the calculations of the monomer structure [40] (bond lengths in the molecule, bond and torsion angles), electron density on atoms, and spectral characteristics showed the inequality of the protons in the methyl group of the methanol molecule. The characteristics of the proton situated in the *trans*-position with respect to the hydrogen of the hydroxy group are distinguished from the characteristics of two other protons of the methyl group. The energy barrier to the rotation of the CH_3-group in the gas phase amounts approximately to 4 kJ mol^{-1} [40]. However, the assumption on a large barrier to rotation in the condensed phase of methanol is not valid for in liquid and solid phase the energy of each interaction decreases due to the coordination effect and larger number of the interactions with the molecules of the nearest surrounding. Therefore for liquid and solid methanol with its characteristic hydrogen bonds and high capability to dimerize in vapor, the energies of the interactions of low stability under consideration should be lesser than 4 kJ mol^{-1}. Thermodynamic analysis of the hydrogen bonds in various classes of organic compounds [41] in the liquid state showed that this type of interactions is approximately tenfold stronger that the nonspecific interactions. For the substantiation of formed interactions, we shall use the information on the participation in complexing of the compounds in question.

The important problem for the thermodynamic consideration of formamide, *N*-methylformamide, and *N*,*N*-dimethylformamide is the understanding of their coordination ability which provides the information on the number of bond vacancies, types of the hydrogen bonds, and the specific interactions. In this connection, the ability of these compounds to form complexes with hydrogen fluoride and hydrogen chloride [42] is especially interesting.

The energy of H-bonds in complexes (HF)*m*DMF at the mixing enthalpy of initial components ~20.9–188.4 kJ mol^{-1} varies in a wide range and is governed by the reciprocal position of the molecules. Among the complexes containing equal number of molecules, the most stable are the complexes with a single molecule of dimethylformamide. The addition to it of more HF molecules occurs with the highest energy gain. The results of performed studies using IR spectroscopy, quantum-chemical calculations, and densimetry are well consistent [43]. The comparison of these studies with those performed for the system HCl–DMF by spectral methods and calculations shows that the results obtained are analogous. These results allow drawing conclusion, that DMF forms exclusively stable heterocomplexes with hydrogen fluoride and chloride [43-45] in all range of concentrations where the coordination number of DMF was 4, 5, and 6. Consequently, the two unshared electron pairs of the oxygen possess two bond vacancies and are involved into the interactions with the molecules of hydrogen fluoride or chloride forming stable hydrogen bonds $-O\cdots H{-}F$. Less stable interactions are formed by the carbon atoms of the methyl groups attached to nitrogen, N–CH_3. In this connection, let us turn to the electron population in the ammonia molecule (N: s = 0.515; $2p_x$, $2p_y$ = 1.210; $2p_z$ = 1.867; H: 0.734) and in $N(CH_3)_3$ (N = +0.024; C = −0.113; H = +0.035). Note also that the hydrogen bond in the liquid ammonia ($DN\cdots H{-}N{-} = 3.9\ kJ\ mol^{-1}$) at a higher positive charge on the hydrogen atom has reduced stability compared with the specific interaction ($DN{-}CH_3{\leftarrow}N = 4.20\ kJ\ mol^{-1}$) in the liquid $N(CH_3)_3$. This means that the donor properties of the nitrogen atom and the electron density granted to the hydrogen atoms of ammonia are essentially reduced. At the same time in the molecule $N(CH_3)_3$ the positive charge on the nitrogen atom and the negative charge on the carbon correspond to the distribution of the electron density with the participation of the reverse dative bond and giving over of the electron density from the p_z-orbital of the nitrogen. Consequently, the essentially unshared $2s^2$-electron pair of the carbon atom formed a more stable specific interaction than the hydrogen atom of the ammonia molecule. This shows directly that the methyl group forms a specific interaction of the stability higher than the hydrogen bond in the liquid ammonia. The presence of a negative charge on the carbon atom of the methyl group (−0.113) and of a positive charge on the nitrogen atom (+0.023) indicates directly that the positively charged hydrogen atom of the HF molecule predominantly interacts with the carbon atom of the methyl group.

Among the existing complexes, the more stable is the complex containing a single DMF molecule. The most stable are the quasisymmetric hydrogen bridges $O\cdots H{-}F$ formed by the carbonyl oxygen atom. The main trend in complexing in the system HF–DMF consists in the addition to the cyclic fragment (HF)*m*DMF (m = 1–7) of successive HF molecules. The cyclic complexes have both purely cyclic and branched structures. The latter complexes are cycles with added to them "tails" of hydrogen fluoride molecules. In the formed four-nuclear complexes, the quasisymmetric hydrogen bridges $O\cdots H{-}F$ occupy two bond vacancies and form the cyclic fragment (HF)*m*DMF (m = 1–7), and the remaining four vacancies form with the "tail" HF molecules the intermolecular interactions of lower energy. Inasmuch as the oxygen atoms of the carbonyl groups (C=O) of the *N,N*-dimethylformamide (Fig. 2.12), *N*-methylformamide, and formamide possess enhanced negative charge and two bond vacancies of two unshared electron pairs, the oxygen atom is capable of the formation of two more stable hydrogen bonds or specific interactions. Therefore, the common feature of these compounds is the ability to form a stable hydrogen bond $-H{-}C{=}O\cdots H{-}C{=}O$.

The presence in the molecule of *N,N*-dimethylformamide of two methyl groups results in the formation of two specific interactions with coordinated molecules of the hydrogen fluoride. The fifth coordination bond originates from the nitrogen atom forming it with the hydrogen of the HF molecule. The last coordination bond is formed by the hydrogen of the fragment (O=)C–H possessing higher

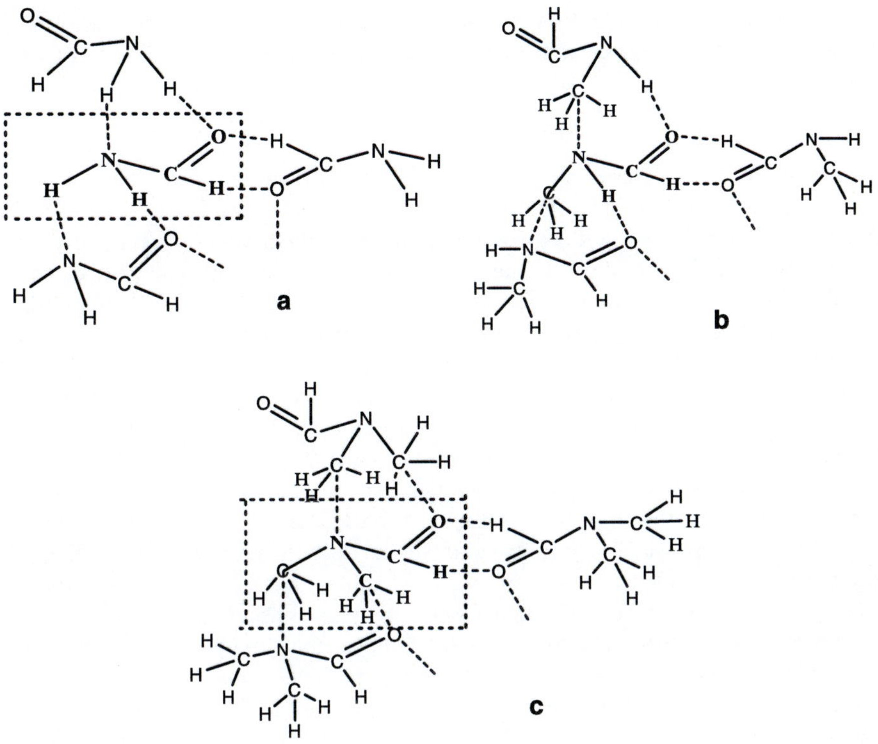

Fig. 2.12 Schematic pictures of hydrogen bonds and specific intermolecular interactions in liquid FA (**a**), NMF (**b**), and DMF (**c**)

positive charge than the hydrogen atoms of the methyl groups. A similar sequence of hydrogen fluoride molecules addition corresponding to the regular decrease in the stability of formed interactions exists in the complex formation with *N*-methylformamide and formamide. Consequently, the liquid functional solvents FA, NMF, and DMF are complex systems which utilizing six bond vacancies form various types of hydrogen bonds and specific interactions with characteristic energy values.

Thus, the sequence of the hydrogen fluoride coordination in its complexing with *N,N*-dimethylformamide shows the enhanced stability of the specific interaction formed by the essentially unshared $2s^2$-electron pair of the carbon atom of the methyl group and the unshared electron pair of the oxygen atom $N–CH_3 \leftarrow O{=}C(H)$. The second type of the specific interactions with lower stability is formed by the analogous methyl group and the unshared electron pair of the nitrogen atom $N–CH_3 \leftarrow N$. The schematic picture of the structure of liquid FA, NMF, and DMF with the coordination of three molecules shown in Fig. 2.12 depicts three types of six hydrogen bonds and specific interactions forming cycles linked into a network by interactions of various stability.

The formamide molecule with two NH_2 groups and six bond vacancies forms three types of hydrogen bonds in the liquid state: H–C=O···H–C=O (two), N–H···N (two), and N–H···O (two) (Fig. 2.12a). One such type of hydrogen bonds forms in liquid *N*-methylformamide, another in *N,N*-dimethylformamide.

The molecule of the *N*-methylformamide coordinates three molecules and forms two types of hydrogen bonds, H–C=O···H–C=O (two) and N–H···O=C–H (two), and two specific interactions *N*-CH_3←N (Fig. 2.12b). In keeping with the curve of the radial distribution, the second type of hydrogen bonds possesses a reduced stability as compared with the hydrogen bonds in the liquid formamide. The hydrogen atom of the NH-group forms the third type of the hydrogen bond N–H←N of lower stability with the nitrogen atom of the molecule in the nearest surrounding. The distinguishing feature of the *N,N*-dimethylformamide with two methyl groups is the formation in the liquid state of two types of the specific interactions through the methyl groups, H–C=O→CH_3–N (two), N–CH_3←N (two), and of two hydrogen bonds H–C=O···H–C=O (Fig. 2.12c). A significant distinguishing feature of liquid *N*-methylformamide and *N,N*-dimethylformamide is the formation of the specific interactions N–CH_3←N existing in the liquid trimethylamine (4.25 kJ mol^{-1}) with the energy exceeding that of the hydrogen bond in the liquid ammonia (3.89 kJ mol^{-1}). Note that it was shown by X-ray electron spectroscopy that the replacement of hydrogen in ammonia by a methyl group leads to insignificant decrease in the electron density shift to the nitrogen atom [46, 47]. It means that the hydrogen atom in the NH_2 and $NH(CH_3)$ groups forms less stable interactions than that formed by the methyl group in $N(CH_3)$, $N(CH_3)_2$, therefore the hydrogen bond in the liquid ammonia DN–H···N should be less stable than the specific interaction DN–CH_3←N. This reasoning permits considering the energies of the hydrogen bonds DN–H···N in the liquid ammonia and the formamide to be the same, equal 3.89 kJ mol^{-1}, and respectively the specific interactions N–CH_3←N in the liquid trimethylamine, *N,N*-dimethylformamide, and *N*-methylformamide equal 4.25 kJ mol^{-1}. The energy of the specific interaction DH–C=O→CH_3–N we take equal to the energy of this interaction in the liquid dimethyl ether (5.83 kJ mol^{-1}) (Chap. 3). Evidently, the electron density distribution in a molecule is governed by the properties of the atoms in its composition, therefore the substitution of the functional groups results in unequal changes in the energies of the intermolecular interactions formed by the bond vacancies of the molecule. Yet the above assumptions are valid for the estimation of the energies of the specific interactions in the complicated system of the functional solvents with a carbonyl oxygen connected through a carbon atom with a nitrogen atom securing a high negative charge (−0.368).

The mentioned types of the six hydrogen bonds existing in the liquid functional solvents (Fig. 2.12) and the specific interactions, whose energy and number constitute the vaporization enthalpy [15] make it possible to describe their contributions into this characteristic of the liquid *N,N*-dimethylformamide by the equation:

$$\Delta_{vap}H^0\,(298) = (2D{=}O\cdots H{-}C{=}O + 2D{=}O \rightarrow CH_3\text{-}N(5.83 \times 2) + 2DN\text{-}CH_3 \leftarrow N\text{-}CH_3(8.40)) = 46.89\,\text{kJ mol}^{-1}. \quad (2.1)$$

Subtracting the energy of the specific interactions DN–CH_3←N and D=O→ CH_3–N from the vaporization enthalpy we estimated the energy of the hydrogen bond D = O···H–C = O at 13.40 kJ mol^{-1} (Table 2.4).

Analogous equations describe the energy contributions of the hydrogen bonds and specific interactions in the *N*-methylformamide and the formamide, respectively:

$$\Delta_{vap}H^0\,(298) = 2D{=}O\cdots H{-}C{=}O + 2D{=}O\cdots H{-}N{-} + 2DN{-}CH_3 \leftarrow N(8.50) = 56.25\,\text{kJ mol}^{-1}(\text{NMF}), \quad (2.2)$$

$$\Delta_{vap}H^0\,(298) = 2D{=}O\cdots H{-}C{=}O + 2D{=}O\cdots H{-}N{-} + 2DN\cdots H{-}N{-}\,(7.78) = 64.98\,\text{kJ mol}^{-1}(\text{FA}). \quad (2.3)$$

Table 2.4 Energies (kJ mol^{-1}) of hydrogen bonds and specific intermolecular interactions in liquid formamide, *N*-methylformamide, and *N*,*N*-dimethylformamide

Compound	$\Delta_{vap}H^0$ (298 K)	Energies of specific interactions		
FA	64.98	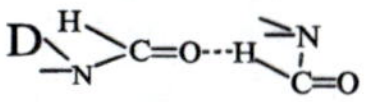 D=O•••H-C 16.50	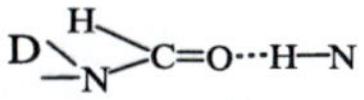 D=O•••H-N 12.00	D–H–N···H–N– 3.90
NMF	56.25	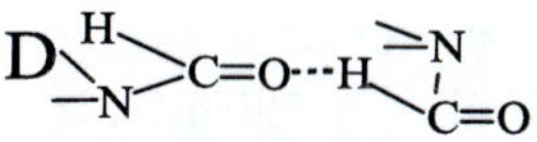 D=O•••H-C 13.40	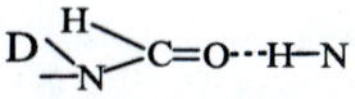 D=O•••H-N 10.55	D–N–CH_3←N 4.25
DMF	46.89	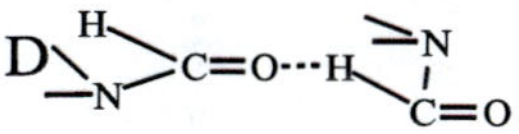 D=O•••H-C 13.40	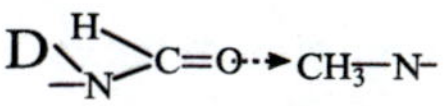 D=O→CH_3-N 5.83	D–N–CH_3←N 4.25

The energies of the hydrogen bonds are as follows:

$$2D = O \cdots H\text{–}C = O + 2D = O \cdots H\text{–}N\text{–} = 47.75\,\text{kJ mol}^{-1}(\text{NMF}), \quad (2.4)$$

$$2D = O \cdots H\text{–}C = O + 2D = O \cdots H\text{–}N\text{–} = 57.20\,\text{kJ mol}^{-1}(\text{FA}). \quad (2.5)$$

The inequality (2.6) following from (2.1), (2.4), and (2.5) demonstrates the increasing energy contribution at replacing the specific interactions DH–C=O→CH_3–N in *N,N*-dimethylformamide by a more stable hydrogen bond 2DH–C=O···H–N– in the *N*-methylformamide.

$$\begin{aligned} 2D &= O \cdots H\text{–}C = O + 2D = O \cdots H\text{–}N\text{–}\,(57.15)(\text{FA}) > 2D \\ &= O \cdots H\text{–}C = O + 2D = O \cdots H\text{–}N\text{–}\,(47.75)\cdot(\text{NMF}) > 2D = O \cdots H\text{–}C \\ &= O + 2D = O \rightarrow CH_3\text{–}N\text{–}\,\left(38.49\,\text{kJ mol}^{-1}\right)(\text{DMF}). \end{aligned} \quad (2.6)$$

Taking the energy of the hydrogen bond $D = O \cdots H - C = O$ equal to that in the liquid *N*, *N*-dimethylformamide (13.40 kJ mol^{-1}) we obtain the value of the energy of the H-bond $D = O \cdots H\text{–}N\text{–}$ equal 10.55 kJ mol^{-1}. The subsequent increase in the overall value of the energy corresponding to the same two types of the hydrogen bonds in the formamide shows their considerable stabilization taking into account the little difference in the hydrogen bond of low stability DN···H–N– and the specific interactions DN–CH_3←N. The energy of these two types of hydrogen bonds may be estimated if, firstly, the energy of the hydrogen bond $D = O \cdots H\text{–}N\text{–} =$ 10.55 kJ mol^{-1} should be taken constant in the *N*-methylformamide and the formamide and, secondly, to take the energy of the hydrogen bond $D = O \cdots H\text{–}C =$ $O =$ 13.40 kJ mol^{-1}. The performed calculation of the energy of the hydrogen bond $D = O \cdots H\text{–}C = O$ resulted in the values 18.00 and 15.20 kJ mol^{-1}, respectively. The more realistic value of this type of hydrogen bond is the average value 16.50 kJ mol^{-1} and the value of the energy of the second type of hydrogen bond $D = O \cdots H\text{–}N\text{–}$ equal 12.00 kJ mol^{-1}. The estimated energies of these two types of hydrogen bonds form two regular series of stabilization:

$$\begin{aligned} D = O \cdots H\text{–}C = O(16.60)(\text{FA}) &> D = O \cdots H\text{–}C = O(13.40)(\text{NMF}) \\ &\geq D = O \cdots H\text{–}C = O(13.40\,\text{kJ mol}^{-1})(\text{DMF}), \end{aligned} \quad (2.7)$$

$$D = O \cdots H\text{–}N\text{–}\,(12.00)(\text{FA}) > D = O \cdots H\text{–}N(10.50\,\text{kJ mol}^{-1})(\text{NMF}). \quad (2.8)$$

Note that analogous series of the hydrogen bond stabilization in the liquid formamide, *N*-methylformamide, and *N*, *N*-dimethylformamide were suggested in [5] based on the investigation by the method of the RDF. The assumptions taken in the thermodynamic analysis of the energy of the hydrogen bonds and the specific interactions in the liquid functional solvents do not introduce considerable errors. This is confirmed by the energy of the hydrogen bonds in the liquid water (10.99 kJ mol^{-1}). On the contrary,

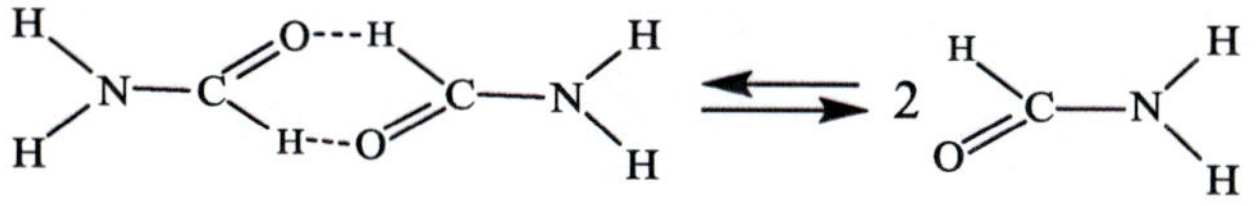

Fig. 2.13 Dissociation process of the dimer molecule of formamide in the vapor phase

the high energy of the hydrogen bond obtained by the quantum-chemical calculations [5] for the chain (67.8 kJ mol^{-1}) and ring (63.2 kJ mol^{-1}) structure of the liquid formamide in comparison with the vaporization enthalpy (64.98 kJ mol^{-1}) does not exclude the presence in vapor of dimeric molecules.

The trend in the decrease of the hydrogen bond energy in the series of formamide derivatives differing 1.5 times at the extremes of this series make it possible also to suggest the presence of dimers in the vapor of the formamide and the absence of dimers in the vapor of its derivatives.

$$\begin{aligned} &D = O\cdots H\text{–}C = O(16.50)(FA) > D = O\cdots H\text{–}C = O(13.40)(NMF)\\ &\quad \approx D = O\cdots H\text{–}C = O(13.40)(DMF) > D\\ &\quad = O\cdots H\text{–}N(12.00)(FF) > D = O\ldots H\text{–}N\ (10.55\,\text{kJ mol}^{-1})(NMF). \qquad (2.9)\end{aligned}$$

The energy of the hydrogen bonds of the liquid formamide does not fundamentally contradict to the presence of dimers in its vapor since the reduction to two of the number of coordinated molecules and the decrease in the number of the intermolecular interactions from six to two in the dimer secures the stabilization of the hydrogen bonds (Fig. 2.13). The fulfillment of these conditions is necessary for the dimerization in the vapor as is observed at the analogous relation between the vaporization enthalpy and dissociation energy of the triethylaluminum dimer [20, 21].

2.4 Energy of Specific Interactions in Liquid Acetonitrile and Dimethyl Sulfoxide

The acetonitrile does not form hydrogen bonds [5, 32, 48–51] and in the nearest surrounding in the liquid state are situated two molecules oppositely directed [5, 51]. This pattern is an unjustified simplification of the structure of the liquid acetonitrile suggesting the formation of weakly bound dimers of low stability having the vaporization enthalpy of 32.7 kJ mol^{-1}. This concept [5, 51] does not provide an understanding of the specificity of its intermolecular interactions in the liquid state. Therefore, applying this concept we have suggested that the acetonitrile molecule has two bond vacancies originating from the unshared electron pair of the nitrogen and from the essentially unshared $2s^2$-electron pair on the carbon atom of the methyl group (Fig. 2.14), forming as a result chain and ring structures in the liquid acetonitrile (Fig. 2.14a, b). Consequently, the acetonitrile molecule has two specific

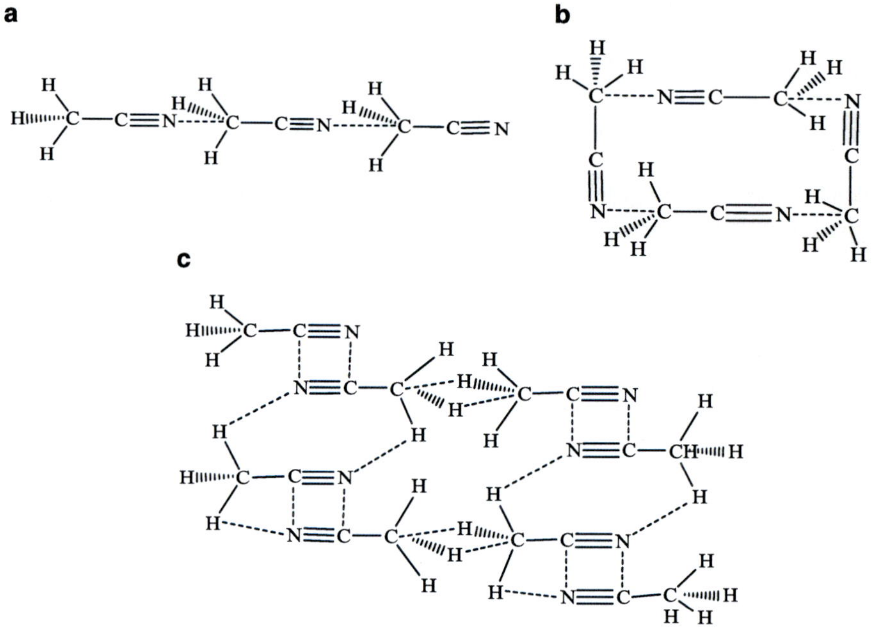

Fig. 2.14 Schematic picture of the liquid structure of acetonitrile (**a**) chain, (**b**) ring (carbon atom of the fragment –C ≡ N does not take part in the intermolecular interaction), and (**c**) network-chain structure

interactions –C–H$_3$C←N≡C–, whose energies equal D–C–H$_3$C←N≡C– = $\Delta_{vap}H^0$ (298 K)/2 = 32.7/2 = 16.35 kJ mol^{-1}. The energy of this type of specific interactions is approximately four times larger than the energy of the interaction DN–CH$_3$←N– (4.20 kJ mol^{-1}) existing in the liquid trimethylamine, and they are 1.5-fold more stable than the hydrogen bonds in water.

It is more realistic to assume that the more stable specific interactions are formed by the unshared electron pair of the nitrogen and by the essentially unshared 2s^2-electron pair of the carbon atom with a triple bond. The remaining bond vacancy of the carbon atom from the methyl group forms an interaction with the hydrogen of the methyl group H$_3$C→H–CH$_2$ of the contiguous molecule (Fig. 2.14c) analogously to methanol [38, 40] (Chap. 1). The RDF of the acetonitrile [5] is well consistent with the schematic picture of its liquid state shown in Fig. 2.14. Thus, the acetonitrile molecule with four bond vacancies forms two interactions –C≡N→C≡N of high stability and two specific interactions DH$_3$C→H–CH$_2$ of low stability with the estimated energy of 0.60 kJ mol^{-1} (Chaps. 1 and 3). The energy of the specific interactions D–C≡N→C≡N is estimated from the difference between the vaporization enthalpy and the overall energy contribution of the interactions of low stability (2.10):

$$DC \equiv N \rightarrow C \equiv N = \frac{\Delta_{vap}H^0\,(298) - DH_3C \rightarrow H - CH_2}{2}. \qquad (2.10)$$

The high energy of this type of specific interactions (15.75 kJ mol^{-1}) indicates the large difference of the charges on the carbon and nitrogen atoms due to the shift of the electron density in the triple bond of acetonitrile.

The thorough investigation by the diffraction method of the solutions of acetonitrile and *N,N*-dimethylformamide in 1:1 ratio showed that the molecules are antiparallel oriented [5, 51] with a short distance between the molecules of acetonitrile and *N,N*-dimethylformamide (316 pm), while the distance between the acetonitrile molecules in the liquid state is 330 pm. Therefore, it is presumable that the existing in the solution specific interactions $=O\rightarrow CH_3–C–N$; $–N–(H_3)C\leftarrow N–$; $–C–(H_3)C\leftarrow N–C–$ are more stable than the interactions in the liquid initial compounds, and the following inequality is valid (2.11):

$$D{=}O \rightarrow CH_3–C–N > D–C–H_3C \leftarrow N \equiv C\text{-}\left(15.75\,\text{kJ}\,\text{mol}^{-1}\right)$$
$$(\text{AN}) > D{=}O \cdots H–C{=}O\left(13.5\,\text{kJ}\,\text{mol}^{-1}\right)(\text{DMF}). \qquad (2.11)$$

The negative value of the mixing enthalpy (−200 kJ mol^{-1}) for the solutions in 1:1 ratio [52] confirms the higher stability of the specific interactions in the solution than in the liquid acetonitrile and *N,N*-dimethylformamide.

The dimethyl sulfoxide has a structure of a trigonal pyramid. The carbon and oxygen atoms are located in the plane of the pyramid base which should be distorted. The electron density distribution in the molecule (Fig. 2.15) [5, 53] results in a high positive charge on the sulfur atom (0.212), a negative charge on the oxygen atom (−0.281), and a significantly lesser negative charge on the carbon atoms of the methyl groups (−0.096). The dimethyl sulfoxide molecule has four bond vacancies, two due to the oxygen atom, and two, from the essentially unshared $2s^2$-electron pairs of the carbon atoms of the methyl groups. Four specific interactions of the same type $D–S{=}O\rightarrow CH_3–S{=}$ are formed. The radial distribution curve for dimethyl sulfoxide contains two peaks at 400–700 pm [5] less pronounced than the analogous peaks of the formamide (350–550 pm). These data allow to suppose, that the dimethyl sulfoxide forms a hydrogen bond more stable than the formed specific interactions and which has insignificant distinction from the hydrogen bond in the liquid *N,N*-dimethylformamide. Four bond vacancies of the dimethyl sulfoxide take part in the interactions in the liquid state with two molecules of the nearest surrounding and in the formation of the chain structure of the liquid state (Fig. 2.16) involving specific interactions $D–S{=}O\rightarrow CH_3–S{=}$ whose energies are determined from the vaporization enthalpy $\Delta_{vap}H^0$ (298 K) = 52.88/4 = 13.22 kJ mol^{-1}.

The energy of this type of specific interactions has the insignificantly decreased value compared with the energy of the hydrogen bonds $D{=}O\cdots H–C{=}O$ (13.40 kJ mol^{-1}) in the *N*-methylformamide and *N,N*-dimethylformamide and the essentially reduced value compared to that of the hydrogen bond energy of the

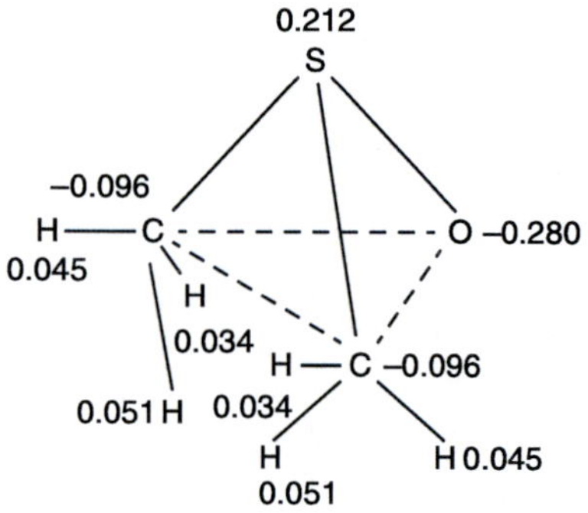

Fig. 2.15 Structure and charge distribution in molecule of dimethyl sulfoxide

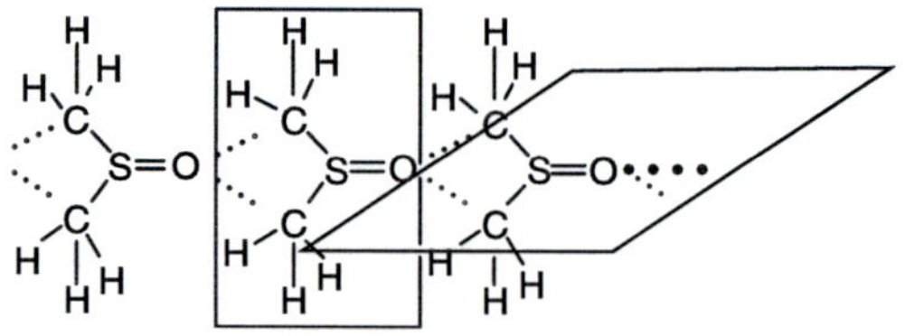

Fig. 2.16 Schematic picture of the liquid structure of dimethyl sulfoxide

formamide (16.60 kJ mol^{-1}). This succession in the energy value of the hydrogen bonds D=O · · · H–C=O and of the specific interaction D–S=O→CH_3–S= in the dimethyl sulfoxide is in complete agreement with the conclusions from the radial distribution curves.

2.5 Hydrogen Bonds and Specific Interactions of *N*, *N*-Dimethylformamide in Solvents

The conclusions on the six-coordination state of the *N*,*N*-dimethylformamide, *N*-methylformamide, and formamide can be supplemented with the data [52] on the complex formation with various solvents. In this connection, we shall discuss the influence of the composition of the binary system *N*,*N*-dimethylformamide–phosphoric acid on the dynamic viscosity of the system [54]. Note that there is no correct equation describing the viscosity of the ideal solutions proceeding from the properties of the individual components. The existing semiempirical equations describing the viscosity isotherms of certain systems are quite unsuitable for the description of the other systems. The concept of the positive deviation on the isotherms of viscosity and isotherms of excessive viscosity is generally recognized. The positive deviation from the additivity is commonly ascribed to the formation in solutions of more stable interactions than in the initial compounds [54]. The experimental data of [54] on the viscosity of solutions in the system *N*,*N*-dimethylformamide–phosphoric acid measured with the accuracy up to 1% show the positive deviation in the total range of the compositions (Fig. 2.17).

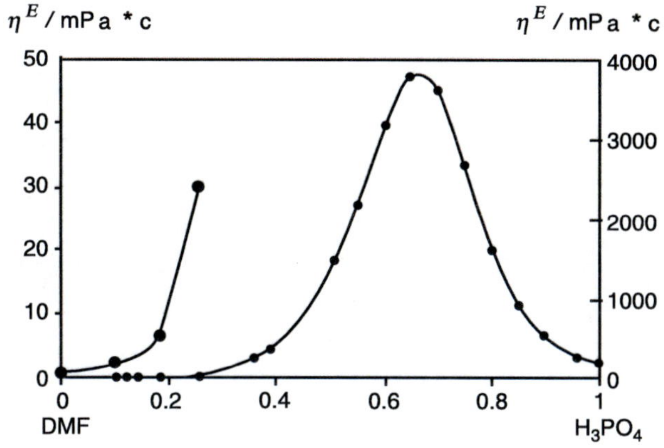

Fig. 2.17 Dependence of the isotherm (298 K) of dynamic viscosity (η/mPa s) on concentration of solution in system *N,N*-dimethylformamide–H_3PO_4: curves (1) (0–25) and (2) (0–100 mol% H_3PO_4) are shown in different scale

This character of viscosity variation with the solution composition is conserved in the temperature range 298–338 K where the viscosity decreases 12-fold.

The maximum on the isotherm in question at 66.7 mol% H_3PO_4 indicates the formation of a complex of the composition $(H_3PO_4)_2{\cdot}DMF$ [54–56] between the phosphoric acid dimer and the *N,N*-dimethylformamide. The dimers of the acid exist in the water solutions, and their presence in the *N,N*-dimethylformamide was substantiated in [57]. The excessive molar viscosity of solutions (η^E_M) in this system repeats the character of the dependence of the dynamic viscosity (η^E) confirming the formation of a stable complex in the solution [54]. According to Fadeeva et al. [58, 59] the narrowness of the peak and the form of the maximum on the curve viscosity–concentration correspond to the stability of the complex formed in the solution [58, 59]. In the discussed case, the maximum illustrates the high stability and consequently the low dissociation degree. In this connection, the primary viscosity growth at dissolution of the phosphoric acid in DMF in the range 0–25 mol% of H_3PO_4 (Fig. 2.18) changed by steep increase in the viscosity from 30 to 260 (η/mPa s) at the content of the phosphoric acid 26.5 mol% shows the finished process of the destruction of the structure of liquid DMF and the end of the formation of the solvate of the composition 3:1 = DMF:H_3PO_4 with coordinating center on the phosphoric acid molecule. This means that the dimers of the phosphoric acid completely dissociated in the solvent and formed a solvate structure of relatively low viscosity about 20 times greater than that of the *N,N*-dimethylformamide. It is therefore presumable that the specific intermolecular interactions formed in the solvate are hardly different in the stability from the interactions existing in liquid DMF. The low values of the dynamic viscosity of solutions in the composition range 0–25 mol% of H_3PO_4 and the positive deviation on the viscosity isotherms show that the energy of dissociation of the phosphoric acid dimers is partially or completely compensated by the newly formed specific interactions.

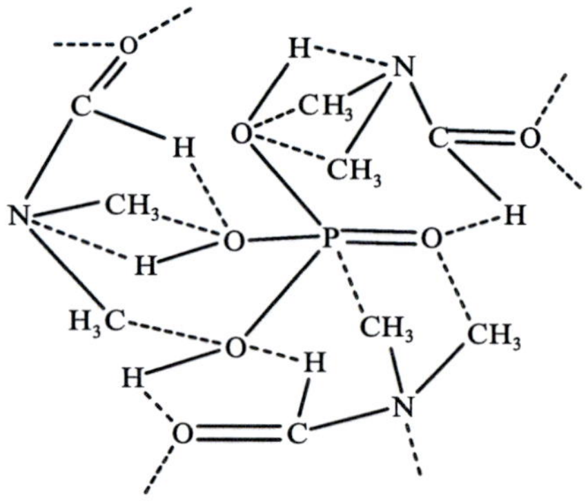

Fig. 2.18 Schematic picture of a complex of *N,N*-dimethylformamide with phosphoric acid

The molecule of the phosphoric acid with 12 bond vacancies is capable to coordinate in the nearest surrounding two and more molecules of *N,N*-dimethylformamide.

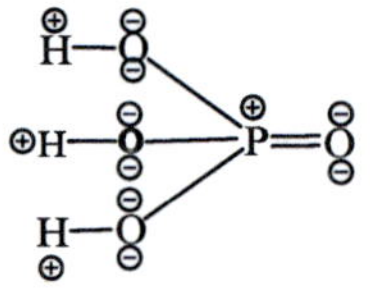

However, the presence in the near surrounding of three DMF molecules is preferable since each molecule having six bond vacancies forms a solvate of the 3:1 composition and with the remaining two bond vacancies cross-links them into the network structure of the solution (Fig. 2.18). Further increase in the phosphoric acid content results in the 25-fold increase in the dynamic viscosity of the solution, approximately from 29 to 3,800 (η/mPa s) ensuring the complex formation, and then the subsequent increase in the phosphoric acid concentration leads to the decrease in the dynamic viscosity finally to the value characteristic of the pure acid.

The growing viscosity in solutions with the concentration of the phosphoric acid in the range 25–66.7 mol%, occurs, on the one hand, because of the destruction of the solvate structure with the molecule of the phosphoric acid as the coordination center, and on the other hand, due to the processes at the enhanced concentration of the phosphoric acid increasing the viscosity of the solution. The understanding of these processes can be provided by the spectral investigations [59] of liquid DMF, phosphoric acid, and their solutions.

According to Corbridge [60] the 100% phosphoric acid undergoes autoprotolysis:

$$2H_3PO_4 = H_4PO_3^+ + H_2PO_4^-.$$

However, this process in the DMF solution is hampered by the formation of the specific intermolecular interactions between the molecules of the solvent and the solute. The spectral study of the phosphoric acid dissolution in diphenylamine showed that at the primary dissolution the intensity ratio of the main absorption bands of the components changed, and a strong background absorption was observed [55] indicating the formation of the specific intermolecular interactions between the components. This means that the protonation of the phosphoric acid did not occur. However, even at the initial dissolution of the phosphoric acid in DMF in the spectrum of the DMF solution a band of the stretching vibrations P–O–P (889 cm^{-1}) is observed whose position remains virtually unchanged at the increasing concentration of the phosphoric acid. The presence of this band shows that even at the low concentration of H_3PO_4 the dehydrated form of the phosphoric acid is present in the solution [6, 55, 60]. At the same time in the solutions of this system, the overlapping occurs of the bands of the stretching vibrations of groups P=O (1,145 cm^{-1}) of the phosphoric acid and the bending vibrations of the group N–C–H (1,092 cm^{-1}) of DMF molecules which in its turn should be assigned to the formation of the specific intermolecular interactions between the molecules of DMF and phosphoric acid. In the spectra of solutions in the range of phosphoric acid content 0.0–0.7 mol fractions, the band of the stretching vibrations of the C–N bond (1,387 cm^{-1}) of the DMF molecule appears and shifts (7 cm^{-1}) to the low frequency region with the simultaneous decrease in its intensity. These data show the stabilization of the bond presumably due to the interactions of the free bond vacancy of the nitrogen atom with the hydrogen atoms of the phosphoric acid providing a hydrogen bond of low stability with the energy similar to that of the hydrogen bond in the liquid ammonia. After reaching the maximum on the plot of the dynamic viscosity of the solution versus its composition (Fig. 2.17) at 0.7 mol fraction of H_3PO_4, the band of the stretching vibrations of the C–N bond (1,387 cm^{-1}) of the DMF molecule disappears and lacks in the region of the compositions 0.7–1.0 mol fraction of the acid. At the same time at the initial addition of the acid and the subsequent increase in its concentration, the low-frequency wing of the stretching vibration band of the C=O bond (1,674 cm^{-1}) of DMF suffers widening and the shift to the low frequency side. This suggests that the free bond vacancies of the carbonyl oxygen form specific intermolecular interactions of low stability. At the acid content of 0.7 mol fraction a band at 1,709 cm^{-1} appears lacking at lower concentration of the acid and characteristic of the stretching vibration of the C=N bond in the protonated DMF [61].

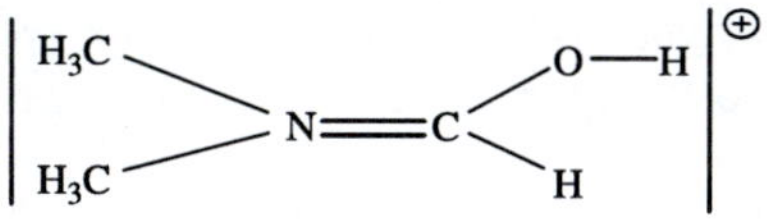

Consequently, a steep decrease occurs in the dynamic viscosity of the solution in the range 0.7–1.0 mol fraction of the acid due to the protolytic reaction and general decrease in the H_3PO_4 concentration (Fig. 2.17). It should also be noted that at reaching 0.2 mol fraction of the phosphoric acid in the solution the intensity of the band of the bending vibrations of the group O=C–N (658 cm^{-1}) of the DMF molecule sharply decreases, and in the DMF spectrum appears a strong band at 672 cm^{-1} which should be assigned to the formation by this group of the specific interactions in the solvate structure. Just in this region the destruction of the structure of liquid DMF is finished, and the solution acquires the solvate structure (Fig. 2.18) with the components ratio 3:1. Further a structure begins to form where the coordinating center is the DMF molecule, but with the growing acid concentration gradually the DMF protonation occurs. At the H_3PO_4 concentration of 0.7 mol fraction, the band at 672 cm^{-1} also disappears confirming the existence of the DMF in the protonated form. Therefore, the high dynamic viscosity of solutions beginning from 40 to 50 mol% of the acid is due to the involvement into the interactions of $H_4P_2O_7$, deprotonated forms of the acid, and the protonated forms of DMF accompanied with the formation of a complex with stable hydrogen bonds.

The dependence of the dynamic viscosity on the concentration shown in Fig. 2.17 has the same character for the phosphoric acid solutions in water and in organic solvents butyl formate, formamide, diisopropyl ether. The solutions of these systems are characterized by the position of the maximum deviation from the additivity of the viscosity curve in the range of the phosphoric acid concentration 0.6–0.7 mol fraction. The height of the maximum peak decreases in the series [61, 62]:

$$\text{DMF[62]} > \text{Diisopropyl ether} > \text{Formamide} > \text{Butylformate} > \text{water[63]}. \quad (2.12)$$

The existing correlation between the height of the maximum and the complex stability indicates the decrease of the hydrogen bonds stability in this series. Therefore, we are able to conclude that the six-coordinate state in the solutions under consideration is due to the bond vacancies of oxygen atom and the hydrogen atom linked to the carbonyl group, to the same oxygen atom and the carbon atom of the methyl group, to the nitrogen atom and the carbon atom of the methyl group, or the hydrogen atom of the amino group, respectively.

The results of the realized untraditional approach to the thermodynamic analysis and the established laws of the variation in the energy of the hydrogen bonds and the specific interactions in the liquid acetonitrile, dimethyl sulfoxide, *N,N*-dimethylformamide, *N*-methylformamide, and formamide illustrate the validity of this approach and the adequate interpretation of the interactions nature. Thus, the well-grounded rejection of the model of the sp^3-hybridization allowed for the first time the application of the concepts of the theoretical chemistry to the understanding of various types of specific interactions and hydrogen bonds in organic compounds and to the estimation of their energies.

References

1. Krestov GA, Vinogradov VI, Kesle YuM (1986) Modern problem of solution chemistry. Nauka, Moscow
2. Alferova VA, Liogonkaja TI, Podolskaja NI, et al (1991) Investigation of possibility of complexation in system dimethylcadmium – methyl derivates of VI group elements by methods of spectroscopy. In: Book of abstract. Properties and application of compounds with molecular crystal structure for new technology. N. Novgorod, USSR, p 33

3. Baev AK (1996) Thermodynamics and specific interactions in binary system of dimethylcadmium–dimethylselenium. Russ J Inorg Chem 41:1214–1219
4. Baev AK (1999) Thermodynamic properties of solutions diethylzinc–diethyltellurium. Russ J Phys Chem 73:1157–1162
5. Ohtaki H, Ishiguro SI (1994) In: Mamontov G, Popov AI (eds) Chemistry of non-aqueous solution. Current progress. VCH, New York, pp 179–226
6. Frisch MA, Trucks GW, Schigell HR, Johnson BG, Robbs MA Gaussian 94, Revision G.3
7. Ohtaki H, Funaki A, Rode BN, Reibnegger GJ (1985) The structure of liquid formamide studied by menus of X-ray diffraction and ab initio LCGO-MO-SCF calculations. Bull Chem Soc Jpn 56:2116–2121
8. Ohtaki H, Iton S (1985) Has liquid formamide a linear-chain structure of ring dimer structure. Z Naturforsch 40a:1357
9. Krestov GA, Beresin BD (1986) Based notions of contemporary chemistry. Chemistry, Leningrad, Russia
10. Ericson S (1968) Effect environment at examination of a reactivity adding of trimethylamine to trimethylborium theory. Chem Acta 10:193–209
11. Baev AK (2001) Thermodynamic parameters of solutions of binary system dimethylzinc–trimethylgallium. Russ J Phys Chem 75:1065–1067
12. Baev AK (2002) Liquid-vapor equilibrium in systems of methyl compounds of elements sulfur subgroup. Russ J Phys Chem 76:614–619
13. Baev AK (2001) Thermodynamic vaporization of alkyl compounds of elements second-six groups. Russ J Izv Vischich Ucheb Zaved Chem Chemo Technol 44:3–13
14. Baev AK (2001) Thermodynamic of solutions of trimethylaluminium and trimethylgallium with tetramethylstanum. Russ J Coord Chem 27:637–640
15. Baev AK (1972) Complex formation ability halogenides of second – six groups elements of periodical system. General and applied chemistry, vol 5. Vaisheshika Educating, Minsk, pp 35–51
16. Baev AK (2003) Novel approaches to the structure of functional solvents and elementorganic compounds with saturated hydrocarbon ligands and their solvates in binary systems. In: Book of abstract. 27th international conference on solution chemistry. Debrecen, Hungary, p 79
17. Baev AK (2004) Insolvency of sp^3-hibrydization model of carbon atom and chemistry of non-electrolytes solutions. In: Book of abstract. IXth international conference. The problems of solution and complex formation in solution. Ploys, Ivanovo, Russia, pp 62–64
18. Baev AK (2004) Lack of co-ordination of thermodynamic properties of elementorganic compounds with sp^3-hibrydization model of carbon atom and its insolvency. In: Book of abstract. Russian research symposium on thermochemistry and calorimeters. Nizhny Novgorod, Russia, pp 40–41
19. Vilkov LV, Mastrukov VC, Sadova NI (1978) Structure determination of geometrical structure of free molecules. Chemistry, Leningrad, USSR
20. Baev AK, Vitalev SE (1989) Thermodynamic vaporization and association of triethylaluminium. Vesti Akademii Navuk Belarusii 2:58–63
21. Vitalev SE, Baev AK (1993) Thermodynamic vaporization and association of trimethylaluminum. J Izv Vischich Ucheb Zaved Chem Chemo Technol 36:36–40
22. Baev AK, Schischko MA (2002) Thermal conduct of triarylaluminiums. Russ J Appl Chem 75:26–32
23. Albert S, Repmeester JA (1978) Methyl group rotation and bonding in solid hexamethyldialuminum, $Al_2(CH_3)_2$. J Chem Phys 70:722–725
24. Levison KA, Parkins PG (1969) Bonding in hexamethyl-dialuminium and related compounds. Discuss Faraday Soc 47:183–189
25. Kviste S, Rytter E (1984) Isolation of monomeric trimethylaluminum (Me_3Al) in argon matrixes. Spectrochim Acta 40A:939–957

26. Jursic BS (1998) Hybrid density functional theory quadratic complete basis set ab initio computation study of structural properties, vibrational spectral and bond dissociation energy for methyl aluminum and methyl gallium. J Mol Struct Theochem 428(1–3):61–66
27. Vranka RG, Amma EL (1967) About crystal structure of trimethylaluminum. J Am Chem Soc 89:3121–3126
28. Pauson PL (1967) Organometallic chemistry. Edward Arnold, London
29. Sheldrick GM, Sheldrick WS (1970) Crystal structure of trimethylthallium. J Chem Soc A Inorg Phys Theoret Chem 28–30
30. Davidson G, Ebsworth EAV, Sheldriek GM, Woodward LA (1965) Vibrational spectra of trisilylphosphine: evidence for planar PSi_3 structure. Chem Commun 7:122–123
31. Nefedov VI, Vovna VI (1989) Electronic structure of organic and elementorganic compounds. Nauka, Moscow
32. Baev AK (1996) Pentacoordinated carbon atom at intermolecular interactions. Russ J Coord Chem 22:399
33. Baev AK (1992) Structural-energetic aspects of association interactions of alkalis elements of I–VI groups. Russ J Chem Thermodyn Thermochem 2:143–158
34. Ohtaki H, Itoh S, Rode BM (1986) The structure of liquid *N*-methylformamide by X-ray diffraction and ab Initio LCGO (linear combination of Gaussian orbitals) – MO-SCP calculation. Bull Chem Soc Jpn 59:271
35. Legon AC, Roberts BP, Wallwork AAL (1980) Equilibrium conformation of four- and five-membered cycle molecules in the gas phase determination and classification. Chem Phys Lett 19:97
36. Yuchnevich GV, Tarakanova EG (1998) Hydrogen bond C—H···O in liquid methanol. J Mol Struct 447:257
37. Hobza P, Sponer J (1999) Structure energetic and dynamical of the nuclear acid pairs: nonempirical ab initio calculations. Chem Rev 99:3247–3276
38. Krupin SV, Puchovskyi YuN (2004) Investigation of new configuration hydrogen-bonded of methanol's dimer by method of molecular orbital's and functional density. In: Book of abstract. IXth international conference. The problems of solvation and complex formation in solution. Ivanovo, Russia, p 136
39. Hobza P, Havlas Z (2000) Blue-shifting hydrogen bonds. Chem Rev 100:4253–4264
40. Ivanov EV, Puchovskie YuP, Abrosimov VK (2004) About formation of weak hydrogen bonds CH···O in liquid methanol. In: Book of abstract. IX international conference. The problem of solvation and complex formation in solutions. Plyos, Russia, p 345
41. Novikov GI (2002) To the problem of hydrogen bonds at the series methane, ammonium, water and hydrogen fluoride. Izv Ser Chem Sci NAS Belarus 3:11–17
42. Yuchnovich GV, Tarakanov EG, Suschko NI, Suchovarchov VF (2004) Complication in system of HF-DMF. In: Book of abstract. IXth international conference. The problems of solvation and complex formation in solution. Ivanovo, Russia, p 90
43. Tarakanova EG, Yuchnovich GV (2005) Structure and stability of complexes formed by molecules of hydrogen fluoride and *N,N*-dimethylformamide. Russ J Struct Chem 46:26–31
44. Maiorov VD, Cisoeva SG, Temkin ON, Kiskina IS (1993) Ion–molecular interactions in system DMF–HCl on the results of IR-spectroscopy. Izv Akad Nauk Ser Chem Russ 7:1577
45. Tarakanova EG, Yuchnovich GV, Zoi ON, Librovich NB, Kislina IS (2004) Structure of complexes (DMF)m(HCl)n ($m = 2$, $n = 3$–4). Russ J Struct Chem 46:18–25
46. Karmon TA (1981) Photoelectron and oze-spectroscopy. L., p 258
47. Nefedov VI (1984) Infrared – spectroscopy of chemical compounds. Chemistry, Moscow
48. Reichell G, Weidner JU, Zimmerman H (1974) Rentgenostructural studies of liquid tetrachloric carbon in framed radial model of fluids. Ber Bunsenges Phys Chem 78:1054–1064
49. Bertagnolli DH, Zeidler M (1978) Nuclear spin reaction in fluids. J Mol Phys 36:177
50. Steinhanser O, Bertagnolli DH (1981) Molecular pair correlation function of liquid acetonitrile derived from perturbation theory with a computer-generated reference function. Chem Phys Lett 78:555–559

51. Radnai T, Itoh S, Ohtaki H (1988) Liquid structures of *N,N*-dimethylformamide, acetonitrile and their 1:1 molar mixture. Bull Chem Soc Jpn 6:3845
52. Fadeeva AYu (2003) IR-spectroscopy investigations of phosphoric acid – *N,N*-dimethylformamide. Russ J Coord Chem 29:551–555
53. Itoh S, Ohtaki H (1987) A study of the liquid structure of dimethylsulfoxide by the X-ray diffraction. Z Naturforsch 42A:858
54. Fadeeva AYu, Shmukler LE, Safonova LP (2005) Intermolecular interactions in system of phosphoric acid–N, N-dimethylformamide on the data of viscosimetry. Izv Akad Nauk Russ Ser Chem 10:2252
55. Anosov VYa, Ozerova MI, Fialkov YuYa (1976) Bases of physico-chemical analysis. Nauka, Moscow
56. Dulova VI, Gordienko NS, Kazan NA, Kalinichenko LG, Breze AI, Gribenkina VN (1979) Differentiated influence solvent on force of acids and base of different compounds and correlations between of these properties and the stability of some complexes of transitional metals, vol 21. Izvestya University of Charkov, Ukraine, p 216
57. Postnikova NN (1970) Thermal phosphorous acid. Chemistry, Moscow, p 304
58. Fadeeva JA, Shmucler LE, Safonova LP (2001) In: Abstract of papers, 27th international conference on solution chemistry. Vaals, Netherland, p 33
59. Fadeeva YuA, Demina LI, Gorbunova AYu, Shmukler LE, Safonova LP, Tsivadze AYu (2003) Orthophosphoric acid–*N,N*-dimethylformamide system: IR study. Russ J Coord Chem 29:515–518
60. Corbridge D (1980) Phosphorus: an outline of its chemistry, biochemistry, and technology. Elsevier, Amsterdam
61. Krestov GA, Afanasiev VN, Efremova LS (1988) Physico-chemical properties of binary solvent. Chemistry, Leningrad
62. Burdin VV, Kislin IS, Maiorova VD et al (1998) The structure of complexes of HCl with *N,N*-dimethylformamide on data of polarized changes in vibration specter. Izv Akad Nauk USSR Ser Chem 12:2484
63. Postnikova NN (1970) Thermally produced orthophosphoric acid: chemistry and technology. Chemistry, Moscow

Chapter 3
Types and Energies of Specific Intermolecular Interactions in Ethers and Organic Cyclic Oxides

3.1 The Structure and Specific Intermolecular Interaction in Liquid Ethers

In molecules of ethers in keeping with [1], two HOMOs are localized on a basic oxygen. In $O(CH_3)_2$, the contribution of the n–σ-conjugation into the HOMO destabilization amounts to 1.5 eV compared to the water molecule providing the oxygen atom with a reduced electron density. As a result, the energy of the specific intermolecular interaction should be considerably lower that the hydrogen bond energy in water, 10.94 kJ mol^{-1}.

The angular structure of the ether molecule with two lone electron pairs of the oxygen atom and terminal methyl groups governs the structure of its liquid and solid state.

The ether molecule has four vacancies, two of which are supplied by the oxygen atom and the other two by the carbon atoms of the terminal methyls of the alkyl groups. The molecules of ether of angular structure form a chain of cycles (Fig. 3.1) where every two successive terminal methyl groups are located on perpendicular planes, or a network in which four specific intermolecular interactions of the same kind –C–O $\rightarrow$ CH_3–O–(–C–O $\rightarrow$ CH_3–O–) are present formed essentially by lone $2s^2$-electron pairs of oxygen with two proper terminal methyl groups and two contact molecules [2–5]. The second characteristic feature consists in the presence of two molecules in the close surrounding. In their turn, the relatively low enthalpy characteristics of ether vaporization suggest the absence of dimerization in the vapor in good agreement with the experimental findings [6]. The existence of the ethers in vapor solely as monomer molecules simplifies the thermodynamic calculations of the energy of specific intermolecular interactions. It is therefore presumable that ethers are convenient compounds to be used in the additive method of the thermodynamic analysis. The energy of this interaction is estimated from the vaporization enthalpy of the corresponding ethers with the symmetric alkyl groups [7]. These values manifest the stabilization of the specific intermolecular interaction D–O $\rightarrow$ CH_3–$(CH_2)_n$– with the growing number of carbon atoms in the alkyl chain

A.K. Baev, *Specific Intermolecular Interactions of Organic Compounds*,
DOI 10.1007/978-3-642-21622-0_3, © Springer-Verlag Berlin Heidelberg 2012

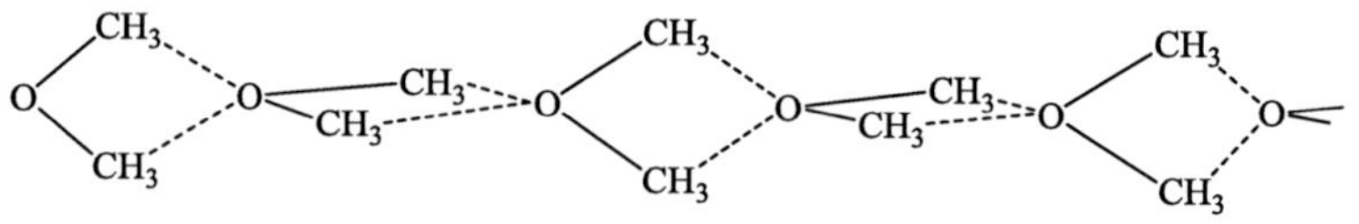

Fig. 3.1 Schematic picture of the structure of chain formed of cycles in liquid ether

Table 3.1 Vaporization enthalpies and energies (kJ mol^{-1}) of specific intermolecular interactions in liquid ethers (298 K)

Compounds	Formula	$\Delta_{vap}H^0(T)$ [8]	Energy of specific interactions D–O → CH_3–	D–O → C_2H_5–	D–O → C_3H_7–	$\Delta_{vap}H^0(T)$ Calculated
Dimethyl ether	C_2H_6O	22.54	5.63	–	–	–
Diethyl ether	$C_4H_{10}O$	27.14	–	6.78	–	–
Dipropyl ether	$C_6H_{14}O$	35.68	8.92	–	8.92	–
Dibutyl ether	$C_8H_{18}O$	44.99	11.25[a]	–	–	–
Dipentyl ether	$C_{10}H_{22}O$	(54.4)[b]	(13.6)[a]	–	–	–
Dihexyl ether	$C_{12}H_{26}O$	64.1 ± 0.1	(16.02)[b]	–	–	–
Methyl ethyl ether	C_3H_8O	(24.80)	5.38	6.78	–	24.32
Propyl ethyl ether	$C_5H_{10}O$	31.39	–	6.78	8.92	33.46
Methyl propyl ether	$C_4H_{10}O$	27.57	5.38	–	8.92	30.60
Methyl butyl ether	$C_5H_{12}O$	29.55	5.38	–	11.23[c]	33.23
Butyl ethyl ether	$C_6H_{14}O$	36.32	–	6.78	11.23[c]	36.06
Decyl methyl ether	$C_{11}H_{24}O$	62.30	5.38	–	8.92	

[a]Calculated per one alkyl group/
[b]Established by interpolation/
[c]R-*n*-Bu

(Table 3.1). The low value of the specific interaction for the dimethyl ether –O → CH_3– is due to the largest contribution of the reverse dative bond into the dissociation energy of the C–O bond and consequently

the reduction of the negative charge on the oxygen atom and of the positive charge on the methyl group carbon. The increased number of carbon atoms in the ethyl group of the diethyl ether decreases the displacement of the electron density from the CH_2-group to oxygen and loosens the effect of the reverse dative bond on the terminal methyl group, therefore the oxygen atom possesses enhanced donor properties

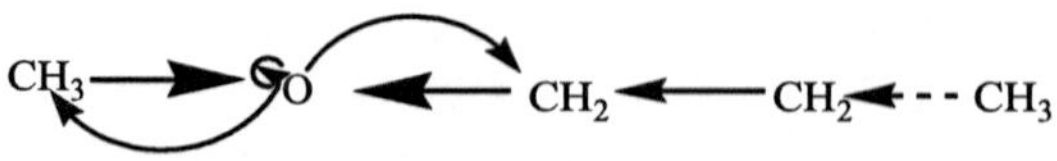

with insignificant diminishing of the acceptor capacity of the terminal methyl group and consequently it results in a more stable specific intermolecular interaction. The effect of the dative bond is observed to the propyl group inclusive (Chap. 1) and further remains virtually constant. Therefore, the energy of the specific intermolecular interaction in ethers with the symmetric alkyl groups attains practically the highest stability also in the dipropyl ether. The energy of this interaction in the dipropyl ether with a normal chain (8.92 kJ mol^{-1}) is lower by 2.08 kJ mol^{-1} than the energy of the hydrogen bond in water (10.94 kJ mol^{-1}) in agreement with the conclusion following from the quantum-chemical calculations [1]. At the same time, the equivalent contribution of the reverse dative bond in the dipropyl and dibutyl ethers of the normal structure suggests either the equal energy or at least insignificant difference in their specific intermolecular interactions. Therefore, the calculated value of the dibutyl ether energy exceeding the energy of the specific interaction of the dipropyl ether in the condensed state corresponds to the additional energy contribution from the CH_2-group into the enthalpy characteristics. Namely, the lacking effect of the reverse dative bond on the terminal methyl group in the dibutyl ether underlies the interaction of hydrogen and carbon atoms of the CH_2-groups and the contribution of the energy of their specific interactions into the enthalpy characteristics of the vaporization processes of the other ethers and organic compounds with alkyl chain exceeding three carbon atoms. The evaluation of the energies of the specific intermolecular interactions in liquid dimethyl, diethyl, and dipropyl ethers makes it possible to estimate the vaporization enthalpy of ethers with different alkyl groups based on the principle of contribution additivity. The vaporization enthalpies listed in Table 3.1 show a good agreement between the experimentally obtained values and those calculated by the additive method for ethyl methyl, propyl ethyl, and butyl ethyl ethers. The difference between the experimental values and those calculated by the additivity method is observed for propyl methyl (1.60 kJ mol^{-1}) and butyl methyl (4.20 kJ mol^{-1}) ethers; the corresponding difference is given in parentheses. This difference may originate from, firstly, the large contribution of the reverse dative bond into the property of the methyl group retaining its individuality; and, secondly, deformation of the cycles in the cyclic chain structure due to the dissimilar length of the alkyl groups under the condition of the presence of two molecules in the close surrounding.

This suggests that the growing difference in the length of alkyl groups in unsymmetrical ethers should result in the increased number of molecules in the close surrounding up to four and in the formation of branched cyclic structure (Fig. 3.2), for instance, in the liquid methyl propyl ether.

Taking as the energy of the specific intermolecular interaction DCH_3–O → CH_3–O = 5.63 for the methyl and the energy D–O → CH_3–$(CH_2)_2$ = 8.92 kJ mol^{-1} as the limiting energy value for the propyl group, we can evaluate the energy value contributed by nine CH_2-groups into the vaporization enthalpy of the decyl methyl

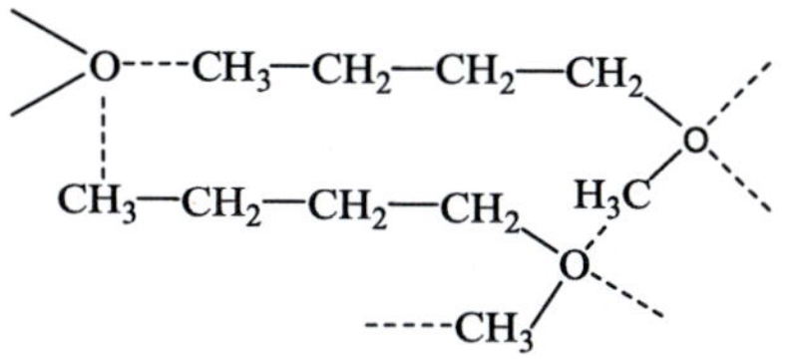

Fig. 3.2 Schematic picture of the liquid structure of ether with different length of alkyl groups

ether (Table 3.1) at 33.2 kJ mol^{-1}, or 3.7 kJ mol^{-1} per one group. However, a single CH_2-group of the symmetric dibutyl ether contributes to the vaporization enthalpy 4.65 kJ mol^{-1}. As a result, we obtain an opportunity to estimate the vaporization enthalpy of a number of ethers with different alkyl groups.

Note that the data on ether vaporization enthalpies in [6, 8] are mutually supplementing. Therefore, it is reasonable to follow the dependence of the enthalpy characteristics on the number of carbon atoms in alkyl groups. These dependences based on the scanty existing data on the vaporization enthalpies of symmetric (1) and unsymmetrical ethyl (2) and methyl (3) alkyl ethers are presented in Fig. 3.3.

Nominally, the successive increase in the number of carbon atoms in the alkyl groups of ether molecules should not lead to a considerable deviation from the straight line corresponding to this dependence. The gradually growing influence of the cooperative interactions might result in a certain deviation from the description of the experimental values of the enthalpy characteristics only at a large number of compounds belonging to the same homologous series.

The plots in Fig. 3.3 describe the dependence in question by two straight lines crossing at the enthalpy value for the vaporization of the methyl propyl ether. This fact manifests the essential changes in the specific intermolecular interaction originating from the reduced influence of the reverse dative bond in the first three members of this homologous series on the energy of the interaction caused by the terminal methyl of the corresponding alkyl group. The description of vaporization enthalpies of compounds in the series methyl butyl ether–decyl methyl ether permits a conclusion that the successive equivalent growth in the vaporization enthalpy value originates from the additive contribution of the growing number

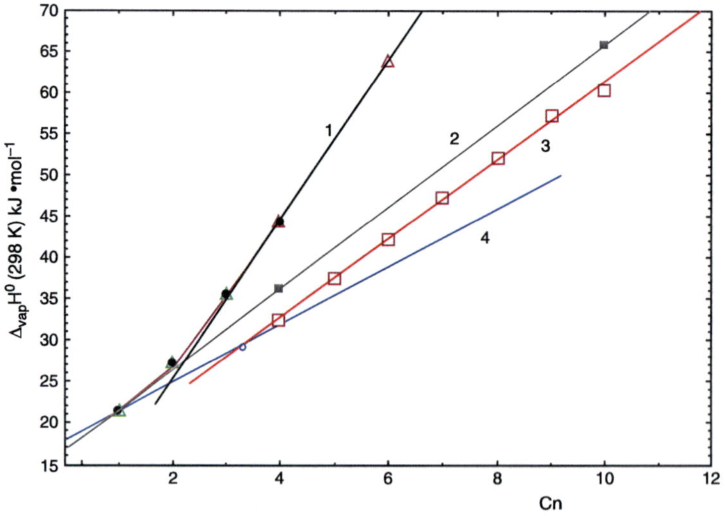

Fig. 3.3 Dependence of $\Delta_{vap}H^0$(298 K) (kJ mol^{-1}) of ethers vaporization on the number of carbon atom in alkyl groups (R) (1) symmetric ethers; (2) ethyl R ethers; (3) methyl R ethers

of the CH_2-groups. The dependence (1) for the three first representatives of the symmetric ethers, dimethyl–dipropyl ethers, reveals clearly the effect of the reverse dative bond on the energy of the specific interaction caused by the same terminal methyl group. The presence of two equal alkyl groups in each ether of this series results in the double decrease in the effect of the reverse dative bond with the successive growth of the number of carbon atoms in the alkyl fragment in the series dimethyl–dipropyl ethers. Accordingly the vaporization enthalpies of these compounds do not fit a straight line, and only approach such dependence starting with the dipropyl ether. Hence, the description of the dependence under consideration of the vaporization enthalpy by two straight lines crossing at the value of the vaporization enthalpy of propyl ether is an experimental proof of the existence of the reverse dative bond in ethers belonging to various homologous series.

A special interest attracts ethers with a branched structure of alkyl groups. It was shown in Chap. 2 that the methyl groups located in the side chain do not take part in the satiric effects but affect the displacement of the electron density to the oxygen; consequently, affect the reverse dative bond –O → $CH_3(CH_3)$–CH.

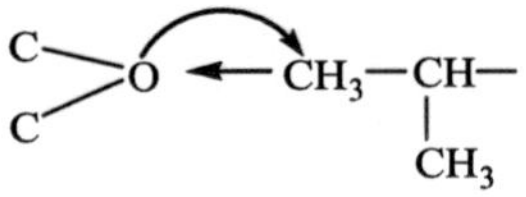

Taking into account that the number of the specific interactions of two molecules of the close surrounding involves a twofold increase in the number of these bonds, and thus in this surrounding operate four isostructural methyl groups. Therefore, the difference in vaporization enthalpies of the dipropyl and diisopropyl ethers (3.56 kJ mol^{-1}) is due to the destabilizing effect of two isostructural CH_3-groups and each one destabilizes the bond by 1.80 kJ mol^{-1} or by 0.90 kJ mol^{-1} calculated per one methyl group of two interacting molecules in the close surrounding (Table 3.2). In the symmetric diisopropyl ether each alkyl chain $CH(CH_3)$–CH_3 contains the same number of carbon atoms as an ethyl group in the diethyl ether. It is therefore presumable that their specific intermolecular interaction would be of similar or equal value. Hence, the difference in the vaporization enthalpy of diisopropyl and diethyl ethers (5.00 kJ mol^{-1}) is the contribution of the energy of the specific interaction originating from the isostructural methyl groups, and the contribution of each group equals 1.25 kJ mol^{-1}. At the same time, the difference between the vaporization enthalpy of the isopropyl ethyl and diethyl ethers (2.90 kJ mol^{-1}) is the overall energy of specific intermolecular interactions provided by two isostructural methyl groups, or the energy of this interaction corresponding to one such group is 1.45 kJ mol^{-1}.

In these calculations, we take into account that two molecules located in the close surrounding form four specific intermolecular interactions D–O → CH_3–(CH_3)–CH and four CH_3-groups belonging to the two molecules form less stable bonds whose energies are evaluated from the difference of the enthalpy characteristics of the vaporization processes. The chain of the *tert*-butyl alkyl group in the *tert*-butyl ether also consists of a two-carbon chain like in the ethyl

Table 3.2 Energies (kJ mol^{-1}) of specific intermolecular interactions of liquid ethers with branched alkyl groups

Ethers	Formula	$\Delta_{vap}H^0$(298 K) [6, 8]	DO → CH_3–	D–O → CH_3–CH_2	D–O → CH_3–CH_2–CH_2–	D*iso*-CH_3	D*iso*-CH_3/2
Diisopropyl	C_3H_80	32.12	–	6.78	–	2.50	1.25
Di-*tert*-butyl	$C_4H_{10}0$	29.82	–	6.78	–	2.68	0.67
Isopropyl ethyl	$C_5H_{12}0$	30.04	–	6.78	–	2.90	1.45
Isopropyl methyl	$C_6H_{14}0$	26.42	5.63	6.78	–	1.60	0.80
Methyl *tert*-butyl	$C_5H_{12}0$	30.4 ± 0.1	5.63	6.78	–	2.8	1.4
tert-Butyl ethyl	$C_6H_{14}0$	33.5 ± 0.1	–	6.78	–	3.2	1.6
Isopropyl *tert*-butyl	$C_7H_{16}0$	34.5 ± 0.1	–	6.78	–	2.46	1.23
tert-Butyl isobutyl	$C_8H_{18}0$	39.2 ± 0.3	–	6.78	8.92	2.6	1.30
Butyl *tert*-butyl	$C_8H_{18}0$	41.6 ± 0.2	–	6.78	8.92 DCH_2 = 4.30	2.88	1.44

group with two methyls attached thereto in the isostructural position; therefore, the difference between the vaporization enthalpy of the former compound and the diethyl ether corresponds to an overall value (2.66 kJ mol^{-1}) of the energy of specific interaction of four such groups, consequently, 0.66 kJ mol^{-1} per one group. Hence, the increased number of isostructural methyl groups in going from diisopropyl to di-*tert*-butyl ether resulted in a considerable decrease in the electron density on the carbon atoms of the groups in question.

The values of the contributions into the vaporization enthalpy of compounds (Table 3.2) of the isostructural methyl groups involved into specific interaction of low stability determined experimentally accurately to the second decimal place form the following series:

$$\text{Di}\textit{tert}\text{ - butyl ether } (0.68) < \text{Isopropyl methyl ether } (0.80) < \text{Diisopropyl ether } (1.25) < \text{Isopropyl ethyl ether } (1.45\ \text{kJ mol}^{-1}).$$

The comparison of these data indicates that in the case of the location of the isostructural methyl group at the first carbon atom, the minimum contribution into the vaporization enthalpy is produced by the methyl groups of the *tert*-butyl fragment. It means that the attachment of two isostructural methyl groups most significantly affects the electron density distribution and consequently on the carbon atoms of both isostructural methyl groups arise diminished negative charges as compared to the unsymmetrical isopropyl methyl ether containing a methyl group at the oxygen. The replacement of the latter by isopropyl in the diisopropyl ether makes it possible to assume that the negative charge on the carbon atoms of the isostructural methyl groups tends to grow increasing the low-stability specific interactions formed by these groups.

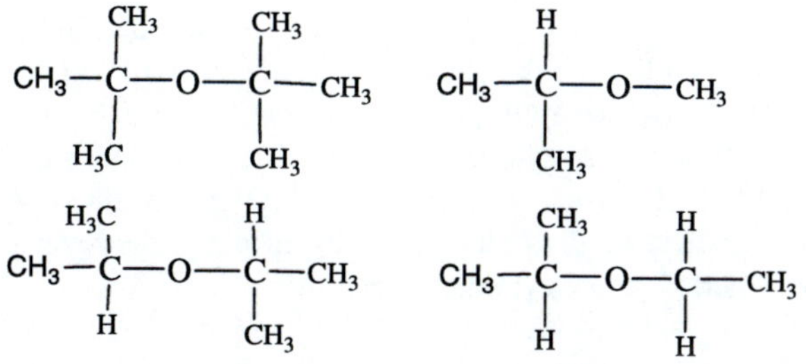

The replacement of the methyl group in the isopropyl methyl ether by an ethyl in the isopropyl ethyl ether also favors the increase in the stability of the specific interaction of the isostructural methyl of the isopropyl group. The results of the performed calculations of the contribution of the isostructural methyl groups in the propyl and butyl derivatives are within the error limits of the experimentally measured enthalpy characteristics, therefore it is possible to state the average value of their contribution at 1.4 kJ mol^{-1}. The results of the thermodynamic analysis show that the determined energies of the specific intermolecular interactions in liquid dimethyl, diethyl, and dipropyl ethers regularly grow with the length of the alkyl chain and reach the limiting value for the dipropyl ether due to the termination of the influence of the reverse dative bond in the alkyl chain of the propyl group. This conclusion is consistent with the quantum-chemical calculations substantiating the reduced electron density on the oxygen in ethers compared to the water molecule [1] and therefore the lower energy of the specific intermolecular interactions in the ethers.

3.2 Energy of Specific Intermolecular Interaction in Solid Ethers

The chemical nature of the phase transitions was revealed in [9, 10]. It was substantiated that the polymorphic transition, melting, sublimation, and vaporization were dissociation processes and were kindred to the dissociation processes in the gas phase of dimeric and more complex molecular forms, to complex formation, disproportionation by the steps of substitution and oxidation, to conproportionation, and the entropy of these processes was the characteristic of the structural transformations. This statement is valid for compounds with ionic, covalent bond types, with hydrogen bonds, specific intermolecular interactions involving a pentacoordinate carbon atom. The presence in the ethers vapor of monomer molecules and the formation of a single type of the specific intermolecular interaction excludes their participation in the polymorphic transitions. The melting enthalpy [6] amounts to one-fourth of their value of the vaporization enthalpy (Table 3.3) indicating significant alterations in the crystal structure [9].

The published enthalpy characteristics of the ethers sublimation (Fig. 3.4) manifest their dependence on the number of carbon atoms in the alkyl chain similar to that discovered for the vaporization process. Hence it is possible to state that the variation in the enthalpy characteristics of vaporization and sublimation as a function of a number of carbon atoms in the alkyl group are described by the same dependences, reflecting the reduction of the effect of the reverse dative bond in going from the methyl to propyl ether derivatives.

Therefore, we are able to use interpolation and extrapolation methods for estimating enthalpy characteristics of vaporization and sublimation of a number of yet unstudied ethers to obtain the values within the current accuracy of the experimental methods.

The above considered calculation of the energy of the specific interactions of liquid symmetric ethers is also valid for the estimation of the energies of the same interactions in crystals. The results of the calculations (Table 3.3) show the stabilization of the specific interactions in solid ethers by 2.0–2.7 kJ mol^{-1} compared to the same interactions in liquid ethers.

Table 3.3 Energies (kJ mol^{-1}) of specific intermolecular interactions in solid ethers

Ethers	$\Delta_{melt}H^0(T)$	$\Delta_{sub}H^0$(298 K) [8]	D–O → CH_3	D–O → CH_3CH_2	D–O → $CH_3(CH_2)_2$	DCH_2 group	DCH_3 isogroups
Dimethyl C_2H_6O	4.94	26.45	6.59	–	–	–	
Diethyl $C_4H_{10}O$	7.19	35.33	–	8.83	–	–	
Dipropyl $C_6H_{14}O$	10.77	46.45	–	–	11.61	–	
Dibutyl $C_8H_{18}O$	12.7	57.4	–	–	11.61	5.5 × 2	
Dipentyl $C_{10}H_{22}O$		68.6	–	–	11.61	5.6 × 4	
Propyl methyl $CH_3O(CH_2)_2CH_3$	7.67	35.24	6.59	–	11.61	–	
Propyl ethyl $CH_3(CH_2)O(CH_2)_2CH_3$	8.40	39.79	–	8.83	11.61	–	
Butyl methyl $CH_3(CH_2)_3OCH_3$	10.85	43.24	6.59	–	11.61	5.80	
Isopropyl methyl $CH_3OCH_2(CH_3)_2$	5.85	32.26	6.59	8.83	–	–	0.70
tert-Butyl methyl $(CH_3)_3COCH_3$	7.60	37.41	6.59	8.83	–	–	1.64
Diisopropyl $(CH_3)_2CHOCH(CH_3)_2$	12.04	44.16	–	8.83	–	–	2.20
Methyl ethyl $CH_3OC_2H_5$	–	30.8[a]	6.59	8.83	–	–	
Pentyl methyl CH_3 $(CH_2)OCH_3$	–	48.2[a] ± 0.5	6.59	–	11.61	5.8 × 2	
Diethyl $C_2H_5OC_2H_5$	–	36.0[a] ± 0.1	–	9.0			
Dihexyl $C_6H_{13}O$ C_6H_{13}	–	79.5[a] ± 0.5	–	–	11.61	5.8 × 6	
Diheptyl $C_7H_{15}O$ C_7H_{15}	–	90.0[a] ± 0.5	–	–	11.61	5.8 × 8	

[a]Values determined by extrapolation and by independent method of additive contributions of energies of specific interactions

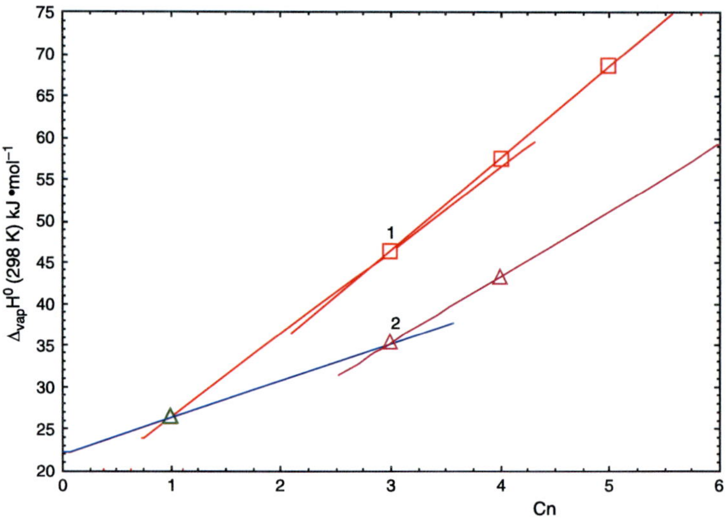

Fig. 3.4 Dependence of sublimation enthalpy on the number of carbon atoms in the alkyl group (R) of symmetric ethers (1) and methyl derivatives (2)

The obtained values of the energy of the specific intermolecular interactions should be applied to the test of the validity of the concept of the additive energy contribution of the specific intermolecular interaction O → CH_3–$(CH_2)_n$ into the sublimation enthalpy of methyl propyl and propyl ethyl ethers.

In the calculation of the sublimation enthalpy of methyl propyl and propyl ethyl ethers, the energy of specific intermolecular interaction was used measured for the crystalline dimethyl (6.59 kJ mol^{-1}), diethyl (8.83 kJ mol^{-1}), and dipropyl (11.61 kJ mol^{-1}) ethers, respectively. The obtained values of the sublimation enthalpy (36.40 and 40.88 kJ mol^{-1}) for the methyl propyl and propyl ethyl ethers differ from the experimental findings by +1.20 and −1.10 kJ mol^{-1}, respectively (Table 3.3).

As shown above, the effect of the reverse dative bond is observed on the alkyl chains with the number of carbon atoms not exceeding three. In this connection, the butyl group providing the maximum value of the energy of the specific interaction (11.61 kJ mol^{-1}) contributes additionally by the CH_2-group to the enthalpy of vaporization and sublimation of compounds. Further increase in the number of CH_2-groups in the pentyl, hexyl, and heptyl ethers also is accompanied by the additional energy contribution corresponding to the number of methylene groups in the alkyl chain (two, three, etc.). Estimated from the enthalpy characteristics, the contribution of the methylene group of the butyl substituent into the sublimation and vaporization enthalpy of ethers amounts to 5.80 and 4.80 kJ mol^{-1} for crystalline and liquid state, respectively.

Using the energies of specific intermolecular interactions and applying the additivity principle, we estimated the sublimation enthalpy of unsymmetrical ethers. Thus obtained enthalpy characteristics of methyl propyl (36.42 kJ mol^{-1}) and propyl ethyl (40.88 kJ mol^{-1}) ethers exceed the experimental data by 1.18 and 1.09 kJ mol^{-1}. In the calculation of sublimation enthalpy of butyl methyl ether, the contribution of the CH_2-group was taken into account. The obtained value of the enthalpy characteristic was 42.0 kJ mol^{-1}, lower than the experimental value by 1.24 kJ mol^{-1}. Thus the calculated value of the sublimation enthalpy at fluctuating error in all events deviated from the experimental data by 1.09–1.24 kJ mol^{-1}. This means that the error in the calculation of the energy of the specific intermolecular interaction in the case of four alkyl groups in two molecules of the close surrounding amounts to 0.26–0.30 kJ mol^{-1} per one alkyl group.

The performed thermodynamic calculations made it possible to start the evaluation of the energy contributed by the isostructural methyl groups depending on their number and the position in the alkyl chain of the ethers (Table 3.3). In discussing the influence of the isostructural groups, we should take into account that their number is doubled since in the intermolecular-specific interactions are involved four alkyl groups of the two molecules of the close surrounding (Fig. 3.1).

Let us turn now to the consecutive decrease in the sublimation enthalpy in the series dipropyl (46.45 kJ mol^{-1}), diisopropyl (44.16 kJ mol^{-1}), and diethyl (35.33 kJ mol^{-1}) ethers. The decrease in the sublimation enthalpy in the first two compounds corresponds to the destabilizing effect at the replacement of hydrogen in the CH_2-group of the alkyl by a methyl group. It should be however taken into consideration that the alkyl chain in the molecule of diisopropyl ether involved into the specific intermolecular interaction is shortened by one CH_2-group, and as a result the energy of the interaction of the terminal methyl group decreases due to the appearance of the reverse dative bond corresponding to the ethyl chain which lacked in the analogous methyl group in the di(*n*-propyl) ether. In their turn, the isostructural methyl groups in the side chain take part in the intermolecular interaction and provide a certain contribution to the sublimation enthalpy of the diisopropyl ether. As was already shown, taking the contribution to the sublimation enthalpy of the two-carbon chains of the ethyl group and the isopropyl group as equivalent we obtained from the difference of the enthalpy characteristics of these compounds the overall value of the energy of the specific intermolecular interactions corresponding to four isostructural methyl groups of two contact molecules of the close surrounding:

$$4DCH_3 = \Delta_{sub}H^0\,(298\,K)d.i.p. - \Delta_{sub}H^0\,(298\,K)d.e. \tag{3.1}$$

Similarly from the difference of the sublimation enthalpy of dipropyl and diisopropyl ethers, we obtain the energy value of the destabilizing effect at replacement of specific interactions of four propyl groups by isostructural methyl groups and the same number of ethyl groups:

$$Ddest.\ E = \Delta_{sub}H^0(298\,K)d.p. - \Delta_{sub}H^0(298\,K)d.i.p. \tag{3.2}$$

The overall contribution of isostructural methyl groups (8.83 kJ mol^{-1}) shows that the energy of the specific interaction corresponding to this group equals 2.20 kJ mol^{-1}. Therefore, we conclude that the reduced value of the sublimation enthalpy of the diisopropyl ether compared to that of the dipropyl ether has nothing in common with the steric effect. This value is due, firstly, to the difference in the specific interactions between a propyl and an ethyl groups, respectively; and, secondly, by the effect of the isostructural methyl group on the electron density distribution in the molecule of the diisopropyl ether.

Analogous interpretation approach is required in considering the sublimation and vaporization enthalpy:

$$\text{Dibutyl}\,(57.4) < tert\text{ - Butyl} < \text{Diethyl}\,(35.33\,\text{kJ}\cdot\text{mol}^{-1}),\ \text{Butyl methyl}\,(43.24\,\text{kJ}\cdot\text{mol}^{-1}) < tert\text{ - Butyl methyl}\,(37.41\,\text{kJ}\cdot\text{mol}^{-1}) < \text{Ethyl methyl ether}\,(30.8\,\text{kJ}\cdot\text{mol}^{-1}).$$

As already shown, the effect of the reverse dative bond in the interaction of an alkyl group containing a large number of methylene units ends at the propyl group, therefore in dibutyl and butyl methyl ethers at equal contributions into the specific interactions from the propyl groups it is necessary to take into account the contribution to the enthalpy characteristic of two or one CH_2-group, respectively. At the same time in solid and liquid butyl methyl and *tert*-butyl methyl ethers the specific interactions are similar to propyl group and ethyl group, respectively. Methyl substituents in these compounds provide the specific interactions of equivalent energy. Consequently, the difference in the enthalpy characteristics of butyl methyl and *tert*-butyl methyl ethers is due to the energy difference in the energies of the specific interactions originating, on the one hand, from the propyl group, and on the other hand, from the ethyl group and the specific interactions of the isostructural methyl groups.

The energy contribution from the isostructural methyl groups of the *tert*-butyl may be calculated taking

$$8\text{DCH}_3 = \Delta_{\text{sub}}H^0(298\,\text{K})\text{t.b.} - \Delta_{\text{sub}}H^0(298\,\text{K})\text{d.e.} \tag{3.3}$$

and the specific interactions of their two-carbon chain alkyls as equal to ethyl groups. Using the additivity principle, we conclude that the energy contribution of the methyl groups of *tert*-butyl methyl ether is equal to the difference between its sublimation enthalpy and the energy of the specific intermolecular interactions:

$$4\text{DCH}_3 = \Delta_{\text{sub}}H^0(298\,\text{K})\text{t.b.m.} - 2\text{DO} \rightarrow \text{CH}_3 - \text{CH}_2 - -2\text{DO} \rightarrow \text{CH}_3. \tag{3.4}$$

This contribution of four isostructural methyl groups amounts to 6.65 kJ mol^{-1}. Thus, the energy of specific interaction corresponding to one isostructural methyl group equals 1.65 kJ mol^{-1}. This energy value is lower than analogous energy contributed by the isostructural methyl group into the enthalpy characteristic of the diisopropyl ether.

The energy of specific interactions originating from isostructural methyls of the isopropyl group in the isopropyl methyl ether at the contact between the molecules of the close surrounding in the crystalline state we obtain from the difference of the sublimation enthalpy of this compound and dimethyl ether (1.46 kJ mol^{-1}), or 0.73 kJ mol^{-1} per one methyl group.

The characteristic distinction of isopropyl methyl, *tert*-butyl methyl, and diisopropyl ethers is the position of the isostructural methyl group at the carbon atom of the two carbon chain directly bound to the oxygen atom. The comparison of the energy of specific interactions originating from the isostructural methyl groups shows that the replacement of the isopropyl in the diisopropyl ether by a methyl destabilized the energy of this specific interaction in *tert*-butyl methyl ether three-fold and in isopropyl methyl ether 1.5 times. Consequently, the methyl group bound to the oxygen atom significantly affects the distribution of the electron density in the molecules of these compounds and therefore the distribution of charges on the isostructural methyl groups.

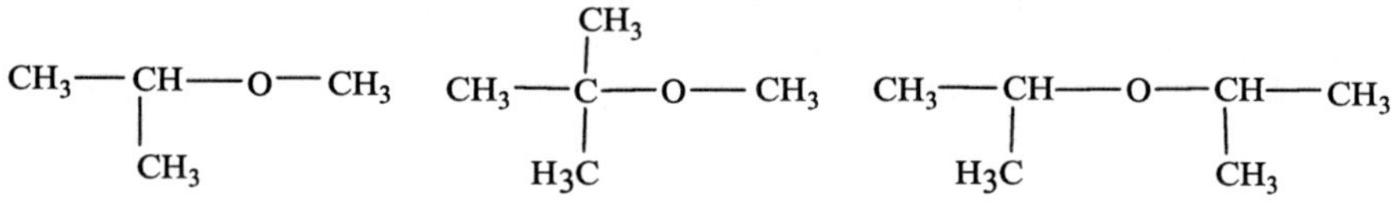

0.70; 32.26 kJ•mol^{-1} 1.65; 37.41 kJ•mol^{-1} 2.20; 44.16 kJ•mol^{-1}

Although the number of the isostructural methyl groups in the molecules of diisopropyl and *tert*-butyl methyl ethers is identical, their energy contribution into the enthalpy characteristic is different. Inasmuch as two isostructural methyl groups belong only to the *tert*-butyl, it is presumable that their location at the same carbon atom of the two-carbon alkyl chain hampers the displacement of the electron density analogous to the ethyl group and as a result the charges on the carbon atoms of these methyls are lesser than that on the carbon atoms of analogous groups in the diisopropyl ether.

The attention is attracted by the high melting enthalpy of the diisopropyl ether (12.04 kJ mol^{-1}) compared to that of the dipropyl ether (10.77 kJ mol^{-1}). This fact shows that under the effect of the crystal field the specific interactions of the four isostructural groups undergo stabilization.

The mentioned data on the decrease in the sublimation enthalpy of the ether having in the alkyl group a methyl in the α position to the oxygen testify to the suggestion that the all-explaining steric effect not only does not explain the thermodynamic properties but on the contrary conceals the causal connection between the thermodynamic characteristics and the distribution of the electron density in the molecule, and impedes the understanding of structural and energy correspondence. The thermodynamic analysis shows that independent of the number of isostructural methyl groups in molecules each of them to the identical degree diminishes the vaporization and sublimation enthalpy of ethers with respect to the ethers of normal structure, in particular, of dipropyl ether. Their role consists in the redistribution of the electron density in the alkyl chain, the redistribution of atomic charges in the molecule.

The high reliability of the considered experimentally measured enthalpy characteristics of vaporization and sublimation (to the second decimal place and in some cases with the accuracy up to 0.1–0.3 kJ mol^{-1}) ensure the validity of the rules of their variations.

The second characteristic feature of saturated ethers is the fact that every successive CH_2-group equally contributes to the enthalpy characteristic of ethers up to the dodecyl methyl ether and at least to dihexyl and hexyl vinyl ethers. This means that the specific feature and the energy of the interaction of the essentially unshared $2s^2$-electron pair of the carbon atom and the positively charged hydrogen atoms remains constant in a large number of ethers in each homologous series. As already shown before, the energy of the specific intermolecular interaction reaches the limiting value in compounds with the propyl alkyl chain. The results of performed thermodynamic calculations of the energies of the specific intermolecular interactions with accounting for the contribution of CH_2-groups proved the validity of this statement in all ethers with experimentally measured and estimated by interpolation and extrapolation enthalpy characteristics.

3.3 Energies of Specific Intermolecular Interaction in Vinyl Ethers

Similarly to liquid and solid saturated ethers, the dependence of the vaporization enthalpy of ethers with the minimal unsaturated alkyl on the number of the carbon atoms in the saturated alkyl chain shown in Fig. 3.5 is described with two straight lines crossing at the point corresponding to the vaporization enthalpy of vinyl propyl ether. This is due to the end of the reverse dative bond effect on the energy of the specific interaction originating from the terminal methyl of the propyl group.

These dependences permit the estimation of the unmeasured enthalpy characteristics of the vinyl ethyl, vinyl heptyl, and vinyl octyl ethers by interpolation and extrapolation. The estimated values of the vaporization enthalpy of these ethers and the experimentally measured data of the vinyl ethers are listed in Table 3.4. Same as in the preceding cases, the method of the additive chemical thermodynamics for the evaluation of the energies of the specific intermolecular interactions is based on the accounting for the energy contribution into the value of the vaporization enthalpy characteristic of all existing bond vacancies in the liquid and the solid state.

The most simple for the thermodynamic calculation of the energy of the specific intermolecular interaction is the divinyl ether. The structure of its liquid and solid state is similar to the structure of the diethyl ether (Fig. 3.6). It is also characterized by the formation of four specific intermolecular interactions $–O \rightarrow CH_2{=}CH–O–$

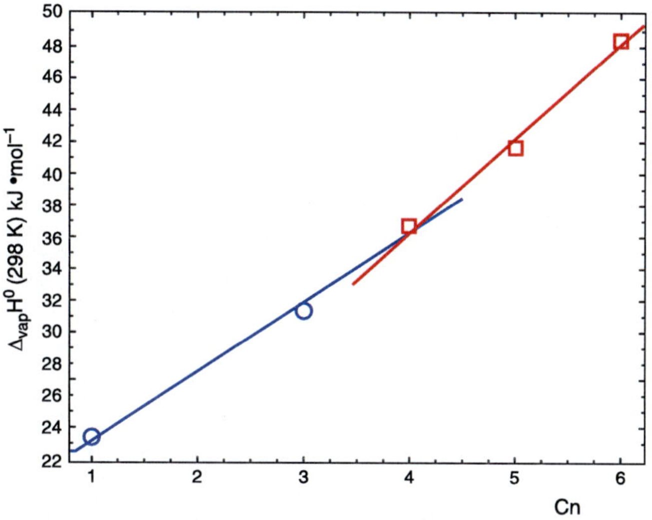

Fig. 3.5 Dependence of $\Delta_{vap}H^0$(298 K) (kJ mol^{-1}) for vaporization of vinyl ethers on the number of carbon atom at in alkyl

Table 3.4 Energies of specific intermolecular interactions (kJ mol^{-1}) of liquid unsaturated ethers

Ethers	Formula	$\Delta_{vap}H^0(T)$ [8, 11]	T (K)	D–O → $CH_3(CH_2)_n$	$D(CH_2)_n$	D–O → CH_2=CH–
Diethyl	$C_4H_{10}O$	35.33	298	6.78	$n = 0$	
Divinyl	C_4H_6O	26.1	301	–	$n = 0$	6.52
Divinyl	C_4H_6O	29.2	268	–	$n = 0$	7.3
Methyl vinyl	$C_3H_{16}O$	23.4	298	5.63	$n = 0$	6.07
Ethyl vinyl	C_4H_8O	29.5	238	7.45	$n = 0$	7.30
Ethyl vinyl	C_4H_8O	27.3[a]	298	6.78	$n = 0$	6.87
Propyl vinyl	$C_5H_{10}O$	31.4	298	8.83	$n = 0$	6.87
Isobutyl vinyl	$C_6H_{12}O$	37.4	281	8.83	$n = 0$	7.02
Butyl vinyl	$C_6H_{12}O$	36.7	298	8.83	$n = 1$ 4.9	7.07
Pentyl vinyl	$C_7H_{14}O$	41.6	298	8.83	4.9 × 2	7.07
Hexyl vinyl	$C_8H_{16}O$	48.4	298	8.83	4.9 × 3	7.04
Heptyl vinyl	$C_9H_{18}O$	54.0[b]	298	8.83	5.7 × 4	6.97
Octyl vinyl	$C_{10}H_{20}O$	59.8[b]	298	8.83	5.7 × 5	6.82

[a]Estimated by interpolation[b]Estimated by extrapolation

originating from the essentially unshared 2s^2-electron pair of the carbon and by the free electron pair of the oxygen. Consequently, the energy of each of the four bonds is equal to the quotient of the division of the vaporization enthalpy of the divinyl ether by the number of bond vacancies, namely equal 6.52 kJ mol^{-1} at 301 K:

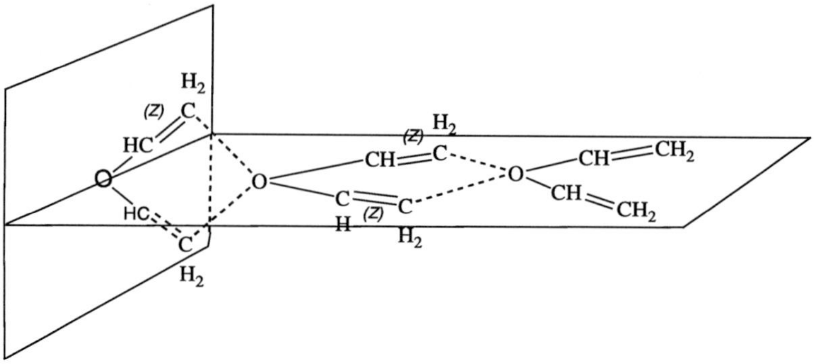

Fig. 3.6 Schematic picture of the liquid structure of divinyl ether

$$D - O \rightarrow CH_2 = CH - O- = \Delta_{vap}H^0(298\,K). \tag{3.5}$$

The energy of the specific intermolecular interaction evaluated from vaporization enthalpy of the divinyl ether established at 268 K (7.3 kJ mol^{-1}) reflects the stabilization of the liquid ether at the temperature decreasing by 30 K. In event of the diethyl ether, the energy of analogous specific interaction at the standard temperature (6.78 kJ mol^{-1}) is only by 0.26 kJ mol^{-1} more stable than that of the bond in the liquid divinyl ether D–O $\rightarrow$ CH_3–CH_2– > D–O $\rightarrow$ CH_2=CH–. Taking this difference as invariable at 238 K, we obtain the energies of the specific intermolecular interactions D–O $\rightarrow$ CH_3–CH_2– and D–O $\rightarrow$ CH_2=CH– in the liquid vinyl ethyl ether equal 7.45 and 7.25 kJ mol^{-1}, respectively. Accounting for the energy of the same interactions at the standard conditions, we obtain the standard vaporization enthalpy of the vinyl ethyl ether at 26.7 kJ mol^{-1}. Taking also the energy of the specific intermolecular interaction in the liquid dimethyl ether (5.63 kJ mol^{-1}) as equal to the energy of the same interaction in the liquid vinyl methyl ether, we evaluated the value of the standard vaporization enthalpy of the latter compound at 24.4 kJ mol^{-1}, by 1.0 kJ mol^{-1} more than the experimentally determined value.

This difference between the experimentally measured vaporization enthalpy and the value calculated from the additive contribution of the energies of the specific interactions into the enthalpy characteristic illustrates their mutual influence. This is due to the dissimilar effect of the reverse dative bond on the energy of the specific interaction of liquid dimethyl ether, on the one hand, and liquid diethyl, divinyl, and methyl vinyl ether, on the other hand.

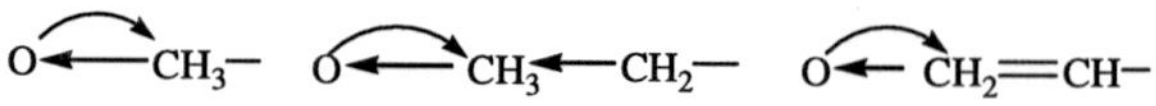

Therefore, the energy of the specific intermolecular interaction O $\rightarrow$ CH_3 in liquid dimethyl and methyl vinyl ethers should be the most proper regarded as invariable. This condition results in the energy of the specific interaction DO $\rightarrow$ CH_2=CH–O– equal 6.07 kJ mol^{-1}. The known rule of the reduction of the effect of the reverse dative bond in the series methyl–propyl group should lead to the stabilization of the specific intermolecular interaction DO → CH_2=CH–O– in liquid propyl vinyl, isobutyl vinyl, and butyl vinyl ethers and to the invariability of the energy of the specific interactions D–O → CH_3–$(CH_2)_n$– in saturated and unsaturated ethers. In the calculation of the energy of the specific interactions, the contribution of the CH_2-groups (4.9 kJ mol^{-1}) into the vaporization enthalpy is taken into account growing with the number of these groups in the saturated alkyl chain. The performed thermodynamic calculations (Table 3.4) manifest the trend to stabilization of the bond energy D–O → CH_2=CH– in the series

$$\begin{aligned}&\text{Methyl vinyl (6.07)} < \text{Ethyl vinyl (6.59)} < \text{Propyl vinyl (6.87)} < \text{Butyl vinyl (7.07)}\\&\approx \text{Pentyl vinyl (7.02)} \approx \text{Hexyl vinyl (7.04); Heptyl vinyl (6.97)}\\&- \text{Octyl vinyl } (6.82\,\text{kJ mol}^{-1})\end{aligned}.$$

In methyl vinyl–propyl vinyl ethers and further in butyl vinyl–hexyl vinyl ethers (7.04 kJ mol^{-1}), the scatter of the value is within the experimental error. The energy of the same bond in compounds where the vaporization enthalpy was evaluated by extrapolation indicated the mean error value at ±0.1 kJ mol^{-1}.

The attention is attracted by the enhanced value of the enthalpy of isobutyl vinyl ether (37.4 kJ mol^{-1}) measured experimentally at 281 K as compared to the enthalpy characteristic of butyl vinyl ether (36.7 kJ mol^{-1}). If based on the experimentally determined values of vaporization enthalpy of the divinyl ether at 301 and 281 K, the increase in the enthalpy characteristic at the decrease in the temperature by 1 K is taken as 0.1 kJ mol^{-1} then we obtain the value of the vaporization enthalpy of the isobutyl vinyl ether equal 35.7 kJ mol^{-1}. Inasmuch as the contribution of the isostructural methyl group is equal to the difference between the vaporization enthalpy of the isobutyl vinyl and propyl vinyl ethers, taking into account the involvement into the interaction of two molecules in the close surrounding,

$$2\text{DCH}_3 = \Delta_{\text{vap}}H^0(298\,\text{K})\text{isob.v.} - \Delta_{\text{vap}}H^0(298\,\text{K})\text{p.v.} \tag{3.6}$$

we obtain the energy of the contribution of two isostructural methyl groups equal 4.3 or 2.15 kJ mol^{-1} per one methyl. This value significantly exceeds the energy contribution in the corresponding saturated ethers. The difference consists in the fact that in the unsaturated ether the isostructural methyl group is located at the second carbon atom of the isobutyl fragment whereas these groups in the saturated ether are bound to the carbon atom linked directly to the oxygen. Therefore, we conclude that the energy contribution of the isostructural methyl group can change depending on its position in the alkyl chain.

The structure of liquid isobutyl vinyl ether is shown by a schematic picture (Fig. 3.7) that depicts the interaction of contacting molecules with the formation of the specific interactions of the isostructural methyl groups with the vinyl

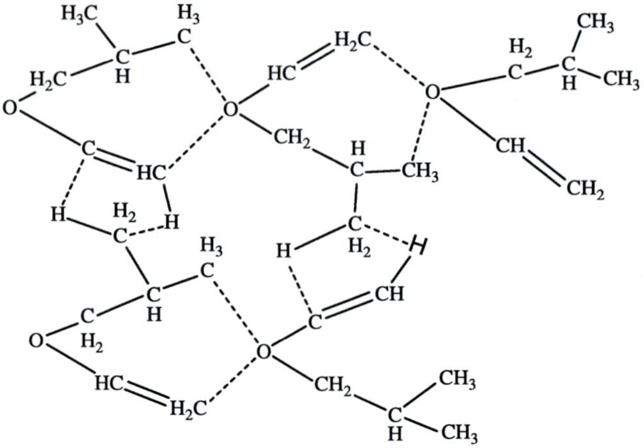

Fig. 3.7 Schematic picture of the liquid structure of isobutyl vinyl ether

hydrocarbon chain, on the one hand, formed by the essentially unshared $2s^2$-electron pair of the carbon atom, and on the other hand, by a hydrogen atom located in the *trans* position of the same isostructural methyl group with =CH– and –CH_2-groups.

The energy value of the arising specific intermolecular interaction of the side-chain methyl group at the second carbon atom of the alkyl chain (2.15 kJ mol^{-1}) is nearly 1.5 times greater than the energy of the interaction of the analogous group in the liquid isopropyl ethyl ether (1.45 kJ mol^{-1}). This energy difference of the interaction is due to the difference in the negative and positive charges on carbon and hydrogen atoms of the isostructural methyl group and the atoms of the vinyl group in the isobutyl vinyl ether.

The nature of the specific intermolecular interaction of all discussed crystalline and liquid ethers is governed by the participation of the pentacoordinate carbon atom of the terminal methyls and isostructural CH_3-groups. Consequently, between the energies of the specific interactions and more complex enthalpy characteristics in various homologous series a correlation should always exist. This can be illustrated by the example of the correlation dependences of vaporization enthalpy on some enthalpy characteristics of ethers from three homologous series taken from Ref. [11] (Fig. 3.8). The correspondence of this correlation to straight lines suggests a conclusion that in each homologous series the influence of the reverse dative bond observed in the ether molecules of the first three compounds of each series is leveled and the subsequent contribution of each CH_2-group with their growing number in the alkyl chain remains invariable due to the same reason, the invariability of the nature of the specific interactions.

The high reliability of the experimentally measured enthalpy characteristics of vaporization and sublimation (to the second decimal place and in some cases with the accuracy up to 0.1–0.3 kJ mol^{-1}) and of the rules of their variations makes it

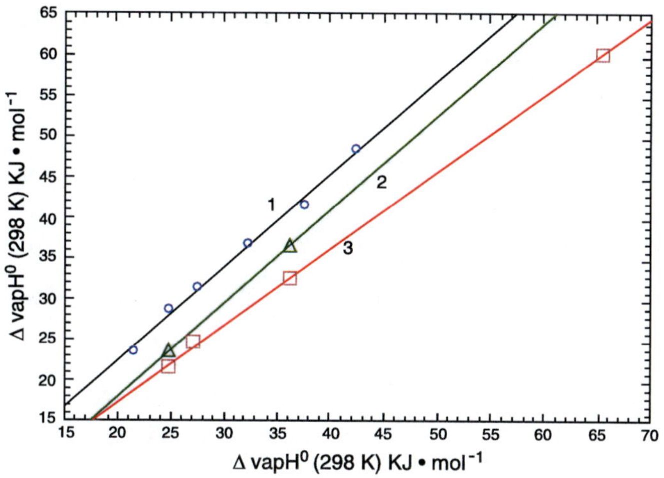

Fig. 3.8 Relation between vaporization enthalpies of ethers (1) methyl-R versus vinyl-R; (2) ethyl-R versus vinyl-R; (3) ethyl-R versus methyl-R

Table 3.5 Vaporization enthalpy of ethers established by extrapolation and interpolation (kJ mol^{-1})

Ethers	Formula	$\Delta_{vap}H^0$(298 K)	Ethers	Formula	$\Delta_{vap}H^0$(298 K)
Pentyl ethyl	$C_7H_{16}O$	41.5	Nonyl ethyl	$C_{11}H_{24}O$	60.7
Hexyl ethyl	$C_8H_{18}O$	46.6	Undecyl ethyl	$C_{11}H_{24}O$	70.4
Heptyl ethyl	$C_9H_{20}O$	51.3	Methyl ethyl	C_3H_8O	24.7
Octyl ethyl	$C_{10}H_{22}O$	56.0	–	–	–

possible to use the interpolation and extrapolation methods for the evaluation of vaporization enthalpies of not yet investigated ethers (Table 3.5).

3.4 Energies of Specific Intermolecular Interaction in Methoxy, Ethoxy, and 1,2-Epoxy Compounds

In Sect. 3.2, we showed that the vaporization enthalpy of unsymmetrical ethers, methoxy ethers, and ethoxy ethers can be determined with a great accuracy from the energy of the specific intermolecular interactions evaluated from the vaporization enthalpy of the symmetric ethers. This can be demonstrated by an example of the 1-ethoxypropane. Inasmuch as every interacting molecule of the ether generates four specific interactions, among which two are formed by ethyl groups (D–O → CH_3–CH_2 = 6.78 kJ mol^{-1}) and the other two by propyl groups (D–O → CH_3–CH_2–CH_2 = 8.92 kJ mol^{-1}) the sum of their energies amounts to 31.40 kJ mol^{-1} whereas the experimentally measured value is 31.39 kJ mol^{-1} (Table 3.6). However, starting with 1-methoxybutane the energy contribution of

Table 3.6 Energies of specific intermolecular interactions (kJ mol^{-1}) of liquid methoxy, dimethoxy, and ethoxy ethers

Ethers	Structure	$\Delta_{vap}H^0(T)$ [8, 11]	T (K)	D–O → CH_3	D–O → $CH_3(CH_2)_n$
1-Ethoxypropane $C_5H_{12}O$	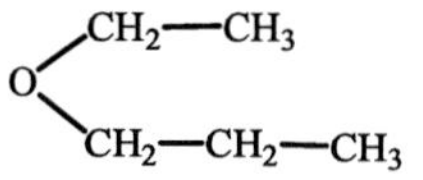	31.39	298	6.20 6.78	8.92
1-Ethoxyisopropane $C_5H_{12}O$	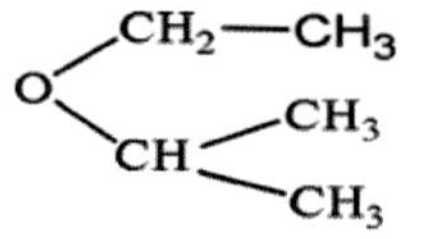	30.0	298	6.78	6.78, $2DCH_3 =$ 3.70
1-Methoxybutane $C_5H_{12}O$	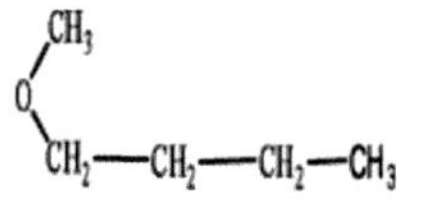	32.5 ± 0.1	298	5.63	10.62
1,2-Dimethoxyethane $C_4H_{10}O_2$	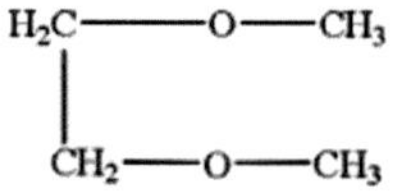	36.8 ± 0.1	298	4.60	–
1,1-Dimethoxyethane $C_4H_{10}O_2$	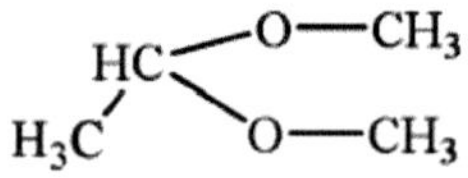	36.4 ± 0.1	298	4.55	–
1-Ethoxy-2-methoxyethane $C_5H_{12}O_2$	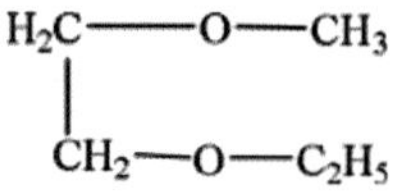	39.8 ± 0.1	298	4.60 × 6	$n = 1$ 6.20
1,2-Diethoxyethane $C_6H_{14}O$	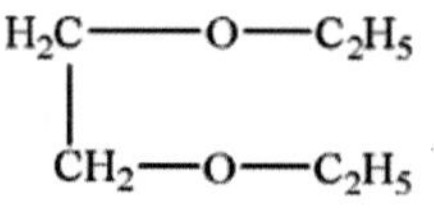	43.2 ± 0.1	298	4.60 × 4	$n = 1$ 6.20
1,1-Diethoxyethane $C_6H_{14}O_2$	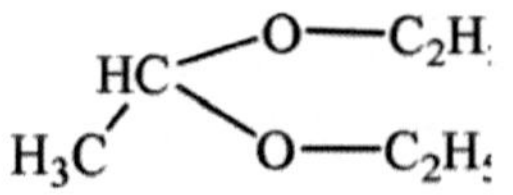	43.2	298	4.60 × 4	$n = 1$ 6.20

(continued)

Table 3.6 (continued)

Ethers	Structure	$\Delta_{vap}H^0(T)$ [8, 11]	T (K)	D–O → CH_3	D–O → $CH_3(CH_2)_n$
1,1-Dimethoxy-butane $C_6H_{14}O_2$	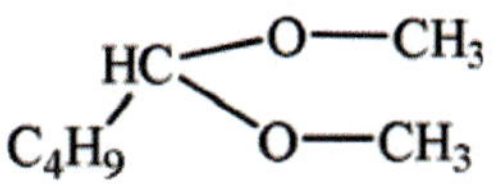	41.2	317	4.60 × 4	5.70
1,1-Dimethoxy-butane $C_6H_{14}O_2$	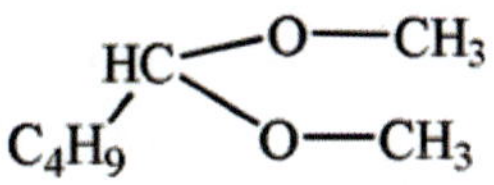	39.8	412	4.60 × 4	5.35
1,1-Dimethoxy-3-butene $C_6H_{12}O_2$	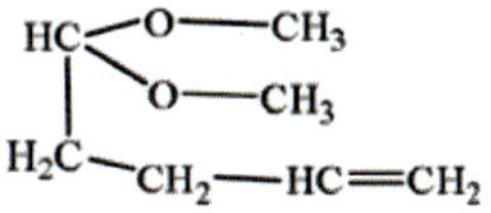	42.0	320	4.60 × 4	5.90
1,1-Dimethoxycyclobutane $C_6H_{12}O_2$	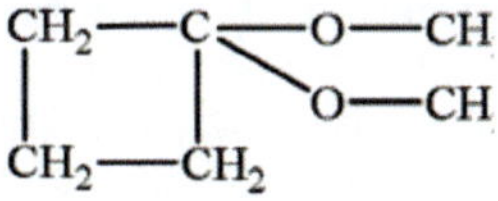	42.0 ± 0.3	298	4.60 × 4	5.90

CH_2-group should be taken into account for it interacts with the molecules of the close surrounding, whereas the influence of the reverse dative bond ends at the third carbon atom of the propyl chain.

In the case of dimethoxy compounds the growing number of oxygen atoms in the molecule results in the enhanced number of bond vacancies and, consequently, of specific interaction; however, it does not find an adequate manifestation in the energy. The molecule of a dimethoxy compound with two oxygen atoms and the same number of terminal methyl groups and the CH_2-groups of the ethane fragment have eight bond vacancies that are involved into the formation of the specific intermolecular interactions (Fig. 3.9). In the molecule of 1,2-dimethoxyethane all interacting alkyl groups contain a single carbon each, therefore the specific interaction D–O → CH_3, D–O → CH_2 are identical, and their energies are determined from the vaporization enthalpy $\Delta_{vap}H^0(T)/8$.

This statement is also valid for the 1,1-dimethoxyethane. As seen from the data of Table 3.6, the energies of the specific interactions in liquid compounds are practically identical and the difference is within the error limits of the calorimetry experiment [7].

Thus it is possible to state that the energy of the specific interaction of the methoxy group is independent of its position in the ethane fragment of 1,2-dimethoxyethane and 1,1-dimethoxyethane.

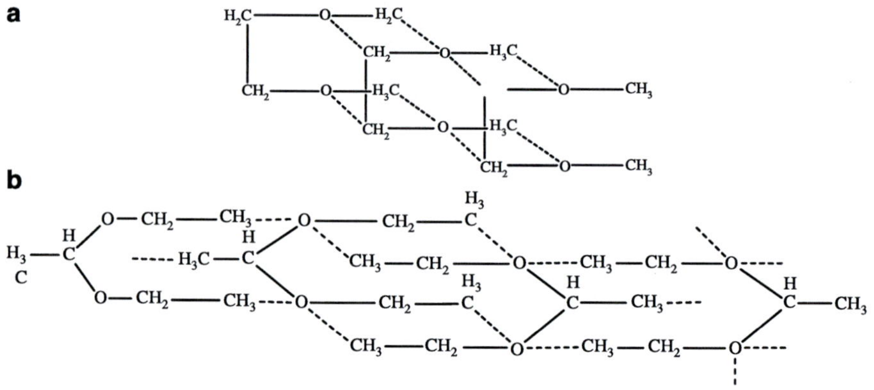

Fig. 3.9 Schematic picture of the liquid structure of 1,2-dimethoxyethane (**a**) and 1,1-diethoxy ethane (**b**)

In liquid 1-ethoxy-2-methoxyethane in the close surrounding form two specific interactions of ethyl groups, two of methyl groups, and four of the ethane fragment. Hence, overall six D–O → CH_3, D–O → CH_2 interactions and two D–O → CH_3–CH_2 bonds are formed. The energy of the latter specific intermolecular interactions is determined from the difference in the vaporization enthalpy and the sum of the energy of the specific interaction of the methyl and the CH_2-groups

$$\begin{aligned} &2D - O \rightarrow CH_3 - CH_2 \\ &\quad = \Delta_{vap}H^0(T) - (4D - O \rightarrow CH_2 + 2D - O \rightarrow CH_3). \end{aligned} \tag{3.7}$$

The established value of the energy of this specific interaction (6.20 kJ mol^{-1}) is lower than in the liquid diethyl ether (6.78 kJ mol^{-1}). The increase in the number of ethoxy groups to two in the molecule of 1,2-diethoxyethane ensures the equal number of specific interactions, 4DO → CH_2 and 4DO → CH_3–CH_2, therefore the energy of the latter bond is determined from the equation

$$2D - O \rightarrow CH_3 - CH_2 = \Delta_{vap}H^0(T) - (4D - O \rightarrow CH_2), \tag{3.8}$$

also at 6.20 kJ mol^{-1}.

Based on the above we conclude that the energies of the specific interactions originating from the methoxy group in the liquid 1,2-dimethoxyethane, 1,1-dimethoxyethane, 1-ethoxy-2-methoxyethane, and from the ethoxy group in the liquid 1-ethoxy-2-methoxyethane and 1,2-diethoxyethane may be used for the establishment of the vaporization enthalpy of yet unstudied compounds. This can be shown by an example of 1,1-diethoxyethane (Table 3.6).

The real and correctly agreed with the experiment pattern of the structural changes in the compounds under consideration makes it possible to estimate the energies of the specific interactions in the liquid 1,1-dimethoxybutane (317 K),

1,1-dimethoxybutane (412 K), and 1,1-dimethoxy-3-butene. Here, we do not take into account the temperature dependence of the energy of the specific interaction generated by the methoxy group and thus introduce a certain error into the calculations for the experimentally measured values of the characteristics determined at temperatures differing from the standard one. Each of the compounds in question form in the liquid case by their methoxy, butane, and butene groups, respectively, four corresponding specific interactions. Therefore, the energy of the specific intermolecular interaction of one among two types should be obtained from the equation where the D–O → CH_3–CH_2 is one of two bonds formed by the terminal methyls of the butyl groups:

$$D - O \rightarrow CH_3 - CH_2 = \frac{\Delta_{vap}H^0(T) - (4D - O \rightarrow CH_3)}{4}. \quad (3.9)$$

Taking the energy of the specific interaction involving the methoxy group at 4.60 kJ mol^{-1}, we obtain the energies of the specific intermolecular interactions of the other methoxy compounds (Table 3.6). It should be noted that the established energies of the specific interaction with a butane fragment in the 1,1-dimethoxybutane (5.70 kJ mol^{-1}, 317 K) and butane fragment in the 1,1-dimethoxy-3-butene (5.90 kJ mol^{-1}, 320 K) differ by 0.2 kJ mol^{-1}. The enhanced stability is characteristic of the specific intermolecular interaction in the compound with the unsaturated fragment. In the liquid, 1,1-dimethoxycyclobutane exists with the same types of the specific interactions. The equal number of these bonds permits the use for the calculation of the energies of the specific intermolecular interactions by the same equation (3.6). The energy of specific interaction with the CH_2-groups of the butane ring (Table 3.6) has the same stability as the bond formed by the butane fragment but diminished compared with the ethyl group.

The compounds with an epoxy group should have different electron density on the terminal methyl and CH_2-group in keeping the position of the oxygen atom in the alkyl chain. Since the fragment is a cycle with two electron pairs on the oxygen, consequently, the charges on the mentioned terminal groups and the displacement of the electron density along the chain depend on the distance of the CH_2-group from the epoxy fragment.

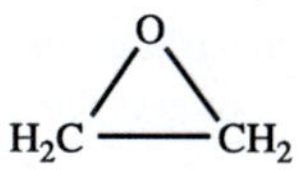

This means that the specific intermolecular interactions formed by these groups should differ in the energies. Their estimation is obviously interesting for it permits a sufficiently adequate judgment on the architecture of the structure of the liquid and the crystalline state and on the strength of the specific intermolecular interactions in the close surrounding and unit volume. The existence of these strongly differing features of the terminal methyl and epoxy groups of the alkyl chain permits the presentation of the liquid state of epoxyalkyls in the form of two

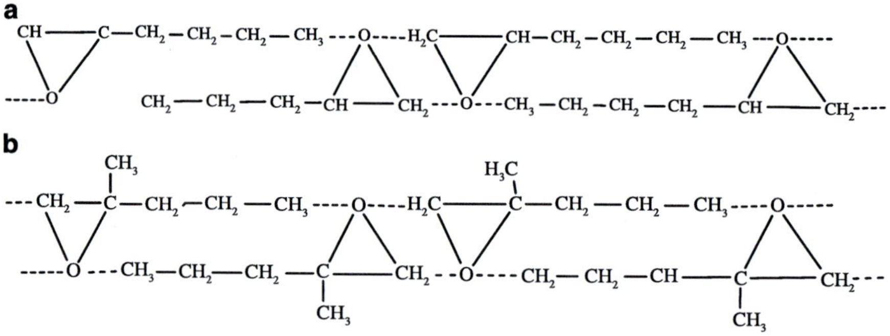

Fig. 3.10 Schematic picture of the liquid structure of 1,2-epoxyhexane (**a**) and 2-methyl-1,2-epoxypentane (**b**)

chains where the molecules are bound by four specific interactions of two dissimilar types (Fig. 3.10). The chains are connected by weak intermolecular interaction, in particular, with the involvement of the side methyl group in the case of 2-methyl-1,2-epoxypentane whose contribution into the vaporization enthalpy may be taken equal to that of the isostructural group in the liquid isopropyl ether, 0.70 kJ mol^{-1}.

Regarding the methyl group in the molecule of methyloxirane as equivalent to isostructural groups with the contribution into the vaporization enthalpy of oxirane as such group of 1.12 kJ mol^{-1}, we should take into consideration the significant influence of the epoxy fragment on the decrease in the electron density and consequently in the charge on the carbon atom of this group. This charge sharply changes the acceptor properties of the carbon atom in this group and the energy of the forming specific interaction with the participation of the essentially unshared $2s^2$ electron pair of the carbon atom. The growing number of carbon atoms in the alkyl chain results in the reduction of the effect of the oxygen in this fragment on the terminal methyl group. This effect comes to the end at the presence of three carbon atoms in the alkyl chain. Therefore, the energy difference of the propyl chain in the 2-methyl-1,2-epoxypentane and 1,2-epoxyhexane should be taken as equal to the energy of the specific intermolecular interactions in the liquid dipropyl ether (8.92 kJ mol^{-1}). Accounting for it and, in the case of 2-methyl-1,2-epoxypentane, for the contribution into the vaporization enthalpy of the side methyl group, and in the case of 1,2-epoxyhexane for the contribution of the third CH_2-group we conclude that the energies of the specific intermolecular interactions with the oxygen atom in this ring DO $\rightarrow$ CH_2–CH– are equal, respectively, to $D - O \rightarrow CH_2 - CH- = (\Delta_{vap}H^0(T) - 2D - O \rightarrow CH_3 - CH_2 - CH_2 - DCH_2)/2$ in 1,2-epoxyhexane and

$D - O \rightarrow CH_2 - CH- = (\Delta_{vap}H^0(T) - 2D - O \rightarrow CH_3$
$-CH_2 - CH_2 - 2D - CH_3)/2$ in 2-methyl-1,2-epoxypentane.

The energy contribution of the CH_2-group into the vaporization enthalpy taken analogously to the case of unsymmetrical ether equal 4.8 kJ mol^{-1}. The performed thermodynamic calculations showed that the energies of the specific intermolecular interactions in the liquid 1,2-epoxyhexane (10.25 kJ mol^{-1}) and 2-methyl-1,2-epoxypentane (10.60 kJ mol^{-1}) are of close value, therefore they should be taken as the limiting value for all epoxy compounds whose vaporization enthalpy is different only because of the energy contribution of all successive CH_2-groups. Hence the estimated values of the vaporization enthalpy of 1,2-epoxyheptane, 1,2-epoxyoctane, and 1,2-epoxynonane amount to 47.9, 52.7, and 57.6 kJ mol^{-1} respectively.

3.5 The Energies of Hydrogen Bonds and Specific Intermolecular Interaction of Liquid and Solid Organic Cyclic Oxides

3.5.1 Saturated Organic Cyclic Oxides

The structural features of the molecules of ethers and cyclic saturated and unsaturated oxides are the oxygen with two free bond vacancies and an essentially unshared $2s^2$(c)-electron pair of carbon atom forming donor–acceptor interactions. The presence of a cycle in the saturated oxide creates in the system a certain rigidity that impedes the displacement of the electron density from carbon atom of the central CH_2-group in oxetane and two analogous groups in oxolane to the oxygen atom. The growing vaporization enthalpy in the series from ethylene oxide to oxetane and to 1,4-dioxane (Table 3.7) is a consequence of the growing displacement of the electron density with the growing number of the CH_2-groups from two to four in the molecule reflecting the stabilization of the specific interactions in this series. The comparison of the structures of the molecule of the ethylene oxide and the methyloxirane at the equal number of the free bond vacancies suggests that the difference in their vaporization enthalpy (1.12 kJ mol^{-1}) corresponds to the energy contribution of the isostructural methyl group into the enthalpy characteristic. In this case it should be taken into consideration that the replaced hydrogen atom does not participate in the interaction. Therefore, the energy value in question that is the difference between the vaporization enthalpy of ethylene oxide lacking these methyl groups and methyloxirane with a single CH_3-group is a sum of the compensated energy contributed by the replaced hydrogen atom, on the one hand, and of the energy contributed by the mentioned methyl group, on the other hand. Inasmuch as the molecule takes part in the interaction not only by its own isostructural methyl groups, but contacts with analogous groups of another molecule of the close surrounding, the number of weak specific interactions formed by these groups is doubled for each molecule. Therefore, the discussed difference between the vaporization enthalpy of ethylene oxide lacking these methyl groups and methyloxirane with a single CH_3-group should be assigned to the contribution of one such group, and the second group interacting with the neighboring molecule compensates the energy of the interaction of the substituted hydrogen atom. Similar pattern in the changes in molecular structure and values of the vaporization

Table 3.7 Energies of specific intermolecular interactions (kJ mol^{-1}) of liquid saturated cyclic oxides (298 K)

Compound	Structure	$\Delta_{vap}H^0$ (*T*) [8]	D–O → CH_2–	$\sum DCH_3iso$ DCH_3=
Ethylene oxide	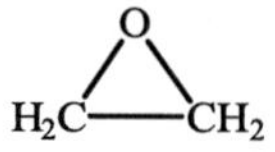	26.80	6.70	–
Oxetane trimethyleneoxe	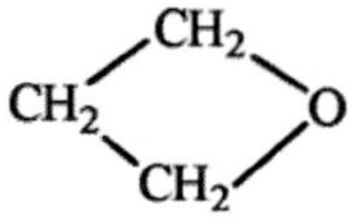	30.78	7.65	–
Oxolane tetrahydrofuran	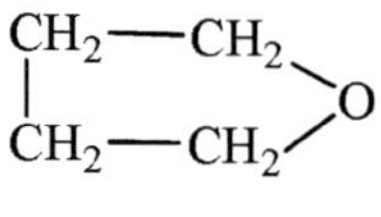	32.90	8.25	–
1,4-Dioxane	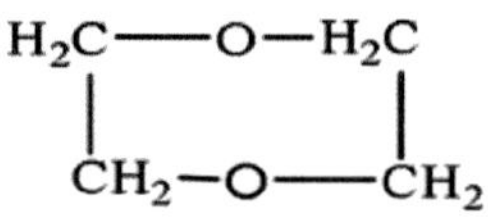	38.64	4.83	–
1,3-Dioxane	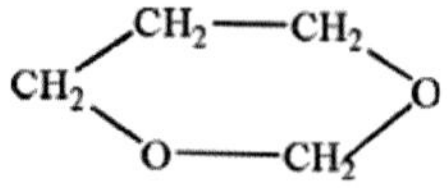	39.10	4.90	–
3,5,7-Trioxacyclononane	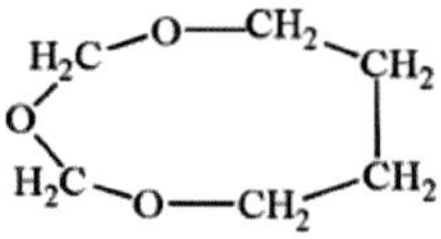	44.7	3.70	–
Methyloxirane propylene oxide	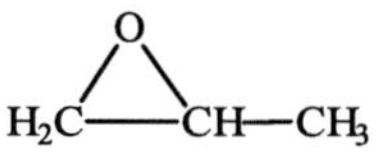	27.92	6.42	$\sum DCH_3iso$ = 2.24 DCH_3=1.12
3,3-Dimethyloxolane	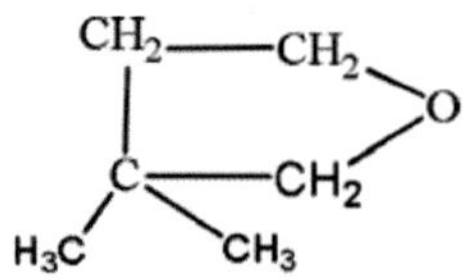	33.9	7.46	$\sum DCH_3iso$ = 4.00 DCH_3 = 1.00

(continued)

Table 3.7 (continued)

Compound	Structure	$\Delta_{vap}H^0$ (*T*) [8]	D–O → CH_2–	$\sum DCH_3 iso$ DCH_3=
2-Methyloxolane	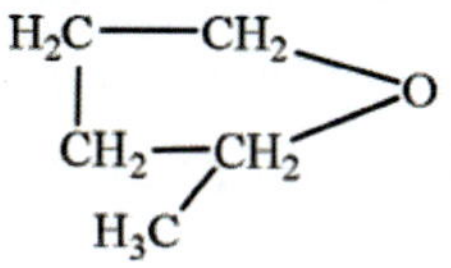	34.00	7.95	$\sum DCH_3 iso = 2.20$ $DCH_3 iso = 1.10$
2,5-Dimethy ltetrahydrfuran	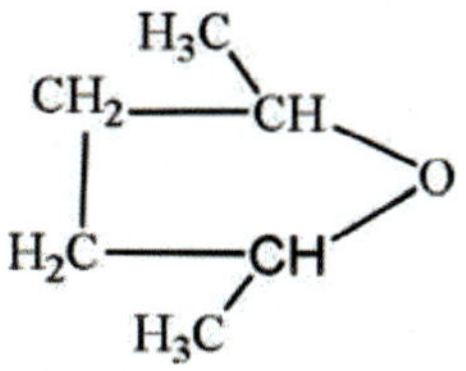	35.4	7.60	$\sum DCH_3 iso = 5.0$ $DCH_3 iso = 1.25$

enthalpy exists in the oxolane (32.90 kJ mol^{-1}) and the 2-methyltetrahydrofuran (34.00 kJ mol^{-1}) (Table 3.7). Consequently, the difference in the vaporization enthalpy of these compounds (1.10 kJ mol^{-1}) is the contribution of two methyl groups of the interacting molecules of the close surrounding. Therefore, it is possible to state that each methyl group takes part in the intermolecular specific interaction in liquid saturated oxides with an energy equal 1.10 kJ mol^{-1} and an overall energy value of 2.2 kJ mol^{-1}. Consequently, the overall value of the energy of stable specific intermolecular interaction in liquid oxolane amounts to 32.90 kJ mol^{-1} and in 2-methyltetrahydrofuran to 31.80 kJ mol^{-1} (Table 3.7).

The presence of a single oxygen atom in the molecule of cyclic oxides oxolane, furan, 2,3-dihydrofuran governs the formation of the cyclic chain structure in the liquid state of these compounds (Fig. 3.11).

In the cyclic chain structure four specific intermolecular interactions –O → CH_2–CH_2– are present. Therefore, the energies of the specific intermolecular interactions are evaluated by dividing the vaporization enthalpy by the number of bond vacancies in the molecule (Table 3.7) [2]. Here the nonspecific interactions, whose low energies are comparable to the error in the experimental measurements are not taken into account. From the established energy values of the specific intermolecular interaction DO → CH_2 follows that they regularly grow in the series:

Ethylene oxide < Oxetane < Tetrahydrofuran

with the growing number of carbon atoms in the cycle. This means that in going from a compound with two methylene groups to compounds with three and four

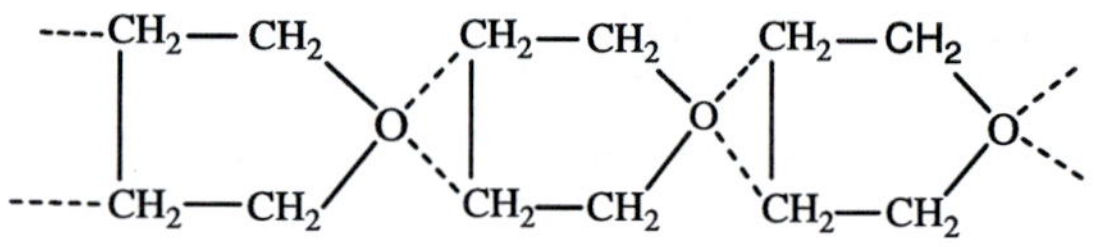

Fig. 3.11 Fragment of schematic pattern of the liquid structure of saturated oxides

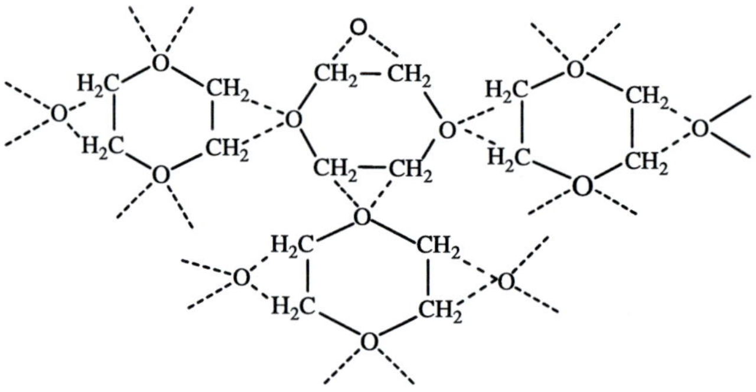

Fig. 3.12 Schematic picture of the network structure of liquid 1.4-dioxane

CH_2-groups in the respective cycles grows the shift of the electron density to the oxygen atom and consequently increase the donor properties of the oxygen and acceptor properties of the carbon atoms.

In the second group of oxide compounds of the dioxane series, the number of bond vacancies is characteristically increased: four of them form two oxygen atoms and four CH_2-groups. The regular position of oxygen atoms and two ethylene groups in the structure of 1,4-dioxane molecule defines the formation of the net structure of the liquid state (Fig. 3.12). Four molecules should be located in the close surrounding, each forming eight bonds of identical type (Fig. 3.12). Consequently, the vaporization enthalpy originates from the rupture of eight specific intermolecular interactions. Therefore at the large overall number of bond vacancies in the 1,4-dioxane molecule compared to ethylene oxide, but equal calculated per one oxygen atom and one $–CH_2–CH_2–$ fragment and at equal number of CH_2-groups per one oxygen in the molecule, it is presumable that the energies of the specific intermolecular interactions of these compounds have similar values. The performed thermodynamical calculations show the deviation in the energy values in the second decimal place. Rounding off the numbers we obtain 4.80 and 4.90 kJ mol^{-1}, respectively, in liquid 1,4-dioxane and 1,3-dioxane. Thus the fragment $–CH_2–CH_2–$ possesses an enhanced rigidity and reduced ability to electrons transmission to oxygen atoms as compared to the $–CH_2–CH_2–CH_2–$ fragment. Consequently, the specific interactions in liquid 1,4-dioxane are somewhat less stable.

Hence, the formation of eight specific interactions by the molecules of 1,4-dioxane and 1,3-dioxane destabilized each bond reducing the energy nearly twofold compared to the specific intermolecular interactions existing in the liquid tetrahydrofuran.

Another interesting question is the effect of the place taken by the CH_2-group in the molecule on the energy of the specific interaction on condition of the equal number of oxygen atoms and equal number of the methylene groups. Just this point is illustrated by the performed thermodynamical calculations (Table 3.7) of the energies of the specific intermolecular interactions in 1,4-dioxane and 1,3-dioxane.

In liquid 1,4-dioxane and 1,3-dioxane each molecule of these compounds contacts with four molecules in the close surrounding. The increased number of oxygen atoms and CH_2-groups at their remaining ratio in the molecule of 3,5,7-trioxacyclononane should result in certain change in the electron density on the oxygen and carbon atoms, and to altered number of contact molecules in the close surrounding. In the 3,5,7-trioxacyclononane molecule, the fragment of four CH_2-groups located between oxygen atoms evenly shifts the electron density to each of the oxygens. The cyclic structure of the molecule should strengthen the role of the dative bond, therefore the energy of the specific intermolecular interaction $=O \rightarrow CH_2=$ should be less stable than in the liquid 1,4-dioxane and 1,3-dioxane. The flat structure of the 1,3,5-trioxane molecule makes it possible to expect the formation of net structure in the solid and liquid state (Fig. 3.13), where in the close surrounding six molecules coordinate so that oxygen atoms interact by their lone electron pairs with two CH_2-groups. The 12 bond vacancies of the molecule provide the energy of the specific intermolecular interaction $-O \rightarrow CH_2$. Therefore, the energy of this specific interaction is evaluated from the vaporization enthalpy $\Delta_{vap}H^0(T)/12$ (3.70 kJ mol^{-1}) [2]. As already mentioned, the energies of the nonspecific interactions are within the error limits of the experimental measurements and do not affect the determined energy value.

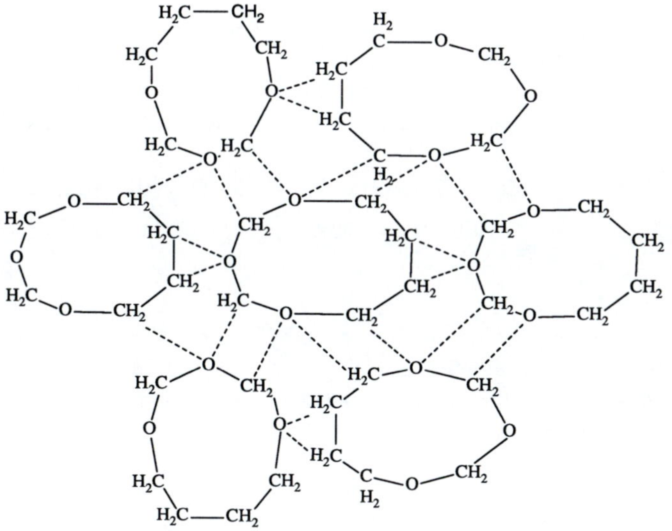

Fig. 3.13 Schematic picture of the 3,5,7-trioxacyclononane in condensed state

The lower stability of the same type of the specific intermolecular interaction in the liquid state of 3,5,7-trioxacyclononane compared with 1,3-dioxane and 1,4-dioxane 1,3-dioxane (4.90) ≈ 1,4-dioxane (4.80) > 3,5,7-trioxacyclononane (3.70 kJ mol^{-1}) at equal ratio of oxygen atoms to the number of CH_2-groups in the molecules should be accounted for by the larger number of coordinated molecules in the close environment of the liquid 3,5,7-trioxacyclononane.

To establish the energy of specific interaction in the liquid methyloxirane, the contribution of the methyl group into the vaporization enthalpy of this compound should be evaluated. This value equals the difference between the vaporization enthalpies of the methyloxirane and ethylene oxide (1.12 kJ mol^{-1}). Since the interaction of two molecules of the close surrounding involves each of the methyl groups, their summary contribution to the vaporization enthalpy amounts to 2.24 kJ mol^{-1}. This means that the rest of the energy is uniformly distributed along the residual four bond vacancies forming the specific interactions DO → CH_2, namely,

$$\mathrm{D-O\rightarrow CH_2} = \frac{\Delta_{\mathrm{vap}}H^0(T) - \sum \mathrm{DCH_3}}{4}. \tag{3.10}$$

From the difference in the vaporization enthalpy of 2-methyltetrahydrofuran and tetrahydrofuran, we obtain the value of the methyl group contribution into the vaporization enthalpy at 1.10 kJ mol^{-1} or its double value corresponds to the overall contribution from two such groups of two interacting molecules of the close surrounding, 2.20 kJ mol^{-1}. Similarly from the difference of the vaporization enthalpies of 2,5-dimethyltetrahydrofuran and tetrahydrofuran, 3,3-dimethyltetrahydrofuran and tetrahydrofuran, the contributions of two methyl groups and, respectively, four these groups of the interacting molecules of the close surrounding were established at 5.00 and 4.00 kJ mol^{-1}, respectively, for 2,5-dimethyltetrahydrofuran and 3,3-dimethyltetrahydrofuran.

The performed thermodynamic calculations (Table 3.7) show that the energies of the specific intermolecular interactions form a series of decreasing stability:

$$\begin{aligned}&\text{2 - Methyltetrahydrofuran } (7.95\,\mathrm{kJ\,mol^{-1}}) > \text{2,5 - Dimethyltetrahydrofuran}\\&(7.60\,\mathrm{kJ\,mol^{-1}}) > \text{3,3 - Dimethyloxolane } (7.46\,\mathrm{kJ\,mol^{-1}}) > \text{Methyloxirane}\\&(6.42\,\mathrm{kJ\,mol^{-1}}).\end{aligned}$$

Firstly, due to the growing number of methyl groups: 2-methyltetrahydrofuran (7.95 kJ mol^{-1}) and 2,5-dimethyltetrahydrofuran (7.60 kJ mol^{-1}).

Secondly, because of the dissimilar positions of the methyl groups in the ring: 2,5-dimethyltetrahydrofuran (7.60 kJ mol^{-1}) and 3,3-dimethyloxolane (7.46 kJ mol^{-1}). Therewith the condition is fulfilled that the location of two methyl groups in the 2,5-position, i.e., at the oxygen atom, provides an enhanced value of the energy of the specific intermolecular interaction. Thirdly, the energy of the specific interactions under consideration decreases with the diminishing number of carbon atoms in the ring.

3.5.2 *Unsaturated Organic Cyclic Oxides*

An interesting correlation is observed between the structure of the molecules of oxolane, furan, and 2,3-dihydrofuran and their vaporization enthalpies: 32.90, 28.49, and 30.80 kJ mol^{-1}, respectively [3]. The replacement of two ordinary bonds in the former one by two double bonds in the second compound results in the decrease in the enthalpy characteristic by 4.40 kJ mol^{-1}. Consequently, the replacement of one ordinary bond by a double bond should lead to the decrease in the vaporization enthalpy by 2.20 kJ mol^{-1} on condition of equal number of the specific intermolecular interactions in liquid compounds (Table 3.8). We obtain a

Table 3.8 Energies of specific intermolecular interactions (kJ mol^{-1}) of liquid unsaturated cyclic oxides (298 K)

Compound	Structure	$\Delta_{vap}H^0$ (*T*) [8]	D=CH–O → CH=	$\sum DCH_3iso$ DCH_3iso
Furan	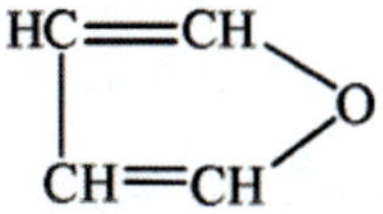	28.49	7.12	–
2,3-Dihydrofuran	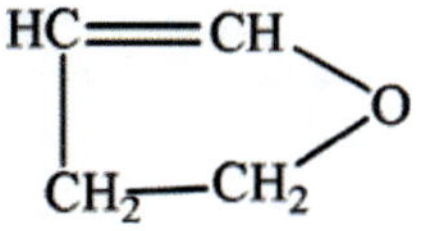	30.80	7.12 D CH_2–O → CH_2– 8.28	–
2-Methylfuran	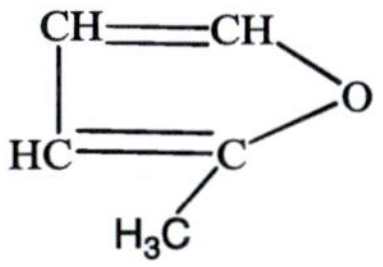	32.20	7.12	$\sum DCH_3 = 3.70$ $DCH_3iso = 1.85$
2.5-Dimethylfuran	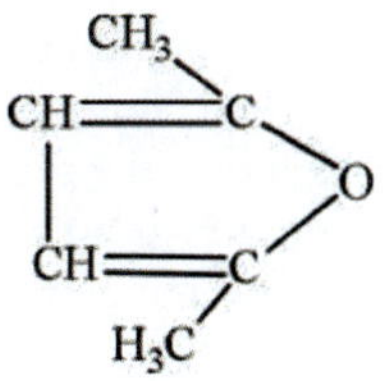	31.80	7.12	$D_1CH_3 = 3.70$ $D_2CH_3 = -0.4$ $DCH_3 = 1.65$
2-*tert*-Butylfuran	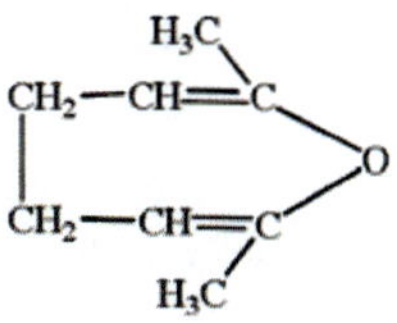	38.1	DO → CH_2–CH_2 = 7.88	$\sum CH_3 = 6.60$ $DCH_3 = 1.65$

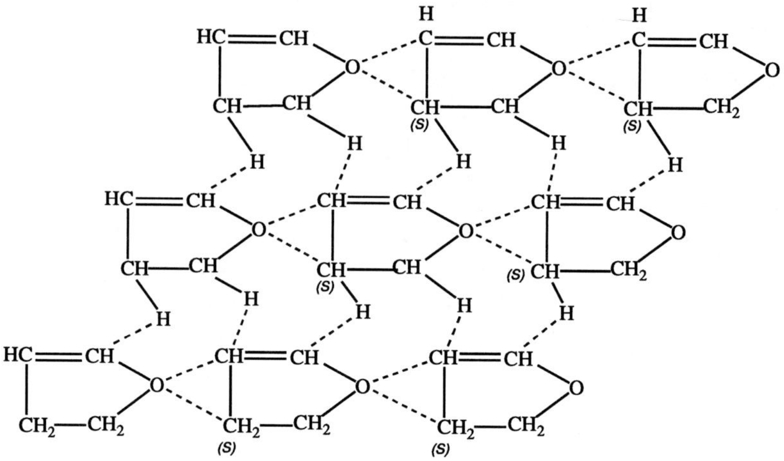

Fig. 3.14 Schematic picture of the liquid structure of 2,3-dihydrofuran

value close to the above (2.10 kJ mol^{-1}) from the difference of the vaporization enthalpies of oxolane and 2,3-dihydrofuran, namely, at replacement of one ordinary bond by a double bond.

The similarity of the structures of these compounds, the equal number of bond vacancies, of ordinary and double bonds permits the representation of their structure in the liquid state by the schematic pattern shown in Fig. 3.14. This structure is a net consisting of chains that are connected reciprocally by weak specific interactions of hydrogen atoms with the carbon atoms involving the essentially unshared $2s^2$(c)-electron pairs. Disregarding the weak dispersion interactions and considering the specific intermolecular interactions formed between the oxygen atom and CH_2–(DO → CH_2–) or CH-groups (DO → CH=) as equal, we conclude that the energy of the specific intermolecular interactions in liquid furan can be estimated directly from the vaporization enthalpy (Table 3.8).

The energy of the specific interaction in the liquid 2,3-dihydrofuran containing an ordinary and a double bond should be taken as including the energy of the interaction DO → C–H = C–H involving the carbon atom at the double bond equal to that existing in the liquid furan (7.12 kJ mol^{-1}). As a result, the energy of the specific interaction of the oxygen with the carbon atom at the ordinary bond DO → C–H = C–H in the compound under consideration is determined by the equation:

$$\mathrm{D-O\rightarrow CH_2}=\frac{\Delta_{\mathrm{vap}}H^0(T)-2\mathrm{DH-C\rightarrow H-C}}{2}. \qquad (3.11)$$

The obtained value of the energy of the specific intermolecular interaction at the ordinary bond in the molecule of the 2,3-dihydrofuran equals 8.28 kJ mol^{-1}. This value completely coincides with the energy of the specific interaction in liquid tetrahydrofuran with the characteristic for this compound ordinary bonds (8.25 kJ mol^{-1}). This energy value provides a good agreement of the vaporization enthalpy of 2,3-dihydrofuran (30.74 kJ mol^{-1}) with the experimentally measured value of the enthalpy characteristic (30.80 kJ mol^{-1}) [8].

The contribution of the methyl group to the vaporization enthalpy of the 2-methylfuran at the replacement of a hydrogen atom in furan amounts to 3.70 kJ mol^{-1} (Table 3.8). This fact demonstrates a certain redistribution of the electron density in the molecule of the 2-methylfuran. However, the essential contribution to the vaporization enthalpy of the first methyl group changes to the negative value (−0.40 kJ mol^{-1}) at the substitution of the second hydrogen in the furan by a methyl group in the 2.5-dimethylfuran. It shows that the position of the second methyl group in the position 5 at a double bond and an oxygen atom not only does not favor the shift of the electron density to the oxygen, but even somewhat impedes it.

The replacement in the oxolane ($\Delta_{vap}H^0(T)$ = 32.90 kJ mol^{-1}) of the ordinary bonds by two double bonds in furan ($\Delta_{vap}H^0(T)$ = 28.49 kJ mol^{-1}) (Table 3.8) creates a rigid system accompanied with the destabilization of the specific interaction =CH–O → CH= by 1.13 kJ mol^{-1} at the retention of the number of carbon atoms in the ring. This is a striking example of the destabilizing effect of the double bond on the specific interactions with the retention of the molecular structure, the structure of the liquid state, of the type of interaction, and of a number of specific interactions. In the compounds under consideration, this effect reduces the energy of the specific intermolecular interaction from 8.25 kJ mol^{-1} in oxolane to 7.12 kJ mol^{-1} in furan.

The presence of a single hydrogen atom at the carbon atom in furan significantly affects its interaction ability whose energy would evidently be close to the cooperative interaction. Therefore, the difference in the vaporization enthalpy of 2-methylfuran (32.20 kJ mol^{-1}) and furan (28.49 kJ mol^{-1}), 2.5-dimethylfuran (31.80 kJ mol^{-1}) and furan corresponds to the two methyl groups of the interacting molecules in the close surrounding. Consequently, the energy of the specific intermolecular interaction is calculated by the above presented equation (3.11). The performed calculations (Table 3.8) demonstrated the total consistency of the values of the energy of the specific interactions DO → CH=CH = 7.12 kJ mol^{-1} in the furan, 2-methylfuran, and 2.5-dimethylfuran. This energy of the specific interactions should be assigned to a fragment containing two carbon atoms of each part of the ring in the furan and its methyl derivatives.

In this connection, a special interest attracts the 2-*tert*-butylfuran possessing the vaporization enthalpy of 38.1 kJ mol^{-1} and a double bond in propylene fragment (Table 3.8). The structure of the liquid 2-*tert*-butylfuran with two double bonds and methyl groups is analogous to that of 2.5-dimethylfuran. Each part of the ring contains three carbon atoms, therefore the reverse dative bond does not affect the energy of the specific intermolecular interaction. Hence its value should be higher than in the furan, 2-methylfuran, and 2.5-dimethylfuran. Inasmuch as in the presence of two double bonds in the molecule of 2.5-dimethylfuran each methyl group contributes 1.65 kJ mol^{-1} into the vaporization enthalpy, in the presence of two such groups two interacting molecules of the close surrounding contribute 6.60 kJ mol^{-1} into the enthalpy characteristic. Thus the difference ($\Delta_{vap}H^0(T) - \sum$ DCH$_3$) is the sum of four specific intermolecular interactions DO → CH$_2$–CH=CH– where each one equals 7.90 kJ mol^{-1}. Hence, the energy of the specific intermolecular interaction in a ring with an exocyclic double bond closed by an oxygen atom is stabilized at replacement of an ethylene fragment (7.12 kJ mol^{-1}) by a propylene fragment. It means that the reverse dative bond in the case of three carbon atoms in the propylene fragment decreases to the minimum, and the energy of the specific interaction attains the maximum value and then remains constant in compounds with the butylene and pentylene fragments.

The structure of the liquid 2-*tert*-butylfuran with two double bonds, methyl groups, and a –CH$_2$–CH$_2$– fragment is analogous to that of the 2.5-dimethylfuran. Each part of the ring contains three carbon atoms, therefore the reverse dative bond does not affect the energy of the specific intermolecular interaction, and its value should exceed that of the 2.5-dimethylfuran. Since the energy contributed by each of two symmetrically located in the molecule methyl groups at the presence in the close surrounding of two molecules amounts to 1.65 kJ mol^{-1}, their overall contribution to the vaporization enthalpy equals 6.60 kJ mol^{-1}. The difference between the vaporization enthalpy and the overall contribution into the enthalpy characteristic from the energy of bonds formed by all four methyl groups ($\Delta_{vap}H^0(T) - \sum$ DCH$_3$) constitutes the value of the energy of the specific intermolecular interaction. The obtained value of the energy of the specific interaction in the liquid 2-*tert*-butylfuran DO → CH$_2$–CH=CH– (7.88 kJ mol^{-1}) illustrates the enhanced stability of this interaction as compared with the same interaction in the liquid 2.5-dimethylfuran and furan (7.12 kJ mol^{-1}).

Interesting information follows from the energies of interactions formed by methyl groups in saturated and unsaturated oxides.

Firstly, one and two methyl groups in the molecules of 2-methylfuran (1.85 kJ mol^{-1}) and 2.5-dimethylfuran (1.65 kJ mol^{-1}) provide a higher contribution to the vaporization enthalpy of these compounds than any of the methyl groups of the molecules of 2-methyltetrahydrofuran (1.10 kJ mol^{-1}), 2,5-methyltetrahydrofuran (1.25 kJ mol^{-1}), or of 3,3-methyloxolane (1.00 kJ mol^{-1}). They consequently form the specific interactions of elevated stability. This fact shows that the methyl group in unsaturated cyclic oxides strongly affects the displacement of the electron density to oxygen atom than in the saturated oxides. As a result on the one hand the enhanced electron density on the oxygen atom is ensured underlying its elevated donor properties, and on the other hand the acceptor properties of the carbon atom are strengthened.

Secondly, the energy of the specific intermolecular interactions in the liquid unsaturated organic oxides increase with the number of carbon atoms in the cycle from four D–O → CH=CH– (7.12 kJ mol^{-1}) to six D–O → CH$_2$–CH=CH– (7.88 kJ mol^{-1}). The latter bond is more stable than the analogous specific interaction in the divinyl ether D–O → CH$_2$=CH– (6.60 kJ mol^{-1}).

3.5.3 *Energies of H-bonds in 1,4-Dioxane-2,5-dione, Ethylene Ozonide, Diethyl Peroxide, and Furaldehyde*

Especially interesting is 1,4-dioxane-2,5-dione whose structure contains two carbonyl groups and two endocyclic oxygen atoms (Table 3.9). Therefore, the reliable interpretation of the energy effects and structural features of the molecule underlying them in this compound requires a consideration of additional thermodynamic

Table 3.9 Energies of hydrogen bonds (kJ mol^{-1}) of liquid 1,4-dioxane-2,5-dione and 2-furaldehyde (298 K)

Compounds	Structure	$\Delta_{vap}H^0$ (*T*) [8, 11]	D CH_2–O → CH_2–	D − O··· H − C − H
1,4-Dioxane-2,5-dione	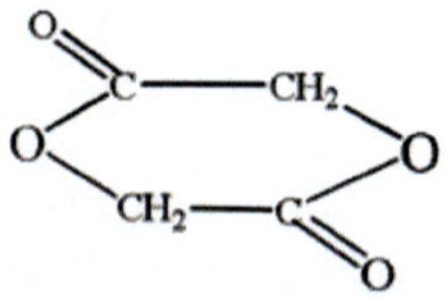	50.4	4.83	2.90
Ethylene ozonide	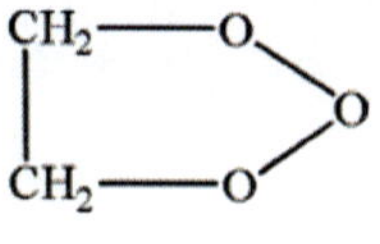	34.8	4.83	1.82
Diethyl peroxide	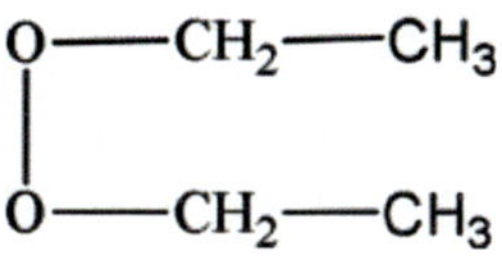	29.0	3.60	–
2-Furaldehyde	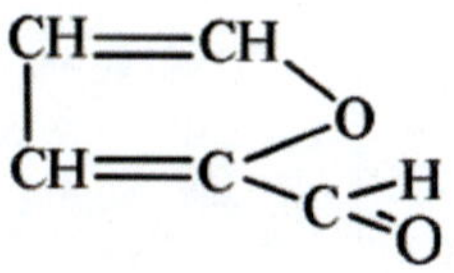	43.24	7.12	5.20 $DCH_2 = 4.9$
		$\Delta_{sub}H^0(T)$	DO → CH solid	–
2-Furaldehyde	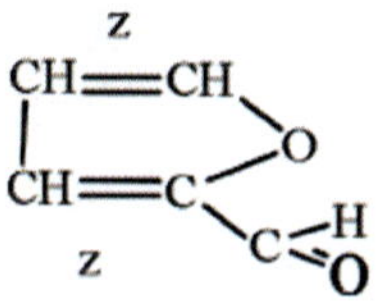	57.61 $\Delta_{melt}H^0$ $(T) = 14.87$	8.06 $DCH_2 = 5.8$	10.30

properties. In this regard, the bond vacancies of the molecule as a whole should be taken into account which we neglected due to their insignificant contribution into the vaporization enthalpy against a background of highly stable specific interactions. Four oxygen atoms with their unshared electron pairs suffer a deficit of electron donors and acceptors. Therefore, the hydrogen and carbon atoms of the CH_2-groups are obliged to provide a significant part of their electron density and considerably increase their acceptor ability. As a consequence the most stable specific interaction in the molecule in question is formed by the essentially unshared $2s^2$(c)-electron pair of the carbon atom of the CH_2-group. Less stable donor–acceptor interactions originate from the hydrogen atoms of the CH_2-group located in the *cis*- and *trans*-positions with respect to the same CH_2-group. In keeping with [12, 13], the *cis*-located hydrogen forms the least stable interactions. The presence in the molecule of 1,4-dioxane-2,5-dione regularly placed two carbonyl groups and two other oxygen atoms leads to a significant change in the distribution of the electron density due to the reciprocal influence of these atoms. The exocyclic oxygen atom of the carbonyl group suffers the least influence on its negative charge compared to the endocyclic oxygen capable of affecting the shift of the electron density of the endocyclic atoms, first of all, from the contiguous carbon atom of the CH_2-group thus favoring the increase in its acceptor properties and thus of the hydrogen. Therefore, the CH_2-group generates the specific intermolecular interaction of the essentially unshared $2s^2$(c)-electron pair of the carbon atom and of the hydrogen located in the *trans* position.

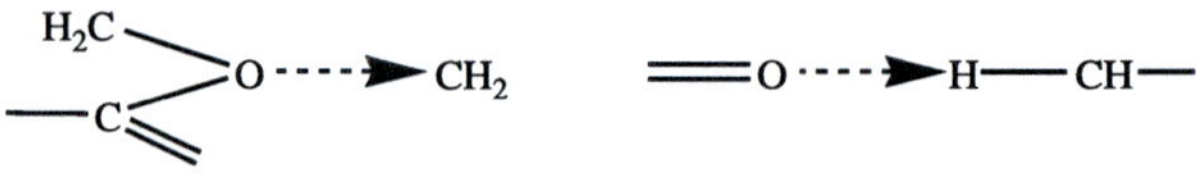

This reasoning makes it possible to represent the structure of the liquid 1,4-dioxane-2,5-dione by a schematic picture shown in Fig. 3.15. The net structure

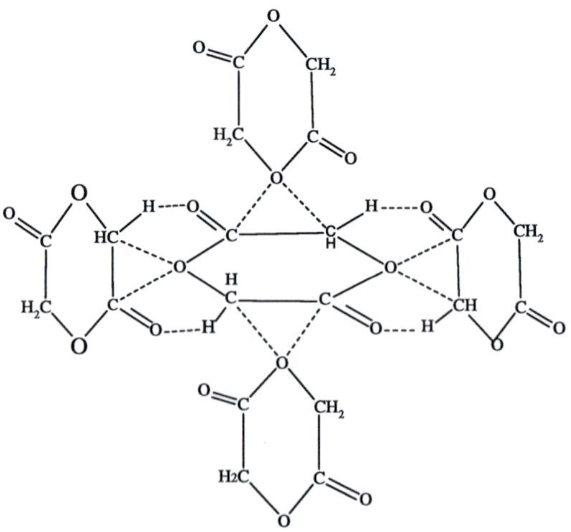

Fig. 3.15 Schematic picture of the liquid structure of 1,4-dioxane-2,5-dione

depicts the coordination of four molecules in the close surrounding with eight specific intermolecular interactions with the participation of the pentacoordinate carbon atom 4D–O $\rightarrow$ CH_2, 4D–O $\rightarrow$ C=O and of four H-bonds $D-O\cdots H-CH$ of low stability. Based on the equal number of carbon and oxygen atoms in the rings of 1,4-dioxane-2,5-dione and 1,4-dioxane and on equal number of the specific intermolecular interactions, we assume the energy of the specific interactions D–O $\rightarrow$ CH_2 and D–O $\rightarrow$ C=O equal to the energy of analogous interactions existing in the liquid 1.4-dioxane (4.83 kJ mol^{-1}). The validity of this assumption is confirmed by the data reported above. This approach provides an opportunity to establish the energy of the hydrogen bond $D-O\cdots H-CH$ arising in the liquid 1,4-dioxane-2,5-dione between the hydrogen atom in the *trans*-position of the CH_2-group and the carbonyl oxygen atom.

The energy of this type of hydrogen bond is equal to the difference between the vaporization enthalpy of the 1,4-dioxane-2,5-dione and the sum of the energies of the specific intermolecular interactions divided by the number of the H-bonds.

$$D-O\cdots H-C-H=\frac{\Delta_{\mathrm{vap}}H^0(T)-4D-O\rightarrow CH_2-4D-O\rightarrow C=O}{4}. \tag{3.12}$$

Results of performed thermodynamic calculations (Table 3.9) indicate the low stability of the H-bond whose energy (2.90 kJ mol^{-1}) is less than the energy of the specific interaction D–O $\rightarrow$ CH_2 (4.83 kJ mol^{-1}). The small value of the energy of the H-bond is an indirect confirmation of the extremely low stability of the interaction of the hydrogen atom situated in the *cis*-position of a methyl and consequently of a CH_2-group.

The effect of the growing ratio of the number of oxygen atoms to the CH_2-groups in a molecule on the energy of a hydrogen bond is illustrated by the ethylene ozonide. The structure of its liquid state may be depicted by the schematic pattern shown in Fig. 3.16. The three oxygens in the molecule of the ethylene ozonide indicate the presence of six bond vacancies by the oxygen atoms that form the specific intermolecular interactions with the carbon atoms of the CH_2-groups and with the hydrogen atoms of these groups. Since each of the molecules present in the close surrounding possesses the same number of the bond vacancies, the transition of a molecule into vapor involves the cleavage of four specific intermolecular interactions and of eight H-bonds of low stability. Consequently, the summary value of the H-bond energy equals

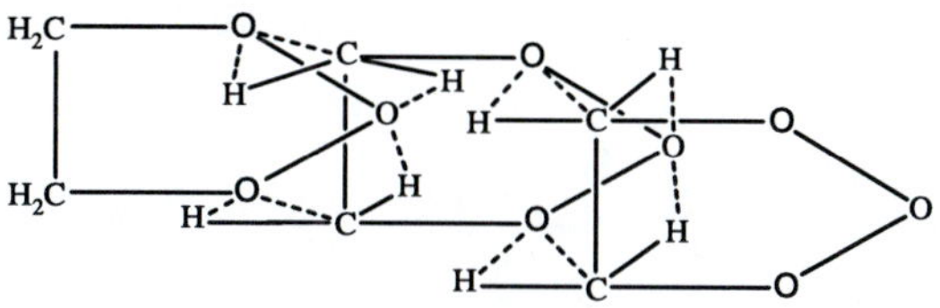

Fig. 3.16 Schematic picture of the liquid structure of ethylene ozonide

$$8D - O \cdots HC_2 = (\Delta_{vap}H^0(T) - 4D - O \to CH_2) \text{ or } D - O \cdots HC_2 = \frac{\Delta_{vap}H^0(T) - 4D - O \to CH_2}{8}. \quad (3.13)$$

In the 1,4-dioxane-2,5-dione having four oxygen atoms in the molecule, among which two are the carbonyl oxygens, the energy of the specific interaction D–O → CH_2 = 4.83 kJ mol^{-1} is considerably less than the energy of analogous bond in the liquid dimethyl ether (5.63 kJ mol^{-1}). The presence of three oxygen atoms in the molecule of the ethylene ozonide with the same number of CH_2-groups between oxygens as in the molecule of 1,4-dioxane-2,5-dione suggests that we can assume the energy of the same type of the specific interaction to be of the same value (4.83 kJ mol^{-1}). The established magnitude of the energy of the H-bond (Table 3.9) amounts to nearly a half of the value of the energy of the analogous specific interaction in the liquid 1,4-dioxane-2,5-dione.

The obtained values of the energy of the specific interactions DO → CH_2 form the series

$$1,4\text{ - Dioxane - }2,5\text{ - dione}\,(2.90\,\text{kJ}\,\text{mol}^{-1}) > \text{Ethylene ozonide}\,(1.82\,\text{kJ}\,\text{mol}^{-1}) > \text{Alcohol}\,(D - O \cdots H - CH_2 = 0.60\,\text{kJ}\,\text{mol}^{-1})$$

where the least value corresponds to the energy of the H-bond formed by the hydrogen in the *trans*-position in the methyl group of methanol.

The diethyl peroxide is a representative of oxides with an open structure of a molecule. The peroxide bridge favors the electron density shift from the terminal methyl groups along the chain providing enhanced negative charges on the oxygen atoms (Fig. 3.17). However their enrichment with negative charge in its turn favors the transfer of a part of the electron density to the carbon atom interacting with this oxygen, namely, a reverse dative bond is operating and decreases the acceptor property of the carbon and consequently destabilizes the specific intermolecular interactions.

Inasmuch as the effect of the reverse dative bond decreases with growing number of carbon atoms in the alkyl chain, the stabilization of the specific interaction should strengthen. However it is impossible to obtain the information on the effect of the reverse dative bond on each carbon atom in the alkyl chain because of

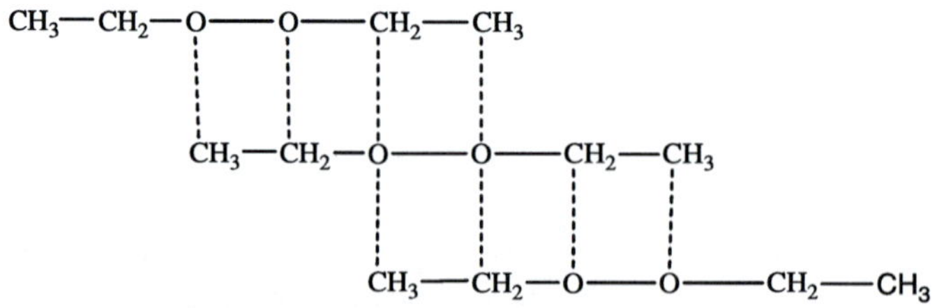

Fig. 3.17 Schematic picture of the liquid structure of diethyl peroxide

the lack of additional thermodynamic data, therefore in the thermodynamic calculation an average energy value of CH_2- and CH_3-groups is taken that they contribute into the vaporization enthalpy of diethyl peroxide. This value makes it possible to reveal the trend to the stabilization of bonds, and based on the sharp difference in energies of the compounds under consideration to establish the law of the appearance of the reverse dative bond.

The schematic drawing of the liquid structure of diethyl peroxide shown in Fig. 3.17 illustrates the chain–net structure of the liquid state of this compound where each oxygen atom forms two specific intermolecular interactions with carbon atoms of CH_2– (D–O → CH_2) or CH_3-groups (DO → CH_3). The overall value of these energies equals the vaporization enthalpy of diethyl peroxide, therefore the mean energy of the specific interaction is $\Delta_{vap}H^0(T)/8 = 3.60$ kJ mol^{-1}. It also follows from the data of Table 3.9 that the energies of the specific intermolecular interactions of liquid oxygen-containing compounds compose the following series of the decreasing interactions stability:

$$\begin{aligned}&\text{Dimethyl ether}(5.63) > 1,4\text{ - Dioxane - }2,5\text{ - dione}(4.83)\\&\quad = \text{Ethylene ozonide } C_2H_4O_3(4.83) > \text{Diethyl peroxide}\left(3.60\,\text{kJ mol}^{-1}\right).\end{aligned}$$

The least stable specific interaction is characteristic of the liquid diethyl peroxide with an open chain of the molecule. Such distribution of the structural and energy parameters of the intermolecular specific interaction in the compounds under discussion corresponds:

Firstly, to the growing number of oxygen atoms in the molecule with the formation of a rigid cyclic structure accompanied with the redistribution of the electron density leading to decreased negative charge on the oxygen atom.

Secondly, at equal number of carbon atoms in linear molecules of dimethyl ether and diethyl peroxide the oxygen atoms of the peroxide bridge in the latter compound like those in the ozonide bridge in the ethylene ozonide possess reduced acceptor ability due to the stable interaction between the oxygen atoms. Consequently the carbon atoms transmit lesser electron density to the oxygen atoms, and because of the growing tendency of the latter to take part in the reverse dative bond with the carbon atom the specific intermolecular interactions formed in the liquid diethyl peroxide are of lower stability than in the liquid dimethyl ether.

The comparison of the structures of furan and 2-furaldehyde molecules (Tables 3.8 and 3.9) shows that their rings are quite identical and therefore they provide an equal contribution into the vaporization enthalpy. The schematic picture of the structure of the liquid 2-furaldehyde shows the existing specific interactions and hydrogen bonds (Fig. 3.18), whose number is 4 and 2, respectively. Equal number of the forming specific intermolecular interactions in the liquid furan and 2-furaldehyde and equal energy contribution of their rings into the vaporization enthalpy correspond to the equal energy of these bonds (7.12 kJ mol^{-1}). It means also that the contribution of the aldehyde group into the enthalpy characteristic is equal to the difference between the vaporization enthalpies of 2-furaldehyde and

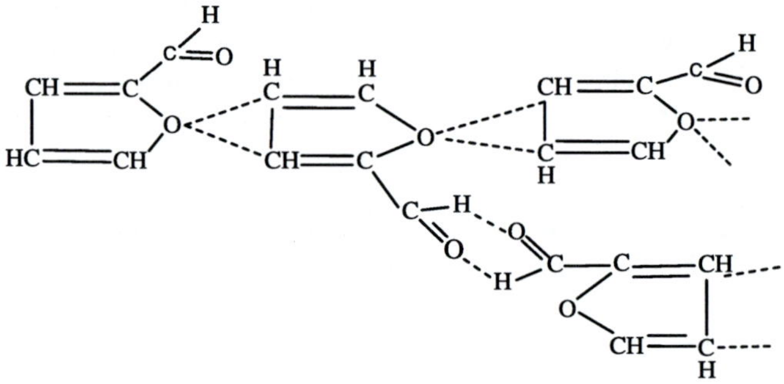

Fig. 3.18 Schematic picture of the condensed structure of 2-furaldehyde

furan. As the aldehyde group alongside oxygen and hydrogen atoms involved into the formation of two hydrogen bonds contains also a carbon atom, to estimate the energy of H-bonds we should take into consideration the contribution of the aldehyde carbon atom into the enthalpy characteristic of 2-furaldehyde. The energy contribution of the carbon atom should be taken equal to the energy contribution into enthalpy characteristic of the CH_2-group of an ether mentioned in Sect. 3.2 and equal 4.9 kJ mol^{-1}. Therewith the energy of interaction (0.60 kJ mol^{-1}) formed by the hydrogen atom in the *trans*-position of the CH_2-group should be taken into account; this hydrogen atom is lacking in the 2-furaldehyde molecule. Hence, the energy of the H-bond of the liquid 2-furaldehyde is calculated by the equation

$$\mathrm{D{=}O\cdots H{-}C} = \frac{(\Delta_{\mathrm{vap}}H^0(T)\mathrm{f.d.} - \Delta_{\mathrm{vap}}H^0(T)\mathrm{fur.}) - (\mathrm{D{-}CH_2{-}D{-}C{\rightarrow}H{-}CH})}{2}. \tag{3.14}$$

The performed thermodynamic calculations showed that the energy of the hydrogen bond between the hydrogen of the aldehyde group and the carbonyl oxygen of the adjacent molecule of the close surrounding is 5.20 kJ mol^{-1}.

The total identity of the ring structure in the 2-furaldehyde and furan makes it possible to use the energy of the specific intermolecular interactions in the furan crystal in the thermodynamic calculations of the energy of the hydrogen bond between the hydrogen of the aldehyde group and the carbonyl oxygen of the adjacent molecule. The calculation was carried out using the same equation (3.14). The calculations utilized the sublimation enthalpy of 1,4-dioxane and furan [3, 4] and the energy contribution into the enthalpy characteristic of the carbon atom of the aldehyde group equal to the contribution of CH_2-group (5.8 kJ mol^{-1}) minus the energy of the interaction of the hydrogen atom in the *trans*-position of the same group. Results of performed thermodynamic calculations (Table 3.9) demonstrate the significant stabilization of the hydrogen bond in solid 2-furaldehyde (10.30 kJ mol^{-1}) compared to the energy of H-bond in the liquid state (5.20 kJ mol^{-1}).

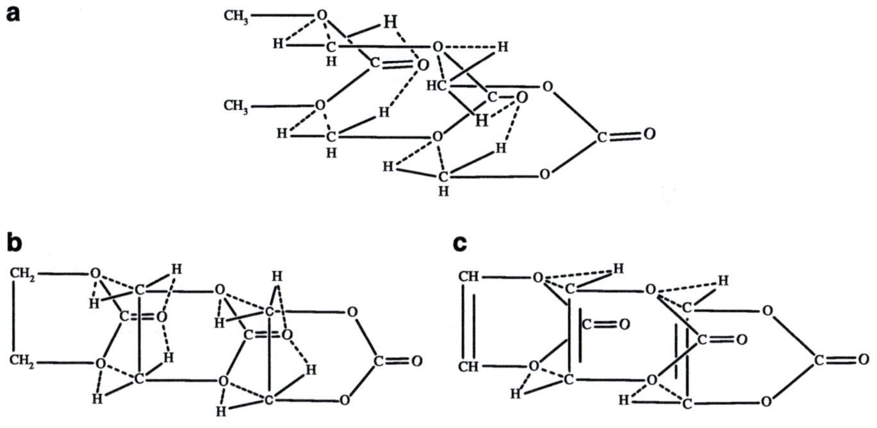

Fig. 3.19 Schematic picture of the liquid FA (**a**), NMF (**b**) and DMF (**c**)

Alkyl carbonates are interesting objects for revealing the influence of the structure of compounds with the open chain and rings, saturated and unsaturated alkyl chains on the energy of specific intermolecular interactions. The molecule of dimethyl carbonate possesses six bond vacancies of three oxygen atoms and two bond vacancies of the carbon atoms of the methyl groups. The lacking four vacancies are provided by the hydrogen atoms of the same methyl groups, therefore in the liquid state in the interaction with the molecules of the close surrounding H-bonds should form of low stability. The schematic picture of the structure of the liquid state (Fig. 3.19) shows four specific intermolecular interactions DO → CH_3 and eight H-bonds formed in the interaction of two molecules of the close surrounding.

It was shown above that the energy of this type of hydrogen bond equals 0.60 kJ mol^{-1}. Therefore, the sum of the energy of the specific intermolecular interaction with the participation of the essentially unshared $2s^2$-electron pair of the carbon atom of the methyl group equals to the difference between the vaporization enthalpy of the dimethyl carbonate and the overall energy of the existing H-bonds

$$4\mathrm{D}-\mathrm{O}\rightarrow \mathrm{CH_3}=\Delta_{\mathrm{vap}}H^0(T)-8\mathrm{D}-\mathrm{O}\cdots\mathrm{H}-\mathrm{CH_2}. \qquad (3.15)$$

The value of the energy (Table 3.10) of the specific interaction (8.10 kJ mol^{-1}) established using the thermodynamic data from Ref. [11] considerably exceeds the energy of the same type of interactions in the liquid dimethyl ether (5.65 kJ mol^{-1}). The reason of this considerable difference in the energies of the specific interactions (2.45 kJ mol^{-1}) of compounds in question is the significant shift of the electron density from the carbon and hydrogen atoms of the methyl groups under the influence of the oxygen atoms of the carbonate fragment. Consequently the carbon atoms bear a slight negative charge, and the oxygen atoms, by contrast, have a high

Table 3.10 Energies of specific intermolecular interactions and hydrogen bonds (kJ mol^{-1}) of liquid alkyl carbonates

Compounds	Structure	$\Delta_{vap}H^0$ (*T*) [11]	T(K)	$D-O\cdots H-C$	D–O → CH_2
Dimethylcarbonate	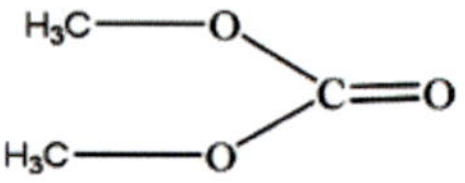	37.1 ± 0.3	298	0.60 Σ=4.8	8.10
Diethylcarbonate	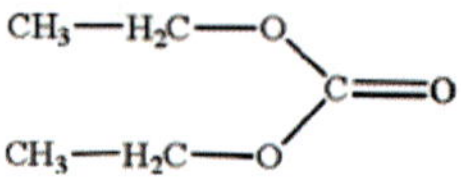	43.6 ± 0.2	298	0.60 Σ=4.8	9.70
Ethylenecarbonate	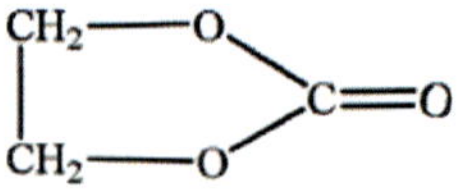	60.3	383	0.60 Σ=4.8	13.9
Vinylenecarbonate	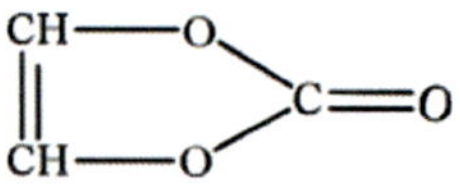	46.9	328	0.60 Σ=2.40	11.1
Propylenecarbonate	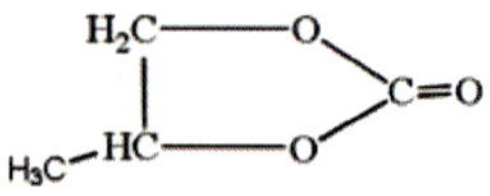	57.8	383	0.60 Σ=3.60 2DCH$_3$iso=3.70	13.1

negative charge and high donor properties. Ethylene carbonate of cyclic structure of the molecule is characterized by about 1.5-fold increase in the experimentally established vaporization enthalpy at 383 K compared with the enthalpy characteristic of the dimethyl carbonate, and accounting for the temperature dependence the vaporization enthalpy of the former is larger nearly two times than that of the latter. This demonstrates a significant stabilization of the energy of specific interaction of the same type D–O → CH_2. In the liquid state of this compound (Fig. 3.19), four specific interactions and eight H-bonds of low stability are formed. The established energy of specific interaction in the liquid ethylene carbonate (Table 3.10) also is nearly twice greater that the energy of the same type interaction in the liquid diethyl ether (6.78 kJ mol^{-1}).

In the liquid phase of vinylene carbonate having a molecule of analogous structure, the same number of specific intermolecular interactions and four H-bonds of low stability are formed. Therefore, the energy of the specific interaction D–O → CH is calculated by the equation:

$$4\mathrm{D-O} \rightarrow \mathrm{CH_2} = \frac{\Delta_{\mathrm{vap}}H^0(T) - 4\mathrm{D-O}\cdots\mathrm{H-CH}}{4}. \qquad (3.16)$$

The value of the energy of the specific interaction D–O → CH (Table 3.10) also demonstrates the high stability of the specific interaction.

In the framework of the described procedure of the thermodynamic analysis of alkyl carbonates we conclude that in the liquid propylene carbonate forms two specific interactions D–O → CH_2 involving CH-groups, two analogous interactions D–O → CH–(CH_3)– and two intermolecular interactions of the low stability marked up isostructural methyl group. This means that the specific intermolecular interactions formed by the carbon atoms should have practically equal values and therefore they are calculated from

$$4\mathrm{D-O} \rightarrow \mathrm{CH} = \Delta_{\mathrm{vap}}H^0(T) - 6\mathrm{D-O}\cdots\mathrm{H-CH} - 2\mathrm{D-CH_3}\mathit{iso}. \qquad (3.17)$$

The obtained energy value of this interaction (Table 3.10) reveals the enhanced stability of the specific interaction compared to the liquid diethyl ether (6.78 kJ mol^{-1}). The data listed in Table 3.10 indicate that the energies of the specific interactions fit to the series:

Ethylene carbonate(13.9)>Propylene carbonate (12.6)>Vinylene carbonate (11.1)>Dimethyl carbonate (8.1 kJ mol^{-1}).

It follows from this series that

- The least stable specific intermolecular interaction in the liquid alkyl carbonates occur in a compound with an open chain. This means that the carbon atoms of the methyl groups in this compound are less capable to shift the electron density even to such strong acceptors like the oxygen atoms of the carbonate group.
- The most stable specific interactions form in the ethylene carbonate containing a saturated dimethylene group in the cyclic structure of the molecule. This means that the dimethylene group in a molecule of cyclic structure exhibits a high ability to supply the electron density, first of all, from the carbon atoms.
- Vinylene carbonate with a double bond in the vinylene group exhibits considerably lower ability to supply the electron density from carbon atoms to the oxygens of the carbonate group.

3.5.4 Crystalline Saturated Organic Cyclic Oxides. Energies of H-bonds

The enthalpy of melting of 1,4-dioxane is the highest among all compounds listed in Table 3.11. This corresponds to the deeper transformation of the structure of its

Table 3.11 Enthalpy of phase transitions and energies of specific intermolecular interactions (kJ mol^{-1}) of solid cyclic oxides (298 K)

Compounds	Structure	$\Delta_{melt}H^0(T)$	$\Delta_{evap}H^0(T)$ [8]		D–O → CH_2–CH_2		DCH_3 iso
			Solid	Liquid	Solid	Liquid	
Methyloxirane	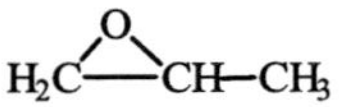	6.53	34.56	27.98	7.89 DCH_3 = 1.50	6.70 DCH_3 = 1.12	
Oxolane	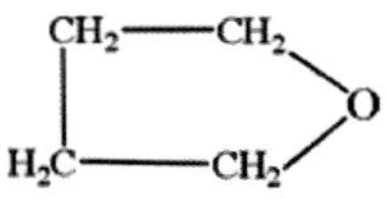	8.54	40.55	32.01	10.16	8.25	–
1,4-Dioxane	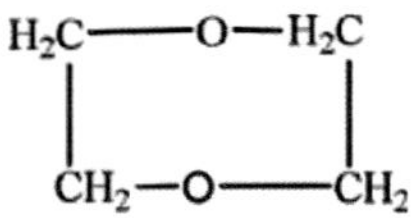	2.33; 12.63	51.27	38.64	6.40	4.83	–
1,3-Dioxane	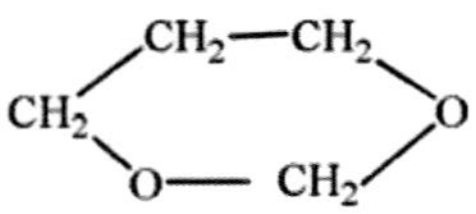	12.70	51.80*	39.10	6.40	4.89	–
4-Methyl-1,3-dioxane	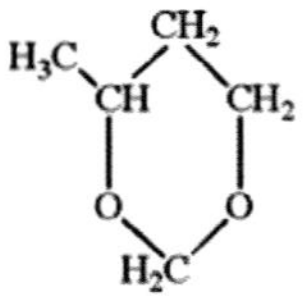	–	–	43.2	–	5.10	1.25 ∑2.50

2,4-Methyl-1,3-dioxane	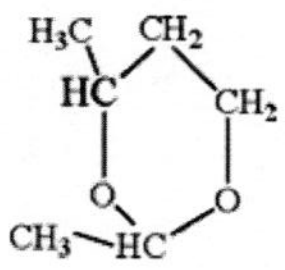	–	–	44.4	–	4.92	1.25 ∑5.00
cis-4,5-Dimethyl-1,3-dioxane	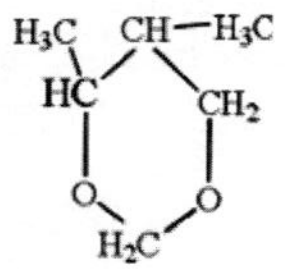	–	–	48.5 (368 K)	–	5.4 0	1.25 ∑5.00
trans-4,5-Dimethyl-1,3-dioxane	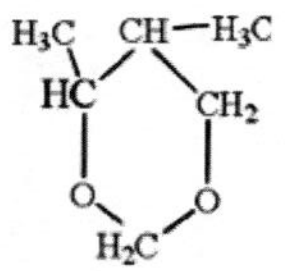	–	–	39.1 (368 K)	–	4.30	0.90 ∑3.60

solid state during melting [9, 10]. On the contrary, the least melting enthalpy is observed for furan since in melting the structure of its solid phase suffered insignificant changes because of the rigid structure of its molecule containing two double bonds. Therefore, the specific intermolecular interactions in furan crystal and in the liquid state should be of close energies. According to our concepts on the chemical nature of the phase transitions [9, 10], the thermodynamic calculations of the energy of the specific intermolecular interactions for various compounds in the solid state are carried out in the same manner as for the energies of the same bonds in the liquid phase. Therefore we can state that the energy of the specific intermolecular interactions in solid oxolane, 1,4-dioxane, 1,3-dioxane, and furan are evaluated from the sublimation enthalpy divided by the number of bond vacancies [7]. Based on the similarity of the energy contribution into the vaporization enthalpy of methyl groups of methyloxirane and isopropyl methyl ether (1.12 kJ mol^{-1}), we take as identical the contribution of the same groups in the solid state (1.40 kJ mol^{-1}). The obtained energies of the specific intermolecular interactions of solid organic oxides (Table 3.11) exceed by 1.2–2.0 kJ mol^{-1} the stability of these interactions in the liquid state. A somewhat enhanced difference in the energies of the specific interactions in the solid and liquid phases is characteristic of 1,4-dioxane, and it is the smallest for the furan. However the enthalpy characteristics of sublimation processes even measured at the standard conditions, as well as the data on the melting process, are not an unambiguous criterion for the proper conception on the degree of the transformation of the structures of the solid state.

A certain understanding of the structural transformations should be obtained from the characteristics of the corresponding states. As this characteristics we used the ratio between the enthalpy of melting or vaporization, sublimation and the energy of the specific intermolecular interactions [9, 10]. The obtained dimensionless quantities describe better the corresponding states and evaluate the depth or the degree of the transformation of the structure of the solid state in the course of melting or the other phase transition. These values (Table 3.11) corresponding to the degree of the transformation of the structures illustrate the regular series of the depth of this process:

$$\text{Furan} < \text{Methyloxirane} < \text{Oxolane} < 1,4\text{ - Dioxane.}$$

The most significant transformation of the structure of the solid state among the substances listed in Table 3.11 occurs in 1.4-dioxane possessing a large number of less stable specific interactions. It means that the depth of the structure transformation at melting is related to the number and the energy of ruptured specific intermolecular interactions and can have a greater magnitude at a large number of less stable bonds [7].

The sublimation of 1,4-dioxane occurs with the transition into vapor of monomer forms of the molecule and with the rupture of eight specific interactions $–O \rightarrow CH_3$ of the energy 6.40 kJ mol^{-1} and of sublimation enthalpy 51.27 kJ mol^{-1}. The sublimation of oxolane with four bond vacancies in the molecule and with the rupture of each bond having the energy $DO \rightarrow CH_2 = 10.16$ kJ mol^{-1} is characterized by enthalpy of 40.55 kJ mol^{-1} that is only by 10.7 kJ mol^{-1} less than the same characteristic of 1,4-dioxane. Analogous conclusion follows from the analysis of the enthalpy characteristic of the vaporization of this compound. Thus these examples additionally support the validity of the concept of the nature of phase transitions substantiated previously in Refs. [9, 10] and of the established rule on the connection of the enthalpy (entropy) characteristic of the phase transition with the number and energy of the ruptured bonds [7].

The chair shape structure of 1,4-dioxaten and planar structure of 1,3-dioxane molecules and the presence of equal number of oxygen atoms and CH_2-groups in each of them (Table 3.7, 3.11) simplifies the thermodynamic calculations. The presence in the molecules of these compounds of equal number of donors and acceptors and CH_2-groups indicates their equal ability to form specific interactions $-O \rightarrow CH_2$ and analogous structure of their liquid state with nearly the same energies of the interactions formed by the essentially unshared $2s^2$-electron pair of the carbon atom. Proceeding from these data we estimated the enthalpy of vaporization and melting of 1,3-dioxane (Table 3.11).

Cyclic molecules of 1,3,5-trioxane and 1,3,5,7-tetraoxane of planar structure are close analogs capable of forming in the solid and liquid states layer–net structure. The schematic picture of this structure in the condensed state of 1,3,5-trioxane is presented in Fig. 3.20. The molecule of this compound containing three oxygen atoms and three CH_2-groups possesses 12 bond vacancies, of which six belong to the lone electron pairs of three oxygen atoms, three essentially unshared $2s^2$-electron pairs of three carbon atoms, and three hydrogen atoms of CH_2-groups. Therefore in the condensed state form six specific intermolecular interactions $D{-}O \rightarrow CH_2$ and six H-bonds $D - O \cdots H - CH$ of low stability.

Since the existing information of the energy of the specific intermolecular interactions involving the pentacoordinate carbon atom is sufficiently valid, it is interesting to obtain data on the energy of the hydrogen bonds in solid 1,3,5-trioxane and 1,3,5,7-tetraoxane. In this connection the energy of the specific intermolecular interactions $-O \rightarrow CH_2$ in solid 1,4-dioxane and 1,3-dioxane (6.40 kJ mol^{-1}) should be used, for in the latter compounds an analogy is present

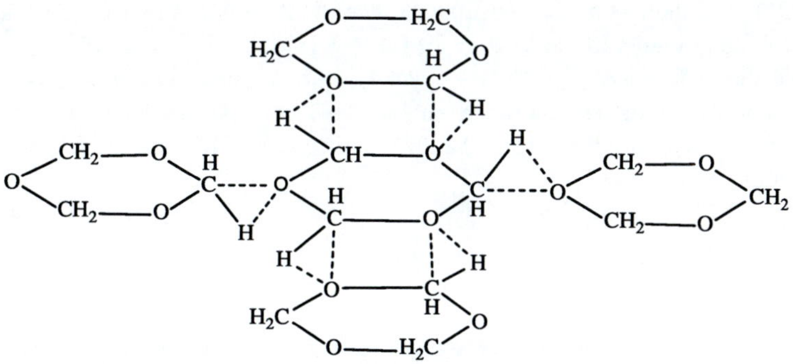

Fig. 3.20 Schematic picture of the condensed state of 1,3,5,-trioxane

in the interaction of CH_2-group with oxygen atoms with those in solid 1,3,5-trioxane and 1,3,5,7-tetraoxane. The molecule of 1,3,5,7-tetraoxane with four oxygen atoms and four CH_2-groups forms eight H-bonds and the same number of highly stable specific intermolecular interactions.

Results of performed thermodynamic calculations (Table 3.12) applying the enthalpy characteristics of the vaporization process taken from Refs. [11, 14] lead to a conclusion that the hydrogen bonds are stabilized in going from 1,3,5-trioxane (2.97 kJ mol^{-1}) to 1,3,5,7-tetroxane (3.55 kJ mol^{-1}) with the growing number in the molecule of CH_2-groups providing some increase in the negative charge on the oxygen atom and consequently the stabilization of the hydrogen bonds.

As shown by the data of Table 3.12, the energy of hydrogen bonds of low stability formed by hydrogen atoms of CH_2-groups and oxygen atom of the oxide compounds are stabilized with growing number of CH_2-groups in the ring

$$1,3,5\text{ - trioxane }(2.97) < 1,3,5,7\text{ - tetraoxane }(3.55\text{ kJ mol}^{-1}).$$

The results of the thermodynamic calculations are attractive for they ensure the unambiguous solution of the problem of evaluation of the energy of hydrogen bonds and specific intermolecular interactions whose quantitative value up to now remained nonestimated even approximately.

3.5.5 *Specific Intermolecular Interaction in Liquid 1,3-Dioxolanes*

The molecule of 2,2-dimethyl-1,3-dioxolane possesses four bond vacancies of two oxygen atoms and the same number of vacancies of four CH_2-groups. Therefore, two methyl groups of each molecule interact with hydrogens of the contact molecules forming weak interactions whose energy equals 1.10 kJ mol^{-1} (Table 3.13). Consequently each molecule forms four such specific interactions (Fig. 3.21), and their contribution into the vaporization enthalpy of 2,2-dimethyl-1,3-dioxolane amounts to 4.4 kJ mol^{-1} (Table 3.13).

Hence the difference $\Delta_{vap}H^0(T) - \sum DCH_3$ corresponds to eight specific intermolecular interactions originating from four free bond vacancies of two oxygen atoms and from essentially unshared $2s^2$-electron pairs of four carbon atoms of the CH_2-groups

$$D = O \to CH_2 = \frac{\Delta_{vap}H^0(T) - \sum DCH_3}{8}. \tag{3.18}$$

The obtained value of the energy of the specific intermolecular interaction (4.60 kJ mol^{-1}) is lower than in the 1,4-dioxane (4.83 kJ mol^{-1}). This fact obviously is due to the effect of the methyl groups. It should be marked here that

Table 3.12 Energies (kJ mol^{-1}) of specific intermolecular interactions and H-bonds of oxanes and 2-methoxy-1,3-dioxolane (T = 298 K)

Compound	Structure	$\Delta_{evap}H^0(T)$		D–O → CH_2		D = O···HCH
		Solid [14]	Liquid [11]	Solid	Liquid	
1,3,5-Trioxane	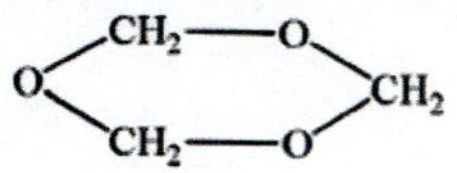	56.2	–	6.40	–	2.97
1,3,5,7-Tetraoxane	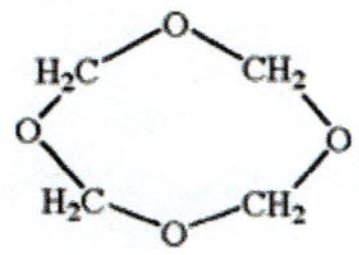	79.6	–	6.40	–	3.55
2-Methoxy-1,3-dioxolane	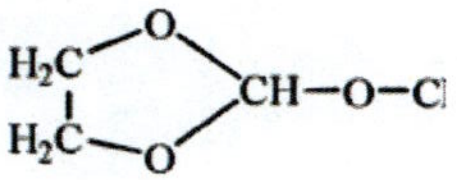	–	46.4	–	5.63 × 2 = 10.76 7.9[a]	D – O···H – CH = 0.60

[a]Estimated by the thermodynamic method

Table 3.13 Vaporization enthalpy and energies of specific intermolecular interactions (kJ mol^{-1}) of dimethyl-1,3-dioxolanes and trimethyl-1,3-dioxolane

Ethers	Structure	$\Delta_{vap}H^0$ (*T*) [11]	*T* (K)	DCH_3	$DO \rightarrow CH_2$
2,2-Dimethyl-1,3-dioxolane $C_6H_{12}O_2$	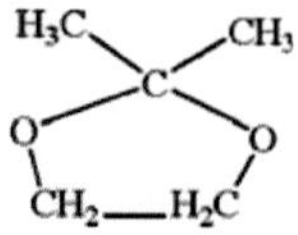	41.1	298	1.10 Σ4.4	4.60
2,2,4-Trimethyl-1,3-dioxolane $C_7H_{14}O_2$	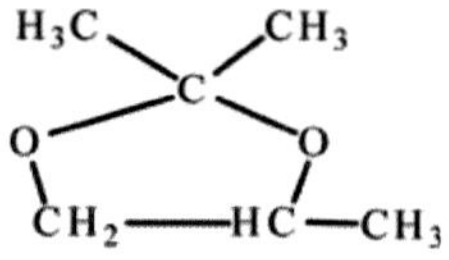	41.9	298	Σ6.2	4.46
2,4-Dimethyl-1,3-dioxolane $C_6H_{12}O_2$	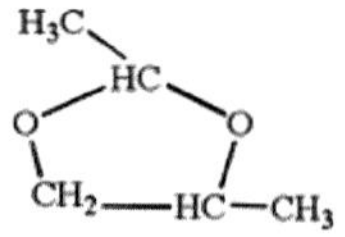	44.9 ± 0.6	298	1.1 Σ4.4	5.06
4,4-Dimethyl-1,3-dioxolane $C_6H_{12}O_2$	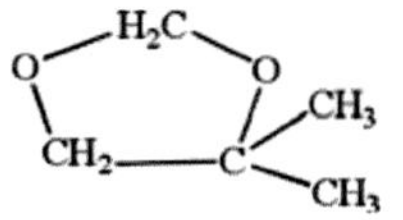	37.1	348	0.80 Σ3.20	4.14
2,4,5-Trimethyl-1,3-dioxolane $C_7H_{14}O_2$	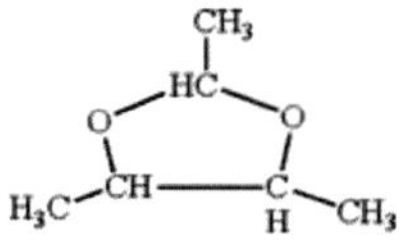	41.5	388	0.90 Σ5.4	4.44
cis-4,5-Dimethyl-1,3-dioxane $C_6H_{12}O_2$	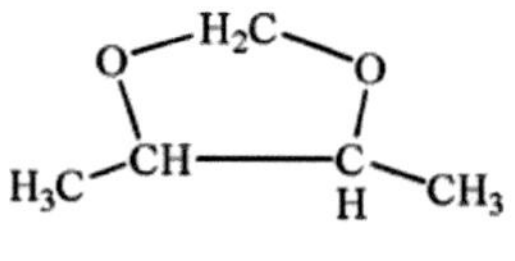	48.5	368	0.90 Σ5.4	5.60
trans-4,5-Dimethyl-1,3-dioxane $C_6H_{12}O_2$	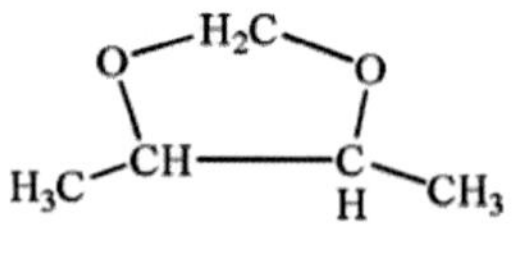	39.1	368	0.90 Σ5.4	4.40

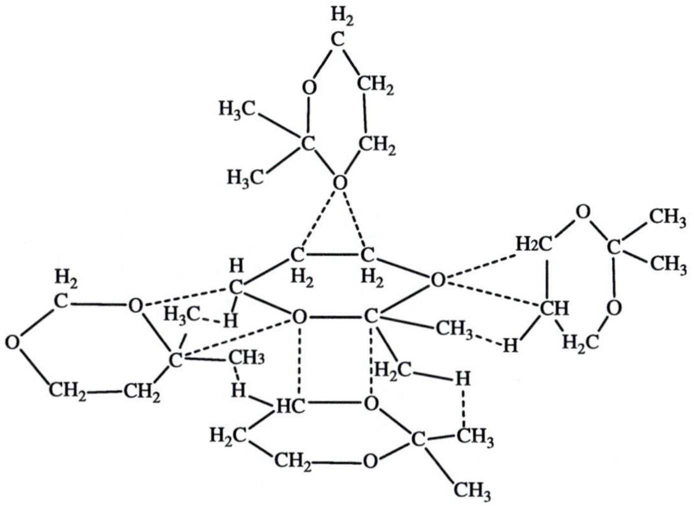

Fig. 3.21 Schematic picture of the liquid structure of 2,2-dimethyl-1,3-dioxolane

the difference in the vaporization enthalpies of 1,3-dioxane and 2,2-dimethyl-1,3-dioxolane amounts to 2.0 kJ mol^{-1} or 1.0 kJ mol^{-1} per one methyl group. About the same energy corresponds to the other two analogous bonds formed by the methyl groups of the molecules of the close surrounding.

The third methyl group of the 2,2,4-trimethyl-1,3-dioxolane molecule provides an additional contribution into the vaporization enthalpy (41.9 kJ mol^{-1}) whose value should be evaluated from the difference from the vaporization enthalpy of 2,2-dimethyl-1,3-dioxolane (41.1 kJ mol^{-1}). This contribution equals 0.80 kJ mol^{-1}, or for two CH_3-groups of the two interacting molecules of the close surrounding, 1.60 kJ mol^{-1}. Based on this difference and on the energy contributed by two other methyl groups of the 2,2,4-trimethyl-1,3-dioxolane their overall contribution into the enthalpy characteristic of vaporization amounts to 6.20 kJ mol^{-1}. Thus the difference between the vaporization enthalpy of 2,2,4-trimethyl-1,3-dioxolane and $\sum DCH_3$ corresponds to the energy of the specific intermolecular interactions existing in the liquid state of the compound under consideration. The energy of the specific interaction DO → CH_2 is determined from (3.18) and is 4.46 kJ mol^{-1}. The obtained values of the energy of the specific interactions in the liquid 2,2-dimethyl-1,3-dioxolane and 2,2,4-trimethyl-1,3-dioxolane suggest that the growing number of the methyl groups from two to three in the molecules of compounds in question is accompanied with the decrease in the energy of interactions DO → CH_2 from 4.83 to 4.60 kJ mol^{-1}.

The comparison of the vaporization enthalpy of 2,4-dimethyl-1,3-dioxolane (44.9 kJ mol^{-1}) and 2,2-dimethyl-1,3-dioxolane (41.1 kJ mol^{-1}) shows that the introduction of the second methyl group in the position 4 results in a considerable increase in the enthalpy characteristic and in the stabilization of the specific intermolecular interaction D–O → CH_2. This permits a statement that the positions of two methyl groups in the 2,4-dimethyl-1,3-dioxolane result in the increased negative charge on the oxygen atoms and decreased negative charge on the carbon atoms of the CH_3-groups considerably changing the respective donor and acceptor properties compared with the charges in the molecule of the 2,2-dimethyl-1,3-dioxolane. Taking the energy contribution of the methyl group located in the position 4 into the vaporization enthalpy of the 2,4-dimethyl-1,3-dioxolane at 1.10 kJ mol^{-1} (Table 3.13), we obtain from the already mentioned equation the energy of the specific interaction DO → CH_2 in the liquid state of the compound equal 5.06 kJ mol^{-1}. Thus the established values of the energy of the specific interactions DO → CH_2 in the liquid 2,2-dimethyl-1,3-dioxolane and 2,4-dimethyl-1,3-dioxolane reveal the stabilization of the considered interactions in the presence of two methyl group in the positions 2,4 of 1,3-dioxolane.

Let us consider now the 4,4-dimethyl-1,3-dioxolane and 2,4,6-trimethyl-1,3-dioxolane whose enthalpy characteristics have been measured at 348 and 338 K, respectively. The deviation from the standard conditions of the determined enthalpy characteristic by 50 and 40 K, respectively, does not impede the estimation of the energy contribution of each methyl group into the vaporization enthalpy of these compounds at 0.80 and 0.90 kJ mol^{-1}, or for two methyl groups 3.2 kJ mol^{-1}, for three groups 5.4 kJ mol^{-1}. Taking into account these contributions, we calculated by (3.18) the energy of the specific intermolecular interaction in the liquid 4,4-dimethyl-1,3-dioxolane and 2,4,6-trimethyl-1,3-dioxolane being equal 4.14 and 4.44 kJ mol^{-1}. The obtained values of the energy of the specific interaction (Table 3.13) obviously show the stabilization of these bonds in event of the location of the methyl groups at different carbon atoms in the dioxolane molecules contiguous to the oxygen atom. On the contrary, the location of the methyl groups at the same carbon atom in the 4,4-dimethyl-1,3-dioxolane destabilizes the specific intermolecular interaction DO → CH_2. Thus we again conclude that the location of the methyl groups at different carbon atoms in the dioxolane molecules favors the shift of the electron density from carbon atoms of the methyl groups to the oxygen atoms, increasing the negative charge on the oxygen and decreasing the negative charge on the carbon.

The energies of the specific interactions obtained by thermodynamic calculation (Tables 3.11 and 3.13) illustrate the enhanced stability of these bonds at the location of the methyl groups in the *cis*-position (Table 3.11). At the same time, the established energy contributions into the vaporization enthalpy of the methyl groups in the dioxane with four CH_2-groups in the ring provide a possibility to evaluate the energy of specific intermolecular interactions in the related dioxolanes with three analogous groups (Table 3.11). This is necessary not only for increasing the amount of information regarding individual compounds but also for their critical understanding and further development of general conceptions on the nature of the specific interactions in various classes of organic compounds.

An important role in obtaining information on the specific interactions in the molecules of the liquid derivatives of the 1,3-dioxolane belongs to functionalized alkyl groups.

The proper way to get the energy contribution from methyl, ethyl, and propyl groups is the thermodynamic method of analysis of the enthalpy characteristics of the vaporization process. These data lead to the conclusion that the methyl groups make equal contribution into the enthalpy characteristics of various cyclic oxides, in particular, methyloxirane, 3,3-methyloxetane, 2-methyltetrahydrofuran (Table 3.7), and other compounds. The average value of the energy contributions into the vaporization enthalpy of the methyl group amounts to 1.10 kJ mol^{-1}. As shown before, the value of this contribution is doubled due to the interaction with the molecules of the close surrounding. Therefore the difference between the vaporization enthalpy, in particular, of the 2-methyl-1,3-dioxolane, and the sum of the contributions of the methyl groups $\Delta_{vap}H^0(T) - \sum DCH_3(=2.20 \text{ kJ mol}^{-1})$ constitutes the energy value contributed by the ring into the same enthalpy characteristic (Table 3.14). The energy contributions of ethyl and propyl groups equal the difference between the vaporization enthalpies of 2-methyl-2-ethyl-1,3-dioxolane and 2-methyl-1,3-dioxolane, 2-methyl-2-propyl-1,3-dioxolane and 2-methyl-1,3-dioxolane, respectively. The established energy contributions of various alkyl groups are compiled in Table 3.14.

The structures of liquid 1,3-dioxolanes should differ only slightly, and they may be represented by the schematic picture in Fig. 3.22. There are shown the specific interactions of low stability formed by methyl, ethyl, propyl, or isopropyl groups whose energies are already considered and listed in Table 3.14. The interactions in the close surrounding are formed by essentially unshared $2s^2$-electron pairs of carbon atoms with the unshared electron pairs of oxygen atoms of the contacting molecules, D–O → CH_2. Therewith the interaction of one carbon atom situated in the position 2 between the cyclic oxygen atoms may somewhat differ from the energy of two other carbon atoms connected with each other by a covalent bond. The identification of their energy requires more thermodynamic data, and the difference in the influence of the fragments –CH_2–CH_2– on the energy of the specific intermolecular interaction is within the error limits of the calorimetric measurements of the vaporization enthalpy, as demonstrated by the established values of the energy of the specific interactions in 1,4-dioxane and 1,3-dioxane (Table 3.11). Therefore, the sum of the energies of the arising four specific interactions D–O → CH_2 and two other

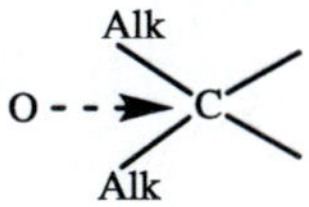

with one or two alkyl groups amount to

$$6\text{D} - \text{O} \rightarrow \text{CH}_2 = \Delta_{vap}H^0\,(T) - \sum \text{DCH}_3. \tag{3.19}$$

Results of performed thermodynamic calculations (Table 3.14) lead to conclusions that the energy of the specific interactions D–O → CH_2 change in a narrow range and are described by a series

Table 3.14 Energies of specific intermolecular interactions (kJ mol^{-1}) of 1,3-dioxolanes (298 K)

Ethers	Formula	Structure	$\Delta_{vap}H^0$ (*T*) [11]	DCH$_3$	D–CH$_3$ (CH$_2$)$_n$	D–O → CH$_2$
2-Methyl-1,3-dioxolane	$C_4H_8O_2$	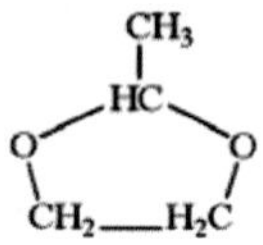	43.0[a]	1.10 ∑2.20	–	6.60
4-Ethyl-1,3-dioxolane	$C_6H_{12}O_2$	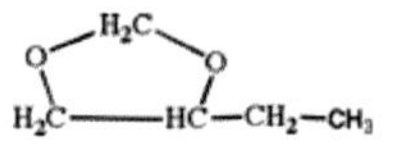	39.3[b]	–	$n = 1$	5.95
2-Methyl-2-ethyl-1,3-dioxolane	$C_6H_{12}O_2$	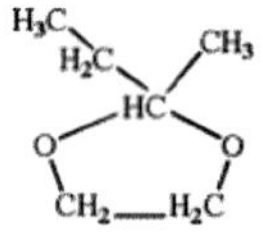	44.8 43.1	1.10 ∑D = 2.20	$n = 1$ 1.80 ∑D=3.60	6.40
2-Propyl-1,3-dioxolane	$C_6H_{12}O_2$	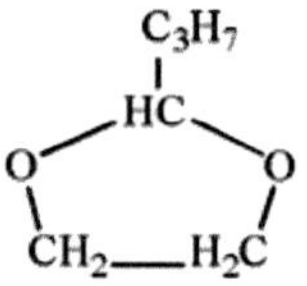	45.3 ± 0.3	–	$n = 2$ 3.30 ∑D=6.60	6.60
2-Methyl-2-propyl-1,3-dioxolane	$C_7H_{14}O_2$	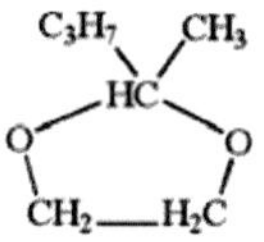	46.3	1.10 ∑D=2.20	$n = 2$ 3.30 ∑D=6.60	6.25
2-Methyl-2-iso propyl-1,3-dioxolane	$C_7H_{14}O_2$	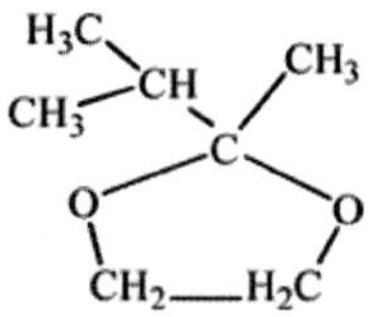	43.9 ± 0.2	1.10 ∑D=2.20	$n = 1$ 0.90 ∑D=1.80	6.60

[a] 308 K
[b] 377 K

2 - Methyl - 1,3 - dioxolane (6.60) = 2 - Propyl - 1,3 - dioxolane = 2 - Methyl - 2 - isopropyl - 1,3 - dioxolane > 2 - Methyl - 2 - ethyl - 1,3 - dioxolane (6.40) > 2 - Methyl - 2 - propyl - 1,3 - dioxolane (6.25 kJ mol^{-1})

and that the difference in the vaporization enthalpy of these compounds originates from the energy contribution of the methyl, ethyl, and propyl groups.

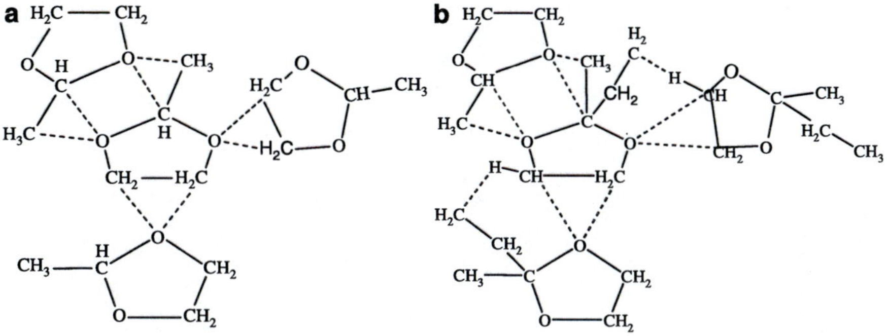

Fig. 3.22 Schematic picture of the liquid structure of 2-methyl-1,3-dioxolane (**a**) and 2-methyl-2-ethyl-1,3-dioxolane (**b**)

3.5.6 Energies of H-bonds and Specific Intermolecular Interaction in Liquid Cyclic Methoxy and Dimethoxy Compounds

Applying the thermodynamic procedure, we can evaluate the energy of the specific intermolecular interaction D–O → CH_2 arising in the liquid 2-methoxy-1,3-dioxolane between the oxygen and carbon atoms of the ring. The structural formula presented in Table 3.15 permits a conclusion that the molecule of 2-methoxy-1,3-dioxolane possesses ten bond vacancies, among which six belong to three oxygen atoms and four to carbon atoms. Based on this assumption, the structure of its liquid state can be schematically drawn as depicted in Fig. 3.23.

This structure includes eight stable specific interactions, two of which are formed by the carbon atoms of the methyl groups and six by the carbon atoms of the CH_2-groups. Besides, in the close surrounding of the coordinating molecule exist also four H-bonds of low stability originating from hydrogen atoms of CH_3- and CH-groups.

Two fragments should be separated in this structure including in their chains different number

of CH_2-groups and different number of oxygen atoms in the direct contact with carbon atoms actually, the carbon atom attached to the methoxy group should suffer the influence of three oxygen atoms drawing off the electron density.

It is therefore probable that the energy of the specific interactions originating from this carbon atom should differ from those formed by the carbon atoms of the –CH_2–CH_2– fragment. If the energy of the specific interaction of the carbon atom

Table 3.15 Energies of specific intermolecular interactions (kJ mol^{-1}) and hydrogen bonds of dimethylmethoxymethane, methoxy-1,3-dioxolane, and cyclic compounds ($T = 298$ K)

Compounds	Structure	$\Delta_{sub}H^0(T)$ [8]	$\Delta_{vap}H^0(T)$ [8, 11]	D–O → CH_2 Solid	D–O → CH_2 Liquid	$D = O \cdots H - CH$
2-Methoxy-1,3-dioxolane $C_4H_8O_2$	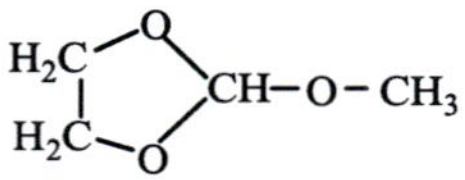	–	46.4	–	5.65 DO → CH_3 = 5.63 × 2 = 11.26	$4DO \cdots H - CH = 4.40$
Dimethylmethoxymethane	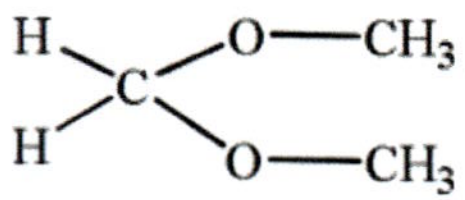	37.26 $\Delta_{melt}H$ $(T) = 8.33$	26.41	5.10	3.60	1.00, solid 0.60, liquid
Cyclopentenyl methyl ether	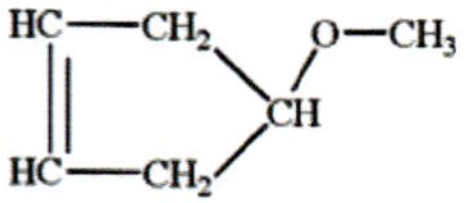	–	42.1 ± 0.6	–	10.25	DO → CH_3 = 5.63 × 2 = 11.26
Cyclohexenyl methyl ether	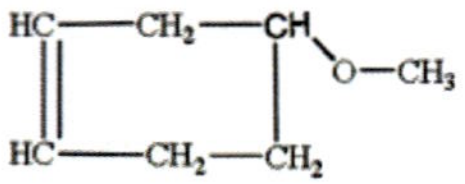	–	43.7	–	10.80	DO → CH_3 = 5.63 × 2 = 11.26
1,1-Dimetoxy cyclohexene	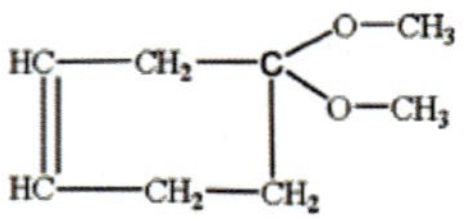	–	48.6	–	8.90	DO → CH_3 = 5.63 × 2 = 11.26

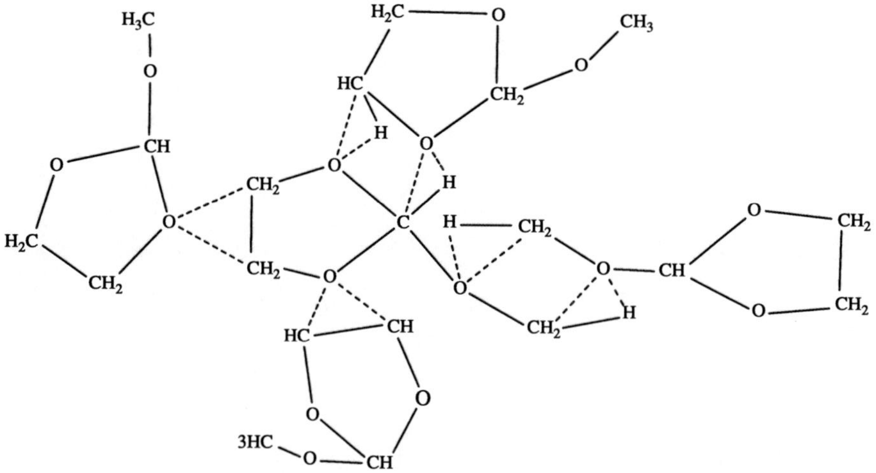

Fig. 3.23 Schematic picture of the liquid structure of 2-methoxy-1,3-dioxolane

from the fragment with two methylene groups with the oxygen atom of the contacting molecule is unknown, by analogy of the bond type with that of the methyl group of the methoxy fragment with the specific interactions in the liquid dimethyl ether the energy of this interaction may be assumed equal 5.63 kJ mol^{-1}. The energy of the H-bond of low stability formed by the hydrogen atom of the CH_3-group in the *trans*-position or by the hydrogen of the CH-group has been already taken equal 1.10 kJ mol^{-1} for the liquid state of the compounds in question. Thus the unknown value is the energy of the specific interaction D–O → CH_2 (or DCH_2–O → CH_2) whose overall value equals the difference between the vaporization enthalpy of 2-methoxy-1,3-dioxolane and the sum of the energy of two other types of the specific interactions DO → CH_3 and of H-bonds

$$6D - O \rightarrow CH_2 = \Delta_{vap}H^0(T) - 2D - O \rightarrow CH_3 - 4D - O \cdots H - CH. \quad (3.20)$$

The obtained value of the energy of the specific interaction DCH_2–O → CH_2 is 5.45 kJ mol^{-1}. This value is sufficiently close to the found energy of the specific interaction formed by the methyl group of the dimethyl ether leading to the conclusion that the charge on the carbon atom linked to the methoxy group and two oxygen atoms of the ring changes insignificantly.

The established value of the energy of the specific interaction in the liquid 2-methoxy-1,3-dioxolane is well consistent with the series describing the dependence on the number of CH_2-groups in the chain: one and a half, three, and four at the number of oxygen atoms in the ring of each compound two, one, and one, respectively.

$$\begin{gathered}2\text{ - Methoxy - }1,3\text{ - dioxolane }(5.45) < \text{Oxetane (trimethylene oxide) }(7.65) < \\ \text{Oxolane (tetrahydrofuran) }(8.25\,\text{kJ mol}^{-1}).\end{gathered}$$

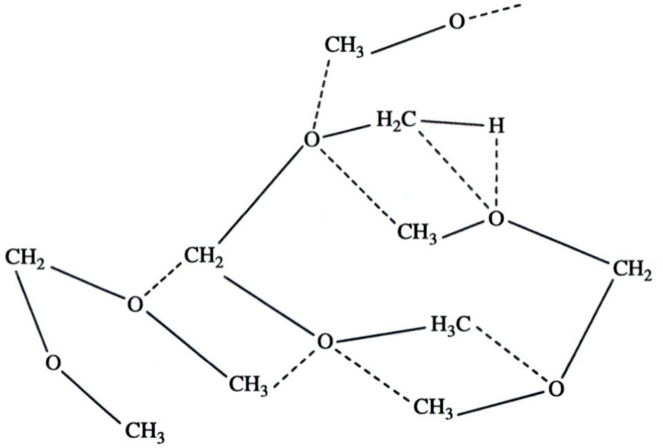

Fig. 3.24 Schematic picture of the liquid structure of dimethylmethoxymethane dimetoxymethane

Hence the growing number of CH_2-groups in the ring provides the stronger shift of the electron density to the oxygen atom of the corresponding fragment.

An interesting compound is dimethylmethoxymethane. The high value of the melting enthalpy of this compound (Table 3.15) with a relatively low vaporization enthalpy demonstrates the significant transformation of the structure of the solid state during melting and considerable destabilization of the specific interactions formed in the liquid state.

The molecule of dimethylmethoxymethane has seven bond vacancies forming strong specific intermolecular interactions, four of which are provided by two oxygen atoms and three remaining vacancies belong to carbon atoms. Having seven bond vacancies, the molecule coordinates four molecules in the solid and liquid state (Fig. 3.24). However three carbon atoms are not enough to fill the four bond vacancies of oxygen, therefore the hydrogen atom in the *trans* position forms an H-bond of low stability whose energy should be taken by analogy with ethers at 0.60 and 1.00 kJ mol^{-1} for liquid and solid state, respectively. The difference between the enthalpy of vaporization or sublimation and the sum of the energies of the H-bonds gives the overall energies of the specific intermolecular interactions $DO \rightarrow CH_3$

$$7D - O \rightarrow CH_3 = \Delta_{vap}H^0(T) - 2D - O \cdots H - CH_2. \tag{3.21}$$

The calculated values of the energies of the specific interactions (Table 3.15) show their low stability as compared with the interactions existing in the liquid and solid methyl ether.

The molecules of the cyclopentenyl methyl ether and cyclohexenyl methyl ether possess an equal number of bond vacancies and form analogous structures in the liquid state. Each molecule coordinates three molecules in the close surrounding (Fig. 3.25) that form two specific intermolecular interactions $DO \rightarrow CH_3$ and three

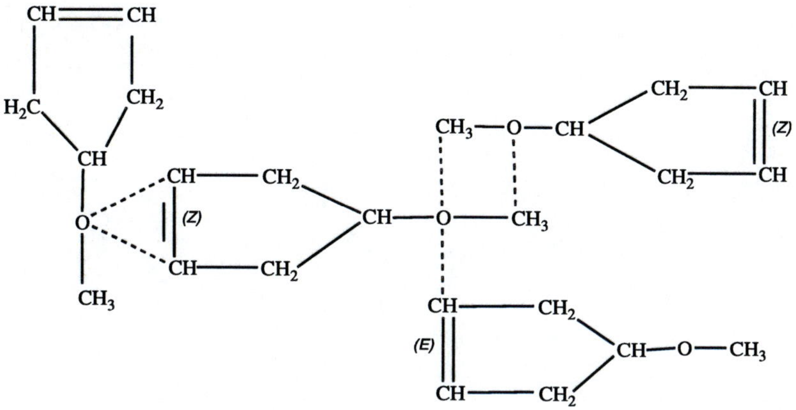

Fig. 3.25 Schematic picture of the liquid structure of cyclopentenyl methyl ether

DO → CH by the methyl and CH-groups of the fragment –HC=CH. Preceding from these data it is possible to write

$$3D - O \rightarrow CH = \Delta_{vap}H^0(T) - 2D - O \rightarrow CH_3. \tag{3.22}$$

Also taking the energy of the specific intermolecular interaction DO → CH_3 = 5.63 kJ mol^{-1} equal to the energy of the same interaction in the liquid dimethyl ether we obtain the energy of DO → CH bonds in the liquid cyclopentenyl methyl ether and cyclohexenyl methyl ether at 10.25 and 10.80 kJ mol^{-1}, respectively. In event of 1,1-dimetoxycyclohexane having eight bond vacancies on a molecule, the structure of the liquid state (Fig. 3.26) contains four specific interactions DO → CH_3 and the same number of DO → CH bonds whose energy contribution into the vaporization enthalpy amounts to

$$4D - O \rightarrow CH = \Delta_{vap}H^0(T) - 2D - O \rightarrow CH_3. \tag{3.23}$$

The performed calculation procedure afforded the energy value of the specific intermolecular interaction under consideration of 8.90 kJ mol^{-1}. This type of energy of the specific interaction has the limiting value for this homologous series, and it remains constant further in 1,1-dimetoxyheptane and 1,1-dimetoxyoctane, whereas the vaporization enthalpy depends on the number of additional CH_2-groups and their energy contribution. In the cases of 1,1-dimetoxyheptane and 1,1-dimetoxyoctane the vaporization enthalpies should be equal to 53.6 and 58.6 kJ mol^{-1}, respectively. The established energy of the specific interaction in the liquid cyclohexenyl methyl ether also is the limiting value for all subsequent members of this homologous series, and their vaporization enthalpy also depends on the number of additional CH_2-groups in the ring and their energy contribution.

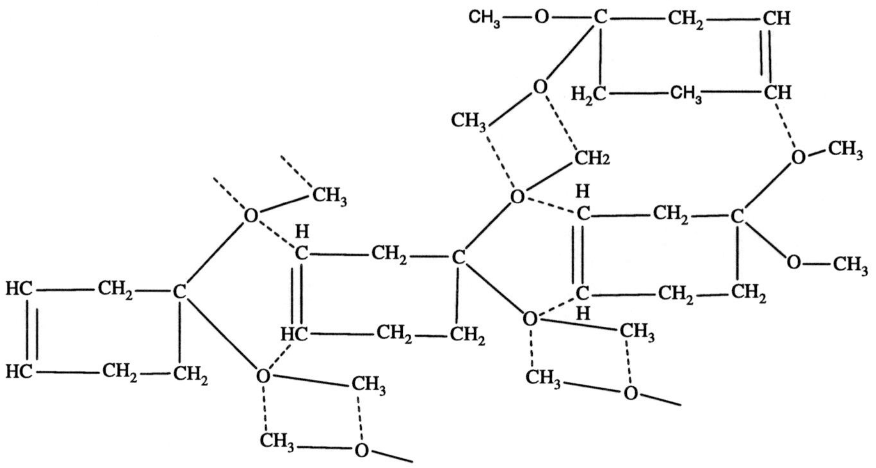

Fig. 3.26 Schematic picture of the liquid structure of 1,1-dimetoxycyclohexene

Therefore, the vaporization enthalpies of cyclopentenyl methyl ether and cyclohexenyl methyl ether should be close to 48.7 and 53.7 kJ mol^{-1}.

3.5.7 The Energies of Specific Intermolecular Interaction in Phenol Ethers

Nowadays, the overlapping of ionization regions of MO of s- and p-type is commonly presumably explained by the high symmetry of benzene that prevents the mixing of the 1b(s)-orbital with the MO of p-type. In event of aliphatic hydrocarbons the overlapping of the ionization regions of MO of s- and p-types is absent, and in the alkyl-substituted benzenes the hyperconjugation is accompanied with the splitting of e_{1g}-MO till 0.5 eV [1].

However, there is no opportunity to get any reliable data on the distribution of the electron density and charges on the atoms.

The molecule of methyl phenyl ether is a complex structure with donor–acceptor properties contained, on the one hand, in the benzene ring, and on the other hand, in the methoxy group. The substitution of a hydrogen atom by a methoxy group produces certain changes in the distribution of the electron density in the benzene ring and consequently in the distribution of charges on the carbon atoms. Methyl phenyl ether should be regarded as an unsymmetrical ether whose benzene ring has replaced a methyl group of the dimethyl ether. It was shown in Sect. 3.2 that the methyl group of unsymmetrical methyl ethers has practically the same properties as in the dimethyl ether and consequently makes the same energy contribution into the enthalpy characteristic of vaporization and sublimation. It is therefore presumable that in the structure of the crystalline and liquid phase of the methyl phenyl ether, versatile specific intermolecular interactions should be present that make an energy contribution in the enthalpy characteristic of the vaporization process inherent to the structure of the groups composing this compound. Therefore the specific intermolecular interactions, same as the hydrogen bonds, are the structure-forming factors for organoelemental and organic compounds. In this sense the most valid is the structural thermodynamic analysis of compounds, group of compounds belonging to the same homologous series or the same class in the establishment of the energy contributions of each type of the specific intermolecular interaction into the enthalpy of vaporization and sublimation. In the case of methyl phenyl ether the similarity of the methoxy group to the same group of the dimethyl ether and virtually constant energy contribution of this group DO → CH_3 into the vaporization enthalpy of unsymmetrical methyl ethers, we can take the energy of this bond in the methyl phenyl ether as equal 5.63 kJ mol^{-1}.

In its turn in the benzene crystal, the flat molecules should form a layer structure where each molecule interacts with two molecules of close surrounding by its CH-groups. Every one of these groups in the benzene molecule forms with a carbon and a hydrogen atom two specific intermolecular interactions H–C → H–C of the same type directed so that the hydrogen atom interacts with the carbon atom of the contact molecule in the same layer, and the carbon atom of the same group interacts with the hydrogen atom of the molecule of the close surrounding from the second layer (Fig. 3.27). Thus six CH-groups of benzene molecule and of molecules from the close surrounding form 12 specific intermolecular interactions DH–C → H–C–. Taking into account that the structure of a unit cell in the crystal is retained at melting and conserves in the elementary volume in the liquid phase and only the bonds of the remote order are broken the energy of the specific intermolecular interaction in the crystal and the liquid phase should be estimated from the enthalpy of sublimation or vaporization, respectively.

$$\mathrm{DH-C\rightarrow H-C} = \frac{\Delta_{\mathrm{vap}}H^0(T)}{12}. \tag{3.24}$$

The calculated values of the energy of the specific interaction in the liquid and crystalline benzene listed in Table 3.16 demonstrate the low stability of these

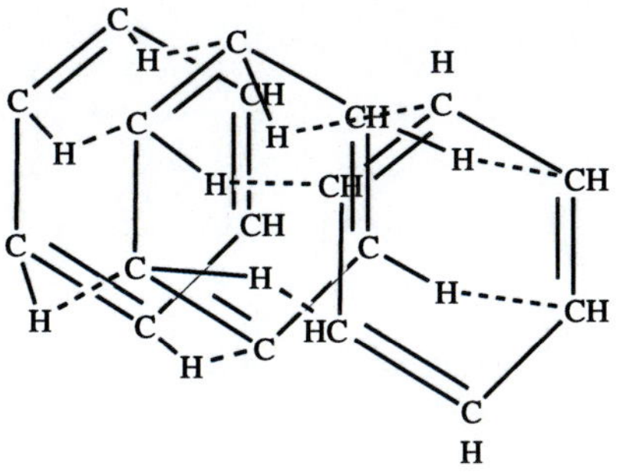

Fig. 3.27 Schematic picture of the structure of benzene in the condensed phase

Table 3.16 Energies of specific intermolecular interactions (kJ mol^{-1}) of liquid phenol ethers (298 K)

Compounds	Formula	$\Delta_{vap}H^0$ (*T*) [8, 11]	DH–C → H–C	D–O → CH_3	D–O → CH
Benzene	C_6H_6	33.8	2.80 3.70 solid	–	–
Methyl phenyl ether	C_7H_8O	46.79	2.80	5.63	3.75
Diphenyl ether	$C_{12}H_{10}O$	67.1	2.80	–	2.80
2,2-Diphenyl-1,3-dioxolane	$C_{15}H_{14}O_2$	82.4[a]	2.80	6.60	–

[a]Estimated with the use of the additivity rule at 413 K

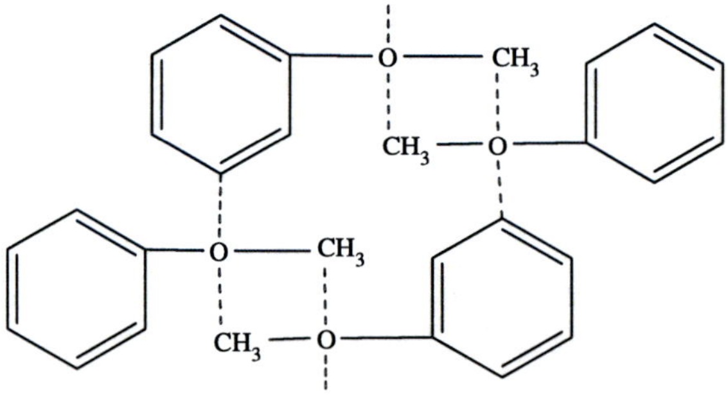

Fig. 3.28 Schematic picture of the liquid structure of methyl phenyl ether

bonds. Just therefore at the large number of the specific interactions (12) and at their low stability, the relatively high value of the vaporization enthalpy corresponds to a low melting and boiling points of benzene.

The schematic picture of the structure of the methyl phenyl ether in the condensed state (Fig. 3.28) illustrates the occupancy of all bond vacancies in the molecule and therefore it depicts the most real pattern of the existing specific interactions and permits getting information on the types and number of these intermolecular interactions. Unoccupied five CH-groups form with the contact molecules of the close surrounding ten specific interactions DH–C → H–C–, methoxy groups form two D–O → CH_3 and two D–O → CH specific interactions. The energies of two among existing three interactions are known, therefore the energy of the bond DO → CH is determined from equation

$$D-O\rightarrow CH=\frac{\Delta_{vap}H^0(T)-10DH-C\rightarrow H-C-2D-O\rightarrow CH_3}{2}. \quad (3.25)$$

The results of preformed calculations (Table 3.16) show the greater stability of this type of specific interactions than those existing in liquid benzene DH–C → H–C–. The specific intermolecular interactions formed in the liquid methyl phenyl ether get weaker in the series

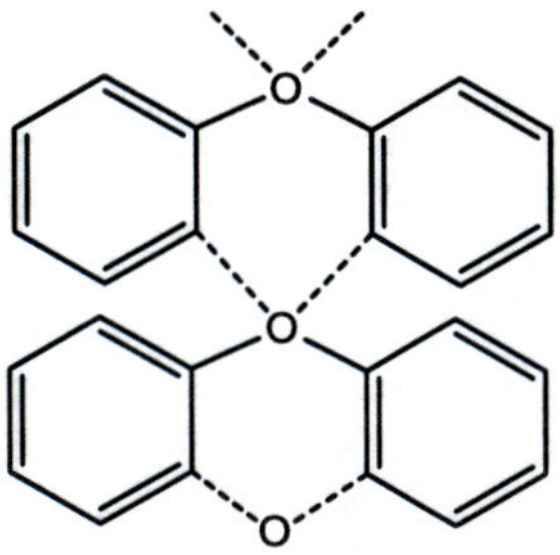

Fig. 3.29 Schematic picture of the liquid structure of diphenyl ether

$$\mathrm{D-O \rightarrow CH_3(5.63) > DO \rightarrow CH(3.75){>}DH-C \rightarrow H-C-\ (2.80\,kJ\,mol^{-1})}.$$

Consequently, the transition of a molecule from the condensed state is governed finally by the rupture of the specific interaction D–O → CH_3.

Compared with the methyl phenyl ether the structure of the condensed state of diphenyl ether seems relatively simple (Fig. 3.29), where form four intermolecular interactions DO → CH and 2DH–C → H–C– involving CH-groups of two benzene rings. It means that the sum of the energies of the specific interaction D–O → CH equals the difference between the vaporization enthalpy and the sum of the energies of the other types of specific interactions DH–C → H–C–, namely interactions between the benzene rings of the molecules of the close surrounding

$$\mathrm{4D-O \rightarrow CH} = \Delta_{\mathrm{vap}}H^0(T) - \mathrm{20DH-C \rightarrow H-C-}\,. \tag{3.26}$$

The energy of this specific intermolecular interaction (2.80 kJ mol^{-1}) practically coincides with the energy of the interaction DH–C → H–C in liquid diphenyl ether.

A very interesting molecule by the structure and the large number of bond vacancies is 2.2-diphenyl-1,3-dioxolane. The structure of the liquid state of this compound (Fig. 3.30) shows the formation of three types of the specific interactions: DH–C → H–C, D–O → CH involving the CH-groups of the benzene rings of overall number 20 and 4, respectively, and four D–O → CH_2 bonds involving the CH_2-groups of the 1,3-dioxolane ring and oxygen atoms of the ring. The experimentally measured at a temperature higher than standard (413 K) vaporization enthalpy of this compound (65.7 kJ mol^{-1}) does not permit the corresponding calculations of the energy of the specific interaction D–O → CH. However, the information obtained on the energies of the specific interactions in the liquid 1,3-dioxolanes and benzene make it possible to estimate the vaporization enthalpy of 2,2-diphenyl-1,3-dioxolane.

The energies of the specific interactions D–O → CH_2 in the liquid 2-methoxy-1,3-dioxolane reported in Table 3.12 suggest to assign to this bond the value of 4.40 kJ mol^{-1}. The overall contribution of four such specific interactions into the vaporization enthalpy amounts to 17.6 kJ mol^{-1}. The energy contribution of two

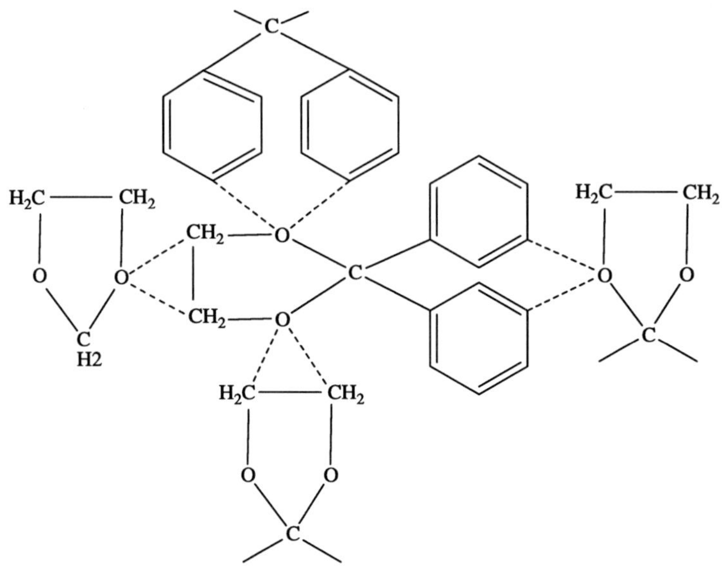

Fig. 3.30 Schematic picture of the liquid structure of 2,2 diphenyl-1,3-dioxolane

benzene rings with 16 specific interactions DH–C → H–C = 2.80 kJ mol^{-1} equals 44.8 kJ mol^{-1}. Four specific interactions D–O → CH formed by the oxygen atom of the diphenyl ether contribute into the vaporization enthalpy 11.2 kJ mol^{-1}. According to the existing energies of the specific intermolecular interactions of the liquid 2,2-diphenyl-1,3-dioxolane, the vaporization enthalpy is 73.6 kJ mol^{-1}.

The performed thermodynamic analysis of cyclic oxides both from the saturated and unsaturated series and the results obtained substantiated the existence of a correlation between the enthalpies of the phase transitions and the structural transformations of the solid and the liquid state of the compounds. The existence of a pronounced correlation between the structure of a molecule and the energies of the specific interactions originating from the essentially unshared $2s^2$-electron pair of the carbon atom are due to the distribution of the electron density in the molecule of the organic oxide and generated therewith charges on the oxygen, carbon, and hydrogen atoms.

A significant influence on the redistribution of the electron density and charges and therefore on the energies of the specific intermolecular interactions D–O → CH_2 and D–O → CH_3 is exerted by the ratio between the numbers of CH_2-groups and oxygen atoms in the ring of the organic oxide. The distribution of the structural and energy parameters of the specific intermolecular interaction of the liquid organic oxides is represented by the series of energy dependences of the forming interactions on the ratio in the ring of the numbers of CH_2-groups and oxygen atoms.

Tetrahydrofuran (8.25) > Oxetane (7.65) > Ethylene oxide (6.70) > 2-Methyl-1,3-dioxolane (6.60 kJ•mol[1]) =

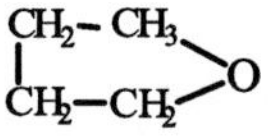
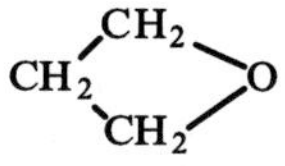
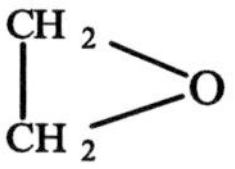
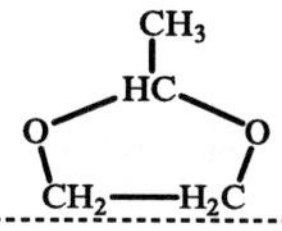

= 2- Propyl-1,3-dioxolane (6.60) = 2-Methyl-2-isopropyl-1,3-dioxolane (6.60 kJ/mol) >

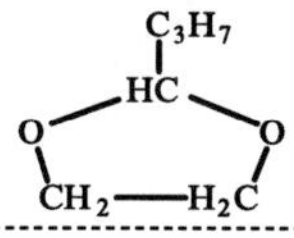
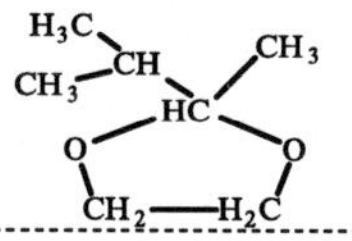

2-Methyl-2- ethyl-1,3-dioxolane (6.40) > 2-Methyl-2-propyl-1,3-dioxolane (6.25 kJ/mol) >

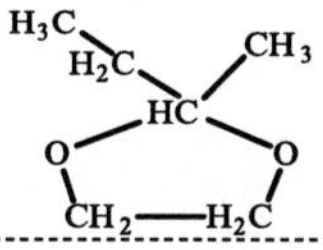
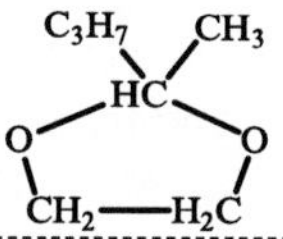

1,3-Dioxane (4.90) ≈ 1,4-Dioxane (4.80) > 3,5,7-trioxacyclononane (3.70 kJ/mol).

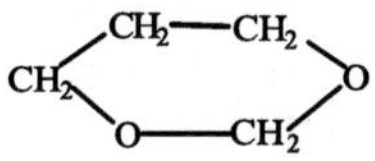
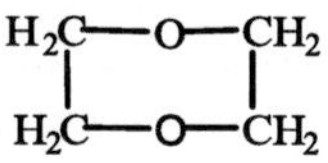
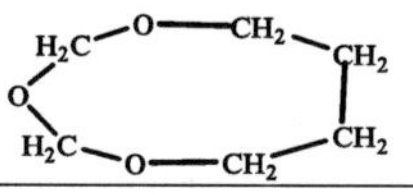

A special interest attracts the revealed H-bonds of low stability in the liquid organic oxides although even the suggestion of their existence in these compounds might be misapprehended. Therefore the energies of these bonds reported in this chapter are interesting for further search for such bonds in various organic compounds, for the development of the problem of hydrogen bonds, structural chemistry of organic compounds in the liquid and crystalline states, the solution theory, the identification of these bonds by spectral methods, and their theoretical understanding from the quantum-chemical standpoint. Here, based on the results of the performed thermodynamic analysis of organic oxides we are able to state:

- H-bonds of low stability in the liquid and crystalline state of this class compounds are formed on condition that the number of bond vacancies of the oxygen atoms in the molecule exceeds the overall number of bond vacancies of the carbon atoms in the alkyl chain or in the cycle of this molecule. In event of insufficient number of the bond vacancies of carbon atoms and remaining unrequited bond vacancies of oxygen atoms the hydrogen atoms in the *trans*-position of CH_3-, CH_2-, or CH-groups or the methyl groups of the other alkyl fragments interact with the oxygen atoms providing the occupancy of all bond vacancies and bond formation with attaining the energy minimum in the system.

The energies of these bonds are commonly significantly less than the energy of the specific intermolecular interactions formed by the essentially unshared $2s^2$-electron pair of the carbon atom $D-O \rightarrow CH_3(CH_2, CH) > D-O \cdots H-CH_2(CH, C)$. The H-bonds of low stability were found in a numerous organic oxides. Their energies in a number of compounds are equal to or close to 0.60 kJ mol^{-1}, and in the liquid compounds their values are even comparable

$$\text{Ethylene ozonide}\ (1.85) > 1,4\text{ - Dioxane - }1,3\text{ - dione}\ (2.90) > 2\text{ - Furaldehyde}\ (5.20\,\text{kJ mol}^{-1})$$

with the energy of the specific intermolecular interaction in the liquid dimethyl ether (5.63 kJ mol^{-1}).

The specific interactions of the CH-groups of liquid benzene merit a special attention. The hypothesis advanced here on the involvement into this interaction of the hydrogen and carbon atoms of the CH-groups giving the specific interaction H–C → H–C seems self-evident, but it requires a confirmation by spectral methods.

One more aspect of the performed thermodynamic analysis of ethers and organic oxides requires a special attention. These are energies of the specific intermolecular interactions formed by isostructural methyl groups and other alkyl groups that replaced hydrogen atoms in the cycles of the organic oxides. First of all the fact should be stressed that the energy contributions were established made by isostructural CH_3-groups into the enthalpy characteristics of vaporization processes of compounds whose molecules do not suffer sterical strains in the structures of their liquid and crystalline states. Yet the energy contributions of these groups into the vaporization enthalpy, namely, in the energy of the specific interactions they formed, can be classed by the compound classes and be ranged in a series along the stability of these interactions

$$\begin{aligned}
&2\text{ - Methyl - 1,3 dioxolane}\ (0.60) < 4,4\text{ - Dimethyl - 1,3 - dioxolane}\ (0.80) < \\
&1,3\text{ - Dioxolane}\ (0.90) = 2,4,5\text{ - Trimethyl - }1,3\text{ - dioxolane}\ (0.90) < \\
&2,2\text{ - Methyl - 1,3 - dioxolane}\ (1.00) = 3,3\text{ - Methyloxetane}(1.00) \leq \\
&2,2,4\text{ - Trimethyl - 1,3 - dioxolane}\ (1.05) < 2\text{ - Methyl - 1,3 - dioxolane}\ (1.10) = \\
&2\text{–Methyltetrahydrofuran}\ (1.10) = \text{Methyloxetane}\ (1.12) < \\
&\text{2,5 - Methyltetrahydrofuran}\ (1.25) < \text{2,5 - Dimethylfuran}\ (1.65) \\
&= 2\text{–}\textit{tert}\text{ - Butylfuran}\ (1.65\,\text{kJ mol}^{-1}).
\end{aligned}$$

Taking into consideration the experimental errors in the measurement of the enthalpy characteristics and understanding that many among the cited values of the energy contributions of the CH_3-groups fell in the limits of these errors, yet a general trend is observed of the stabilizing effect of these groups in the series of 1,3-dioxolane derivatives possessing two oxygen atoms and two CH_2-groups in the ring structure of the molecule, 3,3-methyloxetane with a single oxygen atom in the ring,

2,5-methyltetrahydrofuran with a single oxygen atom in the ring and methyl groups in the 2,5-positions.

$$\begin{aligned}&1,3\text{ - Dioxolane}\,(0.60-0.90) < 3,3\text{ - Methyl oxetane}\,(1.00)\\&= 2\text{ - Methyltetrahydrofuran}\,(1.10) < 2,5\text{ - Dimethylfuran}\,(1.65)\\&= 2-\textit{tert}\text{ - Butylfuran}\,(1.65\,\text{kJ mol}^{-1}).\end{aligned}$$

In this trend to stabilization of the energy contribution of the CH_3-group, the sterical position of these groups does not appear and therefore the effect in question is due exclusively to the influence of the methyl groups on the electron density distribution in the molecule, its stabilization of the molecule. Therewith the methyl group forms a specific interaction of low stability. An important trend consists in the observed increasing contribution of CH_2-, C_2H_5-, and C_3H_7-groups in compounds

$$\begin{aligned}&2\text{ - Methyl}-1,3\text{ - dioxolane}\,(1.10) < 2\text{ - Methyl - }2\text{ - ethyl - }1,3\text{ - dioxolane}\\&(1.80) < 2\text{ - Propyl - }1,3\text{–dioxolane}\,(3.30) = 2\text{ - Methyl - }2\text{ - Propyl - }\\&1,3\text{ - dioxolane}\,(3.30) << 1,2\text{ - Epoxyhexane}\,(10.25),\text{ and }2\text{ - Methyl - }\\&1,2\text{ - epoxypentane}\,(10.60\,\text{kJ mol}^{-1}).\end{aligned}$$

This means that the increased number of carbon atoms in the alkyl chain results in the formation by the terminal methyl group of the propyl fragment of more stable specific interaction. This fact is analogous to the weakening of the effect of the reverse dative bond on the terminal methyl group with the growing number of carbon atoms in the alkyl chain of ethers ensuring the acceptor properties of the carbon atom. It should be noted however that the theoretical understanding of this phenomenon is not a simple problem and requires further development.

The next conclusion requiring a special place in the understanding of the structure of the oxygen-containing organic compounds, of the electron density distribution in molecules, and of the theory of the chemical bond is the stabilization of the specific intermolecular interaction formed by the propyl group of the 1,2-epoxyhexane. The effect of the oxygen atom of this fragment is so high that the interaction energy exceeds the energy of the specific intermolecular interaction in the liquid dipropyl ether (8.92 kJ mol^{-1}). An additional contribution into the stabilization of the same specific interaction makes the isostructural methyl group of the methyl-1,2-epoxypentane. Evidently, these facts should be explained from the position of the oxygen and isostructural methyl group influence on the shift of the electron density resulting in the high acceptor properties of the terminal methyl group and high donor properties of the oxygen atom.

The results of the fulfilled thermodynamic analysis and the performed thermodynamic calculations of the energy of the specific intermolecular interactions and H-bonds of low stability in ethers and organic oxides are used in the next chapters and find further development.

References

1. Nefedov VI, Vovna VI (1989) Electronic structure of organic and elementorganic compounds. Nauka, Moscow
2. Baev AK (2001) Pentacoordinated carbon atom in intermolecular interactions in solutions of alkyl compounds of non-transition elements. In: Book of abstract. Sixth international symposium on carbanion chemistry, Marburg, Germany, p 75
3. Baev AK, Korolkov DV (2002) Phenomenon of dative bonds in elementorganic compounds with alkyl ligands. In: Book of abstract. XVth FECHEM conference on organometallic chemistry, University of Zurich, Zurich, p 350
4. Baev AK (2003) Novel approaches to the structure of functional solvents and elementorganic compounds with saturated hydrocarbon ligands and their solvates in binary systems. In: Book of abstract. International conference on solution chemistry, Debrecen, Hungary, p 79
5. Baev AK (2004) Insolvency of sp^3-hibridization model of carbon atom and chemistry of non-electrolyte solutions. In: Book of abstract. IX international conference. The problem of solvation and complex formation in solutions, Plyos, Russia, pp 62–64
6. Lebedev YuA, Miroschnichenko EA (1981) Thermochemistry of the vaporization of organic compounds. The vaporization and sublimation enthalpy and saturated pressure. Nauka, Moscow
7. Baev AK (1972) Complex formation ability halogenides of second – six groups periodical system. General and applied chemistry, vol 5. Vaisheshika Educating, Minsk, pp 35–51
8. Vasiliev IA, Petrov VM (1984) Thermodynamic properties of oxygen organic compounds. Chemistry, Leningrad
9. Baev AK (1969) To problems of chemical nature of phase transformation general and applied chemistry, vol 1. Vaisheshika Educating, Minsk, pp 197–206
10. Baev AK (1969) Phase condition and complex formation ability of halogenide metals. General and applied chemistry, vol 1. Vaisheshika Educating, Minsk, pp 207–218
11. Chickos JS, Acree WE Jr (2002) Enthalpies of vaporization of organic and organometallic compounds, 1880–2002. J Phys Chem Rev Data 31:519–879
12. Ivanov EV, Puchovskie YuP, Abrosimov VK (2004) About formation of weak hydrogen bonds CH···O in liquid methanol. In: Abstract. IX international conference. The problem of solvation and complex formation in solutions, Plyos, Russia, p 345
13. Krupin SV, Puhovskoi VA (2004) Investigation of new configuration hydrogen bonding dimer of methanol by method of molecular orbital and functional density. In: Book of abstract. IX international conference. The problem of solvation and complex formation in solutions, Plyos, Russia, p 136
14. Chickos JS, Acree WE Jr (2003) Enthalpies of sublimation of organic and organometallic compounds, 1910–2001. J Phys Chem Rev Data 32:537–698

Chapter 4
Specific Intermolecular Interactions of Ketones

4.1 Specific Interactions in Saturated Ketones and Their Isostructural Analogous

In ketones as known from the published data [1], the electrons of the double bond in the carbonyl group C=O are shifted to the more negative oxygen atom. The polarizability of the π-bond of the carbonyl group results in the appearance of a positive charge on the carbonyl carbon atom. The specific interactions of liquid and solid ketones were not discussed.

The quantum-chemical calculations [1] demonstrated that analogously to $HOCH_3$ in the saturated ethers, the contribution of the n–σ conjugation into the destabilization of HOMO as compared to H_2O equals 1.6 eV, and the charge on the oxygen atom is considerably less than in the water molecule. The antibonding character of the n-orbital in the C=O fragment of the formaldehyde molecule $H_2C{=}O$ is confirmed by the parameters of E_{bond}(O1s) exceeding the values of the true n-orbital of water by 1.4 eV. The observed long progression of C=O vibrations is characterized by the frequency by 30% lower than H_2CO in the ground state [1]. These facts indicate the sequence of the decrease in the charges on the oxygen atom in the compounds under consideration in the series $H_2O > H_2CO > O(CH_3)_2$ and in stability of their specific intermolecular interactions with the intermediate values for ketones.

Ketones with an angular structure of the molecule possess four bond vacancies, two formed by the unshared electron pairs of the oxygen, and two by carbon atoms of the terminal methyl groups of alkyl fragments with essentially unshared $2s^2$(c)-electron pairs. Specific intermolecular interactions in liquid ketones form at the expense of the essentially unshared $2s^2$(c)-electron pair involving a pentacoordinate carbon atom. The latter unlike oxygen possesses a positive charge and acceptor properties. Two electron pairs of carbonyl oxygen provide donor properties. The double bond of the carbonyl group results in a planar structure of the ketone molecule, and the four bond vacancies lead to the formation of a complex chain-cyclic structure (Fig. 4.1). The interatomic distance C ↔ O virtually does not affect the intermolecular interactions for the existing steric effect in the angular molecule

A.K. Baev, *Specific Intermolecular Interactions of Organic Compounds*,
DOI 10.1007/978-3-642-21622-0_4, © Springer-Verlag Berlin Heidelberg 2012

Fig. 4.1 Schematic picture of the cyclic-chain structure of liquid ketones

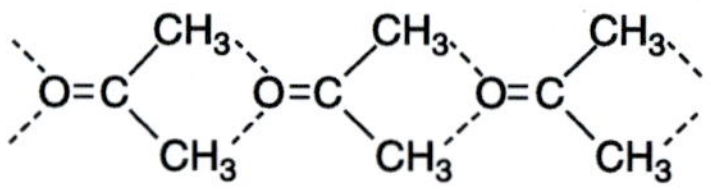

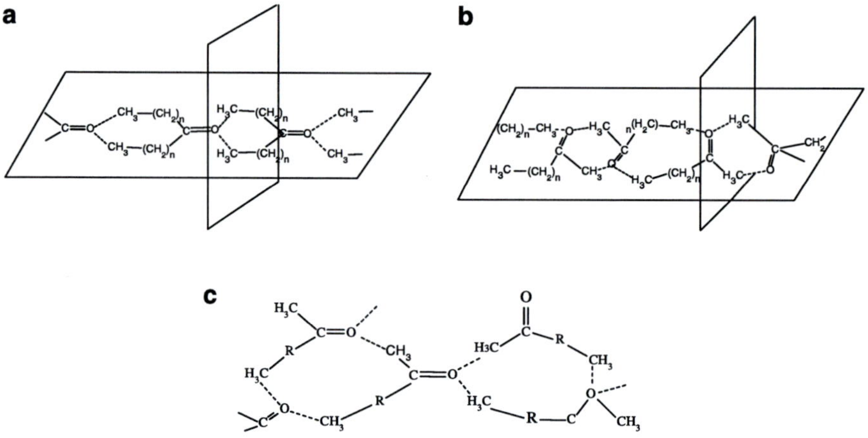

Fig. 4.2 Schematic picture of the cyclic-chain structure of liquid ketones: symmetric (**a**), unsymmetrical ketones (**b**), network structure (**c**)

is insignificant. The four intermolecular interactions =O → CH_3–C– involving the pentacoordinate carbon present in the liquid and solid states of saturated ketones belong to a single type.

The formed bond with the oxygen atom of the contact molecules possesses a donor–acceptor character, arises between a pair of atoms of two molecules, is strictly localized on definite orbitals of the interacting species, and belongs to specific intermolecular interactions [2]. The presence of four bond vacancies in the ketone molecule provides for the participation in the near surrounding of maximum four and minimum two molecules. The latter form the molecular structure with specific intermolecular interactions of the general type =O → CH_3–$(CH_2)_n$–C– in symmetric and unsymmetrical rings resulting in chain-cycle structure in the liquid state. When analogous structure forms from ketone molecule with alkyl groups of different length, an endoeffect of structural distortion may arise that would be absent at the presence of four molecules in the nearest surrounding. In this case the same specific interactions R–C=O → CH_3–C– or CH_3–C=O → CH_3–$(CH_2)_n$–C– exist where R is CH_3–$(CH_2)_n$ (Fig. 4.2). Consequently, in the liquid and solid state a chain structure is possible with two rings on one side: acetone group (2-propanone) CH_3–(C–)=O, and on the other side, of a long carbon chain of the substituent CH_3–$(CH_2)_n$–(C–)=O growing in the series 2-butanone, 2-pentanone, 2-hexanone (Fig. 4.2b). At interaction with four molecules the same bond vacancies form a network structure with a single type and number of specific intermolecular interactions (Fig. 4.2c).

Inasmuch as the vaporization involves the rupture of all bonds, their energy should be evaluated from the vaporization enthalpy $\Delta_{vap}H^0(298)/4$ [3]. In the case of 2-propanone the energy of the specific intermolecular interaction D=O → CH_3–C– equals 7.35 kJ mol^{-1}. The unaccounted nonspecific interactions in liquid ketone give an insignificant error for their energy is low and comparable with the error of the experimental measurement of the vaporization enthalpy (Table 4.1). The obtained values of the energy of the specific interactions (Table 4.1) show their higher stability compared to the interactions in liquid ethers and lower stability than that of the hydrogen bond in liquid water (10.94 kJ mol^{-1}). This is consistent with the quantum-chemical calculations [1] which unambiguously show the increase of the charge on the oxygen atom in the series diethyl ether < acetone < water. In keeping with this sequence of the changing charge on the oxygen atom at the same number of carbon atoms in the alkyl group, the stabilization of the specific intermolecular interaction grows in the series diethyl ether (6.78) < acetone (7.70) < water (10.99 kJ mol^{-1}). Nonetheless, the low values are also due to the influence of the reverse dative bond that decreases the negative charge on the oxygen atom and the shift of the electron density along the alkyl chain to the carbon atom contacting with the oxygen. As a result, the methyl group in 2-propanone has a reduced positive charge and a reduced acceptor property. This type of the intramolecular donor–acceptor bond was substantiated on alkyl compounds AR_k [7–15] (Chap. 2). The effect of the reverse dative bond on the properties of the terminal methyl group gradually decreases with the growing number of carbon atoms in the alkyl chain.

Therefore, the energy of the specific intermolecular interactions in propyl ketones with symmetric alkyls and with different alkyls should be regarded as the limiting value.

Considering the obtained energies of the specific intermolecular interactions involving the pentacoordinate carbon atom from this viewpoint and accounting for the contribution into the stabilizing interaction with the growing number carbon of the alkyl groups in the unsymmetrical methyl ketones

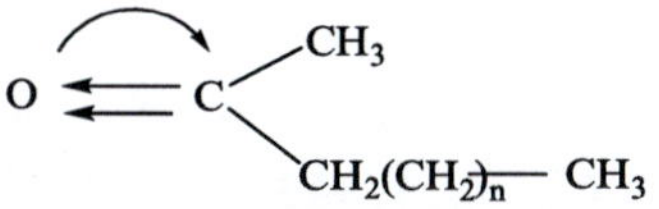

we should estimate the contribution of the specific interaction =O → CH_3– in the unsymmetrical methyl ketones CH_3–C–(=O)–R at 7.70 kJ mol^{-1}. Consequently, the energy of the second type of the specific interaction with the alkyl group containing a large number of carbons in the unsymmetrical ketone should be calculated from the thermochemical equation (4.1):

$$\begin{aligned} D = O &\rightarrow CH_3 - (CH_2)_n - C- \\ &= (\Delta_{vap}H^0(298) - 2D = O \rightarrow CH_3 - C-)/2. \end{aligned} \tag{4.1}$$

The performed calculation of the energy of the specific interaction involving the ethyl fragment of 2-butanone resulted in the value 9.59 kJ mol^{-1}. The energy of the

Table 4.1 Energies (kJ mol^{-1}) of specific interactions in liquid ketones (298 K)

Compounds	Formula	$\Delta_{vap}H^0$(298 K) [4–6]	Energies of specific intermolecular interactions			
			$D{=}O \rightarrow CH_3{-}C{-}$	$DH_3C \rightarrow H{-}CH_2$	$D{=}O \rightarrow CH_3{-}CH_2{-}C{-}$	$D{=}O \rightarrow CH_3{-}(CH_2)_n{-}C{-}$
2-Propanone (acetone)	$CH_3{-}C({=}O){-}CH_3$	30.84	7.70	–	–	–
2-Butanone	$CH_3{-}CH_2{-}C({=}O){-}CH_3$	34.38	7.70	–	9.50	–
3-Methyl-2-butanone	$H_3C{-}CH(CH_3){-}C({=}O){-}CH_3$	34.98	7.70	0.60 (1.20)	9.19	–
3,3-Dimethyl-2-butanone	$CH_3{-}C(CH_3)(H_3C){-}C({=}O){-}CH_3$	37.87 (37.30)	7.70	0.60 (2.40)	$n = 1$ 10.32	–
3-Pentanone (diethyl ketone)	$CH_3{-}CH_2{-}C({=}O){-}CH_2{-}CH_3$	36.15	–	–	9.04	–
2-Methyl-3-pentanone	$CH_3{-}CH_2{-}C({=}O){-}CH(CH_3){-}CH_3$	39.76	–	0.60 (1.20)	9.04 (9.65)	10.25 (9.65)
2,4-Dimethyl pentanone (diisopropyl ketone)	$CH_3{-}CH(CH_3){-}C({=}O){-}CH(CH_3){-}CH_3$	41.58	–	0.60 (2.40)	9.75	–
2-Pentanone (methyl *n*-propyl ketone)	$H_3C{-}H_2C{-}CH_2{-}C({=}O){-}CH_3$	36.13	7.70	–	–	10.32
3-Hexanone	$H_3C{-}(CH_2)_2{-}C({=}O){-}CH_2{-}CH_3$	38.39	–	–	9.04	10.17

3-Methyl-2-pentanone	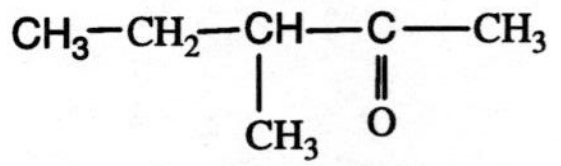	37.18	7.70	0.60 (1.20)	–	10.38
4-Methyl-2-pentanone (isobutyl methyl ketone)	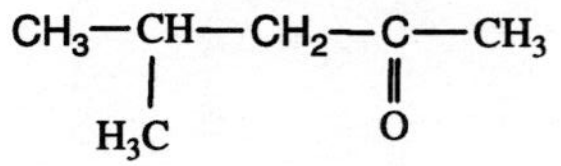	37.69	7.70	0.60 (1.20)	–	10.54
2,6-Dimethyl-4-heptanone	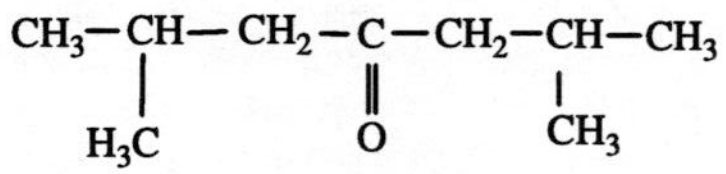	50.92 (51.08)	–	0.60 (2.40)	–	$n = 2$ (11.67)

specific interaction $=O \rightarrow CH_3–CH_2–C–$ in liquid diethyl ketone (3-pentanone) with symmetrical ethyl groups also is established from the vaporization enthalpy $\Delta_{vap}H^0(298)/n$ divided by four bond vacancies giving 9.04 kJ mol^{-1}.

The performed thermodynamic calculation of the energy of the specific interaction in the unsymmetrical 2-butanone demonstrates the increased stabilization of this interaction by 0.55 kJ mol^{-1} compared with the analogous interaction in liquid diethyl ketone (3-pentanone).

Let us discuss now the branched derivatives of 2-butanone. The replacement by a methyl group of a hydrogen in the CH_2-group of this ketone affords a single possible compound 3-methyl-2-butanone having the vaporization enthalpy of 34.98 kJ mol^{-1}, by 0.60 kJ mol^{-1} larger than the vaporization enthalpy of 2-butanone. Thus the contribution of the isostructural methyl group into the enthalpy characteristic of the ketone amounts to 0.60 kJ mol^{-1}, somewhat lower than the contribution (0.70 kJ mol^{-1}) of the analogous group of diisopropyl ether. The similarity of the specific intermolecular interactions in ethers and ketones makes it possible to use this value in evaluation of the energy of intermolecular interactions from the vaporization enthalpy characteristics of the methyl-substituted ketones.

The isostructural methyl forms a specific interaction $H_3C \rightarrow H$ of low stability with a hydrogen of a methyl or methylene group located in the *trans*-position with respect to oxygen.

The replacement of a hydrogen in the CH_2-group requires the addition to the vaporization enthalpy of a compensation contribution, accounting for the interaction of this hydrogen that has been estimated (Chaps. 2 and 3) at 0.60 kJ mol^{-1}. The presence in the near surrounding in the liquid state of two interacting molecules indicates that the summary contribution of these groups into the vaporization enthalpy amounts to 1.20 kJ mol^{-1}. Subtracting this value from the vaporization enthalpy of the 3-methyl-2-butanone and the value of two specific interactions $=O \rightarrow CH_3–$ (15.4 kJ mol^{-1}) accounting for the interaction of the molecules of the near surrounding we obtain the expression (4.2) and the energy of the second type of the specific interaction $D{=}O \rightarrow CH_3–CH_2–C=$ (equal to 9.19 kJ mol^{-1}) (Table 4.1).

$$\begin{aligned} D = O \rightarrow CH_3 - (CH_2)_n - C- \\ = (\Delta_{vap}H^0(298) - 2D = O \rightarrow CH_3 - C - -2D iso CH_3)/2. \end{aligned} \tag{4.2}$$

This value of the interaction energy is less than that of the analogous specific interaction in the liquid 2-butanone by 0.30 kJ mol^{-1}; thus the isostructural methyl group in 3-methyl-2-butanone plays a destabilizing role. The nature of this destabilizing effect consists in the participation of the isostructural methyl group in the electron density distribution in the molecule resulting in the decrease in the negative charge on the oxygen atom and the positive charge of the carbon atom of the terminal methyl group of the chain. The participation of the isostructural methyl group in the redistribution of the electron density in the molecule results in the diminishing of the donor–acceptor properties of the carbon atom of this group providing in this way a possibility for the group to form a specific interaction of low stability.

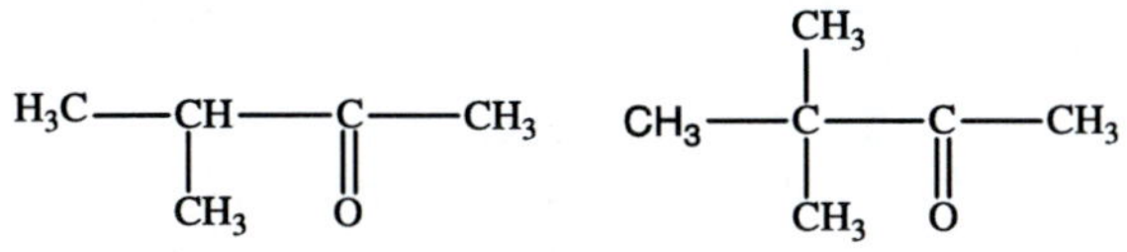

The substitution of the second hydrogen atom by methyl group in the same methylene group in the molecule of 3-methyl-2-butanone is accompanied with the increase in the vaporization enthalpy of 3,3-dimethyl-2-butanone (37.87 kJ mol^{-1}) by 2.98 kJ mol^{-1}. Hence the second methyl group vicinal to the carbonyl group significantly changes the electron density distribution in the 3,3-dimethyl-2-butanone molecule as compared to 2-butanone and 3-methyl-2-butanone providing for an increased negative charge on the hydrogen and a positive charge on the carbon atom of the terminal methyl group of the alkyl fragment O=C–CH–(CH_3)–CH_3. Considering the contribution into the vaporization enthalpy of two methyl groups (1.20 kJ mol^{-1}) and consequently, of four located in the near surrounding (2.40 kJ mol^{-1}) of the liquid 3,3-dimethyl-2-butanone and the contribution of two specific interactions (15.40 kJ mol^{-1}) formed by the fragments with CH_3 groups, we obtained the value of the energy of the second type specific intermolecular interaction D=O → CH_3–CH_2–(–C)=O (4.3) equal to 10.32 kJ mol^{-1}

$$\begin{aligned} & D = O \rightarrow CH_3 - (CH_2)_n - C- \\ & \quad = (\Delta_{vap}H^0(298) - 2D = O \rightarrow CH_3 - C - -DisoCH_3)/2. \end{aligned} \tag{4.3}$$

The results of the performed thermodynamic calculations show that the energies of the specific interactions in liquid 2-butanone and its methyl-substituted derivatives grow in the sequence

$$\begin{aligned} & 3\text{ - Methyl - 2 - butanone } (9.19) < 2\text{ - Butanone } (9.50) \\ & < 3{,}3\text{ - Dimethyl - 2 - butanone } (10.32\,\text{kJ mol}^{-1}). \end{aligned}$$

In the same sequence grows the negative charge on the oxygen of the carbonyl group and the positive charge on the carbon of the terminal methyl group of the long alkyl chain. Thus the introduction of one isostructural methyl group into the structure of 2-butanone results in the change in the electron density distribution in the molecule of 3-methyl-2-butanone leading to the destabilization of the specific intermolecular interaction. At the same time the location of two isostructural methyl groups at the carbon atom of the same alkyl chain led to the stabilization of the same type of the specific interaction.

Based on the above proved conclusion on the effect of the reverse dative bond only in three first representatives of the homologous series of oxygen-containing compounds, we can expect its effect to be finished in 2-butanone and 2-pentanone, symmetric 3-pentanone and 3-hexanone, and their branched derivatives substituted with methyl groups.

Inasmuch as in the molecule of 3-pentanone the alkyl groups contain each three carbon atoms, and the last one is common for both groups, the energy of the specific intermolecular interaction $=O \rightarrow CH_3–CH_2–C–$ is determined directly from the vaporization enthalpy by dividing by four free vacancies capable of forming strong interactions. The obtained value of this interaction for liquid 3-pentanone is 9.04 kJ mol^{-1} (Table 4.1). In the molecule of 2-methyl-3-pentanone one fragment does not contain an isostructural methyl group, and it is therefore expectable that its terminal methyl group forms the specific interaction with the energy equal to that formed in the liquid 3-pentanone, i.e., equal to 9.04 kJ mol^{-1}. Since the second moiety contains the isostructural methyl group, the latter may affect the shift of the electron density at least in this chain. In this connection considering the double contribution of the isostructural methyl groups of the interacting molecules of the near surrounding (1.20 kJ mol^{-1}) we obtain from (4.2) the energy of the specific interaction equal to 10.25 kJ mol^{-1}. This value exceeds by 1.20 kJ mol^{-1} the energy of the analogous interaction formed by the alkyl free of the isostructural methyl group. Whereas no proofs exist that energies of one type of interactions formed by the terminal methyl groups of the alkyl chains of symmetric ketones containing an isostructural methyl group can be different, it is more rightly to assign to both interactions of the 2-methyl-3-pentanone the value of 9.65 kJ mol^{-1} (Table 4.1).

Inasmuch as in the molecule of the symmetric 2,4-dimethyl-3-pentanone (diisopropyl ketone) also two symmetrically located isostructural methyl groups are present, it is possible to establish the energies of the specific interactions and to reveal the effect of these groups on their stabilization. The thermodynamic calculation should take into account the contribution of two isostructural methyl groups of each molecule situated in the near surrounding, and therefore the calculation of the energy of the specific interaction should be performed by (4.4)

$$D = O \rightarrow CH_3 - CH_2 - C- = (\Delta_{vap}H(298) - 4DH_3C - H)/4. \quad (4.4)$$

The obtained energy of the specific interaction under consideration (9.75 kJ mol^{-1}) is close to that calculated for the liquid 2-methyl-3-pentanone. It followed from the data compiled in Table 4.1 that the energy of the specific interactions of compounds in question forms a regular series according to the trend to their stabilization with the growing number of isostructural methyl groups in their molecules originating from the redistribution of the electron density in the molecules.

$$3\text{ - Pentanone(diethyl ketene) } (9.04) < 2\text{ - Dimethyl - 3 - pentanone } (9.65)$$
$$< 2{,}4\text{ - Dimethyl - 3 - pentanone } (9.75\,\text{kJ mol}^{-1}).$$

This leads to a conclusion that the first isostructural methyl group in the presence of two carbon atoms in the alkyl chain provides the maximum stabilization of the specific interaction.

The equal values of the energy of the specific interactions in compounds

$$\text{Methylethyl ketone } (9.04) \approx \text{Diethyl ketone (3 - pentanone) } (9.04)$$
$$\leq 3\text{ - Methyl - 2 - butanone } (9.18\,\text{kJ mol}^{-1})$$

having two carbon atoms in the alkyl moiety show that the only possible position of the isostructural methyl group in the 3-methyl-2-butanone contiguous to the carbonyl group does not affect the reverse dative bond. Therefore, the energy of the specific intermolecular interaction formed by the two-carbon chain remains the same as in the 3-pentanone (diethyl ketone). The second isostructural methyl group vicinal to the carbonyl group in the 3,3-dimethyl-2-butanone affects the dative bond decreasing it and providing for the increase in the electron density on the carbon atom of the terminal methyl group and raising its donor–acceptor properties.

The reverse dative bond in unsymmetrical ketones 2-pentanone (methyl *n*-propyl ketone), 3-hexanone, and 3-methyl-2-pentanone and in symmetric ketones with alkyl chains exceeding three carbon atoms does not affect the donor–acceptor properties of the carbon. Therefore, even the fourth carbon atom (calculating from the carbonyl carbon) provides a certain contribution to the vaporization enthalpy of the corresponding compound. Applying the principle of contributions additivity into the energy of the specific intermolecular interactions we calculated the energy of the specific interaction $=O \rightarrow CH_3–CH_2–CH_2–C–$ in liquid unsymmetrical methyl *n*-propyl ketone (2-pentanone) and 3-hexanone (Table 4.1) at 10.32 and 10.17 kJ mol^{-1}, respectively. These energy values exceed the energy of the specific interaction in 2-butanone having an alkyl chain with three carbon atoms by 0.82 and 0.67 kJ mol^{-1} and by 0.53 kJ mol^{-1} the energy of similar interaction in the liquid 3,3-dimethyl-2-butanone (10.03 kJ mol^{-1}). However, their energies are lower than the energy of the hydrogen bond in water (10.99 kJ mol^{-1}). The above-mentioned passive role of the isostructural methyl group contiguous to the carbonyl group was also found in an unsymmetrical molecule of 3-methyl-2-pentanone, therefore the energy of the specific intermolecular interaction $D{=}O \rightarrow CH_3–CH_2–CH(CH_3)–C–$ (10.38 kJ mol^{-1}) was practically equal to that of the propyl group of the normal structure in the 2-pentanone (methyl *n*-propyl ketone).

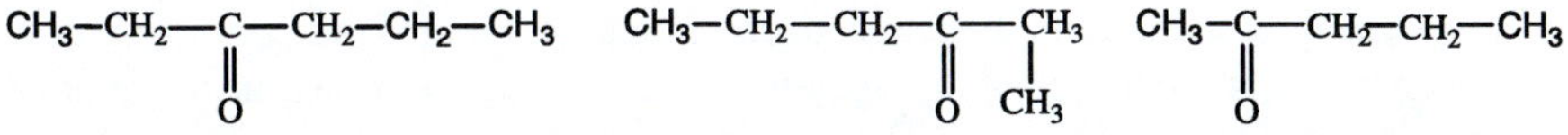

10.17 10.38 10.32 kJ·mol^{-1}

The calculated energies of the specific intermolecular interactions of the liquid 2-pentanone (methyl *n*-propyl ketone) and 3-methyl-2-pentanone with a single isostructural methyl group vicinal to the carbonyl group demonstrate the trend in the stabilization of the specific intermolecular interaction compared to the energy of the specific interactions formed by the molecules of 3-hexanone. It is significant that the removal of the isostructural methyl group from the carbonyl group in 3-methyl-2-pentanone (10.25 kJ mol^{-1}) to the fourth carbon atom in 4-methyl-2-pentanone (isobutyl methyl ketone) (10.54 kJ mol^{-1}) results in the stabilization of the specific intermolecular interaction at the same number of carbon atoms in the alkyl chain. The results of calculations of the energies of the specific interactions presented in Table 4.1 for the liquid symmetric 2,6-dimethyl-4-heptanone show that like the situation in 4-heptanone the involvement of carbon and hydrogen of the methylene groups in the intermolecular interactions is clearly seen. This contribution is not observed in unsymmetrical 4-methyl-2-pentanone (isobutyl methyl ketone) (Table 4.1) that should be explained by the effect of the reverse dative bond which in the molecule of this compound occurs in low degree. These data show that the real values of the energy of the specific interactions with the participation of the pentacoordinate carbon are the energies of the interactions established for ketones with the number of carbon atoms in the alkyl chain no less than four in unsymmetrical ketones. The values of the energy of the specific interactions presented in Table 4.1 for symmetric ketones make it possible to consider the values established for 2,6-dimethyl-4-heptanone as unlikely. Still the thermodynamic calculations for ketones with a large number of carbon atoms in the alkyl chain are useful, for they show the influence of the fourth CH_2-group on the energy of the specific interaction and consequently the necessity to account for this influence and its contribution into the vaporization enthalpy of the compound.

Thus the performed thermodynamic analysis for ketones containing no more than four carbons in the alkyl substituent substantiated the following:

Firstly, the location of the isostructural methyl group of the alkyl moiety contiguous to the carbonyl group in ketones destabilizes the specific interaction (3-methyl-2-butanone), and the location of the isostructural group close to the carbonyl in a symmetric ketone (2-methyl-3-pentanone) shows a stabilizing effect.

Secondly, the increase of the number of the isostructural methyl groups till they are attached to a tertiary carbon atom provides the greatest stabilization of the specific intermolecular interaction (3,3-dimethyl-2-butanone).

Thirdly, the removal the isostructural methyl group further from the carbonyl group in unsymmetrical ketones stabilize the specific intermolecular interaction [3-methyl-2-pentanone and 4-methyl-2-pentanone (isobutyl methyl ketone)].

Fourthly, the maximum value of the energy of the specific interaction in ketones of unsymmetrical structure is attained in 2-pentanone and in symmetric 3-pentanone. The energies of the specific intermolecular interaction of ketones form the following sequence with an average energy value 10.30 kJ mol^{-1}:

$$3\text{ - Pentanone }(10.17) < 2\text{ - Pentanone(methyl } n\text{ - propyl - ketone) }(10.32)$$
$$< 3\text{ - Methyl - 2 - pentanone }(10.38\,\text{kJ}\,\text{mol}^{-1}).$$

The more stable specific interaction $=O \rightarrow CH_3–CH_2–CH_2–C=$ in the methyl *n*-propyl ketone (2-pentanone) (10.32 kJ mol^{-1}) compared with the interaction in the diethyl ketone (3-pentanone) (9.04 kJ mol^{-1}) is due to the increase of the alkyl group to a propyl (or to four-carbon chain taking into account the carbon of the carbonyl group $>C=O$. The energy of this specific interaction is smaller than the energy of the hydrogen bond in water. Just at the four carbon atoms in the alkyl chain of the ketone the reverse dative bond practically does not affect the intermolecular interaction. This fact makes it possible to apply the principle of the additivity of the contributions into the energy of the specific intermolecular interactions and into the enthalpy characteristics of compounds with the equal number of carbon atoms in the alkyl moieties for calculation of the vaporization enthalpy of methyl ethyl ketone (33.48 kJ mol^{-1}) and branched ketones (Table 4.1). The energy of the specific intermolecular interactions found by the thermochemical calculations regularly grows with the growing number of carbon atoms in the alkyl chain and weakening effect of the reverse dative bond, depending on the location of the isostructural methyl groups and their number in the alkyl chain of the symmetric and unsymmetrical ketone molecule.

$$
\begin{aligned}
&2\text{ - Propanone(acetone) }(7.70) < 3\text{ - Pentanone(diethyl ketone) }(9.04)\\
&< 3\text{ - Methyl - 2 - butanone }(9.18) < 3,3\text{ - Dimethyl - 2 - butanone }(10.03)\\
&< 3\text{ - Hexanone }(10.17) < 2\text{ - Pentanone(methyl } n \text{ - propyl ketone) }(10.32)\\
&< 3\text{ - Methyl - 2 - pentanone }(10.38)\\
&< 2,4\text{ - Dimethyl} - 3\text{ - pentanone(diisopropyl ketone) }(10.55\,\text{kJ mol}^{-1}).
\end{aligned}
$$

The energies of the specific interactions of all considered ketones are lower than the energy of the hydrogen bond in the liquid water; for in the latter the larger electron density on the oxygen atom than that on the carbonyl group oxygen of ketones is characteristic which in its turn is higher than the electron density on the oxygen of an ether [1]. Just the difference in the electron density on the oxygen atoms of the mentioned compounds governs the regular decrease in the energy of specific interactions in the series: water > ketones > ethers. This regularity is revealed in the values of the energies of the specific intermolecular interactions of these classes of compounds.

4.2 Energies of Specific Interactions in Liquid Unsymmetrical and Symmetric Ketones

The established values of the energy of the specific interactions in liquid ketones make it possible to turn to acquiring the important information on the energy contribution from each hydrogen of the CH_2-groups into the vaporization enthalpy of ketones with alkyl groups containing more than three carbon atoms.

The regular growth of the values of the ketones vaporization enthalpy in the series

$$
\begin{aligned}
&2\text{ - Propanone} < 2\text{ - Pentanone} < 2\text{ - Hexanone} < 2\text{ - Heptanone}\\
&< 2\text{ - Octanone} < 2\text{ - Nonanone} < 2\text{ - Dodecanone}
\end{aligned}
$$

is described by two intercepting straight lines of dependences on the number of carbon atoms in the alkyl chain of the unsymmetrical molecule (Fig. 4.3). The first part till the interception reflects the decrease in the influence of the reverse dative bond on the shift of the electron density from the terminal methyl group and the

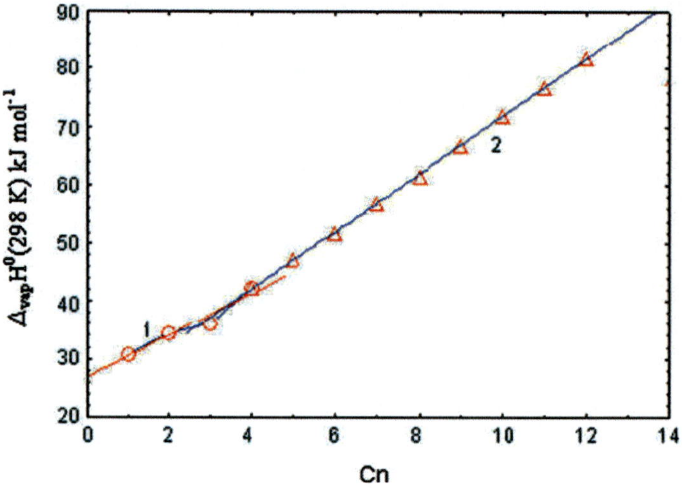

Fig. 4.3 Dependence of vaporization enthalpy on the number of carbon atoms in the hydrocarbon chain of ketones in the homological series of 2-propanone

stabilization of the specific intermolecular interaction formed by ketone molecules. The second part of the line after interception reflects the identical contribution of each next CH_2-group into the enthalpy characteristics. Consequently, this linear dependence describes the overall contributions of the specific interactions involving the carbon and hydrogen atoms of the growing number of the methylene groups in the ketone molecule.

Therefore, the plot of the vaporization enthalpy of unsymmetrical methyl ketones forming two intercepting straight lines demonstrates that the reverse dative bond affects the terminal methyl group of 2-propanone–2-butanone, and its influence is finished at the 2-pentanone where the energy of the specific intermolecular interaction becomes maximal. This means that it remains constant in all liquid unsymmetrical methyl ketones with a larger number of carbon atoms in the alkyl chain, and the growth of the enthalpy characteristics occurs due to the growing contribution of the CH_2-groups.

The character of the dependence of the vaporization enthalpy of symmetric ethers and ketones (Fig. 4.4) on the growing number of carbon atoms in the chain is virtually identical. This dependence shows the end of the influence of the reverse dative bond on the terminal methyl group in the dipropyl ether and 3-pentanone at three carbon atoms in the alkyl chain. The same conclusion follows from the analysis of the correlation presented in Fig. 4.5 between the vaporization enthalpy of symmetric ethers and ketones. The deviation from the straight line of the vaporization enthalpy of 2-propanone and 3-pentanone indicates that the reverse dative bonds in the molecules of these compounds supply different contributions into the stabilization of the specific interactions at the growing number of carbon atoms in the chain up to three. Consequently, the specific interactions in each of the

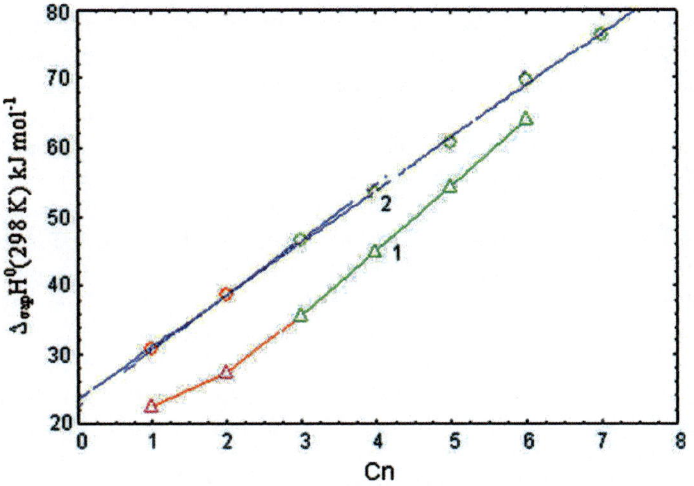

Fig. 4.4 Dependence of vaporization enthalpy of symmetrical ethers (1, 2) and ketones (3, 4) on the number of carbon atoms in the hydrocarbon chains

Fig. 4.5 Correlation between the vaporization enthalpy of symmetrical ethers and the vaporization enthalpy of symmetrical ketones (1) influence of reverse dative bond, (2) influence of CH_2-groups

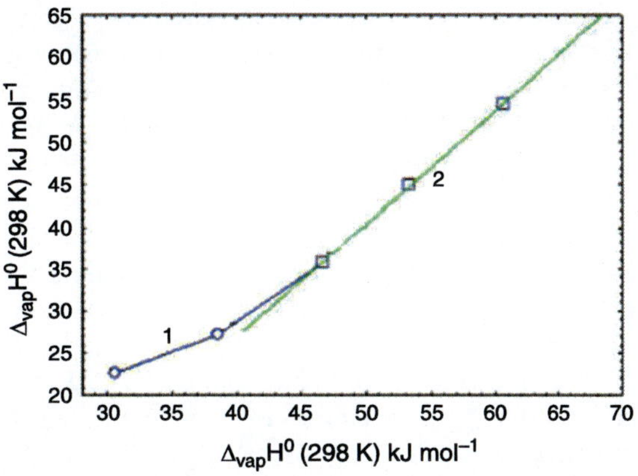

classes in consideration are differently stabilized. Therefore, their contributions into the enthalpy characteristics are not fully compensated.

At the same time the contribution of the CH_2-groups into the enthalpy characteristics is practically identical. Therefore, the specific interaction formed involving oxygen atoms of ethers and ketones essentially differ in the series of compounds dimethyl–diethyl–dipropyl ethers and 2-propanone–3-pentanone, respectively. The constant energy value of the specific intermolecular interactions of compounds with a number of carbon atoms in the chain exceed three results in the linear plot of the correlation $\Delta_{vap}H^0(298)$ethers $= f(\Delta_{vap}H^0(298))$ketones (Fig. 4.5). This shows that the contribution of the methylene groups into the vaporization enthalpy is identical or insignificantly different.

Using the correlation of the vaporization enthalpy with the number of carbon atoms in the long alkyl chain (Figs. 4.3–4.5), we estimated the enthalpy characteristics of a series of previously unstadied ketones (Table 4.2).

Earlier (Chap. 2) we showed that the coordination endoeffect in molecules was observed on the atoms directly linked to the central heteroatom and that the increased mobility of the carbon–carbon chain with the growing number of the CH_2-groups favors the interaction of the positively charged hydrogen atoms with the carbon atoms of the contacting analogous groups in the liquid and solid state of compounds. The comparison of the calculation of the structure of a monomer of methanol, electron density on the atoms, and spectral characteristics led Ivanov et al. [16] to a conclusion on a nonequivalence of the protons of the methanol methyl groups taking part in the interaction with the formation of weak H-bonds. The interactions between the positively charged hydrogen atoms and the carbon atom involving its essentially unshared $2s^2$(c) electron pair are donor–acceptor and not cooperative ones.

The carbon atom of the carbonyl group behaves as a donor–acceptor regulator of the electron density on the carbon atoms linked to it. Therefore, the carbon atom of the carbonyl group giving over its electron density in the interaction with oxygen partially compensates the loss by shifting the electron density from the adjacent carbon atoms of the alkyl chain. The methyl group of unsymmetrical 2-butanone with a decreased positive charge and an acceptor ability forms a more rigid system resulting from the reverse dative bond compared with the carbon atom of terminal methyl group of the second fragment with a larger number of carbon atoms in the chains of 3-pentanone, 4-heptanone. The influence of the reverse dative bond in symmetric ketones is finished on the terminal methyl groups of the 3-pentanone, and further increase in the enthalpy of vaporization and sublimation in the compounds of this homologous series is due to the increasing contribution of the growing number of CH_2-groups. Therefore, the energy of the specific interaction of 2-propanone and 3-pentanone with four equivalent bond vacancies is established from the vaporization enthalpy $\Delta_{vap}H^0(298)/4$ and sublimation enthalpy $\Delta_{sub}H^0(298)/4$. It is therefore possible to state that the energy of the specific interaction $=O \rightarrow CH_3–CH_2–C–$ (9.65 kJ mol^{-1}) attained the maximum value and further in all compounds of this homologous series it will be constant. Consequently, for all further compounds of this series it is possible to evaluate the contribution of CH_2-groups or the overall energy of the two types of the specific interactions $= O \rightarrow CH_3 - CH_2 - C - +2DH_2C \rightarrow H$, and also the energy of $DH_2C \rightarrow H$ interaction.

The value of the energy of this interaction of one CH_2-group we obtain subtracting from the vaporization enthalpy of further member of the series that of the lower member and dividing the difference by the number of methylene groups. Since the difference of the vaporization enthalpy of 2-tetradecanone (81.8 kJ mol^{-1}) and 2-pentanone (36.13) amounts to 45.7 kJ mol^{-1}, the energy contribution of a single CH_2-group is 5.09 kJ mol^{-1}. This value accounts for the interaction of two molecules of the near surrounding in the structure of the liquid state. Therefore, two bond vacancies of the molecule of the unsymmetrical ketone (Fig. 4.2b, c)

Table 4.2 Estimated values of the vaporization enthalpies (kJ mol^{-1}) of the unsymmetrical ketones of the homological series of 2-propanone

Compound	Structure	$\Delta_{vap}H^0(298)$ [4]	Compound	Structure	$\Delta_{vap}H^0(298)$
2-Heptanone	$CH_3-(CH_2)_4-C(=O)-CH_3$	47.2	2-Pentadecanone	$CH_3-(CH_2)_{12}-C(=O)-CH_3$	87.2
2-Octanone	$CH_3-(CH_2)_5-C(=O)-CH_3$	52.0	2-Hexadecanone	$CH_3-(CH_2)_{13}-C(=O)-CH_3$	92.2
2-Nonanone	$CH_3-(CH_2)_6-C(=O)-CH_3$	57.0	2-Heptadecanone	$CH_3-(CH_2)_{14}-C(=O)-CH_3$	97.2
2-Decanone	$CH_3-(CH_2)_7-C(=O)-CH_3$	61.9	2-Octadecanone	$CH_3-(CH_2)_{15}-C(=O)-CH_3$	102.2
2-Indecanone	$CH_3-(CH_2)_8-C(=O)-CH_3$	66.8	2-Nonadecanone	$CH_3-(CH_2)_{16}-C(=O)-CH_3$	107.0
2-Dodecanone	$CH_3-(CH_2)_9-C(=O)-CH_3$	71.8	2-Eicosanone	$CH_3-(CH_2)_{17}-C(=O)-CH_3$	112.2
2-Tridecanone	$CH_3-(CH_2)_{10}-C(=O)-CH_3$	76.8	2-Heneicosanone	$CH_3-(CH_2)_{18}-C(=O)-CH_3$	117.2
2-Tetradecanone	$CH_3-(CH_2)_{11}-C(=O)-CH_3$	81.8	2-Decosanone	$CH_3-(CH_2)_{19}-C(=O)-CH_3$	122.3

forming the specific interaction by the long alkyl chain with the energy of the butyl fragment make an increased contribution into the vaporization enthalpy, and every next CH_2-group supplies additionally 5.09 kJ mol^{-1}.

Inasmuch as this energy is supplied by two CH_2-groups of the molecules of the near surrounding, each methylene group provides 2.55 kJ mol^{-1}. The presence of two hydrogen atoms in the CH_2-group results in the conclusion that each hydrogen atom interacting with the carbon of the neighboring chain contributes into the enthalpy characteristics 1.27 kJ mol^{-1}. Thus the mean value of the energy of the intermolecular interaction of low stability $DH_2C \rightarrow H$ equals 1.27 kJ mol^{-1} in the series 2-pentanone–2-tetradecanone.

Taking into account that the specific interaction formed by CH_3- and C_2H_5-groups of unsymmetrical ketones makes an appropriate contribution into the vaporization enthalpy of 2-butanone and 3-pentanone, respectively, we may refine the values of the energy of interactions $DHC \rightarrow H$ and the contribution of this interaction into the enthalpy characteristics of the unsymmetrical ketones and consequently to estimate the vaporization enthalpy of the unstadied ketones of the corresponding homologous series with an accuracy no worse than that of the experimental methods. Hence we conclude that the vaporization enthalpy of ketones is equal to $DHC \rightarrow H$ and the contribution of this interaction into the enthalpy characteristics of the unsymmetrical ketones and consequently to estimate the vaporization enthalpy of the unstadied ketones of the corresponding homologous series with an accuracy no worse than that of the experimental methods. Hence we conclude that the vaporization enthalpy of ketones is equal to the sum of the energies contributed by all existing interactions (4.5).

$$\Delta_{vap}H^0(298) = 2D{=}O{\rightarrow}CH_3{-}C{-} + 2D{=}O{\rightarrow}CH_3{-}(CH_2)_n{-}C{-} + 2DH_2C{\rightarrow}H. \quad (4.5)$$

The results of the performed calculations of the energy of the specific intermolecular interactions of the unsymmetrical ketones $=O \rightarrow CH_3{-}C{-}$ and $=O \rightarrow C_2H_5{-}C{-}$ and the value of the overall energies of two types of the specific intermolecular interactions $D = O \rightarrow CH_3 - (CH_2)_n - C - +2DH_2C \rightarrow H$, and also energies of low-stability interactions $CH_2 \rightarrow H$ calculated for the growing number of CH_2-groups in the series of ketones are compiled in Table 4.3. The obtained values of the energies of the low-stability interactions show their successive stabilization in the series of unsymmetrical ketones 2-butanone–2-heptanone and the growing contribution from the summary bond characteristics $D = O \rightarrow CH_3 - (CH_2)_n - C - +2DH_2C \rightarrow H$ with the increasing number of the CH_2-groups in the alkyl chain.

The value of these data consists also in the possibility to use them for estimation of the vaporization enthalpy of unstadied ketones, including the unsymmetrical ketones with the number of carbon atoms four and more in the shorter alkyl chain.

The results of the performed thermodynamic calculations for symmetric ketones compiled in Table 4.4 correspond to the specific intermolecular interactions $DH_2C \rightarrow H$ and the summary value of two types of interactions $D = O \rightarrow CH_3$

Table 4.3 Energies (kJ mol^{-1}) of specific interactions in liquid unsymmetrical ketones

Compound	Structure	D=O → CH_3–C–	D=O → CH_3–$(CH_2)_n$–C–	$\Delta_{vap}H^0(298)$ Experiment [4]
2-Butanone	CH_3—CH_2—C(=O)—CH_3	7.70	9.50 ($n = 0$) $n = 0$	34.38
2-Pentanone	CH_3—$(CH_2)_2$—C(=O)—CH_3	7.70	10.37 ($n = 1$) $DH_2C \rightarrow H = 0.44$ $n = 1$	36.13
2-Hexanone	CH_3—$(CH_2)_3$—C(=O)—CH_3	7.70	13.40 ($n = 2$) $DH_2C \rightarrow H = 0.75$ $n = 2$	42.22
2-Heptanone	CH_3—$(CH_2)_4$—C(=O)—CH_3	7.70	15.93 ($n = 3$) $DH_2C \rightarrow H = 1.27$ $n = 3$	47.25
		D=O → CH_3–CH_2–	–	–
3-Hexanone	CH_3–$(CH_2)_2$—C(=O)—CH_2—CH_3	9.04 ($n = 1$)	11.95 ($n = 1$) $DH_2C \rightarrow H = 1.15$ $n = 1$	41.98
3-Octanone	CH_3—$(CH_2)_4$—C(=O)—CH_2—CH_3	9.04 ($n = 1$)	15.93 ($n = 3$) DH C → H = 1.0 $n =_2 3$	(49.94)[a]
4-Octanone	CH_3—$(CH_2)_4$—C(=O)—CH_2—CH_2—CH_3	D=O → $CH_3(-CH_2-)_n$– 10.30 ($n = 2 + 1$)	– 13.40 ($n = 2$)	– –(47.40)[a]

[a]Estimated value of vaporization enthalpy

$-CH_2-C-+2DH_2C \rightarrow H$, whose contribution into the enthalpy characteristics is described by (4.6)

$$DH_2C \rightarrow H = (\Delta_{vap}H^0(298) - 4D = O \rightarrow CH_3 - CH_2 - C-)/8n, \quad (4.6)$$

where n is the number of CH_2-groups exceeding that of 3-pentanone.

The successive growth of the overall energy of the mentioned specific interactions $D{=}O \rightarrow CH_3(CH_2)_n{-}C{-}$ and $2DH_2C \rightarrow H$ indicates the increased contribution of the growing number of the methylene groups, or more exact, the growing number of the specific interactions $H_2C \rightarrow H$ formed by these groups. From the data compiled in Table 4.4 it follows that the obtained values of the specific interactions $H_3C \rightarrow H$ of low stability for symmetric ketones of the series 4-heptanone–8-pentadecanone are identical in the error limits of the experimentally measured vaporization enthalpy of these compounds.

Therefore, the nearly identical contribution of CH_2-group into the vaporization enthalpy of the compounds from the series 4-heptanone–6-undecanone permits the use of extrapolation for the estimation of the vaporization enthalpy of symmetric ketones with the number of carbon atoms in the alkyl chain exceeding five (Table 4.4).

Table 4.4 Energies (kJ mol^{-1}) of specific intermolecular interactions in liquid symmetrical ketones

Compound	Structure	$\Delta_{vap}H^0(298)$ [4, 6]	$D{=}O \rightarrow CH_3(CH_2)_n{-}C{-}$
2-Propanone	$CH_3-C(=O)-CH_3$	30.57	7.64 $n = 0$
3-Pentanone	$CH_3-CH_2-C(=O)-CH_2-CH_3$	38.58	9.65 $n = 1$
4-Heptanone	$CH_3-(CH_2)_2-C(=O)-(CH_2)_2-CH_3$	46.70	11.67 $DH_2C \rightarrow H = 1.02$ $n = 2$
5-Nonanone	$CH_3-(CH_2)_3-C(=O)-(CH_2)_3-CH_3$	53.30	13.33 $D\,CH_2 \rightarrow H = 0.84$ $n = 4$
6-Undecanone	$CH_3-(CH_2)_4-C(=O)-(CH_2)_4-CH_3$	60.67	15.17 $DH_2C \rightarrow H = 0.92$ $n = 6$
7-Tridecanone	$CH_3-(CH_2)_5-C(=O)-(CH_2)_5-CH_3$	(69.7)[a]	(17.43) $DH_2C \rightarrow H = 1.14$ $n = 8$
8-Pentadecan-one	$CH_3-(CH_2)_6-C(=O)-(CH_2)_6-CH_3$	(76.4)[a]	(19.10) $DH_2C \rightarrow H = 0.83$ $n = 10$

[a]Estimated

4.3 Unsaturated Ketones and Diketones

The above cited energy of the specific interactions in 2-propanone plays a key role in the determination of the energy of specific intermolecular interactions in unsaturated symmetric and unsymmetrical ketones. The analogy of the bonds at the carbonyl oxygen in the saturated ketones and methyl vinyl ketone makes it possible to rightfully use the energy of the specific interactions of 2-propanone for the determination of the energy of the intermolecular interaction D=O → CH_2=CH–C– involving the double bond in the methyl vinyl ketone.

The complete analogy in the structures of 2-butanone and methyl vinyl ketone molecules (the same number of bond vacancies, four, and the similarity of the structures of the liquid state) substantiate the identical approach to the determination of the energy of the specific interaction from the vaporization enthalpy. The presence of a methyl group in the molecules of these compounds permits the application of the value of the energy of the specific interaction existing in the liquid 2-propanone =O → CH_3–C– as equal to the energy of the analogous interaction in the liquid methyl vinyl ketone (7.74 kJ mol^{-1}). The energy of the second type of the specific interaction D=O → CH_2=CH–C– is obtained from the difference between the vaporization enthalpy and the sum of the energies of two specific interactions =O → CH_3–C– by (4.7)

$$D = O \rightarrow CH_2 = CH - C- = (\Delta_{vap}H^0(298) - 2D = O \rightarrow CH_3 - C-)/2. \quad (4.7)$$

The obtained energy of this interaction (Table 4.5) in methyl vinyl ketone (8.70 kJ mol^{-1}) demonstrates the reduced stability with participation of the terminal CH_2-group attached to a double bond as compared to the analogous interaction of the methyl group of 2-butanone. The energy of specific interaction in liquid ketones with an equal number of carbons in alkyl substituents is described by a regular series of growing stabilization of the second type interaction including unsymmetrical ketones and symmetric 3-pentanone having the highest energy of this interaction:

Methyl vinyl ketone (8.70) $<$ 2 - Butanone (9.50) $<$ 3 - Pentanone (9.65 kJ mol^{-1}).

The energy of the specific interaction in the methyl vinyl ketone reduced, respectively, by 0.80 and 0.95 kJ mol^{-1} leads to a conclusion that the presence of the double bond in the substituent decreases the acceptor properties of the carbon atom of the terminal CH_2-group and donor properties of the oxygen atom. On the other hand, the effect of the reverse dative bond on the terminal methyl group removed by three carbon atoms from the oxygen in the 3-pentanone is reduced to the minimal value. At the same time the established contribution into the enthalpy characteristics of the CH_2=CH–C– moiety combined with the known contribution of the CH_2-group (Table 4.5) opens the way to evaluate the vaporization enthalpy of yet unstadied unsymmetrical ketones possessing double bonds in the alkyl moiety.

Table 4.5 Energies (kJ mol^{-1}) of specific interactions in liquid unsaturated methyl vinyl ketone and diketones

Compound	Structure	$\Delta_{vap}H^0(298)$ [4]	D=O → CH_3–C–	D=O → CH_2=C–C–
Methyl vinyl ketone C_4H_6O	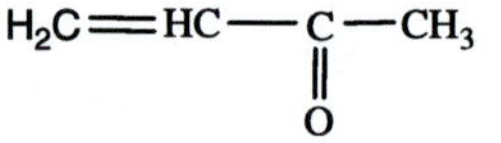	32.9	7.74	8.70
Diketone $C_4H_4O_2$	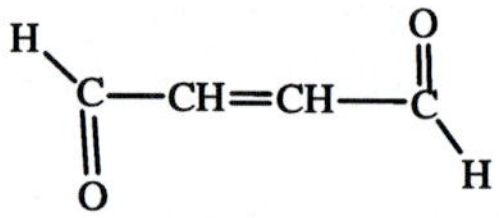	42.9	D=O···H–(C)= 5.5	D=O → CH=CH–C–20.9/4 = 5.22
2,5-Hexanedione $C_6H_{10}O_2$	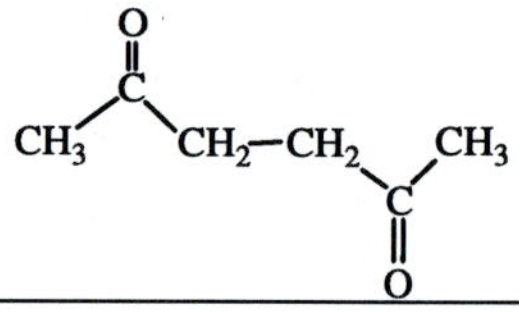	50.1 (T = 401 K)	7.74	D=O → CH_2–CH_2–C– 5.01

The published information on the vaporization enthalpy of diketones corresponds to 2,5-hexanedione (Table 4.5). The molecule of this compound contains eight bond vacancies, among them four belong to two oxygen atoms, and the remaining four belong to two methyl and two CH_2-groups forming the specific interactions through essentially unshared $2s^2$-electron pairs of carbon atoms.

The schematic picture of the structure of liquid 2,5-hexanedione (Fig. 4.6) shows the four equivalent specific interactions involving methyl groups whose energy in each case may be regarded as equal to the energy of the analogous interaction in liquid 2-propanone (7.74 kJ mol^{-1}). Therefore, it is possible to determine the energy of the specific interaction with the CH_2-group of the fragment –CH_2–CH_2– situated between the carbonyl groups whose oxygen atoms are under certain strain with respect to the shift of the electron density. Consequently, the carbon atoms of these groups should possess reduced acceptor properties and should form less stable intermolecular interactions. The value of the energy of the specific interaction D=O → CH_2=CH–C– is determined by (4.8) taking into account the energy contributions of all existing interactions D=O → H_3C–C–

$$D = O \rightarrow CH_2 - CH_2 - C- = (\Delta_{vap}H^0(298) - 2D = O \rightarrow H_3C - C-)/2. \quad (4.8)$$

The obtained value of the energy of the specific interaction $D = O \rightarrow CH_2 - CH_2 - C-$ (Table 4.5) indicates the reduced stability of this type interaction (5.01 kJ mol^{-1}) in agreement with our conclusion. It means that increase in the number of the carbonyl groups to two at the six carbon atoms in the backbone chain considerably decreases the negative charges on the oxygen atoms of the carbonyl groups, their donor properties, and the acceptor properties of the carbon atoms of the CH_2-groups.

The unsaturated dialdehyde attracts a special interest. The structure of this molecule is analogous to the structure of 2,5-hexanedione molecule and therefore the structure of

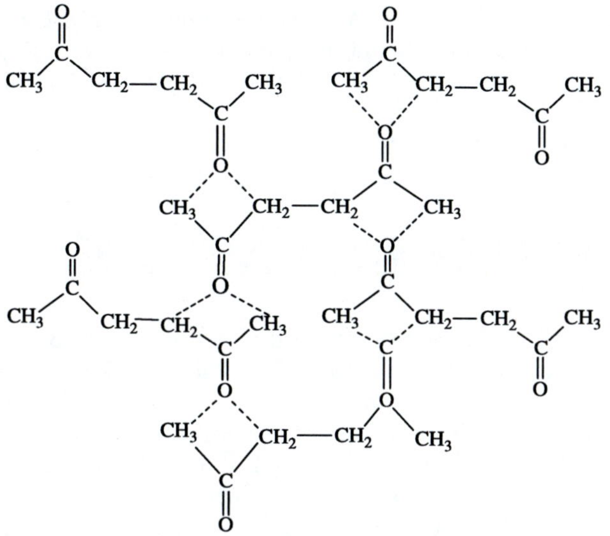

Fig. 4.6 Schematic picture of the chain-cyclic structure of liquid 2,5-hexanedione

their liquid state should be also similar. Therefore, the overall number of the bond vacancies should be also eight, among them four are furnished by the lone electron pairs of oxygen atoms. Two vacancies with acceptor properties are supplied by the carbon atoms of the fragment –CH=CH– possessing a double bond and the two remaining vacancies are furnished by the hydrogen atoms attached to the carbonyl groups. The latter form two H-bonds and two specific interactions $= C = O \cdots H - C -$. Taking the energy of this bond equal to the hydrogen bond in liquid aldehyde (5.5 kJ mol^{-1}) (Chap. 5) we estimated the value of the energy of the specific interaction =O → CH=CH– by (4.9) at 5.22 kJ mol^{-1} (Table 4.5).

$$D = O \rightarrow CH = CH - C- = (\Delta_{vap}H^0(298) - 2D = O \cdots H - C-)/2. \quad (4.9)$$

This type of specific interaction is by 0.2 kJ mol^{-1} more stable than the analogous interaction =O → CH_2–CH_2– in liquid 2,5-hexanedione. It follows from the energies of these two types of specific interactions that the replacement of the methyl group in the 2,5-hexanedione building a more stable interaction (7.74 kJ mol^{-1}) than the hydrogen bond $D = O \cdots H - C -$ formed by the hydrogen atom of the aldehyde results in the redistribution of the electron density and the increase in the charge on the oxygen of the carbonyl group providing the hydrogen atom of this fragment the enhanced positive charge. As a consequence of the enhanced positive charge on the hydrogen atom and the negative charge on the oxygen atom in the dialdehyde as compared to its charge in the molecule of 2,5-hexanedione, the specific interaction D=O → CH=CH–C– has higher stability than the interaction D=O → CH_2–CH_2–C– in liquid 2,5-hexanedione.

This allows a conclusion on the validity of the estimation of the low stability of the specific interaction in the liquid diketone. The performed estimation of the energy of this type hydrogen bond was based on the identity of the bond in liquid diketone and dialdehyde at the equal number of carbon atoms in the bridging fragment between the carbonyl groups.

4.4 Energies of Specific Interactions in Liquid Cyclic Ketones

Cyclic ketones are characterized by enhanced vaporization enthalpy compared with symmetric and unsymmetrical molecular forms of hydrocarbons with the same number of carbon atoms in the molecule (Table 4.6). The difference between vaporization enthalpy of cyclopentane and cyclohexane equal to 4.50 kJ mol^{-1} is due to the different number of the CH_2-groups in their molecules, and the difference in the case of cyclopentanone and cyclohexanone amounting to 2.34 kJ mol^{-1} corresponds to the decreased influence of the reverse dative bond on the specific interaction formed by the carbonyl oxygen with a carbon atom because of the growing number of the methylene groups in the hydrocarbon ring. An interesting series constitute the variations in the energy contribution from the CH_2-group into the vaporization enthalpy of the saturated cyclic hydrocarbon at the standard temperature:

$$C_5H_{10}\,(5.70) \geq C_6H_{12}\,(5.50) = C_7H_{14}\,(5.50) \geq C_8H_{16}\,(5.40) = C_{12}H_{24}\,(5.40\,\text{kJ mol}^{-1}).$$

It should be taken into consideration that in the near surrounding two molecules are present; therefore, this energy contribution belongs to two CH_2-groups. Consequently, the mean value corresponding to a single hydrogen atom is 1.42 kJ mol^{-1} in cyclopentane and cyclohexane, and 1.35 kJ mol^{-1} for all further compounds.

The difference in values 5.70 and 5.40 kJ mol^{-1} in the vaporization enthalpy of cyclic hydrocarbons reflects the difference in the properties of homologs, therefore this difference should be assigned to the participation of hydrogen atoms in the intermolecular interaction. Since the hydrogen atom in the *cis*-position forms less stable interactions [16] the more essential contribution into the enthalpy characteristics belongs to the hydrogen atom in the *trans*-position in two CH_2-groups of the contacting molecules of the near surrounding.

Although on the first sight the influence of the reverse dative bond is hard to follow in cyclic ketones, the dependence drawn in Fig. 4.7 of the vaporization enthalpy on the number of carbon atoms in the ring demonstrates the analogy in the character of the variation of the vaporization enthalpy to that observed for ketones with open structure of the molecules (Figs. 4.4 and 4.5). The plot of the vaporization enthalpy consisting of two intercepting straight lines shows the end of the reverse dative bond influence on the third carbon atom in the cyclohexanone ring.

Table 4.6 Energies (kJ mol^{-1}) of specific interactions in liquid cyclic ketones

Compound	Formula	$\Delta_{vap}H^0(298)$ [4, 6]	T (K)	$\sum 4DH_2C \rightarrow H\text{–}CH_2$	$D{=}O \rightarrow CH_2{=}$
Cyclopropanone	C_3H_4O	35.6[a]	–	6.43	7.30
Cyclopropane	C_3H_6	19.3[a]	–	–	–
Methylcyclopropanone	C_4H_6O	39.4	298		7.30
Cyclobutanone	C_4H_6O	38.4	298	5.8	8.15
Cyclobutane	C_4H_8	23.4[a]		–	–
Cyclopentanone	C_5H_8O	42.72	298	5.66	9.14
Cyclopentane	C_5H_{10}	28.5	298	5.66	–
Cyclohexanone	$C_6H_{10}O$	45.06	298	5.50	9.84
Cyclohexane	C_6H_{12}	33.0	298	5.50	–
Cycloheptanone	$C_7H_{12}O$	50.7	298	5.50	10.10
Cycloheptane	C_7H_{14}	38.5	298	5.50	–
Cyclooctanone	$C_8H_{114}O$	54.2	298	5.40	9.60
Cyclooctane	C_8H_{16}	43.3	298	5.40	–
Cyclononanone	$C_9H_{16}O$	51.4; 59.4[a]	348	5.23	9.60
Cyclodecanone	$C_{10}H_{18}O$	55.2; 64.2[a]	368	5.23	9.60
Cyclodecane	$C_{10}H_{20}$	52.3[a]	358	5.23	–
Cycloundecanone	$C_{11}H_{20}O$	60.3; 68.7[a]	378	5.20	9.50
Cycloundecane	$C_{11}H_{22}$	57.3[a]		5.20	–
Cyclododecanone	$C_{12}H_{22}O$	61.0; 73.7[a]	388	5.10	9.50
Cyclododecane	$C_{12}H_{24}$	62.8; (52.6)	298 (401)	5.10	–
Cyclotridecanone	$C_{13}H_{24}O$	78.6[a]	298	5.00	9.60
Cyclotetradecanone	$C_{14}H_{26}O$	83.3[a]	298	5.00	9.50
2-Cyclohexen-1-one	C_6H_8O	49.5	298	6.70	10.70
2-Cyclohexene	C_6H_{10}	33.5	298	6.70	
Benzene	C_6H_6	33.89	298	–	–

[a]Estimated for 298 K

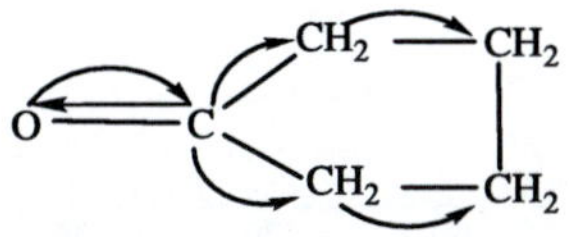

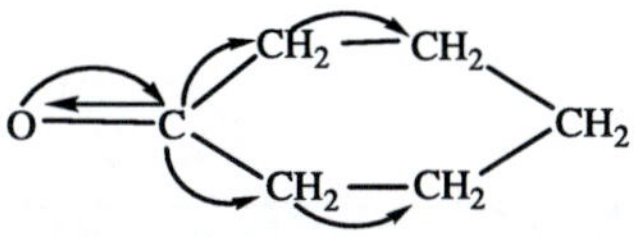

This means that the influence of the reverse dative bond is notable from cyclopropanone to cyclohexanone. The second straight line corresponds to the increase in the enthalpy characteristics with the subsequent increase in the number of carbon atoms in the ring, namely, with the energy contribution of the CH_2-groups of the respective ketones. In the saturated cyclic hydrocarbons no reverse dative bond is operating, therefore the increase in the enthalpy characteristics is due to the contribution of the growing number of the CH_2-groups in the ring. As a result the plot of the vaporization enthalpy of cyclic hydrocarbons from cyclopentane to cyclododecane is a straight line including a large number of compounds (Fig. 4.7). These dependences were used for estimation of the vaporization enthalpy of yet unstadied cyclic ketones and hydrocarbons (Table 4.6).

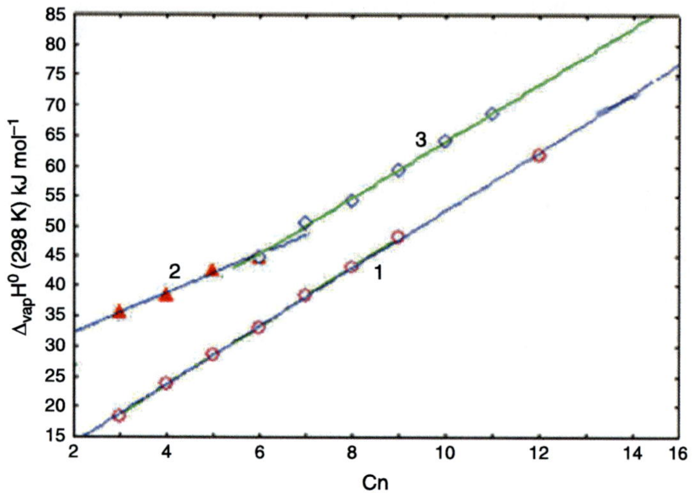

Fig. 4.7 Dependence of vaporization enthalpy of cyclic hydrocarbons (1) and ketones (2) on the number of carbon atoms in the ring

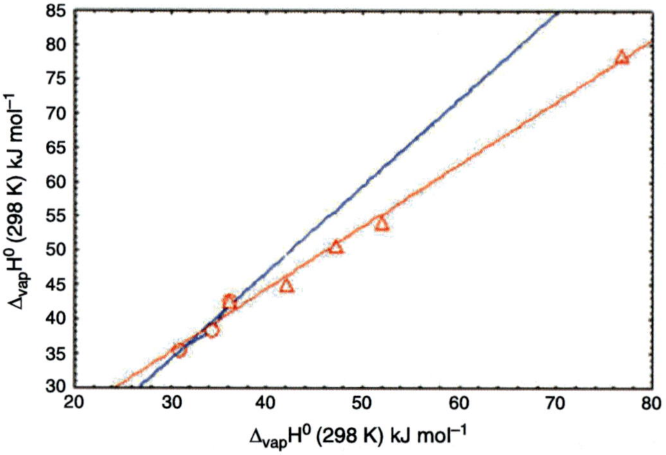

Fig. 4.8 Plot of vaporization enthalpy of cyclic ketones (2) versus $\Delta_{vap}H_0(298)$ of ketones of homological series of 2-propanone (1) the influence of reverse dative bond and (2) the influence of the number of CH_2-groups

The plot of the vaporization enthalpy of cyclic ketones versus that of the homologous series of 2-propanone presented in Fig. 4.8 leads to a conclusion that the reverse dative bond operating in the molecules of the initial compounds of this homologous series does not significantly affect the energies of the specific interactions. It is clear from the sufficiently good description by this plot of the vaporization enthalpy from cyclopropanone till cyclotridecanone. Consequently, the dependence presented in Fig. 4.8 and the combination of other properties make it possible to use also interpolation and extrapolation for estimation of the lacking vaporization and sublimation enthalpy or the other interesting characteristics of yet unstadied ketones.

In the molecule of the cyclic ketone the oxygen atoms with its lone electron pairs provides two bond vacancies. Consequently, two bond vacancies should be supplied by two CH_2-groups of the hydrocarbon ring the least affected by the shift of the electron density caused by the oxygen of the carbonyl group and by the influence of the reverse dative bond. Therefore, it is possible to suggest the schematic picture of the structure of liquid cyclic ketones (Fig. 4.9) where four equivalent specific interaction $=O \rightarrow CH_2–C–$ with the molecules of near surrounding are shown formed between the oxygen atoms and the CH_2-groups of the corresponding hydrocarbon rings. The molecules of cyclic ketones with a planar structure and with lone electron pairs of the oxygen form chain structure in the liquid state with equal interatomic distances (Fig. 4.9).

Cyclic ketones with a lower number of methylene groups, methylcyclopropanone and cyclopropanone with the same structure of the ring and distinguished only by the isostructural methyl group (Table 4.6) have vaporization enthalpies of 39.4 and 35.6 kJ mol^{-1}, respectively. The energy difference is contributed by the isostructural methyl group of the methylcyclopropanone (3.8 kJ mol^{-1}). This methyl group is interacting with the contacting molecule linking different chains. Therefore, each molecule supplies its proper methyl group and interacts with analogous group of the contiguous molecule. Thus the difference in the vaporization enthalpies of ketones in question corresponds to two isostructural methyl groups or 1.9 kJ mol^{-1} per single group.

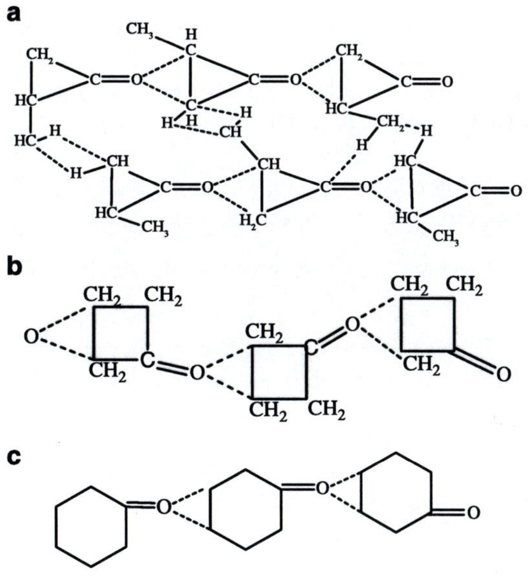

Fig. 4.9 Schematic picture of the chain structure of liquid cyclic ketones (**a**) methylcyclopropanone, (**b**) cyclobutanone, and (**c**) cyclohexanone

In Sect. 4.1 the contribution into the vaporization enthalpy of an isostructural methyl group of ketones was estimated at 0.60 kJ mol^{-1}. The different contribution into the vaporization enthalpy of these methyl groups should be ascribed to a special role of the cyclopropyl ring where the reverse dative bond more than usual decreases the stability of the forming specific interactions. Therefore, in the redistribution of the electron density the isostructural methyl group also takes part, and with the low positive charge on the carbon atom it is capable to form specific interactions of low stability.

A characteristic feature of the cyclic ketone molecules with more than four carbon atoms in the ring is the involvement of the whole cycle into the interaction with the analogous structures of the molecules of the near surrounding. Therefore, the going over into vapor requires the rupture of the specific interactions with the contiguous molecules of the near surrounding, for instance, in cyclopentane, cyclohexane, or cyclic ketones. Since we are interested in the energy of the specific interaction between the carbonyl oxygen and two carbon atoms of the most remote CH_2-groups, the energy contributed by the hydrogen atoms in the CH_2-group may be evaluated from the vaporization enthalpy of cyclopentane, cyclohexane for cyclopentanone, cyclohexanone, respectively. Therefore, the energy supplied by substitution of the hydrogen atoms of the carbonyl group $DH_2C \rightarrow H-$ should be subtracted from the vaporization enthalpy of the corresponding cyclic hydrocarbons. The summary contribution of four such interactions is evaluated from the vaporization enthalpy dividing by the number of the CH_2-groups present in the ring of the corresponding hydrocarbon (Table 4.6). This permits the calculation of the energy of the specific intermolecular interaction $=O \rightarrow CH_2–C–$ from the expression (4.10):

$$\begin{aligned} D &= O \rightarrow CH_2 - (CH_2)n - C- \\ &= (\Delta_{vap}H^0(298) - 4D\,DH_2C \rightarrow H - CH)/4. \end{aligned} \tag{4.10}$$

The results of performed calculations are compiled in Table 4.6.

The results of performed thermodynamic calculations (Table 4.6) illustrate the regular series of changes in the energy of the specific intermolecular interactions existing in the cyclic ketones:

Cyclopropanone (7.30)<Cyclobutanone (8.15)<Cyclopentanone (9.14) <Cyclohexanone (9.84) ≈ Cycloheptanone (10.00) ≈ Cyclooctanone (9.60 kJ mol^{-1}) ≈ Cyclodecanone (9.60) ≈ Cyclododecanone (9.50 kJ mol^{-1}).

These data permit a conclusion that the weakening of the effect of the reverse dative bond in cyclic ketones is accompanied by the stabilization of the specific intermolecular interaction from cyclopropanone till cyclohexanone (9.84) ≈ cycloheptanone (10.00 kJ mol^{-1}), and then a trend is observed of decreasing stability with the growing number of carbon atoms in the ring from cyclooctanone to cyclododecanone.

The energy of the specific interaction in liquid 2-cyclohexen-1-one was determined by (4.10). The energy from the substituted hydrogen atoms at the carbon of the carbonyl group $DH_2C \rightarrow H–CH$ was established from the vaporization enthalpy of 2-cyclohexene equal to 6.70 kJ mol^{-1}. The obtained value of the specific interaction is the mean value of two existing specific interactions $D{=}O \rightarrow CH_2–(CH_2)–C–$ and $D{=}O \rightarrow CH_2–CH{=}CH–C–$ (10.70 kJ mol^{-1}) that indicates their enhanced stability as compared to the analogous interaction $D{=}O \rightarrow CH_2–(CH_2)–C–$ in liquid cyclohexanone.

The specific feature of benzophenone is the rigid structure of the benzene ring. The quantum-chemical calculations showed that even with a CF_3 substituent in the molecule $C_6H_5CF_3$ a strong polarization occurred in the ring with an alternation of the charges on carbons without electron density shift to CF_3 [1]. The value of the vertical potential and E_{bond}(C1s) led to the conclusion on the overall acceptor nature of the OH-group in C_6H_5OH. Therefore, the oxygen atom of the carbonyl group should have far lower charge than on the hydroxy group of the benzyl alcohol. All these facts suggest that the specific interactions between the oxygen of the carbonyl group and the CH-groups of the benzene ring should be unstable. The flat structure of the benzophenone molecule with unshared electron pairs of the oxygen atoms situated at an angle of 90° and the double bond of the carbonyl group substantiates the formation of the flat structure of the benzophenone molecule and the network structure of the liquid state (Fig. 4.10). The overall energy value of four specific intermolecular interactions is equal to the difference between the vaporization enthalpy of benzophenone and the double vaporization enthalpy of benzene. Therewith, it is necessary to exclude the contribution into the enthalpy of vaporization or sublimation of benzene of two substituted hydrogen atoms. The contribution of interactions formed by the hydrogen atom of the benzene ring $DHC \rightarrow H–C$ is established from the enthalpy characteristics of benzene, 6.78 and 8.8 kJ mol^{-1} for vaporization and sublimation, respectively. The energy of the specific interaction is calculated by (4.11)

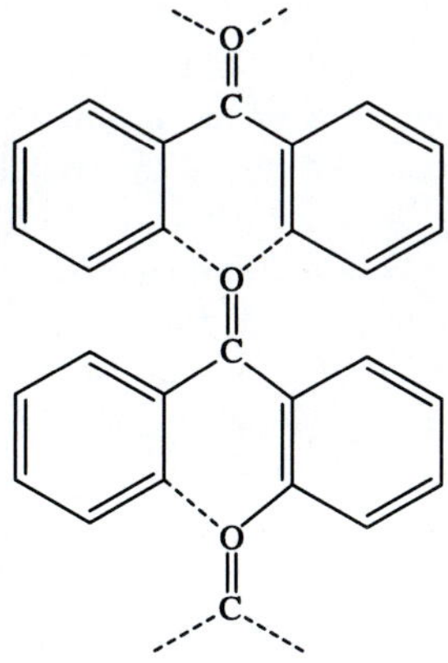

Fig. 4.10 Schematic picture of the liquid structure of benzophenone

$$D = O \rightarrow CH = = (\Delta_{vap}H^0(298)b.f. - 2\Delta_{vap}H^0(298)b. + 2DHC \rightarrow H - C =)/4. \tag{4.11}$$

The energy of the interaction =O → CH= (Table 4.6) under consideration equals 3.6 kJ mol^{-1} and thus it is about three times less than the energy of the specific interactions in liquid cyclopentanone and cyclohexanone.

4.5 Energies of Specific Interactions in Solid Ketones

Symmetric ketones except for 2-propanone suffers low-temperature polymorphic transformation that in the presence of four specific intermolecular interactions of a single type results in their insignificant weakening. The limited data on thermodynamic properties exist mainly on the ketones of an unsymmetrical structure. These data show that only 2-pentanone and 3-hexanone undergo the polymorphic transition with the enthalpy contribution of 0.138 and 0.698 kJ mol^{-1}, respectively. The enthalpy of melting (Table 4.7) is four to five times smaller than the sublimation enthalpy of ketones demonstrating significant transformation of their crystal structures in the course of melting [17, 18]. The measured sublimation enthalpies correspond to high temperature, therefore the heat effects of the polymorphic transformations are not accounted for in the thermodynamic calculations.

Sublimation enthalpies of symmetric ketones 2-propanone, 3-pentanone, and 5-nonanone (3, 4) show the same trend in the dependence of the sublimation enthalpy on the number of carbon atoms in the alkyl chain as seen for unsymmetrical ketones (1, 2) (Fig. 4.11). As seen from this figure, the first members of the series of unsymmetrical ketones 2-propanone–2-pentanone and of symmetric ketones 2-propanone–3-pentanone due to the influence of the reverse dative bond have a reduced ability to the stabilization by the crystal field.

The dependences shown in Fig. 4.12 evidence the differentiating influence of the crystal field on the character of variation of the sublimation enthalpy of symmetric and unsymmetrical ketones because of the difference of the number of carbon atoms in the chains. The plots of sublimation enthalpies versus the number of carbon atoms presented in Fig. 4.12 for the unsymmetrical ketones of the homologous series of 2-propanone are completely analogous to the dependences of vaporization enthalpy: the points are located on two intercepting straight lines, one belonging to 2-propanone and 2-butanone, the other to 2-pentanone, 2-hexanone, and 2-tetradecanone. This character of the dependence reflects the end of the effect of the reverse dative bond in the alkyl chain of 2-butanone, where at the oxygen is linked the third carbon atom C–CH_2–CH_3. Further increase in the enthalpy characteristics is due to the growing number of the CH_2-groups and their involvement into the intermolecular interactions with the molecules of the near surrounding. Analogous dependence of the vaporization and sublimation enthalpy on the number of carbon atoms in the alkyl chain of ketones and including a large

Table 4.7 Energies (kJ mol^{-1}) of specific intermolecular interactions in solid ketones

Ketone	$\Delta_{melt}H^0(T)$ [4]	$\Delta_{sub}H^0(T)$ [4]	$\Delta_{vap}H^0(T)$ [4]	D=O → CH_3C–	D=O → CH_3–CH_2–C–	D$(CH_2)n$
				Sublimation		
2-Propanone	5.73	36.62	30.84	9.16	–	–
3-Pentanone	0.112[a], 0.0098[a] 11.59	47.74	36.15	–	11.94	–
5-Nonanone	0.313[a] 24.98	78.18	53.30	–	11.94	7.32 $n = 4$
2-Butanone	8.365	42.74	34.38	9.16	12.22	
2-Pentanone	0.138[a] 10.68	46.88	36.13	9.16	12.22	4.05
2-Hexanone	14.90	57.12	42.22	9.16	12.22	4.58 $n = 2$
3-Hexanone	0.698[a] 13.43	51.88	38.39	–	12.22	2.92
2-Tetradecanone	49.12	130.9	81.78	9.16	12.22	8.80 $n = 10$
3-Methyl-2-butanone	9.34	44.32	34.98	9.16	12.22	D i-CH_3[b] = 0.80
3,3-Dimethyl-2-butanone	11.33	49.25	37.87	9.16	12.22	D i-CH_3= 1.62
2,4-Dimethyl-3-pentanone	11.18	52.76	41.58	9.16	12.22	D i-CH_3= 1.25

[a]Isomorphous transformation[b]Isostructural methyl group

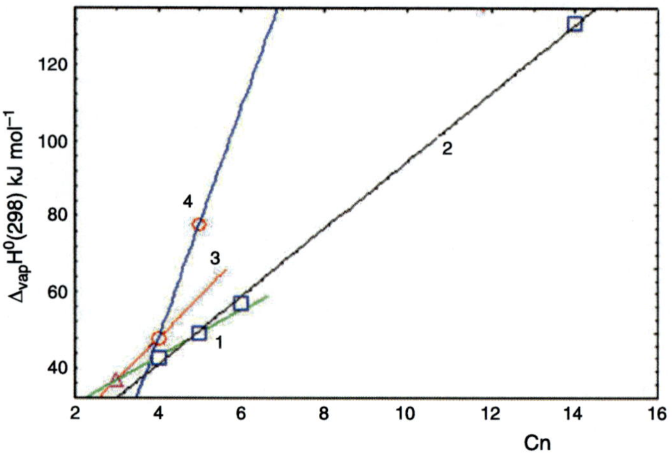

Fig. 4.11 Dependence of sublimation enthalpy of unsymmetrical (1, 2) and symmetric ketones of homological series 2-butanone (3, 4): (1, 3) leveling of the influence of reverse dative bond, (2, 4) the influence of growing number of CH_2-groups in hydrocarbon chains

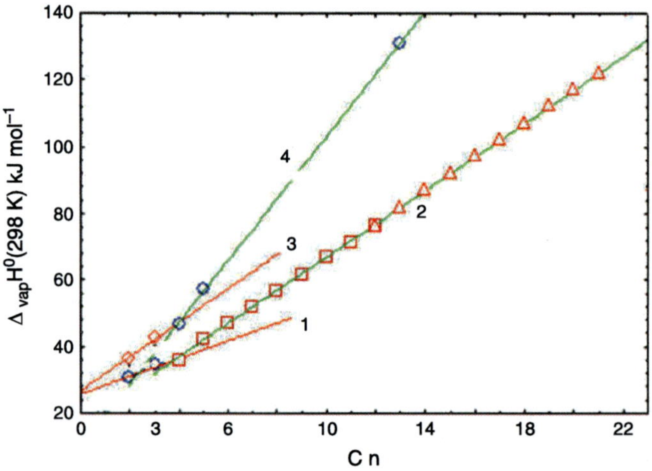

Fig. 4.12 Dependence of vaporization (1, 2), sublimation (3, 4) enthalpy of unsymmetrical ketones of homologous series of 2-propanone on the number of carbon atoms in alkyl chain: (1, 3), leveling of the influence of reverse dative bond, (2, 4), the influence of growing number of CH_2-groups in hydrocarbon chains

number of compounds in the linear plot makes it possible to use interpolation and extrapolation for estimation of enthalpy characteristics of unstadied ketones of symmetric and unsymmetrical structure within the error limits corresponding to the experimental calorimetric measurement. A similar conclusion follows from the correlation of the sublimation enthalpy of unsymmetrical homologs of 2-propanone and symmetric ketones (1) and of correlation between the sublimation and vaporization enthalpy of unsymmetrical ketones from the 2-propanone series (Fig. 4.13). It should be noted that the correspondence of these correlations to straight lines indicates the compensation of the effect of the reverse dative bond on the energy of the specific interactions in the first members of the homologous series. This permits a conclusion that these dependences of enthalpy characteristics provide a reliable evaluation of the thermodynamic properties of a large number of ketones.

Symmetric structure of ketone molecules of 2-propanone, 3-pentanone possessing four bond vacancies permits the use of the sublimation enthalpy $\Delta_{sub}H^0(298)/4$ (Table 4.7) for evaluation of the energy of specific interactions existing in their crystal state. Utilizing the energies of the specific interactions formed by the fragment C–CH_3 (2-propanone) and C–CH_2–CH_3 (3-pentanone) by calculation based on the additive contributions, we obtain the sublimation enthalpy of 2-butanone (42.20 kJ mol^{-1}) underestimated as compared with the experimentally measured value by 0.54 kJ mol^{-1}. This figure indicates the insignificant difference in the contributions of the energy of specific intermolecular interactions in the symmetric and unsymmetrical ketones which should be ascribed to the close to equivalent effect of the reverse dative bond and small difference in the shift of the electron density in the fragments with two and three carbon atoms. This error can be excluded by using in the calculation the energy of the specific interaction existing in 2-propanone, and the energy of the specific interaction involving a fragment with three carbon atoms –C–CH_2–CH_3 should be determined from the sublimation enthalpy of 2-butanone (Table 4.7).

The smallest contribution of the CH_2-group into the sublimation enthalpy is observed in 2-pentanone (4.14 kJ mol^{-1}) and 3-hexanone (6.12 kJ mol^{-1}) with one excessive CH_2-group. It is due to the partial influence of the reverse dative bond on the fourth carbon atom of the alkyl chain. With moving off of the CH_2-group along the chain reduces the influence of the reverse dative bond, and further stabilization occurs only by the effect of the crystal field. With the virtual end of the effect of the reverse dative bond in compounds with three carbon atoms in the alkyl chain finishes the growth of the energy of the specific intermolecular interaction, therefore the

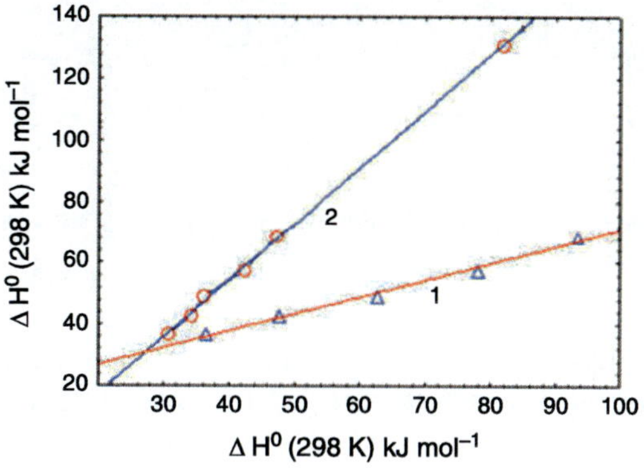

Fig. 4.13 Plot of sublimation enthalpy of the unsymmetrical (1) versus symmetrical ketones and $\Delta_{sub}H^0(298)$ versus $\Delta_{vap}H^0(298)$ (2) of unsymmetrical ketones of homologous series of 2-propanone

difference in the thermodynamic characteristics of sublimation and vaporization of 2-butanone and 2-pentanone should be ascribed to the value of the contribution into these characteristics of the energy of the specific interactions formed by CH_2-groups. The character of dependence of $\Delta_{sub}H^0(298) = f(C_n)$ (Fig. 4.11) demonstrates the growing contribution of these groups into the stabilization of the sublimation process. This indirectly confirms the practically constant energy of the specific intermolecular interaction in ketones with alkyl chain containing more than three carbon atoms. Note that this rule was also valid for ethers (Chap. 3). Thus, the minimum energy contribution of the CH_2-group is 4.14 kJ mol^{-1}, and maximum, in 2-butanone–2-tetradecanone is 8.80 kJ mol^{-1}. The performed calculations of the energy of specific intermolecular interactions in four solid unsymmetrical ketones are listed in Table 4.7.

Let us consider the role of the isostructural methyl group and the importance of its definite position in the alkyl chain on its contribution into the sublimation enthalpy of ketones. Taking into account that the additive contribution of two fragments with different number of carbon atoms practically does not differ from the experimentally measured vaporization enthalpy of unsymmetrical ketones, the difference between the sublimation enthalpy of a ketone possessing isostructural methyl groups and the enthalpy characteristics of the corresponding ketone of normal structure amounts to the contribution of the isostructural methyl groups.

$$n \cdot \mathrm{DCH_3} = \Delta_{sub}H^0(T)iso - \Delta_{sub}H^0(T)\mathrm{k.n.s.}$$

In the case of 2,4-dimethyl-3-pentanone the overall contribution of two isostructural methyl groups is 5.0 kJ mol^{-1} or 2.50 kJ mol^{-1} per one group. However, the position of two interacting molecules in the near surrounding in the crystal increases the number of the isostructural methyl groups, and consequently each of them contributes 1.25 kJ mol^{-1} into the sublimation enthalpy. The influence of the isostructural methyl groups on the sublimation enthalpy becomes obvious at comparison the enthalpy characteristics and structures of molecules depicted below.

$\Delta_{sub}H^0(298)$ kJ•mol^{-1}	44.32	49.25	52.76
Structure	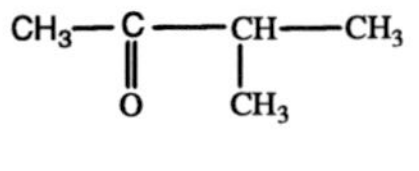	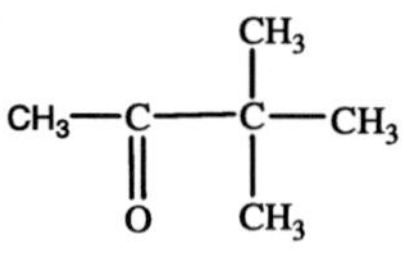	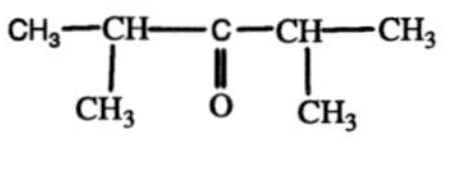
Di-CH_3 kJ•mol^{-1}	1.60	3.25	2.50
	3-methyl-2-butanone	3,3-dimethyl-2-butanone	2,4-dimethyl-3-pentanone

One methyl group in the branched unsymmetrical molecule of 3-methyl-2-butanone contributes 1.60 kJ mol^{-1} that results in the sublimation enthalpy exceeding that of 2-butanone (44.32 kJ mol^{-1}). In the molecule of 3,3-dimethyl-2-butanone two isostructural methyl groups at the same carbon atoms contribute 3.25 kJ mol^{-1} each, and two analogous methyl groups symmetrically located in the molecule of 2,4-dimethyl-3-pentanone provide a decreased contribution into the sublimation enthalpy.

This leads to the conclusion that

Firstly, the effect of the isostructural methyl groups is more significant when two such groups are attached to the same carbon atom in the molecule of 3,3-dimethyl-2-butanone.

Secondly, the presence of two isostructural methyl group symmetrically located in the molecule provide inadequate, reduced contribution into the sublimation enthalpy of 2,4-dimethyl-3-pentanone.

The important feature of the 2,4-dimethyl-3-pentanone is the presence in the molecule of symmetrically situated fragments of three carbon atoms with isostructural methyl groups. This suggests that the formation of fragments with four or maximum five carbon atoms should provide the molecule with isostructural methyl groups the maximum stiffness due to the redistribution of the electron density and charges on atoms.

The analysis of the parameters of ketones substantiated the regular character of variation of their enthalpy of vaporization and sublimation described by the linear dependence $\Delta_{vap}H^0(T) = f(C_n)$ that is used for estimation of the enthalpy characteristics of the unstadied representatives of this class compounds with unsymmetrical structure of molecules (Table 4.8). The errors in the established values for ketones 2-heptanone–2-tridecanone do not exceed $\pm$ 0.1 kJ mol^{-1} and for 2-pentadecanone, 2-hexadecanone do not exceed $\pm$ 0.2 kJ mol^{-1}.

The thermodynamic calculation of the energy of the specific interaction in solid cyclohexanone by the equation given in Sect. 4.2 taking into account bond vacancies results in 4.3 kJ mol^{-1}, and in solid benzophenone, in 6.95 kJ mol^{-1} (Table 4.9).

4.6 Energies of Specific Interactions and H-bonds in Solid Diketones and Oxyketones

The simplest cyclic diketone is cyclobutandione whose characteristic feature is the participation of the CH_2-group in the interaction with one oxygen atom supplying two bond vacancies (Table 4.10). One bond is formed by the carbon atom with the essentially unshared $2s^2$-electron pair, and the second one is formed by the hydrogen atom located in the *trans*-position that in contrast to the hydrogen in the *cis*-position is capable of forming more stable bonds [16]. The planar structure of the cyclobutanone molecule suggests the formation of chain structure in the liquid state with two molecules in the near surrounding (Fig. 4.14). The chains of cycles may be linked by weak intermolecular interactions formed by the hydrogens of the

Table 4.8 Phase transition enthalpies (kJ mol^{-1}) of unsymmetrical ketones

Ketone	$\Delta_{melt}H^0(T)$ [4]	$\Delta_{sub}H^0(T)^a$	$\Delta_{vap}H^0(T)$ [4]	Ketone	$\Delta_{melt}H^0(T)$ [4]	$\Delta_{sub}H^0(T)^a$	$\Delta_{vap}H^0(T)$ [4]
2-Heptanone	18.1	67.7	46.9	2-Dodecanone	39.7	120.3	72.0
2-Octanone	22.2	78.3	52.0	2-Tetradecanone	43.9	142.2	77.1
2-Nonanone	26.7	88.8	57.0	4-Heptanone	–	62.7	–
2-Decanone	31.1	99.3	61.9	6-Undecanone	–	93.3	–
2-Undecanone	35.2	109.8	67.0	7-Tridecanone	–	108.8	–

[a]Estimated

Table 4.9 Energies (kJ mol^{-1}) of specific interactions in solid cyclic ketones (T. middle)

Ketones	Formula	$\Delta_{sub}H^0(T)$ [4]	T (K)	D=O → CH_2–	$\Delta_{vap}H^0(298)$ [4]	D=O → CH_2–
Cyclohexanone	$C_6H_{10}O$	49.3	254	4.3	45.1	3.85
Cyclohexane	C_6H_{12}	37.6	265	–	33.	–
Benzophenone	$C_{13}H_{10}O$	93.8	298	6.95	75.6	3.6
Benzene	C_6H_6	36.4	279	–	33.8	–

CH_2-groups located in the *cis*-position. The presence of two hydrogen atoms in the fragment –CH_2–C=O prevents strong electron density shifts from the next fragment –CH_2–CH_2– to the carbonyl oxygen atom, therefore the oxygen atoms have equal negative charges. The energy of the specific intermolecular interaction =O → CH_2– in the liquid 1,2-cyclobutandione should be taken equal to the same interaction in the liquid 2-propanone (7.70 kJ mol^{-1}) because of the same number of carbon atoms in each fragment. This is confirmed by the low influence of the hydrogen atoms in the terminal methyl group of the 2-propanone on the shift of the electron density to the carbonyl oxygen atom. Evidently the energy of the same specific interaction in the solid 2-propanone (9.16 kJ mol^{-1}) should be regarded as equal to the analogous interaction in the solid 1,2-cyclobutanedione. The performed thermodynamic calculation provided the value of the energy of the hydrogen bond $D = O \cdots H - C$ equal to 8.05 kJ mol^{-1} at 270 K.

The symmetric position of the carbonyl oxygen in the 1,3-cyclobutandione also ensures the equivalent shift of the electron density in the fragments –C–CH_2–C=O, therefore the practically equal negative charges on the oxygen atoms should be expected. Yet this fragment is essentially different from the fragment in the molecule of the 1,2-cyclobutanedione –C–CH_2–CH_2–, therefore the charges on the oxygen atoms in these two cyclobutanediones should be different.

Therefore, we can state that the energy of the formed hydrogen bonds and specific interactions in solid (and liquid) states of these compounds should have certain distinctions. This is also due to the position of methylene groups in the 1,3-cyclobutanedione which are situated between carbonyl groups, therefore they equally participate in the shift of the electron density to each oxygen atom, and the alkyl chain includes three carbon atoms whereas in the ring of the 1,2-cyclobutanedione the fragment –CH_2–CH_2– possessing higher rigidity is to a lesser extent capable of the shift of the electron density, and that especially affects the stabilization of the most strong bond. The presence of an obvious contradiction in the vaporization enthalpy of the compounds under consideration should be attributed to the higher stability of the specific interaction =O → CH_2– compared with =O → CH_2 involving fragments with three and two carbon atoms, respectively, in the 1,3-cyclobutanedione and 1,2-cyclobutanedione. Consequently, the difference in the vaporization enthalpies of 4.5 kJ mol^{-1} should be assigned to the stabilization of the hydrogen bond and the specific interaction =O → CH_2–C–. It is interesting to estimate the energy of the specific interaction in the solid 1,3-cyclobutanedione equal to the energy of the analogous specific interaction in the solid symmetric 3-pentanone (11.94 kJ mol^{-1}) with characteristic even distribution

Table 4.10 Energies (kJ mol^{-1}) of specific interactions in solid cyclic diketones and dioxolanes (T. middle)

Compound	Structure	$\Delta_{sub}H^0(T)$ [4, 19]	T (K)	D=O → CH_2=	$\Delta_{vap}H^0(T)$	D=O → CH_2=
1,2-Cyclobutane-dione $C_4H_4O_2$	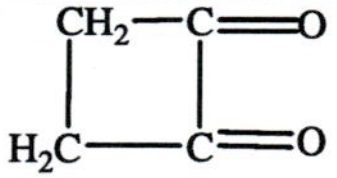	69.1	270	9.16 D = O···H – CH = 8.05	–	–
1,3-Cyclobutane-dione $C_4H_4O_2$	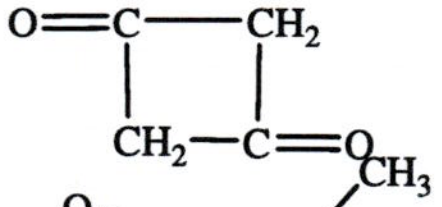	73.6	298	11.96 D = O···H – CH = 6.45	–	–
2,2,4,4-Tetramethyl-1,3-cyclobutanedione $C_8H_{12}O_2$	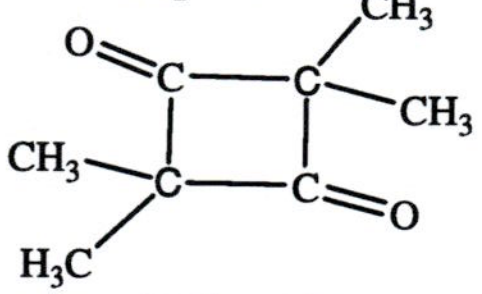	72.4	298	D=O → CH_3 = 9.05	–	–
1,4-Cyclohexanedione $C_6H_8O_2$	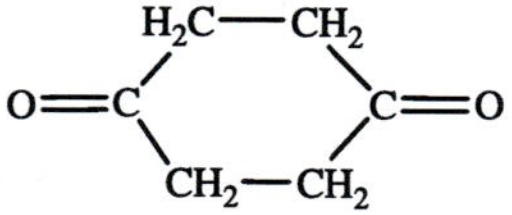	75.6	298	D=O → CH_2= 9.45	–	–
1,3-Cyclohexanedione $C_6H_8O_2$	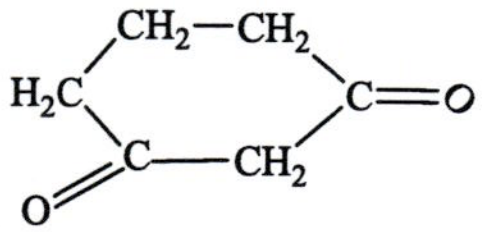	89.2	298	D_1=O → CH_2= 12.23 D_2=O → CH_2= 10.10	–	–
5,5-Dimehyl-1,3-cyclohexanedione $C_8H_{12}O_2$	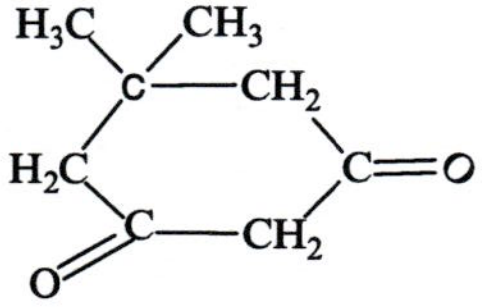	99.8	298	D_2=O → CH_3= 11.63 D_1=O → CH_2= 11.80	–	DCH_2 = 3.76

4,4-Dimehyl-1,3-cyclohexanedione $C_8H_{12}O_2$	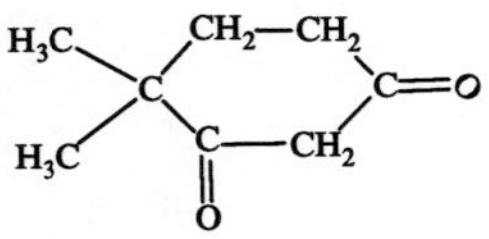	99.2	298	$D_1{=}O \rightarrow CH_2 = 12.23$ $D_2{=}O \rightarrow CH_3 = 11.63$	–	$DCH_2 = 3.76$
1,3-Dioxolan-2-one $C_3H_4O_3$	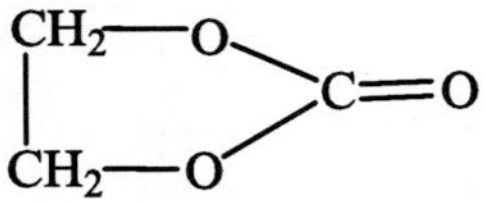	69.6	304.4	9.16 $D = O \cdots H - CH = 4.10$	56.30 (423 K)	7.70 $D = O \cdots H - CH = 3.20$
4-Methyl-1,3-dioxolan-2-one $C_4H_6O_3$	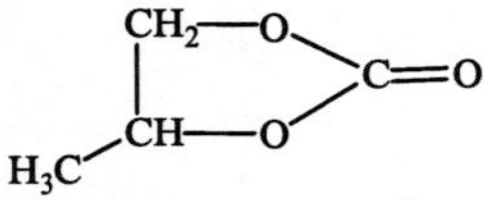	64.82	224.85	9.16 $D = O \cdots H - CH = 4.45$	55.2 (428 K)	7.70, $DCH_3 = 1.1$ $D = O \cdots H - CH = 3.90$
1,4-Dioxane-2,5-dione, $C_4H_4O_4$	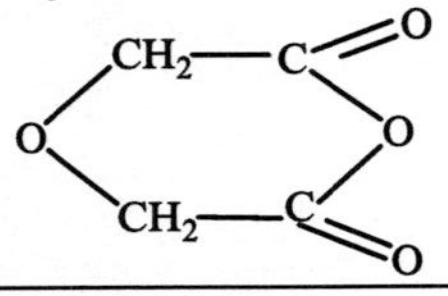	–	–	–	50.4	4.83 $D = O \cdots H - CH = 1.45$

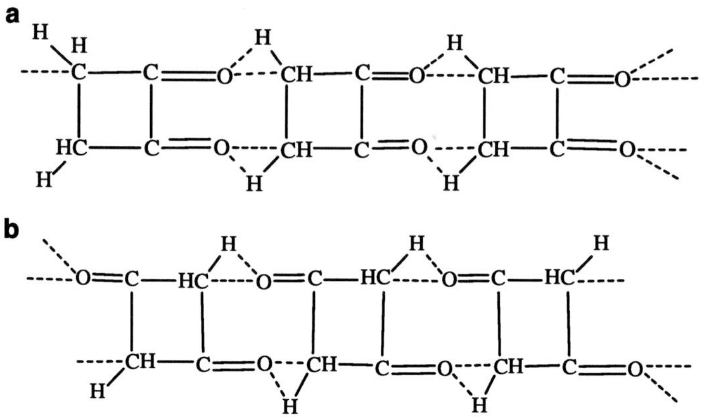

Fig. 4.14 Schematic picture of the liquid structure of diketones (**a**) 1,2-cyclobutanedione, (**b**) 1,3-cyclobutanedione

of the electron density in the fragment, as is observed also 1,3-cyclobutanedione. The obtained value of the hydrogen bond energy equal to 6.45 kJ mol^{-1} shows well its reduced stability as compared to that in the solid 1,2-cyclobutanedione.

In the solid 2,2,4,4-tetramethyl-1,3-cyclobutanedione eight similar specific intermolecular interactions =O → CH_3–C–C– are formed whose energy is determined from the vaporization enthalpy at $\Delta_{sub}H^0(298)/8 = 9.05$ kJ mol^{-1}. The established energies of the specific interactions in solid ketones at the equal number of carbon atoms in the carbon–carbon chain varies in the following regular series:

$$2\text{ - Butanone}\,(D = O \rightarrow CH_3 - CH_2 - C- = 12.23)$$
$$>3\text{ - Pentanone}\,(D = O \rightarrow CH_3 - CH_2 - C- = 11.94)$$
$$>1,3\text{ - Cyclobutanedione}(D = O \rightarrow CH_2 - C- = 9.16)$$
$$>2,2,4,4\text{ - Tetramethyl - }1,3\text{ - cyclobutanedione}$$
$$(D = O \rightarrow CH_3 - C(CH_3) - C- = 9.05) > 1,2\text{ - Cyclobutanedione}$$
$$(D = O \rightarrow CH_2 - C- = 8.05\,\text{kJ mol}^{-1}).$$

In conformity to this series the energy of the specific interactions decrease in going from unsymmetrical to symmetrical ketones and further to cyclic diketones.

The effect of the number of CH_2-groups in the diketone molecules on the energy of the specific interactions in their crystals is revealed in 1,4-cyclohexanedione. The energy of the specific interaction of this compound is also determined directly from the enthalpy characteristic $\Delta_{sub}H^0(298)/8$ (D=O → CH_2 = 9.45 kJ mol^{-1}). In the case of the 1,2-cyclobutanedione the energy of a similar bond (D=O → CH_2 = 8.05 kJ mol^{-1}) in the presence in the molecule of a rigid –CH_2–CH_2– fragment is fundamentally reduced by 1.40 kJ mol^{-1} as compared to solid 1,4-cyclohexanedione due to the interaction of two unshared electron pairs of oxygen with essentially unshared $2s^2$-electron pairs of carbon atoms of two CH_2-groups. This is an important factor illustrating the validity of the concept of equal energies of the specific interactions in solid 2-propanone and 1,2-cyclobutanedione. The molecule of 1,4-cyclohexanedione may be regarded as built from two molecules of 2-propanone with replacing of a hydrogen atom by a bond with a carbon atom. Therewith, the specific interactions formed by the CH_2-groups are more stable (9.45 kJ mol^{-1}) than those involving the CH_3-groups of the 2-propanone (9.16 kJ mol^{-1}). This means that the –C–C– bond in the fragment ensures the increased charge on the oxygen atoms, more positive charge on carbon atoms, and increased acceptor properties. In keeping with this the number of methylene groups increased to three also ensures additional stabilization of the specific intermolecular interactions in the solid 1,3-cyclohexanedione that should be more or equal to the energy observed in the solid 2-butanone (12.23 kJ mol^{-1}) on condition of the equal number of carbon atoms in the substituent of the latter and the ring of the former. Two CH_2-groups at the second carbonyl group are removed from each other to a larger interatomic distance compared to the already mentioned CH_2-groups and thus should form less stable bonds. The results of the thermodynamic calculations are consistent with this conclusion (Table 4.10).

Methyl groups of the 4,4-dimethyl-1,3-cyclohexanedione and 5,5-dimethyl-1,3-cyclohexanedione significantly affect the electron density distribution and considerably increase the charges on the oxygen atoms and the acceptor qualities of the carbon atoms of the methyl and methylene groups. The methyl groups located in the 4,4 and 5,5 positions of the 1,3-cyclohexanediones at the existence of three or four carbon atoms in the chain exclude the influence of the reverse dative bond, and therefore the conditions exist for the formation of stable specific intermolecular interaction (Fig. 4.15).

On the other hand, two methyl groups also take part in the formation of two analogous-specific interactions with an oxygen of the second carbonyl group whose energy should be equal or nearly equal to the energy of the existing interactions in the solid 4,4-dimethyl-1,3-cyclohexanedione (12.23 kJ mol^{-1}). The residual CH_2-group contributes to the enthalpy characteristics an energy whose value can be estimated from the vaporization enthalpy of cyclohexane at $\Delta_{sub}H^0(T)/6 = 3.76$ kJ mol^{-1}. The result of the calculation (Table 4.10) of the energy of the specific interaction in the solid 4,4-dimethyl-1,3-cyclohexanedione demonstrates the high stability of the specific interactions involving both methylene and methyl groups, whereas the latter is characterized by the reduced stability.

It should be stressed that the structural features of the molecules of 4,4-dimethyl-1,3-cyclohexanedione and 5,5-dimethyl-1,3-cyclohexanedione affect the electron density distribution and the variation of the contributions of methyl and CH_2-groups into the enthalpy characteristics of each of these compounds. The energy of the

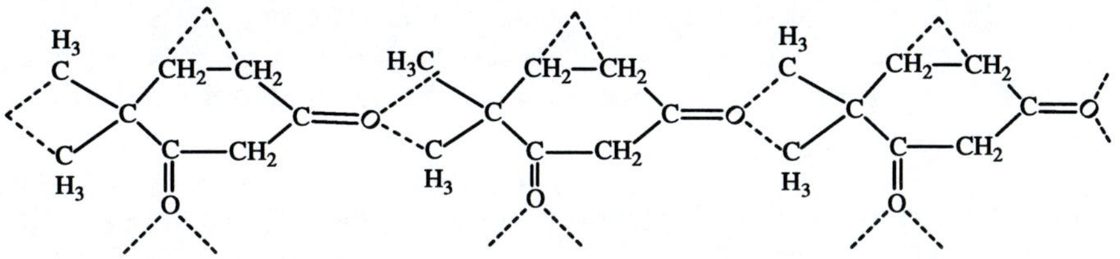

Fig. 4.15 Schematic picture of the solid structure of 4,4-dimethyl-1,3-cyclohexanedione

specific interaction involving the methyl group attains the maximum value in the solid 4,4-dimethyl-1,3-cyclohexanedione. The subsequent increase in the number of carbon atoms in the ring with attached isostructural methyl groups in the solid 5,5-dimethyl-1,3-cyclohexanedione affects only the charge magnitude on the carbon in the CH_2-group and on the oxygen atom. Evidently, this should lead to the subsequent stabilization of the interactions of the CH_2-group situated near the carbonyl group. This leads to a conclusion on the similarity of the energy of the specific interactions of the methyl groups with oxygen atoms in solid 4,4-dimethyl-1,3-cyclohexanedione and 5,5-dimethyl-1,3-cyclohexanedione.

From the thermodynamic analysis and the obtained values of the energy of the specific intermolecular interaction of sold cyclic diketones and oxyketones (Table 4.10), the following conclusions are evident:

Firstly, the insignificant growing of the vaporization enthalpy and the stabilization of the specific interaction in going from the 1.2 cyclobutanone to 1,4-cyclohexanone originates from the virtually constant reverse dative bond at the constant number of carbon atoms (two) in the cycle chain.

Secondly, the changed position of the carbonyl group in the ring of 4,4-dimethyl-1,3-cyclohexanedione in going to 5,5-dimethyl-1,3-cyclohexanedione and the corresponding increase in the number of CH_2-groups favors the shift of the electron density to the oxygen atom, the enhancement of its negative charge, the growth of the acceptor properties of the carbon atoms of the ring, and the stabilization of the specific intermolecular interactions.

Thirdly, the substitution of hydrogen atoms in CH_2-groups by methyl groups is accompanied by the growing vaporization enthalpy of the diketone, and by significant stabilization of the specific intermolecular interaction because of the weakening of the effect of the reverse dative bond on the terminal methyl group of the chain in the presence of three carbon atoms in the chain and the absence of the said effect in the presence of four carbon atoms.

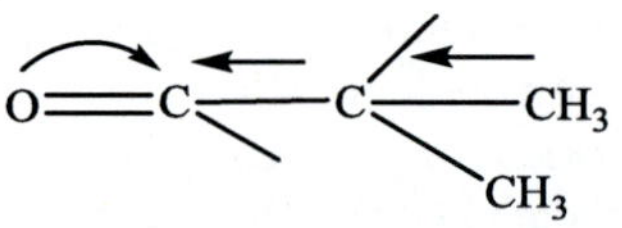

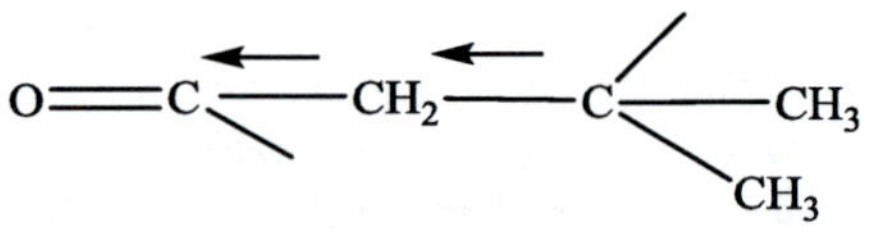

Fourthly, in contrast to the isostructural methyl groups in symmetric and unsymmetrical ketones, the CH_3-groups of the cyclic ketones form stable specific intermolecular interactions with the energy comparable with that of the hydrogen bonds in water.

Fifthly, the hydrogen atoms of the methylene groups of cyclobutanone located in the *trans*-position form stable hydrogen bonds with the carbonyl oxygen atom (6.45 and 8.05 kJ mol^{-1}).

Oxyketones possess reduced values of the vaporization enthalpy and consequently lower stability of the intermolecular interactions. The presence in the oxygen atom of two unshared electron pairs and the presence in the oxyketone structure of two methylene groups indicate that the two bond vacancies of the former should participate in the formation of weak specific intermolecular interactions with the hydrogen atoms located in the *trans*- and *cis*-positions of the CH_2-groups of the molecules of the contacting chains linking them together. The oxygen atom of the carbonyl group also with two bond vacancies forms with the remaining two hydrogen atoms of the CH_2-group of the 1,3-dioxolan-2-one molecule two hydrogen bonds $= O \cdots H - CH$ which should differ in the energy from the other two H-bonds formed by the oxygen atoms of the ring. In its turn every CH_2-group possesses a bond vacancy of the essentially unshared $2s^2$-electron pair of its carbon atom and another bond vacancy of the positively charged hydrogen atom located in the *trans*-position (Fig. 4.16).

Thus the 1,3-dioxolan-2-one molecule forms four specific intermolecular interactions $-O \rightarrow CH_2-$ and four hydrogen bonds $= O \cdots H - CH -$ and four $-O \cdots H - CH -$ of low stability. Taking into consideration that the oxygen forms the specific interaction with the CH_2-group directly linked to the ring oxygen and thus is like 2-propanone, we are able to estimate the energy of this interaction in liquid and solid state of the oxyketone at 7.70 and 9.16 kJ mol^{-1}, respectively. Hence we can calculate the energies of the hydrogen bonds in the 1,3-dioxolan-2-one if the energy of the two types of hydrogen bonds $D = O \cdots H - CH$ and $D\text{–}O \rightarrow CH_2-$ of low stability would be taken as equal. The calculation results based on this assumption are listed in Table 4.10. The vaporization enthalpy of 1,3-dioxolan-2-one and 4-methyl-1,3-dioxolan-2-one (Table 4.10) are given at close temperatures. Therefore, the difference in their vaporization enthalpies (1.1 kJ mol^{-1}) corresponds to the endothermic effect on replacement of the hydrogen atom in the molecule of 1,3-dioxolan-2-one by a methyl group. The influence of the methyl group in the oxyketone is evident and it consists in the redistribution of the electron density in the molecule of the compound. This value originates from the replacement of the H-bonds by the interactions formed by an isostructural methyl group with the exocyclic oxygen that posses lower stability. Consequently, the number of hydrogen bonds in the liquid and crystalline 4-methyl-1,3-dioxolan-2-one is reduced to six. Taking the energy contribution of the isostructural methyl group into the enthalpy of the vaporization and sublimation equal to the contribution of analogous group in 3-methyl-2-butanone, 1.20 and 1.60 kJ mol^{-1}, respectively, we carried out the

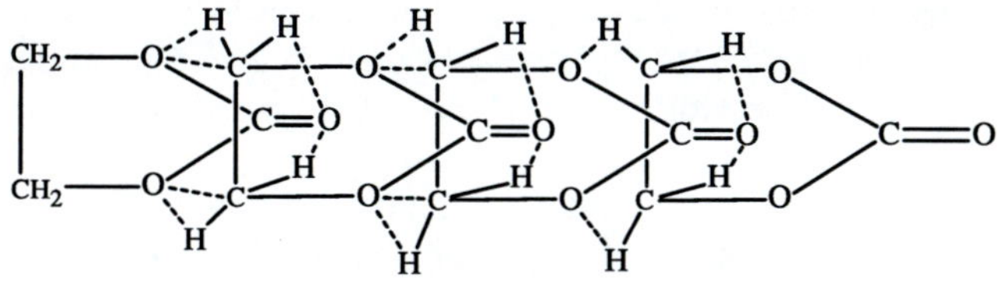

Fig. 4.16 Schematic picture of the liquid structure of 1,3-dioxolan-2-one

corresponding calculations whose results are listed in Table 4.10. The value obtained for the energy of the H-bonds in the 4-methyl-1,3-dioxolan-2-one (3.90 kJ mol^{-1}) is higher than that in the liquid 3-methyl-2-butanone (3.20 kJ mol^{-1}). Thus the decrease in the number of the H-bonds from eight 3-methyl-2-butanone to six in 4-methyl-1,3-dioxolan-2-one results in their stabilization.

The sublimation enthalpy of these compounds (Table 4.10) were measured at sharply differing temperatures, 304 and 224 K, consequently, their difference (4.8 kJ mol^{-1}) should be larger at 298 K. The hydrogen bond energies in solid 1,3-dioxolan-2-one and 4-methyl-1,3-dioxolan-2-one estimated by the thermodynamic calculations are given in Table 4.10.

The low value of the vaporization enthalpy, high volatility, and a large number of the bond vacancies in the molecule of 4-dioxane-2,5-dione indicate the relatively reduced stability of the specific interactions as compared to the liquid 1,3-dioxolan-2-one. The carbon atom of the carbonyl group linked to oxygen like in a carboxy group should possess a relatively high positive charge and high acceptor qualities. Consequently, like the carbon atom of the methylene group it forms a specific interaction –O → C–O–. The molecule of 4-dioxane-2,5-dione, in the near surrounding forms four specific interactions D–C–O → C–O–, four specific interactions D–C–O → CH_2–O–, and eight interactions $D = O \cdots H - CH$ of low stability. Applying the additivity principle, the energy of the specific interaction D–O → CH_2–O– should be taken equal to the energy of analogous interaction existing in the liquid 1,4-dioxane D–O → CH_2–O– (4.83 kJ mol^{-1}) (Chap. 3), and the energies should be equated of the specific interactions D–O → C–O– = D–O → CH_2–O– in the liquid 4-dioxane-2,5-dione. With these assumptions the calculated value of the hydrogen bond of low stability is 1.45 kJ mol^{-1}.

The results of thermodynamic calculations demonstrate the low stability of the H-bonds $D = O \cdots H - CH -$ in liquid and solid cyclic diketones and dioxolanes. These data show also the decrease in the H-bonds stability with their growing number in the oxyketone 1,3-dioxolan-2 one compared with 1,4-dioxane-2,5-dione at the same number of the CH_2-groups in the molecules (Table 4.10).

4.7 The Thermodynamics and Energies of Specific Interactions in Saturated and Unsaturated Ketones C_{13}, C_{18}

In the preceding sections of this chapter it was stated that the thermodynamic characteristics of the vaporization process are more known for saturated ketones with the number of carbon atoms in alkyl chain not exceeding 14, although here also exist significant gaps completed with the use of correlations. However, ketones of unsaturated and saturated series with the number of carbon atoms 13 and more remain unstadied. Some among them as final products of isophytol synthesis possess a practical value and scientific interest due to the special features of the molecular structure and the importance for the development of the theoretical concepts of the intermolecular interactions. However, even the boiling points at normal conditions published in various sources are different. For instance, the published boiling points at normal conditions for saturated C_{18} ketone (6,10,14-trimethylpentadecanone-2) vary in the range from 339 to 438 K [20, 21]. It was stated that pseudoionone (6,10-dimethylundecatrien-3,5,9-one-2) had a non-linear structure [22], and in keeping with [23], the hexahydropseudoionone (6,10-dimethylundecanone-2) possessed a similar structure. A scanty information is available on the unsaturated ketone C_{18} (6,10,14-trimethylpentadecadien-3,5-one-2), and the normal boiling points 383 and 423 K given in [24, 25] are unreliable. Yet, the establishment of valid thermodynamic characteristics of these compounds is necessary from practical reasons and for the thermodynamic analysis of ketones properties with the goal of substantiating the correspondence between the structure and energy and of establishing the energy of the specific intermolecular interactions existing in their liquid state. The latter are especially important for the understanding of the specific intermolecular interactions of compounds with more complex molecules leading to the thermodynamic analysis of peptides and proteins.

These reasons stimulated the thermodynamic investigations of the vaporization process in saturated and unsaturated polyatomic ketones with precision methods on well-purified compounds taking into account their various properties [26, 28] applying the developed and approbated measurement procedure ensuring the results of high accuracy [29].

Aiming at proving the high reliability of the studies performed and the validity of the obtained experimental data we considered it useful to describe some points of the procedure and to give some experimental results.

The study of the vapor–liquid equilibrium was performed by the static tensiometric method with a membrane zero-gauge along a developed procedure approbated on a large number of compounds belonging to versatile classes. The procedure ensures obtaining precision experimental results [29]. The vapor pressure was measured at fixed temperature in the course of heating and cooling of the system with the manometer MChR-3 with an accuracy ± 0.1 mm Hg. The temperature in the liquid thermostat was maintained with the accuracy ± 0.05 K. The investigations were carried out in a wide range of temperatures and pressures.

The vaporization enthalpy of the initial compounds was measured by a calorimeter calibrated by the vaporization enthalpy of methanol and benzene in the temperature range 298.15–343.15 and 298.15–353.25 K, respectively. The experimental findings for the vaporization enthalpy of methanol and benzene at the limiting temperature of the mentioned ranges are well consistent with the most valid published data: for methanol at 298.15 K, 37.31 $\pm$ 0.32 and 37.40 kJ mol^{-1} [30], 37.28 kJ mol^{-1} [31] and at 343.12 K, 34.71 $\pm$ 0.48 and 34.85 kJ mol^{-1} [31]; for benzene at 298.15 K, 33.80 $\pm$ 0.27 and 33.82 kJ mol^{-1} [30], 33.84 kJ mol^{-1} [31] and at 353.25 K, 30.70 $\pm$ 0.38, 30.75 kJ mol^{-1} [30], 30.73 kJ mol^{-1} [31].

Inasmuch as the precision of the studies and the values of the thermodynamic properties essentially depend on the purity of the initial compounds the identified compounds were subjected to thorough purification. The unsaturated ketone C_{18} was distilled repeatedly at the residual pressure 0.027 kPa [29]. The main fraction boiled in the temperature range 377.15–379.15 K. The amount of impurities in the compound under study did not exceed 0.5% from the main substance. The saturated ketone C_{18} was distilled at the residual pressure 0.027 kPa. The main fraction was separated in the temperature range 382.15–385.15 K. The amount of impurities not removed by distillation was 0.4% from the main substance. The dissolved gas was removed from the purified initial compounds in a vacuum at the residual pressure of 6.7×10^{-4} kPa by freeze-thaw method (cooling with liquid nitrogen), and then compounds under study were distilled into glass ampules and stored in sealed ampules in the dark at 278 K. The purity of compounds under study was checked by refraction index (n_D^{298}) [32].

Ketones C_{18} are viscous, and their distilling to ampules due to some heating resulted in the partial decomposition. Therefore, unlike the other compounds the ketones C_{18} were introduced into the glass membrane chamber from open ampules. After evacuation the liquid substances were degassed. In the preliminary experiments with ketones C_{18} we established that their irreversible decomposition started at 546.35 K. For the unsaturated ketone C_{18} the increase in the pressure at this temperature was 346 Pa in 8.5 h, for the saturated ketone, 420 Pa in 1.5 h. In this connection the measurements of the saturated vapor pressure of the unsaturated ketone C_{18} at temperature higher than its decomposition point was performed faster, namely, the temperature intervals of 10° were got over in 30–40 min. The vapor pressure was measured at establishing equilibrium in the system at fixed temperatures. This procedure made it possible to increase the temperature of the measurement of the vapor pressure and to obtain consistent values of the vapor pressure of ketone C_{18} up to 560 K.

The temperature dependences of the saturated vapor pressure of the pseudoionone (6,10-dimethylundecatrien-3,5,9-one-2), hexahydropseudoionone (6,10-dimethylundecanone-2), saturated ketone C_{18} (6,10,14-trimethylpentadecanone-2), and unsaturated ketone C_{18} (6,10,14-trimethylpentadecadien-3,5-one-2) fit well to a straight line plot $\lg P = f(1/T)$ in all studied temperature range. This result additionally proved the purity of the studied compounds [26]. The clear transition into the unsaturated vapor (Fig. 4.17) of all studied substances shows that in the vapor phase processes are absent leading to the change in the number of moles of gas, saturated ketones C_{18}. The correspondence to the Gay–Lussac law on the line of the gas expansion shows that in the vapor monomer molecules of ketones C_{13} and C_{18} are present. This is confirmed by the good agreement between the mean molecular mass of the compounds under study in the vapor phase determined from the unsaturated vapor pressure, mass of the sample, and the volume of the membrane chamber, and that calculated from the empirical formula (Table 4.11) [29].

The initial data of vapor pressures were approximated by Clapeyron–Clausius and Antoine equations [26]. The calculated values of the vaporization enthalpy and entropy of ketones C_{13} and C_{18} are listed in Table 4.12.

The results of measuring vaporization enthalpy of ketones C_{18} by calorimetric method are given in Table 4.13. It is seen that the accuracy of these experimental methods is quite similar. The vaporization enthalpies of ketones C_{18} obtained by calorimetric method are well consistent with the thermodynamic characteristics

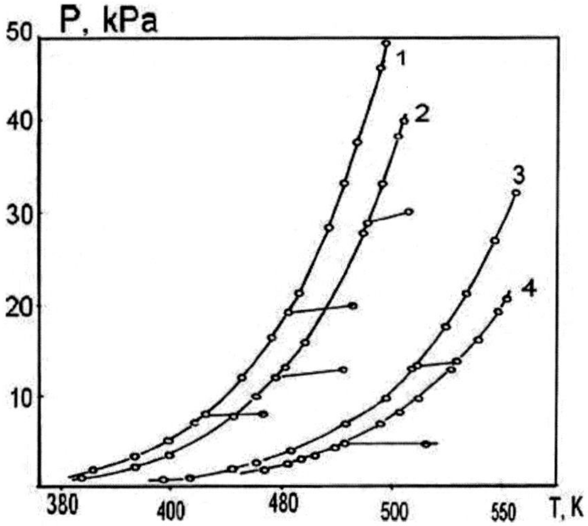

Fig. 4.17 Dependence of saturated and unsaturated vapor pressure on temperature (1) pseudoionone, (2) hexahydropseudoionone, (3) unsaturated, and (4) saturated ketones C_{18}

Table 4.11 Mean molecular mass of ketones C_{13}, unsaturated and saturated ketones C_{18}

Compound	Formula	M formula	M exp
Pseudoionone (6,10-dimethyl-3,5,9-undecatrien-2-one)	$C_{13}H_{20}O$	192.30	191.7 191.5
Hexahydropseudoionone (6,10-dimethylundecane-2-one)	$C_{13}H_{26}O$	198.35	197.8 197.9
Unsaturated ketone C_{18} (6,10,14-trimethyl-3,5-pentadecandien-2-one)	$C_{18}H_{32}O$	264.45	262.1
Saturated ketone C_{18} (6,10,14-trimethylpentadecan-2-one)	$C_{18}H_{36}O$	268.49	265.4

Table 4.12 Vaporization enthalpy (kJ mol^{-1}) and entropy (J mol^{-1} T) of unsaturated and saturated ketones

Compound	Temperature interval (K)	$\Delta_{vap}H^0(T)$	$\Delta_{vap}S^0(T)$
Pseudoionone	365–504	36.84 ± 0.40	82.6 ± 1.0
Hexahydropseudoionone	360–512	37.93 ± 0.42	83.9 ± 1.0
Unsaturated ketones C_{18}	396–560	43.19 ± 0.68	87.0 ± 1.1
Saturated ketones C_{18}	445–556	44.36 ± 0.40	87.8 ± 1.1

Table 4.13 Vaporization enthalpy (kJ mol^{-1}) of unsaturated ketone C_{18} and saturated ketone C_{18} from calorimetric measurements

Compound	Temperature (K)			
	418.55	438.65	458.55	463.45
Unsaturated ketone C_{18}	43.58 ± 0.20	43.51 ± 0.19	43.42 ± 0.20	43.37 ± 0.22
Saturated ketone C_{18}	44.83 ± 0.19	44.68 ± 0.17	44.53 ± 0.20	44.48 ± 0.20

established by tensiometric procedure at the mean values of the same temperature range (Table 4.12). The decrease in the vaporization enthalpy of ketones C_{18} at growing temperature (Table 4.13) is quite regular and is due to the weakening of the specific intermolecular interactions [26, 28].

An impressive result of the performed thermodynamic investigation of saturated and unsaturated ketones C_{13} and C_{18} is their very small values of the vaporization enthalpy compared with the same characteristics of ketones from the saturated series with smaller number of carbon atoms. This experimental result demonstrates the absurdity of the assumption still existing in the scientific literature that the molecules with a large molecular weight possess high vaporization enthalpies (Chap. 1). It is also especially interesting that the enthalpy characteristics of the compounds under considerations are quite similar. The vaporization enthalpy of 3-pentanol measured at 335 K (36.15 kJ mol^{-1}) is close to the vaporization enthalpy of pseudoionone (36.84 kJ mol^{-1}) that in its turn is smaller than the vaporization enthalpy of 3-hexanone (38.39 kJ mol^{-1}). Thus the destruction of the liquid phase structure of saturated and unsaturated ketones C_{13} at the vaporization requires less energy consumption compared with the destruction of the liquid phase structure of 3-hexanone (38.30 kJ mol^{-1}) and 2,4-dimethyl-3-pentanone (diisopropyl ketone) (41.58 kJ mol^{-1}) taking into account the presence in the vapor of each compound of only monomers. This leads to a conclusion that the location of isostructural methyl groups in the backbone carbon chain at every fourth carbon in the saturated and unsaturated ketones C_{13} and C_{18} strongly affects the distribution of the electron density in the hydrocarbon chain and the energies of the specific intermolecular interactions formed by the terminal methyl groups of the chain. It should be mentioned in this connection that the vaporization enthalpy of ketone C_{13} with a normal structure estimated based on the correlation method and thermodynamic data for ketones possessing the highest reliability (Table 4.12) amounts to 76.2 kJ mol^{-1}. At the same time the vaporization of its isostructural (Table 4.12) hexahydropseudoionone requires 37.93 kJ mol^{-1}. This enthalpy characteristics practically equals the vaporization enthalpy of 3,3-dimethyl-2-butanone (37.87 kJ mol^{-1}) whose structure contains fragments resembling those in the structure of ketones C_{13} and C_{18}.

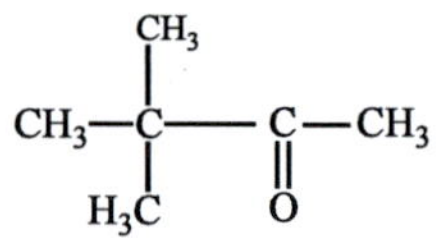

The extremely low values of the vaporization enthalpy connected with the energy of the specific intermolecular interactions reflect the stiffness of the molecular structure at its mobility within the chain or the extremely low capability of the bond vacancies of the molecules under consideration toward the formation of the intermolecular interactions. The similarity of the structures of saturated ketones C_{13} and C_{18} indicates that the difference in their vaporization enthalpies is due to the various numbers of fragments present in their structures. Therefore, the difference in the vaporization enthalpies corresponds to the energy contributed by the fragment below (6.43 kJ mol^{-1}) into this thermodynamic characteristic of the vaporization process.

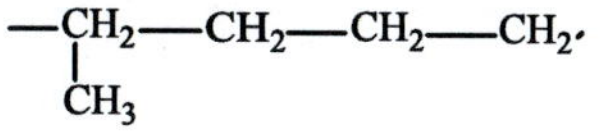

The similarity of the structures of the molecules of unsaturated ketones, pseudoionone and ketone C_{18}, also makes it possible to assign the difference in their vaporization enthalpies to the energy contribution of the fragment below (6.35 kJ mol^{-1}) into the enthalpy characteristics.

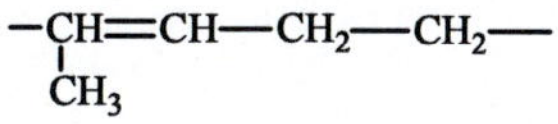

The energy of the contribution of these fragments into the vaporization enthalpy of ketones C_{13} and C_{18} exceeds the contribution of one CH_2-group (5.00 kJ mol^{-1}) into the vaporization enthalpy of the series 2-heptanone–2-octadecanone by 1.35 kJ mol^{-1}. This contribution is comparable with that of one isostructural methyl group. Therefore, it is possible to conclude that each of the five CH_n-groups of these fragments, including the isostructural methyl group, contributes in to the vaporization enthalpy 1.29 and 1.27 kJ mol^{-1}, respectively, for the saturated and the unsaturated carbon chains. Taking into account that the considered contribution into the vaporization enthalpy originates from the interacting fragments of two molecules of the near surrounding the number of interacting CH_n-groups increases twice. This means that the contribution of one such group into the vaporization enthalpy amounts to 0.13 kJ mol^{-1}. Since these groups in the fragments interact involving two hydrogen atoms, this value decreases to 0.07 kJ mol^{-1}. Thus this value is lower than the experimental error in the measurements of the enthalpy characteristics. Therefore, the interactions formed by the saturated and unsaturated fragments of ketones C_{13} and C_{18} should be regarded as cooperative.

Nonetheless, using for each fragment the corresponding energy value we obtain the energy contribution from two or three analogous fragments into the vaporization enthalpy of ketones C_{13} or C_{18} (Table 4.12). The remaining energy value corresponds to the energies of two specific intermolecular interactions formed by the methyl group linked to the carbonyl group (like in acetone) and the terminal methyl group of the long carbon chain (4.12).

$$\Delta_{vap}H^0(T) = n \cdot \text{Dfrag.} + 2\text{D} = \text{O} \rightarrow \text{CH}_3 - \text{R} - \text{C} - +2\text{D} = \text{O} \rightarrow \text{CH}_3 - \text{C} - . \quad (4.12)$$

Regarding this value as constant for every analogous fragment of ketones C_{13} and C_{18} and the energy contributed by the methyl group linked to the carbonyl =C=O group as equal to the energy of the specific interaction in acetone (7.70 kJ mol^{-1}), we are able to estimate the energy of the specific intermolecular interaction formed by the terminal CH_3-group of the carbon–carbon chain R. The obtained values of the energies of the specific intermolecular interactions

Table 4.14 Energies (kJ mol^{-1}) of specific intermolecular interactions with participation of the terminal methyl group of alkyl fragment [26]

Compound	Formula	$\Delta_{vap}H^0(298)$	D=O → CH_3–	∑D Fragments	D=O → CH_3–R′	D = O···CH_3 – R''
Pseudoionone (6,10-dimethyl-3,5,9-undecatrien-2-one)	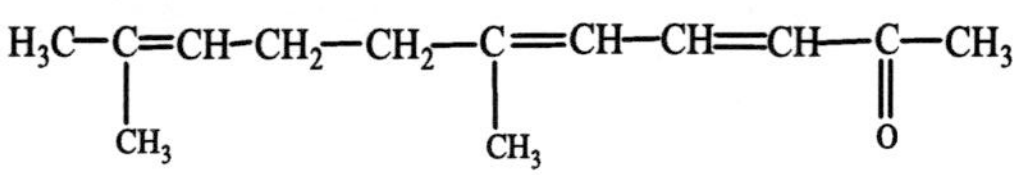	36.84	7.70	12.70	4.40	4.38
Hexahydropseudoionone (6,10-dimethylundecane-2-one)	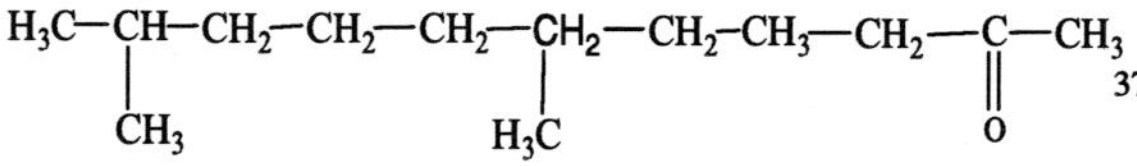	37.93	7.70	12.80	4.90	4.80
Unsaturated ketone C_{18} (6,10,14-trimethyl-3,5-pentadecadien-2-one)	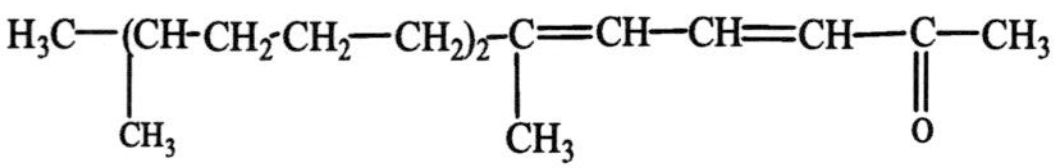	43.19	7.70	19.20	4.30	4.30
Saturated ketone C_{18} (6,10,14-trimethylpentadecan-2-one)	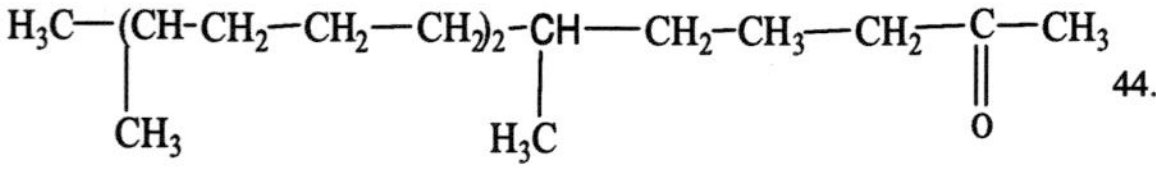	44.36	7.70	19.39	4.90	4.80
Unsaturated ketone C_{23}[a]	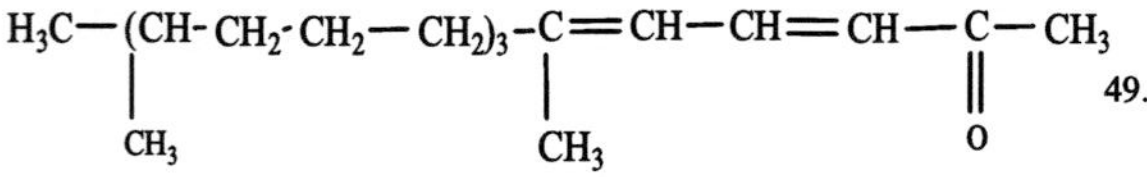	49.54	7.70	25.55	4.40	–
Saturated ketone C_{23}[a]	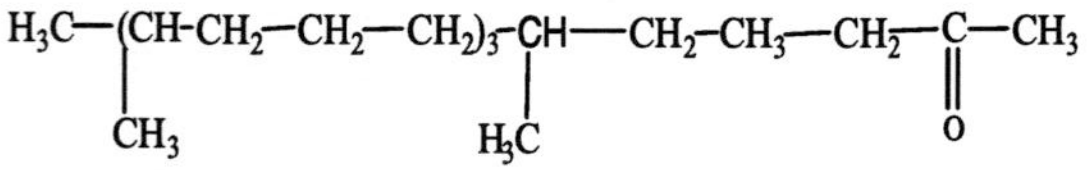	50.76	7.70	25.78	4.90	–
Unsaturated ketone C_{28}[a]	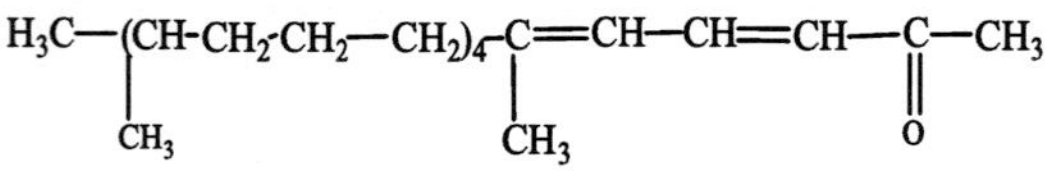	55.85	7.70	31.90	4.40	–
Saturated ketone C_{28}[a]	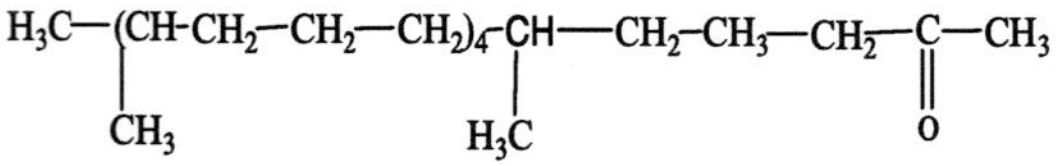	56.59	7.70	32.21		–

[a]Estimated

$=O \rightarrow CH_3$–R–C– formed by the terminal methyl groups of alkyl chain O=C–R–CH_3 differ within the limits of the experimental errors in the measurement of enthalpy characteristics of liquid ketones C_{13} and C_{18} with a saturated R′ and an unsaturated R″ backbone chain, respectively (Table 4.14) [26]. It follows from Table 4.14 that the energies of the specific intermolecular interaction of the terminal methyl group in the unsaturated ketones C_{13} and C_{18} are equal (4.30 kJ mol^{-1}). The equality of analogous interactions is also observed in the saturated ketones C_{13} and C_{18} (4.90 kJ mol^{-1}). These values are significantly lower that the energies of analogous interactions in liquid symmetric ketones, acetone analogs R–C–(=O)–R, and ketones with unsymmetrical structure with normal and branched alkyl chains (Tables 4.3 and 4.4).

References

1. Nefedov VI, Vovna VI (1989) Electronic structure of organic and elementorganic compounds. Nauka, Moscow
2. Krestov GA, Beresin BD (1986) Based notions of contemporary chemistry. Chemistry, Leningrad
3. Baev AK (1972) Complex formation ability halogenides of second – six group's periodical system. Obschaia and Prikladnaia Khimia, vol 5. Vyshaisha Sckola, Minsk, pp 35–51
4. Vasiliev IA, Petrov VM (1984) Thermodynamic properties of oxygen organic compounds. Chemistry, Leningrad
5. Rid R, Prausnic DZH, Shervud T (1977) The properties of gas and liquids. New York
6. Lebedev YuA, Miroschnichenko EA (1981) Thermochemistry of the vaporization of organic compounds. The vaporization and sublimation enthalpy and saturated pressure. Nauka, Moscow
7. Baev AK (1998) Pentacoordinated carbon in Liquid state and solutions of the alkyls main group element. In: Book of abstract of the XVIIth international conference on organometallic chemistry, Munich, Germany, p 18
8. Baev AK (1998) Thermodynamic study of structural changing in system of dimethylcadmium with dimethylmercury. Russ J Inorg Chem 43:1325–1331
9. Baev AK, Cherniak IN (2000) Equilibrium of liquid-vapor in the systems of trimethylaluminiun–trimethylarsin and triethylindium–triethylphosphin. Russ J Phys Chem 74:980–985
10. Baev AK (2000) Thermodynamic parameters of the solutions of the system dimethylzinc–trimethylgallium. Russ J Phys Chem 75:1180–1182
11. Baev AK (2001) Thermodynamics of solutions of trimethylaluminium and trimethylgallium with tetramethylstanum. Russ J Coord Chem 27:637–640
12. Baev AK (2001) Pentacoordinated carbon atom in intermolecular interaction in solutions of alkyl compounds of non-transition elements. In: Book of abstract of VIth international symposium on carbonion chemistry. Marburg, Germany, p 75
13. Baev AK (2001) Thermodynamics and specific intermolecular interactions in solutions of non-electrolytes with pentacoordinated carbon atom. In: Book of abstract, 27th International conference on solution chemistry. Vaals, Netherlands, p 36
14. Baev AK (2001) Specific interactions in solvates of non-electrolyte solutions with isostructural molecular of components and pentacoordinated carbon atom. In: Book of abstract, VIIIth international conference. The problems of solvation and complex formation in solution, Ivanovo, Russia, p 246

15. Baev AK (2002) Thermodynamics of solutions non-electrolytes with five-coordinated carbon atom at specific intermolecular interactions. In: Book of abstract, XIV international conference on chemical thermodynamic, St. Petersburg, Russia, p 135
16. Ivanov EV, Puchovskie YuP, Abrosimov VK (2004) About formation of weak hydrogen bonds CH···O in liquid methanol. In: Book of abstract, IX international conference. The problem of solvation and complex formation in solutions, Plyos, Russia, p 345
17. Baev AK (1969) Phase condition and complex formation ability of galogenide metals. General and applied chemistry, vol 1. Vaisheshika Educating, Minsk, pp 207–218
18. Weichet J, Hodrova J, Kvita V (1957) Studies in the vitamin K and E series. IV. A new synthesis of isophytol and vitamin K. Chem Listy 51:568–571
19. Leets K, Kogerman A, Ammon K (1977) Syntetheses α, β, γ-tocopherole from isoprene. Isv Akad Nauc USSR Chem 28:271–274
20. Bonblik T, Fried V, Hala E (1973) The vapor pressures of pure substances. Elsevier, New York
21. Nazarov NK, Gucev BP, Gunar VI (1957) Full syntheses of Isopropen alcohols. Isv Akad Nauc USSR Chemistry Chem 10:1267–1270
22. Beresovskii D, Pisch M (1973) Chemistry of vitamins. Industry, Moscow
23. Nadirov NK (1986) Tocopheroles (vitaminec of series E) – biological activity compounds. Nauka, Moscow
24. Baev AK (1987) Chemistry of gas-heterogenic systems of elementorganic compounds. Science and Technology, Minsk
25. Baev AA, Baev AK (2011) Thermodynamics of saturated and unsaturated ketones C_{13} and C_{18} and the energies of specific interactions. Russ J Phys Chem 85:823–832
26. Ioffe BV (1983) Refractometric methods of chemistry. Leningrad, Russia
27. Baev AK (1969) Problems of chemical nature of phase transformation. General and applied chemistry, vol 1. Vaisheshika Educating, Minsk, pp 197–206
28. Baev AA, Baev AK (2007) Specific intermolecular interactions of pentacoordinated carbon atom and hydrogen-bonding in liquid alcohols of acetilene series and of vitamin E synthesis' components. XVII international conference on chemical thermodynamics in Russia; X International conferencee on the problems of solvation and complex formation in solutions 2 p. 681
29. Morawetz E (1972) Enthalpies of vaporization of n-alkanes from C12 to C20. J Chem Thermod 4:139–144
30. Majer V, Svoboda V (1985) Enthalpies of vaporization of organic compounds: A critical review and data complication; Blackwell Scientific Publications: Oxford
31. Svoboda V, Vesely F, Holub R, Rick J (1973) Enthalpy data of liquids. I. The dependence of heats of vaporization of methanol, propanol, butanol, cyclohexene and benzene on temperature. Collect Czech Chem Commun 38:3539–3543
32. Wadso J (1966) Calorimetric for a determination of enthalpy vaporization small quantity of compounds at 25 C0. Acta Chem Scand 20:544–552

Chapter 5
Types and Energies of Hydrogen Bonds and Specific Intermolecular Interactions in Alcohols

5.1 Specific Interactions in Saturated Monohydric Alcohols

The high vaporization enthalpies of alcohols (over 40 kJ mol^{-1}) suggest that strong specific intermolecular interactions are operating in their liquid state. The spectral studies in the region 3,100–3,600 cm^{-1} revealed a doublet splitting due to the intermolecular interaction in liquid alcohols [1] with the formation of dimers and more complex associates. The study of the IR spectra of the vapor phase of alcohols made it possible to identify the band γ(OH) characteristic of the H-bonded dimers [2]. N.D. Sokolov [3] substantiated that the nature of the hydrogen bond consists in a donor–acceptor interaction. I.S. Perelygin [4] demonstrated that the formation of a bond through hydrogen of a hydroxy group increased the capability of the interaction via the oxygen atom, and, on the contrary, the bond formation through the oxygen atom increased the proton-donor properties of the hydroxy group. Therefore, the OH group is involved in the formation of the hydrogen bonds simultaneously by both atoms. The hydrogen bond was formulated in [5] as a specific interaction arising between a pair of atoms of two molecules strictly localized on definite orbitals of the reacting species.

The low symmetry of the alcohol molecules complicates the interpretation of the experimental data in the approximation of the multicenter molecular orbitals that results in ambiguous assignment of the vertical ionizing potentials in some publications [6].

It has been already substantiated in Chaps. 1 and 2 that even in the methane molecule the valence $2s^2(c)$ electron pair of the carbon atom remains essentially localized on the C atom, and only two of its valence electrons on its AO $2p_x$, $2p_y$, $2p_z$ effectively take part in the formation of the hypervalent C–H bonds with all four hydrogen atoms [7–13]. The same feature is characteristic of alkyl compounds AR_k.

In the mentioned compounds and methane the sp^3-hybridization model is not valid, and the effective intermolecular interactions constantly observed experimentally in alkyl derivatives of the elements of the main groups [14–16] are underlain by the same electron distribution with the existence of a pseudoatomic

A.K. Baev, *Specific Intermolecular Interactions of Organic Compounds*,
DOI 10.1007/978-3-642-21622-0_5, © Springer-Verlag Berlin Heidelberg 2012

$2s^2(c)$ electron pair. Thus, the reverse dative bond is proven and obtains further extension in the homologous series of alcohols. The same feature of the reverse dative bond appears in the alcohol molecules and consequently it affects the energy of specific

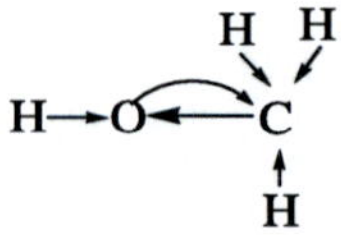

intermolecular interactions in the structures of liquid alcohols [10].

The presence of the essentially unshared $2s^2(c)$ electron pair of the carbon atom provides it with an exceptional previously unknown additional bond vacancy in the alcohol molecule making possible its involvement into the formation of specific interaction and a pentacoordinate state [10, 11]. It is presumable that this property of carbon being unknown and the attractiveness of the sp^3-hybridization model has prevented the accounting for the possibility of the oxygen atom of organic molecules and water to form two specific interactions. Hence it should be stated that the alcohol molecule forms four specific interactions, two among them by the oxygen atom, one is formed by the hydrogen of the hydroxy group, and one by the carbon atom of the terminal methyl group (Fig. 5.1).

Alcohols are a classic example of the formation of a net structure in the liquid state by the hydrogen bonds. Thus, the high coordinating ability of liquid alcohols is due to the presence of the essentially unshared $2s^2(c)$ electron pair of the carbon

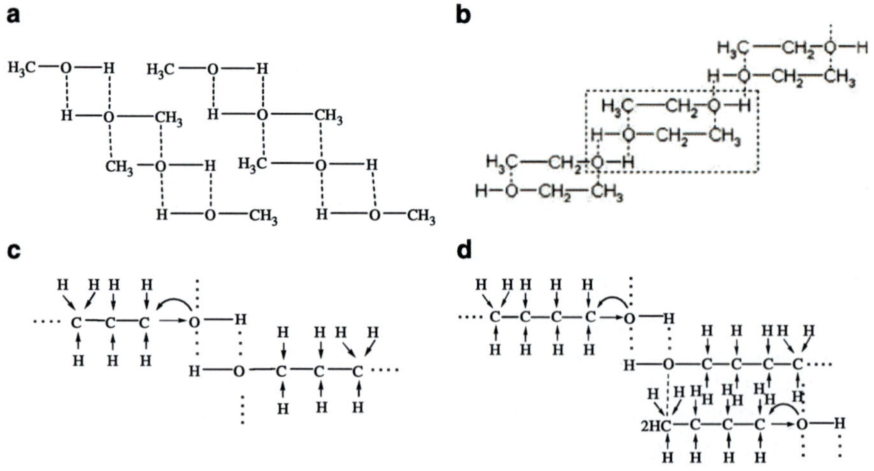

Fig. 5.1 Schematic pictures of the chains formed by saturated normal alcohols: (**a**) methanol, (**b**) ethanol, (**c**) 1-propanol, (**d**) 1-butanol

in the methyl group, of the atoms of oxygen and hydrogen of the hydroxy group. These features provide a possibility for the alcohol molecule to interact with two, three, and even four molecules of the nearest surrounding in the liquid and solid states. In this connection, the alcohol molecules can form a chain and a network structure (Fig. 5.1). The main elements of the structure in question are two hydrogen bonds H–O···H–O and two specific intermolecular interactions with the pentacoordinate carbon atom H–O → CH_3– and H–O → CH_3–CH_2 forming the chain-cyclic structure of the liquid and solid state.

The interatomic distance O···H (278–281 pm) in liquid methanol and ethanol was established from X-ray [17], neutron [18] diffraction, and quantum-chemical calculations [19, 20]. It was shown that the number of the nearest neighbors was 1.77 [20]. The quantum-chemical calculations provided a close value of the molecules in the nearest surrounding (1.85). This value is a strong argument supporting the formation of the chain-cyclic structure in liquid alcohols (Fig. 5.1). These chains are connected by weak interactions formed by hydrogen atoms and the essentially unshared $2s^2$ electron pairs of the carbon atoms of CH_2-groups of the molecules in the nearest surrounding leading to the pentacoordinate state of the carbon atom.

However, the established interatomic distances O↔O cannot prove that the energy of the hydrogen bond does not vary in the series methanol–ethanol–1-propanol. The established interatomic distance O↔O can change insignificantly in the alcohol series for the charges on the oxygen and carbon atoms vary due to the weakening of the reverse dative bond influence on the terminal methyl group affecting the stabilization of the hydrogen bond H–O···H–O.

The decrease in the influence of the reverse dative bond on the energy of the specific interaction with the growing number of carbon atoms in the alkyl chain up to three assures its stabilization and consequently leads to the increase in the vaporization enthalpy in the series

$$CH_3OH\ (37.43) < C_2H_5OH\ (42.26) < n - C_3H_7OH\ (47.48) < n - C_4H_9OH\ (52.33\ \text{kJ mol}^{-1}).$$

This relation is confirmed by the variations in the enthalpy characteristics depending on the length of the alkyl chain of ethers, ketones, and aldehydes whose evaporation consists in going over into the vapor of monomer molecules of these compounds (Fig. 5.2). The different character of the variation in the vaporization enthalpy of aldehydes, unsymmetrical CH_3–O–R and symmetrical R–O–R ethers with the number of carbon atoms in the molecule less than four indicates that the stabilization of the specific interaction formed by the terminal methyl group of the alkyl chain and the weakening of the influence of the reverse dative bond finished after the propyl derivative. These compounds in the respective homologous series are characterized by attaining the maximum energy of the specific interaction 5.63, 6.78, and 8.92 kJ mol^{-1} by ethers, respectively. Further growth in the enthalpy characteristic is due to the contribution of the energy from the growing number of

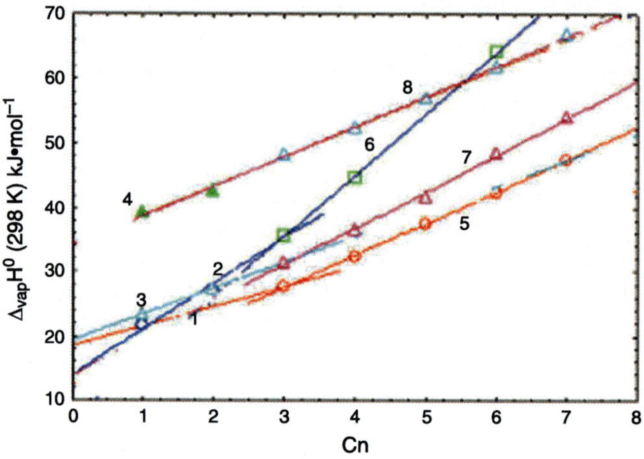

Fig. 5.2 Dependence of vaporization enthalpy of compounds versus the number of carbon atoms in alkyl chain: (1, 5) aldehydes, (2, 6) symmetric ethers, (3,7) unsymmetrical methyl ethers, and (4, 8) 1-alkanols

carbon atom in the alkyl fragment a little exceeding 4 kJ mol^{-1} per one CH_2 group in the molecules. The different slope of the plots $\Delta_{vap}H^0$ (298) $= f$ (Cn) for ethers and aldehydes in the region of 1–3 carbon atoms in the alkyl fragment reflects the difference in the energy of the specific interactions in the respective compounds of these homologous series (Fig. 5.2). The dissimilar slope of the plots at further increase in the number of the carbon atoms in the chain corresponds to the different contribution of the CH_2-groups of compounds in these homologous series into the vaporization enthalpy. Essentially steeper plots (Fig. 5.2) for symmetrical ethers are due to the double increase in the effect of two equal alkyl chains in these compounds.

The feature that distinguished alcohols from ethers and aldehydes is their ability to exist in vapor as stable dimers. Therefore, the vaporization and the sublimation occur with going over in the vapor phase of monomer and dimer molecules. The enthalpy characteristics of dissociation of dimers into single molecules of methanol, ethanol, and further propanol and butanol decreasing in this series unavoidably affects the character of the relation to the number of the carbon atoms in the alkyl chain of alcohols. The high accuracy of the published vaporization enthalpies of alcohols of the normal structure (1-alkanols) [20] obtained by calorimetry permits the assumption that the description by a single straight line of the dependence of the vaporization enthalpy of alcohols on the number of carbon atoms (taking into account the scale interval in Fig. 5.2) is due to the mutual compensation of the effects of the reverse dative bond and the stability of dimers of methanol, ethanol, and 1-propanol.

On approaching the scale interval on the plot to the accuracy of the calorimetric experiment the measured dependence $\Delta_{vap}H^0(298) = f$ (Cn) should be described by three parts including methanol–1-propanol, 1-propanol–1-octanol, and 1-octanol–1-heptadecanol. The first fragment corresponds to the influence of the reverse dative bond and the compensating effect of the going over into vapor dimeric forms of alcohol molecules, and the subsequent fragments illustrate the growing contribution into the vaporization enthalpy of the accumulating in the molecule CH_2-groups. In this case, the dependence under consideration reflects the fairly incomplete compensation of the effect of the reverse dative bond by the dimerization of the molecules of methanol, ethanol, and 1-propanol. Evidently, the compensating effect of the dimerization on the vaporization and sublimation processes should be different in the dissimilar homologous series. The description of the vaporization enthalpy of alcohols in the series 1-propanol–1-octanol by a single straight line $\Delta_{vap}H^0(298) = f\,(\mathrm{Cn})$ is a definite indication that the dimerization degree in 1-propanol is essentially lower than in ethanol and methanol. This assumption is valid, and it is confirmed by the thermodynamic investigation of the gas–liquid equilibrium of aluminum alkyls [14–16]. The thermodynamic properties reported in Table 5.1 correspond materially to going over into the vapor of the dimeric forms of trimethylaluminum molecules whose association is close to 100%. For these forms of molecules, the vaporization occurs with the reduced value of the vaporization enthalpy as compared to the value corresponding to the total process and the vaporization of monomer trimethylaluminum molecules.

The comparison of the enthalpy of vapor formation with dimer molecules of triethylaluminum and of their dissociation energy shows that in the saturated vapor at 400–420 K around 50% of the molecules are associated. The study of the tripropylaluminum demonstrated the presence in the vapor of no more than 3% of dimers [16]. Hence, three processes are involved in the gas-phase equilibrium. The vaporization in the form of monomer molecules requires the rupture of five intermolecular interactions existing in the liquid state whereas the going to the vapor phase of dimers leads to the cleavage of four intermolecular bonds. Therefore, the number of the coordinated molecules essentially increases the dissociation energy of the dimer. Thus, the intermolecular interactions of two molecules in solution are always less stable than in the dimer molecule.

As seen from Table 5.1, the vaporization enthalpy of monomer and dimer forms of molecules is considerably different at the high degree of dimerization in the vapor: the difference is by 3.3 and 0.6 kJ mol^{-1} at 50% degree of association in the vapor of triethylaluminum. Therefore, the energy of the hydrogen bond in methanol might be underestimated by 1–1.5 kJ mol^{-1}. Consequently, we may presume that the same corresponds to the alcohols from the homologous series of 1-alkanols.

Table 5.1 Vaporization enthalpy (kJ mol^{-1}) and dissociation energies (kJ mol^{-1}) of dimer molecules of aluminum alkyls

Process	$\Delta_{vap}H^0(T)$	Process	$\Delta_{vap}H^0(T)$
$\{Al(CH_3)_3\}l = (Al(CH_3)_3)v^a$	41,3 ± 0.2	$\{Al(C_2H_5)_3\}l = (Al(C_2H_5)_3)v^a$	63.7 ± 3.7
$\{Al(CH_3)_3\}l = (Al(CH_3)_3)v$	63.0 ± 1.1	$\{Al(C_2H_5)_3\}l = (Al(C_2H_5)_3)v$	64.3 ± 7.2
$\{2Al(CH_3)_3\}l = (Al_2(CH_3)_6)v$	38.0 ± 1.1	$\{2Al(C_2H_5)_3\}l = (Al_2(CH_5)_6)v$	62.3 ± 2.2
$(Al_2(CH_3)_6)v = 2(Al(CH_3)_3)v$	88.3 ± 0.9	$(Al_2(C_2H_5)_6)v = 2(Al(C_2H_5)_3)v$	66.2 ± 1.1

[a]Total vaporization process

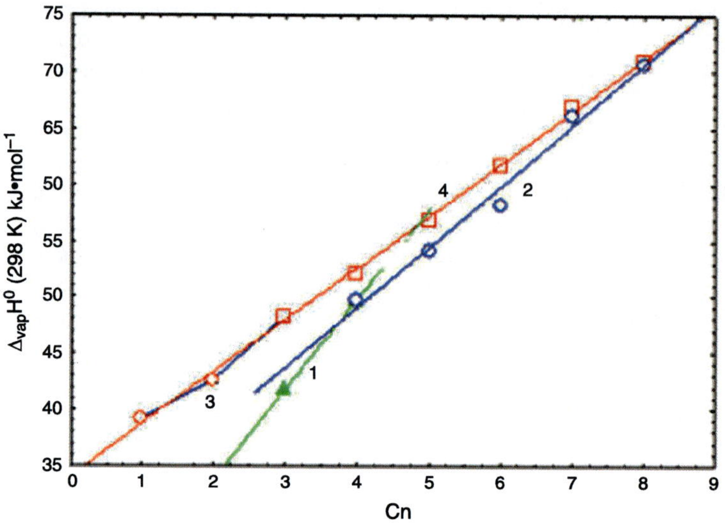

Fig. 5.3 Dependence of vaporization enthalpy of compounds versus the number of carbon atoms in alkyl chain of 2-alkanols (1,2) and 1-alkanols (3,4)

The validity of this assumption is supported by the fact of the incomplete compensation of the effect of the reverse dative bond in ethanol and methanol from this series (Fig. 5.3). This is also supported by the character of the dependence $\Delta_{vap}H^0(298) = f$ (Cn) for the alcohols of the homologous series of 2-alkanols (Fig. 5.3) that is described by two crossing straight lines where the first part corresponds to the decrease in the effect of the reverse dative bond on the enhanced ability of the terminal methyl group in the fragment CH_2–CH_3 of the 2-propanol to form the specific interaction.

Influence of the revers dative bond is completed at 2-butanol with tree carbon atoms in chain CH_2–CH_2–CH_3, and the subsequent growth of the enthalpy characteristic corresponds to the growing contribution of the CH_2-groups in the series 2-butanol– 2-nonanol. This means that the association in the vapor of 2-propanol is not sufficient to compensate the effect of the reverse dative bond and that the dissociation energy of the 2-propanol dimers is essentially lower than in the dimer.

The influence of the reverse dative bond is finished at 2-butanol with three carbon atoms in the series of 1-propanol.

The vaporization enthalpy of the alcohols from the series of 1-alkanols contains the contributions of the energy of two hydrogen bonds and two specific intermolecular interactions H–O → CH_3–CH_2– (Fig. 5.3). In going over into the vapor of monomer and dimer molecules, the enthalpy characteristic obtains the contribution of the cleavage energy of two specific interactions and the partial contribution of the rupture of two hydrogen bonds in the liquid alcohol and of weak nonspecific interactions.

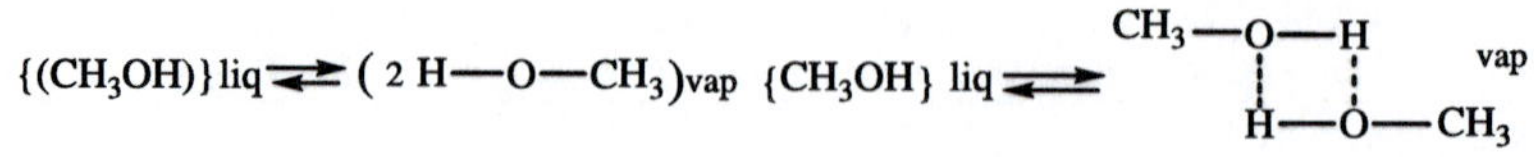

The vaporization of monomer molecules requires the rupture of two specific interactions, two hydrogen bonds, and nonspecific interactions. Consequently, the vaporization of dimer molecules of alcohols occurs with a rupture of two specific interactions and the reduction of the number of coordinated molecules. As a result the hydrogen bonds are stabilized. Therefore, the dissociation energy in vapor of dimers with the hydrogen bonds is higher than the cleavage energy of the hydrogen bonds in liquid alcohols.

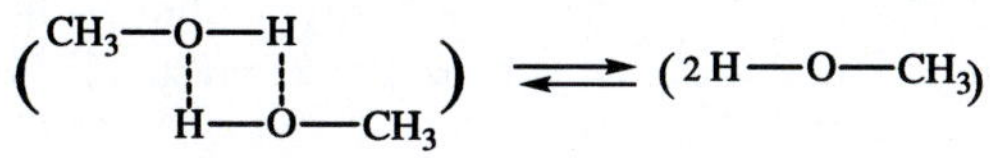

The character of the dependence $\Delta_{vap}H^0(298) = f$ (Cn) for the homologous series of 2-alkanols is analogous to similar dependences for ketones and ethers that do not dimerize in vapor (Fig. 5.4). Therefore like in ketones and ethers, the priority characteristic of these alcohols is the operation of the reverse dative bond in the first compounds of this series. Therefore, the enthalpy dependence on the number of carbon atoms in the alkyl chain reflects the contributions into the enthalpy characteristic from the energy of the most stable specific interactions formed by the pentacoordinate carbon atom and from the hydrogen bond D–O → CH_3– < D=C=O → CH_3– < DH–O···H–O.

The specific interaction H–O → CH_3 formed alongside the hydrogen bond of one oxygen atom (Fig. 5.1) is similar to that observed in ethers, and it may be estimated from the vaporization enthalpy of appropriate ethers [10] (Table 5.1).

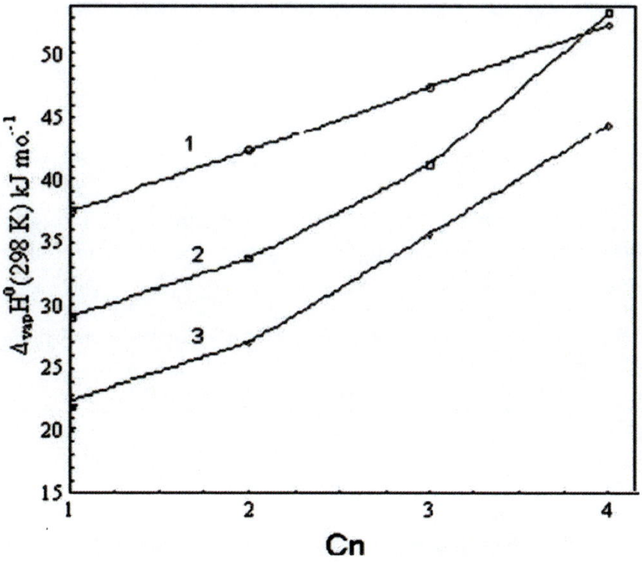

Fig. 5.4 Dependence of vaporization enthalpy of compounds versus the number of carbon atoms in alkyl chain: (1) saturated 1-alkanols, (2) ketones, (3) symmetrical ethers

The calculations utilized the experimentally established by calorimetric method enthalpy characteristics of ethers [21].

Evidently, the obtained energy values of the specific intermolecular interactions can be applied to the estimation of the energy of hydrogen bonds of the other classes of compounds based on the additivity principle. The similarity of the types of specific interactions in liquid ethers and alcohols distinguished by the influence of hydrogen atoms and methyl groups in the alkyl chains makes it possible to account for this similarity in evaluating the energy of hydrogen bonds in liquid alcohols.

Inasmuch as in liquid methanol and ethanol form two specific interactions of the same type

$$\mathrm{DH{-}O \rightarrow CH_3 \text{ and } D{-}O \rightarrow CH_3{-}CH_2} \tag{5.1}$$

and two hydrogen bonds whose cleavage is required for going to vapor of the alcohol molecules we may write the following equation, e.g., for methanol

$$2\mathrm{DH - O \cdots H - O} + 2\mathrm{D - O \rightarrow CH_3} - (11.26) = 37.45\,\mathrm{kJ\,mol^{-1}} \tag{5.2}$$

using the energy of the specific interactions in the liquid dimethyl ether for establishing the energy of the hydrogen bonds.

$$\mathrm{DH{-}O{\cdots}H{-}O} = \left(\Delta_{\mathrm{evap}} H^0(298) - 2\mathrm{D{-}O{\rightarrow}CH_3{-}(CH_2)}\,(n{=}0,1,2)\right)/2. \tag{5.3}$$

In calculating the hydrogen bond energy for 1-butanol and the other alcohols of this homologous series, the energy contribution should be taken into account of the increasing number of the CH_2-groups of the alkyl chain

$$\begin{aligned}\mathrm{DH - O \rightarrow H - O} \\ &= \left(\Delta_{\mathrm{vap}} H^0(298) - 2\mathrm{D - O \rightarrow CH_3 - CH_2 - CH_2} - n\,\mathrm{DCH_2}\right)/2.\end{aligned} \tag{5.4}$$

The hydrogen bond energy in the liquid methanol is 13.10 kJ mol^{-1}. The hydrogen bond energies estimated for four alcohols (Table 5.2) regularly increase

Table 5.2 Energies (kJ mol^{-1}) of specific intermolecular interactions and hydrogen bonds of liquid ethers and alcohols (T = 298 K)

Ethers	D–O → CH_3–$(CH_2)_n$	Alcohols	$\Delta_{vap}H^0(T)$ [18]	DCH_2	D–O–H···O–H
Dimethyl ether C_2H_6O	5.63	Methanol	37.45	–	13.08
Diethyl ether $C_4H_{10}O$	6.78	Ethanol	42.36	–	14.4
n-Dipropyl ether $C_6H_{14}O$	8.92	1-Propanol	47.48	–	15.04 (14.82)
n-Dibutyl ether $C_8H_{18}O$	8.92	1-Butanol	52.39	4.7	14.95

in the series methanol–1-propanol = 1-butanol. The maximum value of this energy is reached in 1-propanol = 1-butanol and equals 15.00 ± 0.05 kJ mol^{-1}. This energy of hydrogen bond exceeds the analogous value for liquid water (11.99 kJ mol^{-1}). The nonspecific interactions have not been taken into consideration for they do not exceed the experimental error in the measurement of the vaporization enthalpy of alcohols.

The obtained values of the hydrogen bond energies permit a conclusion that they adequately explain the nature of the intermolecular interaction, make understandable the experimentally established facts of stable interactions existing in water–organic and organic solutions. The attention should be attracted to the fact that the measurement of high accuracy of the thermodynamic characteristics provides a deeper understanding of the specificity of the interactions. For instance, the difference of vaporization enthalpy in alkyl compounds of metals (Table 5.3) with even and odd number of carbon atoms in the alkyl chain (Δ) has a pronounced regularity. A similar dependence is observed for aliphatic alcohols (Fig. 2.9). This regularity in alcohols makes it possible to suggest that in the molecule of 1-propanol the shift of the electron density of the carbon atom of the terminal methyl group to the oxygen results in the enhanced negative charge on the middle carbon of the propyl group. In compounds with an even number of carbon atoms, for instance, in butane, the enhanced electron density should be observed between two middle carbon atoms as compared with both neighboring groups (Fig. 5.5). Note the enhanced energy of the hydrogen abstraction (379.3 kJ mol^{-1}) from the terminal methyl group compared to the abstraction energy from the CH_2-group at the second carbon atom (371.7 kJ mol^{-1}). Since the fifth carbon atom in the chain, the energy is virtually unchanged within the error of the performed calculations [23].

It was stated [24] that in the hydroxy-containing organic media rather weak complexes might exist with the hydrogen bonds of the type C–H···O. The formation of such interactions is probable and therefore results in somewhat different spectral lines [25]. Based on the analysis of the experimentally observed difference in the spectral lines corresponding to the stretching vibrations of the bonds of various

Table 5.3 Difference (*Δ*) in vaporization enthalpies (kJ mol^{-1}) of compounds with odd and even number of carbon atoms in the alkyl chain [22]

Compound	(*Δ*)	Compound	(*Δ*)	Compound	(*Δ*)	Compound	(*Δ*)
$ZnMe_2$	8.4	$ZnEt_2$	3.3	$ZnPr_2$	8.6	$ZnBu_2$	–
$SnMe_4$	14.0	$SnEt_4$	8.4	$SnPr_4$	13.0	$SnBu_4$	–
$TeMe_2$	5.9	$TeEt_2$	3.9	$TePr_2$	7.9	$TeBu_2$	6.1
CH_3OH.	4.83	C_2H_5OH	5.22	n-C_3H_7OH	4.85	n-C_4H_9OH	–

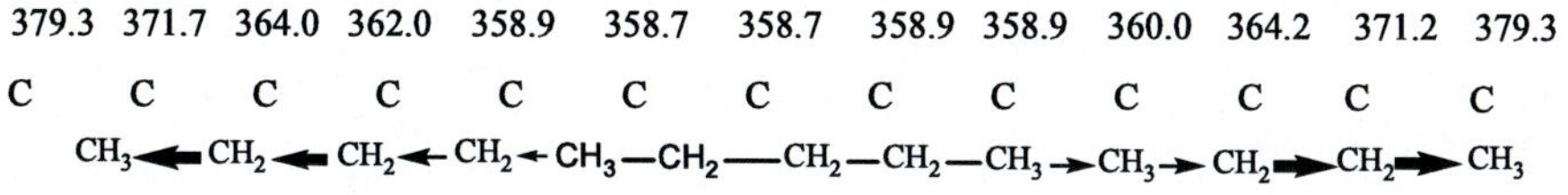

Fig. 5.5 Schematic picture of displacement of the electron density in molecule of hydrocarbon

nature in gaseous and liquid methanol it was concluded [26] that in the condensed phase of this alcohol formed on the average one C–H···O–H bond per each molecule. The calculation of the density functional [27] indicated that the CH···O bond plays an important role in the stabilization of methanol trimers. The possible existence of very weak HCO···HCO intermolecular interactions in the liquid dimethylformamide was stated in [28]. The comparison of the results of the monomer structure, the electron density on the atoms, and the spectral characteristics were interpreted in [29, 30] as nonequality of the protons in the methyl group of the methanol molecule. The said characteristics of the proton located in the *trans*-position with respect to the hydroxy hydrogen atom differ from those of two other protons of the methyl group. Analogous existence of the enhanced charge on the hydrogen atom in the *trans*-position was calculated for the methyl groups of the dimethyl sulfoxide [31]. The presence on the protons of weak positive charged 0.034, 0.051, and 0.045, respectively, indicates the special role of the hydrogen in the *trans*-position in the formation of the weak hydrogen bond. It was shown in [32, 33] that the C–H···O–H type of the hydrogen bond in methanol resulted in the shortening of the C–H bond and in the blue shift in the corresponding spectra; the estimated energy of the interactions did not exceed 4 kJ mol^{-1} [29]. This value cannot be real since it exceeds the hydrogen bond energy in liquid ammonia (3.86 kJ mol^{-1}). It was stated in [26] that this bond should be approximately ten times less stable than the regular hydrogen bond. Note that the transition from the gas into the liquid state is accompanied with the coordination of the molecules of the nearest surrounding and consequently with the sharp decrease in the energy of the hydrogen bonds. Therefore, the more probable seems the estimation of the energy of such bond at 0.60 kJ mol^{-1} presented in Chap. 2. Nonetheless, the positively charged hydrogen of the methyl group of methanol in the *trans*-position to the hydroxy group can interact with the carbon atoms of the bending alkyl chain possessing enhanced electron density (for instance, with the odd carbon atom). These interactions make a contribution into the vaporization enthalpy. However, this contribution is incommensurable with the energy of the specific intermolecular interaction involving the essentially unshared $2s^2$(c) electron pair of the carbon in the pentacoordinate state. The energy of this interaction reaches its maximum value in liquid compounds with a propyl substituent, and at further growing of the chain this value remains virtually constant, and the vaporization enthalpy increases mainly due to the interactions of the positively charged hydrogen atoms of the growing number of the CH_2-groups with the carbon atoms of the chains showing the donor properties.

The increase in the vaporization enthalpy in the series *n*-butanol (52.39)–*n*-octanol (70.98 kJ mol^{-1}) amounts to 18.59 kJ mol^{-1} or 4.65 kJ mol per one CH_2-group. Taking into account that in the nearest surrounding in liquid alcohol four molecules may be present, the energy of the interaction contributed by each hydrogen atom of the four CH_2-groups to the vaporization enthalpy equals 0.58 kJ mol. Thus, this value is incomparable with the values of the energies of the hydrogen bond and the specific intermolecular interaction involving the pentacoordinate carbon atom of the terminal methyl group of the carbon chain. The energy of the interaction under consideration is not of a cooperative character and is the result of a donor–acceptor interaction of the species.

The main characteristic of alcohols is the reverse dative bond [7] affecting the energy of the hydrogen bond and the energy of the specific intermolecular interaction originating from the essentially unshared $2s^2$ electron pair of the carbon atom supplying it with the pentacoordinate state.

The other important characteristic of alcohols, ketones, and ethers is the participation of the hydrogen atoms of the growing number of the CH_2-groups in the molecules in the interaction with the contacting alkyl chains. Their contribution into the vaporization enthalpy of the compound is equal to the difference $\Delta_{vap}H^0(298)$ of two contiguous homologs or of two remote homologs divided by the number of CH_2-groups. The remaining energy value is contributed by two hydrogen bonds and two specific intermolecular interactions. The difference between the vaporization enthalpy of 1-propanol and 1-pentanol gives a mean value of the contribution from the CH_2-group equal to 4.83, and for the series 1-heptanol–1-dodecanol this value equals 5.03 kJ mol^{-1}. Taking the energy of the specific intermolecular interaction of 1-propanol as a constant value we established the energies of the hydrogen bonds of alcohols of normal structure (Table 5.4).

The presence in the oxygen atoms in the molecules of ethers and saturated alcohols of the normal structure of two lone electron pairs and four bond vacancies underlies the analogy in the regular variations in their energy characteristics. This is confirmed by the character of the variation in their vaporization enthalpy and is reflected in the influence of the CH_2-groups on this characteristic. Therefore, for additional proof of the established energy of the hydrogen bonds in liquid alcohols we employed the values of the energy D–O → CH_3–(CH)n of dibutyl ether and decyl methyl ether. The obtained energies of the hydrogen bonds equal 14.93 and 15.00 kJ mol^{-1}, respectively. They are in good agreement with the values (Table 5.2) established from the thermodynamic properties of alcohols.

Table 5.4 Energies (kJ mol^{-1}) of specific intermolecular interactions and hydrogen bonds (kJ mol^{-1}) of liquid 1-alkanols

Alcohol	$\Delta_{vap}H^0(298)$ [21]	D–O → CH_3–CH_2–CH_2	DCH_2	DO–H···O–H
1-Butanol	52.39	8.92	4.7	14.9
1-Pentanol	56.94	8.92	4.83 × 2	14.8
1-Hexanol	61.72	8.92	4.83 × 3	14.4
1-Heptanol	66.81	8.92	4.83 × 4	15.2
1-Octanol	70.98	8.92	5.03 × 5	14.9.
1-Nonanol	76.86	8.92	5.03 × 6	15.5
1-Decanol	81.0[a]	8.92	5.03 × 7	15.5
1-Undecanol	86.5[a]	8.92	5.03 × 8	15.7
1-Dodecanol	91.96	8.92	5.03 × 9	15.4

[a]Established by interpolation

5.2 The Thermodynamics of Saturated Monohydric Alcohols with Branched Structure

The growing demand in the knowledge of the hydrogen bond energies in the alcohols with the normal and branched alkyl chain requires the measurement of their physicochemical characteristics. Special requirements should be addressed to the thermodynamic characteristics of the vaporization processes whose accuracy assures the validity of the magnitudes found in thermochemical calculations and of the conclusions derived therefrom.

Today the publications on the methods and approaches to the measurement of the hydrogen bond energies are not too numerous. This problem for organic compounds is complicated by its connection with the specific intermolecular interactions. The thermodynamic approach firmly gained a foothold and successfully develops in physical organics and solution chemistry. The main advantage of the thermodynamic method remains the possibility of its successful application even in the case of insufficient knowledge of the substance structure. The lack of this knowledge is partially supplemented by the structure–energy compliance. Therefore, the requirements to the validation of the thermodynamic properties of substances are constantly growing. In this respect, the thermodynamic approach to the evaluation of the energy of hydrogen bonds and of specific intermolecular interactions in liquid alcohols with branched alkyl groups requires a special attention. An important place in these studies is occupied by acetylene alcohols and their binary solutions.

The structure of the 2-propanol molecule was studied by electronography [31]. The model fitting best to the experiment has the following bond distances and bond angles: $C_1C_2 = 1.56$ Å, $C_2C_3 = 1.55$ Å, $C_2O = 1.40$ Å, $CH = 1.09$ Å, $C_1C_2C_3 = HCH = 109°28'46''$, and $C_2OH = 105°56'$. Therefore, the 2-propanol molecule should be represented by a formula

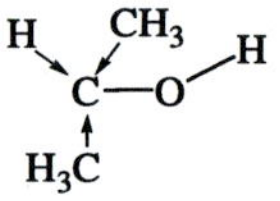

The published data on dependence on the temperature of the isobaric heat capacity of 2-propanol at various pressures were collected and evaluated with respect to the accuracy in [21]. The vapor pressure of 2-propanol was measured by comparative ebulliometry [32, 33]. The direct calorimetric measurements on samples containing 99.9% of 2-propanol are published in [34, 35]. Some experimental data were published in [36, 37], and the measurements of the vapor pressure were presented in [38, 39]. It should be noted that between the results of calorimetric and ebulliometric measurements large discrepancies existed.

The temperature dependence of the saturated vapor of isobutanol was studied by ebulliometry [32], with the use of Gilles device [40], by combining two ebulliometers [41], and also by the calorimetry of the gas flow [42]. The discrepancies in the values of the saturated vapor pressure at high temperature were due to the difference in the procedures of the substance purification and of measurements ignoring the dimerization in vapor of 2-propanol and 2-methyl-1-propanol.

In [43], the temperature dependence of the vapor pressure was measured in a wide range of temperatures and pressures for various initial amounts of substances and with the variation of the volume of the membrane chamber. A good reproducibility of data obtained in different experiments was attained (Fig. 5.6).

The experimentally measured vapor pressure of 2-propanol and 2-methyl-1-propanol fit to Antoine equation [43]

$$\ln P\ (\text{kPa}) = A_1 + B_1/(C + T)$$

and the obtained values of the vaporization enthalpy have a high accuracy (Table 5.5).

The molecular mass of alcohols calculated from the pressure of unsaturated vapor, initial mass of the sample, and the volume of the membrane chamber used in the static method exceeds the molecular mass calculated from the empirical formulas of 2-propanol and 2-methyl-1-propanol (Table 5.6). This fact shows the presence in the vapor of dimers alongside the monomeric forms of the molecules. The decrease in the average molecular mass with the higher temperature of the transition into the unsaturated vapor and the simultaneous increase in the vapor pressure indicate the decrease in the partial contribution from the dimers into the temperature dependence.

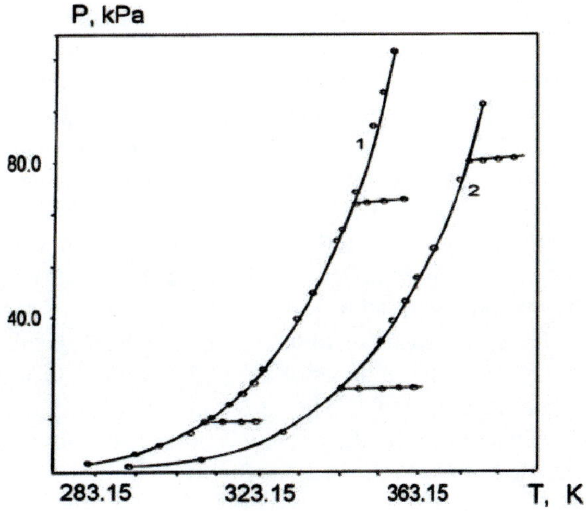

Fig. 5.6 Temperate dependence of pressure of saturated and unsaturated vapor of isopropyl alcohol (1) and isobutyl alcohol (2)

Table 5.5 Enthalpy (kJ mol^{-1}) and entropy (J mol T) of vaporization of 2-propanol and 2-methyl-1-propanol [43]

Compounds	$\Delta_{vap}H^0$(T)	$\Delta_{vap}S^0$(T)	*T*bp. (K)
2-Propanol	41.72 ± 0.4	117.4 ± 1.2	355.3 ± 0.1
2-Methyl-1-propanol	44.2 ± 0.4	116.0 ± 1.1	380.8 ± 0.1

Table 5.6 Average molecular mass of 2-propanol and 2-methyl-1-propanol at the temperature of transition of saturated to unsaturated vapor

Compounds	No	*T* (K)	M exp	M calc
2-Propanol	1	314.43	71.3	60.098
	2	327.33	64.3	
2-Methyl-1-propanol	1	320.95	83.8	74.122
	2	348.95	79.6	

Table 5.7 Vaporization enthalpy (kJ mol^{-1}) of 2-propanol and 2-methyl-1-propanol from the data of calorimetric method

T (K)	$\Delta_{vap}H^0(T)$	
	2-Propanol	2-Methyl-1-propanol
298.25	45.30 ± 0.19	50.78 ± 0.19
311.75	44.42 ± 0.19	49.90 ± 0.20
323.55	43.33 ± 0.18	48.81 ± 0.20
338.35	41.85 ± 0.19	47.30 ± 0.21
353.25	41.00 ± 0.20	45.51 ± 0.20

The calorimetric measurement of the vaporization enthalpy of *i*-PrOH and *i*-BuOH was carried out in a calorimeter and using our procedure described in Chap. 4. The mean values of the enthalpy and entropy of vaporization obtained from a series of experiments at given temperature are listed in Table 5.7.

The vaporization enthalpies measured by the calorimetric method and presented in Table 5.5 correspond to the middle of the studied temperature range. The identity of results of two independent experiments at each temperature attests the reliability of the obtained thermodynamic characteristics of the vaporization of *iso*-PrOH and *iso*-BuOH. The temperature dependence of vaporization enthalpy on temperature (Fig. 5.7) completely coincided with the results of calorimetric measurements performed by other researchers [33, 41, 42, 44] and disagrees with the values published in [45, 46] due to the different purity of the initial compounds. In our experiments, *iso*-PrOH and *iso*-BuOH were subjected to more thorough purification than in the previous publications, and the used amount of initial compounds was about 200–300 times smaller.

The established fact of the increase in the molecular mass in the vapor of *i*-PrOH at 314.45 K approximately by 15.7%, at 327.35 K by 6.7%, and for *i*-BuOH at 320.45 K and 348.95 K by 11.6 and 6.9%, respectively, indicates the proportional contribution of the dissociation enthalpy of dimers into the overall vaporization enthalpy of *iso*-PrOH and *iso*-BuOH.

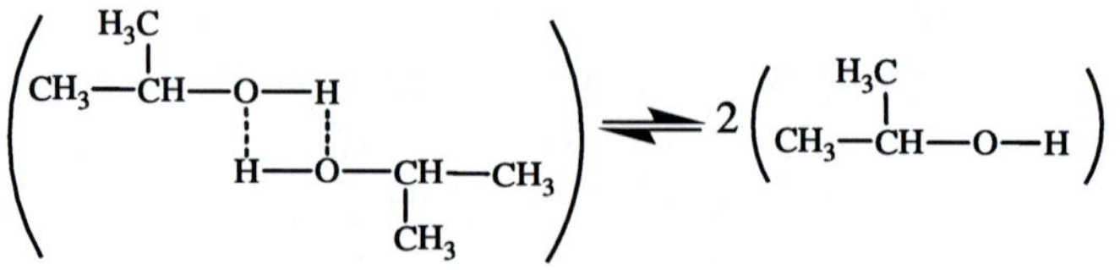

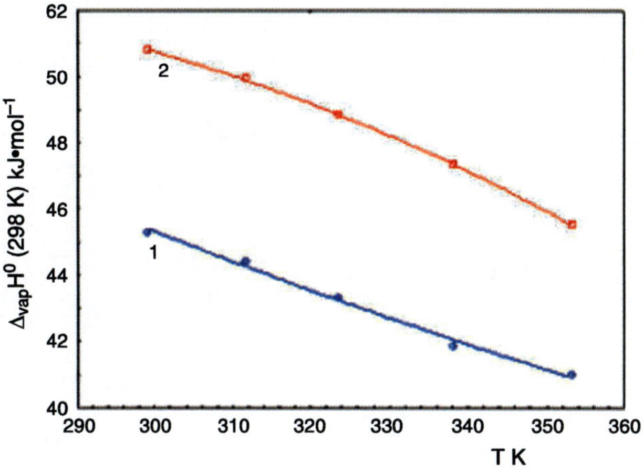

Fig. 5.7 Temperature dependence of the vaporization enthalpy of 2-propanol (1) and 2-methyl-1-propanol (2)

The interatomic distances in the cycles of the chain structure of liquid alcohols possessing isostructural methyl groups also should depend on the energy of the hydrogen bond and of the specific intermolecular interaction –O → CH_3.

The stabilization of the hydrogen bonds in the alcohols of normal structure observed at the growth of the number of carbon atoms to three in the series methanol–1-propanol should also exist in the branched alcohols in the series of 2-propanol and 2-methyl-1-propanol. This is caused by the decrease in the effect of the reverse dative bond with the growth of the alkyl chain.

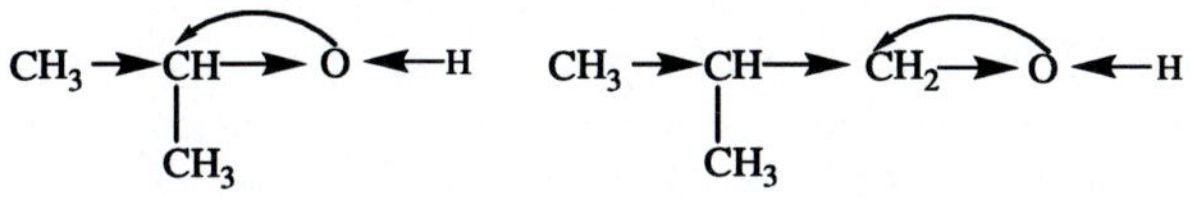

The analogy in the decrease in the association degree of the alcohols of normal and branched structure with the increasing number of carbon atoms in the alkyl chain makes it possible to expect an insignificant difference in the energy of the hydrogen bonds between the ethanol and isopropyl alcohol, 1-propanol and isobutyl alcohol. The isostructural methyl groups replace the hydrogen atom and are localized in the *trans*- or *cis*-position with respect to the hydrogen of the hydroxy group. They form less stable specific intermolecular interactions compared with those of the terminal methyl group of the alkyl chain (Fig. 5.8).

Due to the principle of the energy minimum of the system, the formation of the most stable specific interaction should occur with the participation of the terminal methyl groups with the larger negative charge on the carbon atom. The remaining isostructural CH_3-group and the positively charged hydrogen atom are capable to interact with analogous groups and CH_2-groups of the alkyl chains of the contacting molecules forming a specific interaction H_3C → H of low stability. This interaction

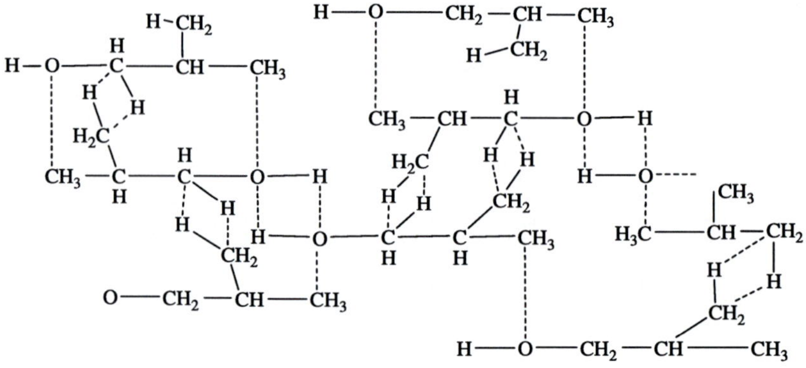

Fig. 5.8 Schematic picture of the network structure of liquid 2-methyl-1-propanol

occurs with the involvement of the essentially unshared $2s^2$ electron pair of the carbon atom that ensures the pentacoordinate state for the carbon atom of the isostructural methyl group (Fig. 5.8).

At equal number of carbon atoms in the ethyl group of ethanol and 2-propanol, the higher value of the vaporization enthalpy of the latter is due to the contribution of the isostructural methyl group (5.5a):

$$D iso - CH_3 = \Delta_{vap}H^0(T)(iso\text{ - p}) - \Delta_{vap}H^0(T)\ (\text{etal.}). \qquad (5.5a)$$

The calculated contribution of the isostructural methyl group into the vaporization enthalpy of 2-propanol equals 2.90 kJ mol^{-1} (Table 5.7), and this value is somewhat underestimated for the enhanced degree of ethanol dimerization is not taken into account. Considering that two interacting molecules occur in the nearest surrounding the obtained value of the energy contribution from the isostructural methyl group amounts to 1.45 kJ mol^{-1}.

In the molecule of isobutyl alcohol (2-methyl-1-propanol), one isostructural methyl group is present possessing a carbon atom with the essentially unshared $2s^2$(c) electron pair providing a free bond vacancy. Therefore, in the liquid state two specific interactions $H_3C \rightarrow H$ of low stability form whose energy is evaluated from the difference of the vaporization enthalpies of 2-methyl-1-propanol and 1-propanol (5.5b):

$$2D iso - CH_3 \rightarrow H = \Delta_{vap}H^0(T)(2 - M - 1 - p) - \Delta_{vap}H^0(T)(1 - p). \quad (5.5b)$$

The energies of two stable specific intermolecular interactions H–O $\rightarrow$ CH_3–CH–CH_2 and two hydrogen bonds H–O···H–O have the same values like in the liquid 1-propanol (Table 5.8).

The carbon skeleton of 2-butanol molecule contains a propyl chain and a terminal methyl group linked to the same carbon atom as the hydroxyl group. Considering that the propyl group forms more stable specific interaction with the

Table 5.8 Vaporization enthalpy, energies of hydrogen bonds, and of specific interactions (kJ mol^{-1}) in liquid branched saturated alcohols related to propanol and butanol (298 K)

Compounds	Structures	$\Delta_{vap}H^0(298)$ [21]	D $H_3C \rightarrow H–CH_2$	D–O → $CH_3(CH_2)n$	D H–O···H–O
Ethanol	$CH_3–CH_2–OH$	42.38	–	6.78	14.40
2-Propanol	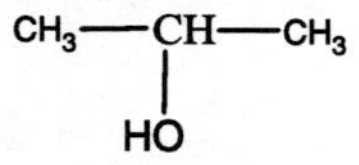	45.30	2.90:2 = 1.45	6.78	14.40
1-Propanol	CH_3-CH_2-CH_2-OH	48.30	–	8.92	15.17
2-Methyl –1-propanol (isobutyl alcohol)	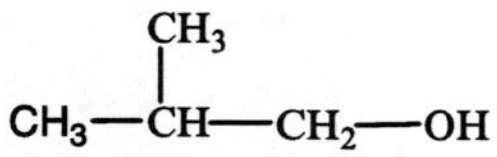	50.79	∑2.49:2 = 1.25	8.92	15.17
2-Butanol (*sec*-butyl alcohol)	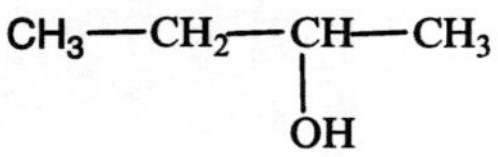	49.74	∑1.44:2 = 0.72	8.92	15.17
2-Methyl –2-propanol (*tert*-butyl alcohol)	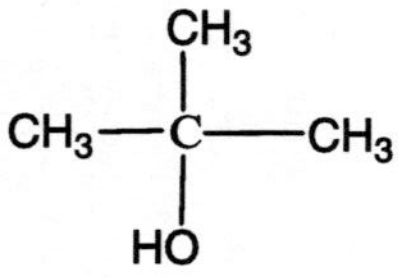	46.81	4.10:4 = 1.03	6.78	14.40
1-Butanol	$CH_3–CH_2–CH_2–CH_2OH$	52.30	–	8.92 DCH[2]	14.80

oxygen atom than the methyl group, the latter with the analogous groups of the interacting molecules of the nearest surrounding acts like an isostructural methyl group forming the interactions of low stability. The methyl group of the propyl fragment possesses enhanced acceptor properties due to the weakened effect of the reverse dative bond. In the near surrounding of 2-butanol and 2-methyl-1-propanol only less active hydrogen atoms are present forming less stable interactions $H_3C \rightarrow H{-}CH_2$. Therefore, it is possible to conclude that the molecules of 2-butanol and 2-methyl-1-propanol possess the same number of bond vacancies. Two of these vacancies form hydrogen bonds, and the other two, stable specific interactions $H{-}O \rightarrow CH_3{-}CH_2{-}CH_2$. Each molecule of 2-butanol utilizing the isostructural functions of methyl group forms with a molecule of the nearest surrounding two specific interactions of low stability $H_3C \rightarrow H{-}CH_2$. This means that in liquid 2-butanol containing in the molecule one isostructural methyl group the main contribution into the vaporization enthalpy originates from two hydrogen bonds H–O···H–O– and the specific intermolecular interactions with the three-carbon chain $DH{-}O \rightarrow CH_3{-}CH_2{-}CH_2{-}$ of the energy 8.92 kJ mol^{-1}. At the same time the overall energy contribution from two methyl groups is equal to the difference between the vaporization enthalpies of 2-butanol and 1-propanol and amounts to 1.44 kJ mol^{-1}, or the energy of the specific interaction formed by this group in the nearest surrounding equals 0.72 kJ mol^{-1}. Similarly, the contribution of the isostructural methyl group in 2-methyl-1-propanol is determined from the difference in the vaporization enthalpies of this compound and 1-propanol (Table 5.8).

Note that the experimentally measured vaporization enthalpies are obtained with a high accuracy and therefore the energy of the discussed specific interaction is also determined with a low error warranting the first figure after the decimal point.

The molecule of 2-methyl-2-propanol (Fig. 5.9) with two isostructural methyl groups also forms in the liquid state the chain-cyclic structure. The isostructural methyl groups take part in joining equivalent and form cpecific interactions of different stability. This is due to the influence of the reverse dative bond from the oxygen on the methyl group located in the *trans*-position with respect to this oxygen (Fig. 5.9b). As a result it possesses higher acceptor properties and therefore it is directly involved in the formation of the stable specific interaction. The molecule of 2-methyl-2-propanol with two isostructural methyl groups forms in the nearest surrounding four specific interactions of low stability $H_3C \rightarrow H$. The main contribution into the vaporization enthalpy of this compound, like in the other alcohols related to 1-butanol, originates from two hydrogen bonds and two specific intermolecular interactions formed by the alkyl chain of two carbons with the oxygen atom of the hydroxy group that has been taken equal to the energy of a similar interaction in the liquid diethyl ether amounting to 6.78 kJ mol^{-1}. From the difference of the enthalpy characteristics of 2-methyl-2-propanol and of ethanol, the contribution of four isostructural methyl groups was evaluated at 4.20 kJ mol^{-1}. Consequently, the energy of one specific interaction of low stability $H_3C \rightarrow H$ contributes 1.05 kJ mol^{-1}.

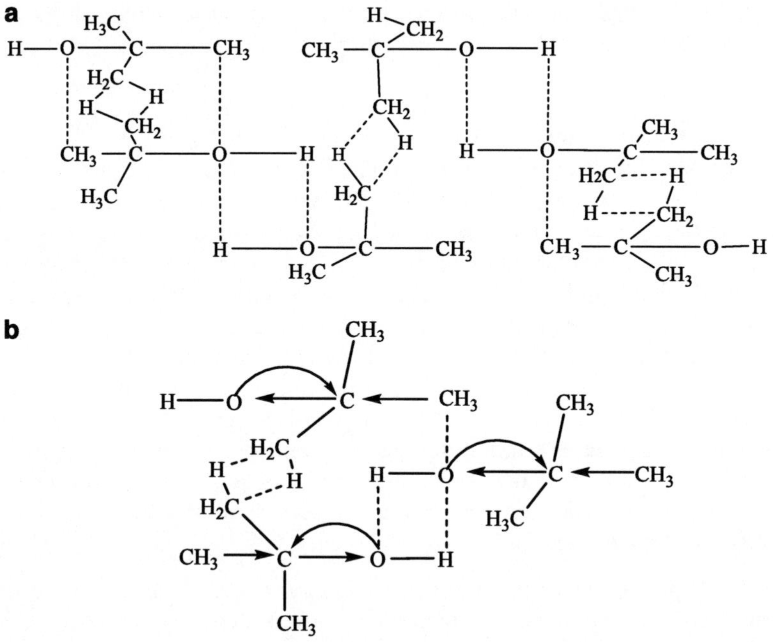

Fig. 5.9 Schematic picture of the liquid structure of 2-methyl-2-propanol (**a**) and reverse dative bond (**b**) in 2-methyl-2-propanol (*tert*-butanol)

The trend of the decreasing dimerization degree in the series of saturated alcohols of normal and branched structure

$$1\text{ - Propanol} > 2\text{ - Methyl - 1 - propanol} > 2\text{ - Propanol} > 2\text{ - Methyl - 2 - propanol}$$

reflected the stabilization of the hydrogen bond in the alcohols with a three-carbon (propyl) chain. The energies of the low-stable specific interactions $H_3C \rightarrow H$ formed by the isostructural methyl group decrease with moving off the methyl from the oxygen atom:

$$2\text{ - Propanol }(1.45) > 2\text{ - Methyl - 1 - propanol }(1.25) >$$
$$2\text{ - Methyl - 2 - propanol }(1.03) > 2\text{ - Butanol}(0.72\,\text{kJ mol}^{-1}).$$

This type of specific interactions is the least stable in 2-butanol, whose methyl group shifts less electron density to the oxygen atom than the isostructural methyl groups of 2-propanol, 2-methyl-1-propanol, or 2-methyl-2-propanol.

Based on this data we conclude that the isostructural methyl group of 2-propanol or two analogous groups of 2-methyl-2-propanol do not create a sterical effect and the decrease of the vaporization enthalpy of these alcohols is due to:

Firstly, to the decrease in the energy of the specific interaction in replacing a propyl chain in the 2-methyl-2-propanol by ethyl chain in 2-propanol and 2-methyl-2-propanol.

Secondly, by the small energy contribution into the vaporization enthalpy by the specific interactions formed by the isostructural methyl groups.

Thirdly, the stability of interactions formed by isostructural methyl groups depends on their place in the alkyl chain of the alcohol molecule. The thermodynamic analysis taking into account the bond vacancies of alcohol molecules and the analysis of their structures provides a possibility to obtain a detailed schematic picture of the liquid state that makes it possible to write the equation for estimation of the energy of hydrogen bond, of specific interactions of high and low stability. The results of this thermodynamic analysis for saturated alcohols related to 1-propanol and 1-butanol (Table 5.8) show the possibility to perform the thermodynamic calculations of the energy of the low stable hydrogen bonds. The schematic pictures of the structures of liquid alcohols presented in Figs. 5.8 and 5.9 give a notion of the specificity of the interactions that form the chain-cyclic structures which are linked into a network by the interactions of the isostructural methyl groups.

The hydrogen bond energies are stabilized in the series of normal and branched alcohols with the growing number of carbon atoms in the longer alkyl chain in the alcohol.

$$\text{Methanol} \rightarrow \text{Butanol and Isopropanol} \rightarrow \text{Isobutanol.}$$

This is due to the decreasing influence on the terminal methyl group of the reverse dative bond between the oxygen atom and contiguous to it carbon atom. The subsequent increase in the number of carbon atoms in the chain does not notably affect the stabilization of the hydrogen bond, and the growing contribution to the vaporization enthalpy provides the increasing number of the CH_2-group which is not insensitive to the interaction with the analogous groups of the contacting molecules.

The conclusions based on the performed thermodynamic analysis can be supplemented by the consideration of the alcohols of 1-pentanol family.

The structure of liquid alcohols related to 1-pentanol is analogous to the above discussed structures of the saturated alcohols. Their difference from 1-propanol consists only in the number of the methylene groups in the alkyl chains and in the location of the isostructural methyl group in the alkyl chain. The isomeric alcohols 2-methyl-1-butanol and 3-methyl-1-butanol (Table 5.7) with an insignificant difference in the vaporization enthalpy (0.3 kJ mol^{-1}) illustrate the weak dependence on the position of the isostructural methyl group at the second and third carbon atom in the alkyl fragment. The contribution of the energies of the low stability specific interactions into the vaporization enthalpy of these compounds equals the difference between their enthalpy characteristics and the vaporization enthalpy of 1-butanol (Table 5.8). The energy of the hydrogen bond is equal to that in the liquid 1-butanol or it should be calculated by the equation taking into account the energy of two specific intermolecular interactions $D\text{–}O \rightarrow CH_3\text{–}CH_2\text{–}CH_2\text{–}$ corresponding to the alkyl chains equal to those in dipropyl ether (8.92 kJ mol^{-1}).

$$\mathrm{DH-O\cdots H-O} = (\Delta_{\mathrm{vap}}\,\mathrm{H}^0(298) - \mathrm{2D-O} \rightarrow \\ \mathrm{CH_3-CH_2-CH_2} - 2\,\mathrm{D}iso - \mathrm{CH_3} - \mathrm{2DCH_2})/2\,. \qquad (5.6)$$

The energy contribution of CH_2-group we obtained as a mean value from the difference between vaporization enthalpies of 1-propanol to 1-octanol (4.75 kJ mol^{-1}).

As already stated, in the interaction of molecules in the nearest surrounding the number of the formed specific interactions involving the isostructural methyl groups is doubled. The molecules of alcohols 2-methyl-1-butanol and 3-methyl-1-butanol are distinguished from the molecule of 1-butanol by the lack of the isostructural methyl group in the latter, therefore obviously the difference in the vaporization enthalpy of the branched alcohol and 1-butanol gives the overall energy contribution of the low stability specific interactions formed by the isostructural methyl groups

$$\mathrm{DH_3C \rightarrow H-CH_2} = \left(\Delta_{\mathrm{vap}}H^0(T)iso.\mathrm{m.\ b.} - \Delta_{\mathrm{vap}}H^0(T)1-b\right)/2. \qquad (5.7)$$

The overall values of these energies in the liquid 2-methyl-1-butanol and 3-methyl-1-butanol amount to 1.70 and 2.00 kJ mol^{-1}, respectively. The high accuracy of the vaporization enthalpy measured by the calorimetry assures the high accuracy of the obtained energy values and consequently of the energies of the interactions in question equal 0.85 and 1.0 kJ mol^{-1}. The results of the calculations (Table 5.9) demonstrate the decrease in the energy of the hydrogen bond nearly by 0.40 kJ mol^{-1} compared to the energy of the H-bond in the liquid derivatives of 1-propanol and 1-butanol (Table 5.8).

It follows from the data of Table 5.8 that the location of the OH-group in the molecule of 3-pentanol and of CH_3-group in the 3-methyl-1-butanol at the same carbon atom of the alkyl chain resulted in close values of their vaporization enthalpy (difference 0.7 kJ mol^{-1}). The vaporization enthalpy of 2-methyl-2-butanol is by 0.54 kJ mol^{-1} less than the vaporization. The molecule of 2-methyl-2-propanol with two isostructural methyl groups forms in the nearest surrounding four specific interactions of low stability $H_3C \rightarrow H$. The main contribution into the vaporization enthalpy of this compound, like in the other alcohols related to 1-butanol, originates from two hydrogen bonds and two specific intermolecular interactions formed by the alkyl chain of two carbons with the oxygen atom of the hydroxy.

It permits an assumption that the propyl fragments in the 3-pentanol also possess different properties in the interactions in the liquid and the crystalline state. The schematic pictures presented in Figs. 5.10 and 5.11 for the liquid and crystalline states of 3-pentanol and 2-methyl-2-butanol reflect the difference in the specific interaction of the alkyl chains of these compounds.

The structure of liquid 3-pentanol consists of chains from small and large cycles formed by the hydrogen bonds and the stable specific intermolecular interactions H–O $\rightarrow$ CH_3–CH_2–CH_2. The chains are linked into a network by specific interactions of low stability formed by the hydrogen atom of the terminal methyl group with the atoms of the methyl groups of the contacting chains.

Table 5.9 Vaporation enthalpy (kJ mol^{-1}) of branched alcohols related to 1-pentanol

Compound	Structure	$\Delta_{vap}H^0(298)$ [21]	D–O → CH_3–$(CH_2)_2$	DH_3C → H–CH_2	DCH_2–	DH–O···H–O
1-Pentanol	CH_3-CH_2-CH_2-CH_2-CH_2-OH	56.94	8.92	–	4.75 × 2	14.80
3-Methyl –1-butanol	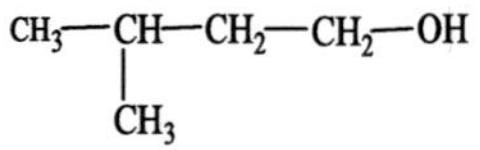	54.3	8.92	2.0:2 = 1.0	4.75	14.85
2-Methyl –1-butanol	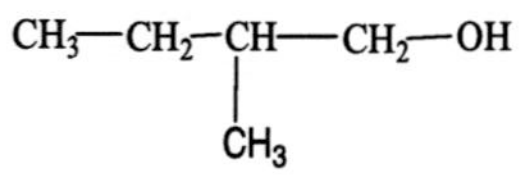	54.0	8.92	1.7:2 = 0.85	4.75	14.85
3-Pentanol	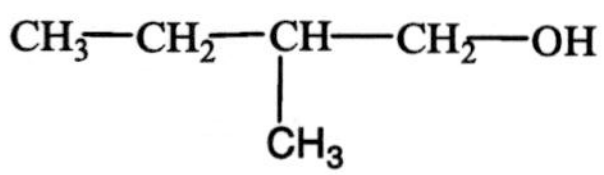	53.6	8.92	2.0:2 = 1.0	4.75	14.85
2-Methyl –2-butanol	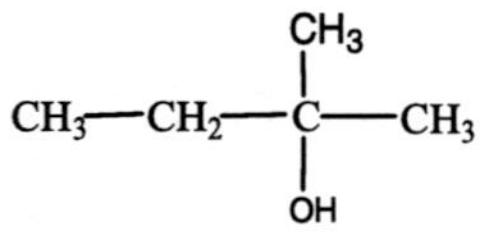	49.2	8.92	0.40 × 2	–	14.85

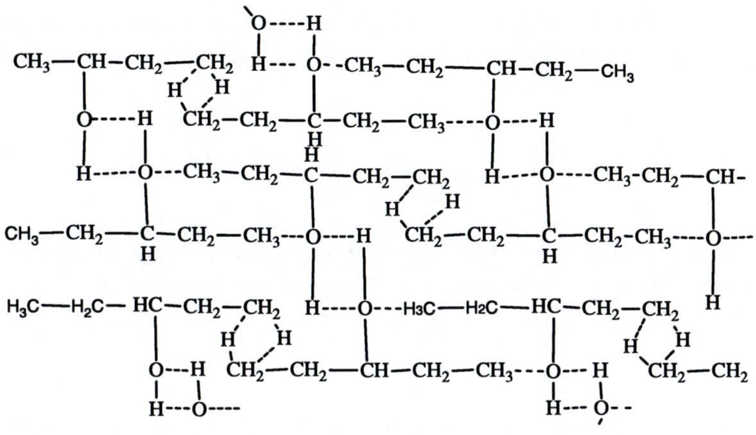

Fig. 5.10 Schematic picture of the liquid structure of 3-pentanol

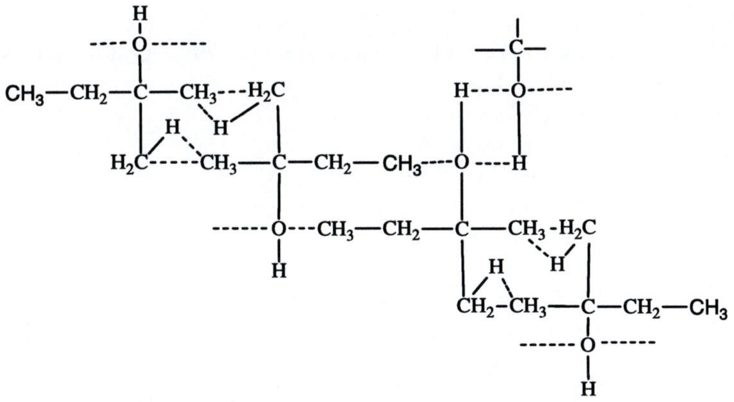

Fig. 5.11 Schematic picture of the liquid structure of 2-methyl-2-butanol

The specific interactions are formed by the second propyl fragment of the 3-pentanol in the same way as it occurs in ethers and alcohols and in the case under consideration at the three carbon atoms in the alkyl fragment the energy of this interaction should be taken equal to 8.92 kJ mol^{-1} as the specific interaction in the dipropyl ether. Taking the energy of the hydrogen bond equal to that in liquid 2-butanol and the contribution of the methylene group into the vaporization enthalpy equal 4.75 kJ mol^{-1} we can write (5.8a)

$$\mathrm{DH_3C \rightarrow H - CH_2} = (\Delta_{\mathrm{vap}}H^0(298) - 2\mathrm{DO - H \cdots}$$
$$\mathrm{O - H} - 2\mathrm{D - O \rightarrow CH_3 - CH_2 - CH_2 - DCH_2})/2\,. \qquad (5.8\mathrm{a})$$

The obtained value of the energy of the specific interaction $H_3C \rightarrow H\text{–}CH_2$ of low stability amounts to 0.65 kJ mol^{-1} (Table 5.9). Taking into account all interactions existing in the liquid 3-pentanol and the energy contribution of CH_2-groups into the vaporization enthalpy we can write an equation analogous to (5.8a)

$$\begin{aligned} &DH_3C \rightarrow H - CH_2 = (\Delta_{vap}H^0(298) - 2DO - H\cdots \\ &O - H - 2D - O \rightarrow CH_3 - CH_2 - CH_2 - DCH_2)/2\,. \end{aligned} \qquad (5.8b)$$

After calculations using the above-mentioned energies of the hydrogen bond and the specific interaction formed by the propyl fragment of the 2-methyl-2-butanol we obtained the values of the energy of the specific interaction of low stability formed by the isostructural and terminal methyl groups of 2-methyl-2-butanol (Table 5.9).

$$\begin{aligned} &DH_3C \rightarrow H \\ &\quad = \left(\Delta_{vap}H^0(298) - 2DH - O\cdots H - O - 2D - O \rightarrow CH_3 - CH_2 - CH_2\right)/4. \end{aligned} \qquad (5.10)$$

As seen from Table 5.8, the calculated energies of the low stable interactions are essentially lower at the location of the methyl and hydroxy groups at the second carbon atom of 2-methyl-2-butanol compared with the energy of the analogous interaction in 2-butanol (0.72 kJ mol^{-1}).

The thermodynamic analysis of structures of liquid alcohols related to 1-pentanol and of the specific intermolecular interactions and hydrogen bonds existing in them made it possible to estimate the versatile features of these interactions, their number, and the contribution of the methylene groups into the vaporization enthalpy. In the performed calculation procedure, we used the notion on the equality of the energies of the specific intermolecular interactions in liquid methanol, ethanol, 1-propanol, branched alcohols on the one hand, and on the other hand in liquid dimethyl, diethyl, or dipropyl ether, respectively. At the number of carbon atoms in the chain exceeding three we additionally accounted for the contribution of the methylene groups into the vaporization enthalpy. Note that the alcohols of the homologous series 2-alkanols starting with 2-butanol do not form dimers in vapor. Therefore, the obtained energies of the hydrogen bonds are real, and the taken condition of its constant value in derivatives of 2-butanol and 3-pentanol in the calculations of the energies of the specific interactions of low stability $H_3C \rightarrow H\text{–}CH_2$ is well founded.

Thus, the performed calculations underlain by the thermodynamic analysis of structures of liquid alcohol related to 1-pentanol and of their vaporization enthalpy yielded the valid value of the energy of the hydrogen bond (14.85 kJ mol^{-1}).

The results of the performed thermodynamic analysis and the calculated energies contributed by specific interactions $H_3C \rightarrow H\text{–}CH_2$ formed by isostructural methyl group show their low stability in the alcohols of 1-propanol family. The interaction energies of the liquid derivatives of 1-propanol have smaller values than those of 1-butanol derivatives. Their common regular series reflects the trend of decrease in their stability

2 - Methyl - 1 - propanol (1.25) > 2 - Butanol (1.15) >
2 - Methyl - 2 - propanol (1.05) > 1 - Propanol (1.00) > 2 - Propanol (0.58) >
2 - Methyl - 1 - butanol (0.50) > 2 - Methyl - 2 - butanol (0.40 kJ mol^{-1}).

5.3 Role of Isostructural Functional Groups in Saturated Alcohols and Specific Interactions

No less interesting is revealing the role of hydroxy group and isostructural methyl groups at the large number of carbon atoms in the alkyl chain. To clear these questions we use the most reliable thermodynamic data obtained by calorimetry and by the static method with the use of membrane zero-manometer and compiled in the most reliable published sources [21, 47, 48]. The shortening by one methylene group of the interacting alkyl fragment with the conservation of the overall number of carbon atoms in the molecule of 2-methyl-1 propanol compared to 1-butanol results in the decrease in the vaporization enthalpy of the former by 2.50 kJ mol^{-1} (Table 5.8). The location at this carbon atom of the hydroxy group in the 2-butanol diminished the enthalpy characteristic by 2.56 kJ mol^{-1}. The lengthening of the carbon skeleton in the 1-pentanol molecule and the attachment of the isostructural methyl group to the third carbon atom results in the decrease of the enthalpy characteristic by 2.60 kJ mol^{-1}, and the presence of a hydroxy group at the same carbon atom in the 3-pentanol leads to the vaporization enthalpy decrease by 3.34 kJ mol^{-1}.

The comparison of the thermodynamic characteristics of vaporization and sublimation is based in the published sources on the traditional approach underlain by the steric effect of the methyl and hydroxy groups.

The presence in the nearest surrounding of two molecules with analogous groups leads to the interaction between these groups which has been considered inert. Both linear and branched organic molecules readily change the position of their chains possessing free bond vacancies forming the specific intermolecular interactions. Therefore, the basis of the condensed state is the interspecies interaction which is not easy to reveal and to establish its physicochemical characteristics. We showed in the preceding chapters that the decrease in the vaporization and sublimation enthalpy in the compounds possessing isostructural methyl groups as compared with the compounds of the normal structure was due to the replacement of a stable specific intermolecular interaction by a similar less stable interaction. For instance, it is observed in the replacement of the specific interaction formed by a propyl group in 1-propanol by that of an ethyl group in 2-propanol and a CH_3-group with the same essentially unshared $2s^2$ electron pair of the carbon atom but with lesser acceptor and donor properties. In turn, the isostructural methyl groups are involved in the intermolecular interactions and provide a contribution into the enthalpy characteristic whose energy also affects the difference in the properties of the compounds in question. A considerable decrease in the enthalpy characteristics is caused by the replacement by isostructural methyl groups with their essentially low energy contribution into the vaporization and sublimation enthalpy of CH_2-groups that each one contributes 4.7 kJ mol^{-1} at their number in the alkyl chain exceeding four.

The alcohols with branched molecules have lower vaporization enthalpy due to the effect of methyl and hydroxy groups on the shift of the electron density in their molecules and on the distribution of the electron density and charges on the atoms involved in the existing intermolecular interactions. Therefore, it is possible to state that the valid interpretation of the thermodynamic properties can be obtained only on the condition of getting information on the energies of the specific interaction. However, the vaporization enthalpy of the majority of the branched alcohols was determined at temperatures differing from the standard conditions. Hence only for a small number among them it is possible to establish the hydrogen bond energies and the energies of the specific interactions of low stability formed by the isostructural methyl groups. In this connection, let us first describe the characteristics of the thermodynamic process of vaporization.

The maximum influence of the isostructural methyl group on the vaporization enthalpy characterizing the energies of specific interactions and hydrogen bonds is observed at the location of these groups at the second carbon atoms in the derivatives of 1-alkanols: 2-methyl-1-pentanol, 2-methyl-1-hexanol, and 2-methyl-1-heptanol.

At the location of the hydroxy group at the second carbon atom in the derivative of 2-hexanol or at the third carbon atom in the derivative of 3-heptanol, the isostructural methyl group causes the maximum decrease in enthalpy characteristic when both the hydroxy and methyl groups are linked to the same carbon atom:

$$\begin{aligned}&2\text{-Methyl-2-pentanol}\ (54.7,\ 298\,\text{K})\\&\quad = 3\text{-Methyl-3-hexanol}\ (54.7(355\text{K})\,\text{kJ}\,\text{mol}^{-1}).\end{aligned}$$

However in the longer chain of octanols, the maximum effect of decreasing the vaporization enthalpy by the isostructural methyl group was observed at its location at the next carbon atom: 3-methyl-2-heptanol (48.0, 351 K), 4-methyl-3-heptanol (48.0, 366 K), or at the previous carbon atom in the 2-methyl-3-heptanol (54.8, 367 K kJ mol^{-1}) when the hydroxy group is attached to the middle carbon atom of the chain.

The presence of two isostructural methyl groups in the alcohol structure produces the maximum decrease in the enthalpy characteristic when they are attached to the carbon atom next to that bearing the hydroxy group, like in the alcohols

$$\begin{aligned}&2,2\text{ - Dimethyl - 1 - butanol}\ (53.7, 313\,\text{K})\\&\quad = 3,3\text{ - Dimethyl - 2 - butanol}\ (53.6\,\text{kJ}\,\text{mol}^{-1}(298\,\text{K}))\end{aligned}$$

or at location of these groups at the carbon atom previous to that linked with the hydroxy group, as in 2,2-dimethyl-3-pentanol (51.4, 338 K kJ mol^{-1}). This regularity is conserved in trimethyl-substituted octanols where the vaporization enthalpy grows in the series:

$$2,2,3\text{ - Trimethyl - 3 - pentanol }(47.3,\ 333\,\mathrm{K})$$
$$<2,2,4\text{ - Trimethyl - 3 - pentanol}(57.1,\ 343\,\mathrm{K})$$
$$<2,4,4\text{ - Trimethyl - 3 - pentanol }(60.6.328\,\mathrm{K\ kJ\,mol^{-1}}).$$

In the alcohols of the normal structure, the electron density shift occurs in a single chain and the specific intermolecular interaction is formed by the terminal methyl groups with the contacting molecules of the nearest surrounding. In the branched alcohols with the location of the oxygen atom, e.g., at the second carbon atom like in 2-butanol the electron density shifts simultaneously from two terminal methyl groups.

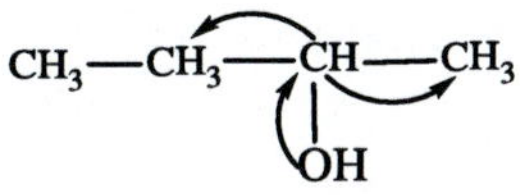

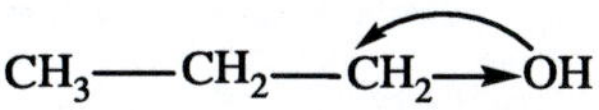

However, the reverse dative bond in the short methyl fragment essentially fills up the electron deficit on the carbon atom of this group. Consequently, this isostructural methyl group forms the specific interaction of low stability (Tables 5.8 and 5.9). In the ethyl fragment the electronic deficit is filled up to a lesser degree, therefore the enhanced charge on the carbon atom of its terminal methyl group makes it possible to form the specific interactions of higher stability than the methyl group (Table 5.9). The reverse dative bond does not affect the methyl group of the propyl fragment of the 2-butanol and moreover the butyl fragment of the 2-pentanol. Therefore, the charges on the carbon atom of the methyl group are virtually unchanged, and they form the specific interactions with the energy equivalent to that of the analogous interaction in liquid dipropyl ether (2DH–O → CH_3–CH_2–CH_2–). As a result the less reactive methyl group attached to the carbon atom bearing the hydroxy forms a specific interaction of low stability with the hydrogen atoms in the *trans*-position in analogous group's 2D*iso*H_3C → H–CH_2.

Due to the same number of carbon atoms in the molecules of 1-butanol and 2-butanol, it is clear that the vaporization enthalpy of the latter decreased by the value of the energy contribution of the methylene group (4.75 kJ mol^{-1}) is incompletely compensated by the specific interaction of low stability. As already mentioned, the overall contribution of two interacting CH_3-groups of the molecules of the nearest surrounding is equal to the difference between the vaporization enthalpies of 2-butanol and 1-propanol or 2-pentanol and 1-butanol, respectively. This condition is valid for liquid and solid alcohols 2-hexanol and 1 pentanol, 2-heptanol and 1-hexanol, etc.

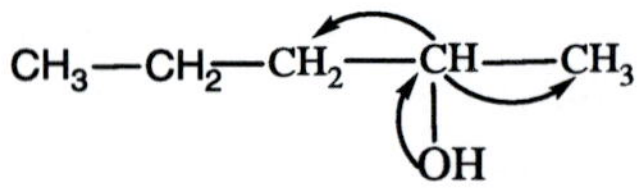

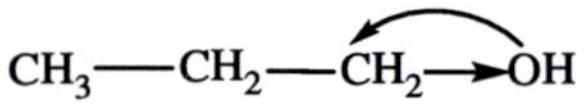

The energy of the formed specific interaction of the terminal methyl group of the propyl group does not differ from the energy of the analogous type of the specific interaction in the dipropyl ether. Therefore, taking into account the contribution into the vaporization enthalpy of the CH_2-group we obtain the equation for evaluation of the energy of the hydrogen bond in liquid 2-butanol or accounting for the larger number of the methylene groups ($n > 1$), for evaluation the energy of the hydrogen bond in 2-pentanol, 2-hexanol, 2-heptanol, etc.

$$\begin{aligned} DH-O\cdots H-O = (&\Delta_{vap}H^0(298) - 2D-O \\ &\rightarrow CH_3-CH_2-CH_2-2DCH_3 \\ &\rightarrow H-CH_2-D(CH_2)n/2. \end{aligned} \quad (5.11)$$

In the molecules of 3-pentanol, 3-hexanol, 3-heptanol, 3-octanol, and the other alcohols of this homologous series the smallest alkyl fragment is the ethyl group, and the overall contribution into the vaporization enthalpy is also equal to the difference between the enthalpy characteristics of 3-pentanol and 1-propanol or 3-hexanol and 1-butanol.

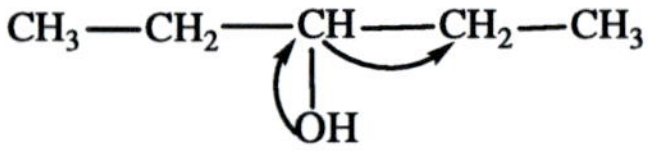

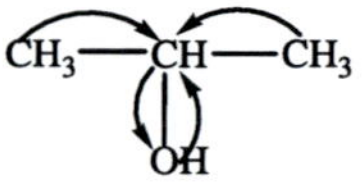

$$2DC_2H_5 = \Delta_{vap}H^0(298)(3\text{–}\Pi) - \Delta_{vap}H^0(T)1\text{–}\Pi p. \quad (5.12)$$

Accounting for the number of CH_2-groups [49] the energy of the hydrogen bond is determined by the equation

$$\begin{aligned} DH-O\cdots H-O = (&\Delta_{vap}H^0(298) - 2D-O \\ &\rightarrow CH_3-CH_2-CH_2-2DH_3C \\ &\rightarrow H-CH_2D(CH_2)n)/2. \end{aligned} \quad (5.13)$$

In 4-octanol, the overall contribution of the propyl fragment is determined from the vaporization enthalpy difference of this alcohol and 1-pentanol, and consequently the energy of the hydrogen bond is calculated by the equation

$$\begin{aligned}\mathrm{DH-O\cdots H-O} &= (\Delta_{\mathrm{vap}}H^0(298) - 2\mathrm{D-O} \\ &\rightarrow \mathrm{CH_3 - CH_2 - CH_2 - 2DCH_3} \\ &\rightarrow \mathrm{H - CH_2 - D(CH_2)}n/2. \end{aligned} \tag{5.14}$$

From the performed thermodynamic calculations of the energies of hydrogen bonds and of specific intermolecular interactions of low stability (Table 5.10), the following conclusions can be made:

– The energy of the hydrogen bonds in liquid alcohols of the normal structure

$$1\text{ - Butanol }(14.80) = 1\text{ - Pentanol }(14.80) = 1\text{ - Hexanol }(14.82)$$
$$\approx 1\text{ - Heptanol }(15.00) \approx 1\text{ - Octanol }(14.70\,\mathrm{kJ\,mol^{-1}})$$

is practically constant and its variations are within the error of the vaporization enthalpy experimentally measured by calorimetric method.

– The energies of the hydrogen bonds in the liquid isomeric alcohols with different position of the hydroxy group

$$1\text{ - Hexanol }(14.80) = 2\text{ - Hexanol }(14.80) = 3\text{ - Hexanol }(14.88\,\mathrm{kJ\,mol^{-1}})$$

are practically equal with the variations within the error of the vaporization enthalpy experimentally measured for this compounds by calorimetric method.

– The energies of the specific intermolecular interactions of low stability formed by the terminal methyl group of the shorter alkyl fragment or by ethyl group are described by the series of their regular variation

$$\mathrm{DCH_3}\text{:}\quad 2\text{ - Hexanol }(0.68) < 2\text{ - Butanol }(0.80) < 2\text{ - Pentanol }(0.95)$$
$$< 2\text{ - octanol }(1.95,\ 268\,\mathrm{K}) < 2\text{ - Heptanol }(2.20,\ 259\,\mathrm{K},\ \mathrm{kJ\,mol^{-1}})$$

$$\mathrm{DC_2H_5}\text{:}\quad 3\text{ - Pentanol }(2.65) < 3\text{ - Hexanol }(3.15) < 3\text{ - Octanol }(4.40,\ 268\,\mathrm{K})$$
$$< 3\text{ - Heptanol }(5.45,\ 259\,\mathrm{K},\ \mathrm{kJ\,mol^{-1}}).$$

The understanding of the influence of the isostructural methyl groups' position in the alkyl chain of alcohols can be derived from the values of the energy of specific interactions of low stability formed by these groups and of the hydrogen bond energies. The energy of these specific interactions can be evaluated by comparison of the properties of the alcohol molecule with a normal chain and an alcohol with the same chain with isostructural methyl group attached to it (Table 5.11), for instance, 1-pentanol and 2-methyl-1-pentanol or 1-hexanol and 3-methyl-1-hexanol.

The difference in the vaporization enthalpy of each of these pairs of alcohols is due to the energy contribution from the isostructural methyl groups, and consequently,

Table 5.10 Energies of hydrogen bonds and specific interactions (kJ mol^{-1}) in liquid alcohols

Alcohol	Structure	$\Delta_{vap}H^0(T)$ [21]	T (K)	$D(CH_2)\cdot n$	DCH_3	$D\text{–}O \rightarrow CH_3(CH_2)n$	$DHO\cdots H\text{–}O$
1-Butanol	$CH_3\text{-}CH_2\text{-}CH_2\text{-}CH_2\text{-}OH$	53.33	298	$n = 1$ 4.75	$\sum = 1.90:2 = 0.95$	$n = 3$ 8.92	14.80
2-Butanol	$CH_3\text{—}CH_2\text{—}CH(OH)\text{—}CH_3$	49.74	298	–	$\sum = 1.57:2 = 0.80$	$n = 3$ 8.92	14.84
1-Pentanol	$CH_3\text{-}CH_2\text{-}CH_2\text{-}CH_2\text{-}CH_2\text{-}OH$	56.94	298	$n = 2$ 4.75×2	–	$n = 3$ 8.92	14.80
2-Pentanol	$CH_3\text{—}CH_3\text{—}CH_3\text{—}CH(OH)\text{—}CH_3$	54.2	298	$n = 1$ 4.75	$\sum = 1.90:2 = 0.95$	$n = 3$ 8.92	14.80
3-Pentanol	$CH_3\text{—}CH_2\text{—}CH(OH)\text{—}CH_2\text{—}CH_3$	53.6	298	$n = 1$ 4.75	$\sum = 2.0:2 = 1.0$	$n = 3$ 8.92	14.85
1-Hexanol	$CH_3\text{-}CH_2\text{-}(CH_2)_3\text{-}CH_2\text{-}OH$	61.72	298	$n = 3$ 4.75×3	–	$n = 3$ 8.92	14.82
2-Hexanol	$CH_3\text{—}(CH_2)_3\text{—}CH(OH)\text{—}CH_3$	58.3	298	$n = 2$ 4.75×2	$\sum = 1.36$ 0.68	$n = 3$ 8.92	14.80

3-Hexanol	$CH_3-CH_2-CH_2-CH(OH)-CH_2-CH_3$	58.6	298	$n = 1$ 4.75	$\sum C_2H_5 = 6.30$ 3.15	$n = 3$ 8.92	14.88
1-Heptanol	CH_3-CH_2-$(CH_2)_4$-CH_2-OH	66.8	298	$n = 4$ 4.75×4	–	$n = 3$ 8.92	14.83
2-Heptanol	$CH_3-(CH_2)_4-CH(OH)-CH_3$	66.1	259	$n = 3$ 4.75×3	$\sum = 4.38$ 2.20	$n = 3$ 8.92	15.06
3-heptanol	$CH_3-(CH_2)_2-CH_2-CH(OH)-CH_2-CH_3$	67.0	259	$n = 2$ 4.75×2	$\sum C_2H_5 = 10.9$ 5.45	$n = 3$ 8.92	14.40
1-Octananol	CH_3-CH_2-$(CH_2)_5$-CH_2-OH	71.1	298	$n = 5$ 4.75×5	–	$n = 3$ 8.92	14.70
2-Octananol	$CH_3-(CH_2)_5-CH(OH)-CH_3$	70.7	268	$n = 4$ 4.75×4	$\sum = 3.90$ 1.95	$n = 3$ 8.92	15.06
3-Octananol	$CH_3-(CH_2)_4-CH(OH)-CH_2-CH_3$	71.6	268	$n = 3$ 4.75×3	$\sum C_2H_5 = 9.80$ 4.90	$n = 3$ 8.92	14.80

Table 5.11 Energies of hydrogen bonds and specific interactions (kJ mol^{-1}) of liquid saturated alcohols with isostructural methyl groups

Alcohol	Structure	$\Delta_{vap}H^0(T)$ [21]	T (K)	$D(CH_2)\cdot n$	$D\text{–}O \rightarrow CH_3(CH_2)n$	$D iso H_3C \rightarrow H\text{–}CH_2$	$DHO\cdots H\text{–}O$
2-Methyl-3-pentanol	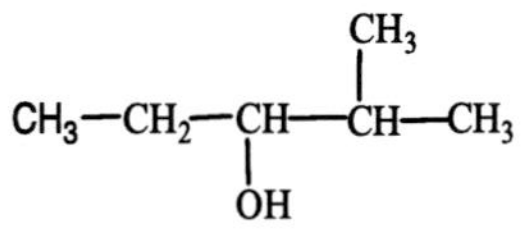	59.4	298	$n = 2$ 4.75 × 2	8.92	$\sum = 2.46$ 1.23	14.80
3-Methyl-1-pentanol	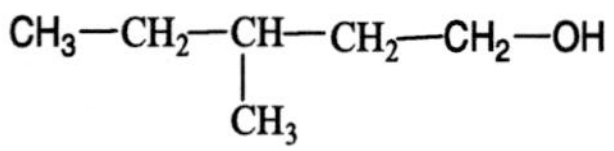	61.7	298	$n = 2$ 4.75 × 2	8.92	$\sum = 4.8$ 2.46	14.80
2-Methyl-2-pentanol	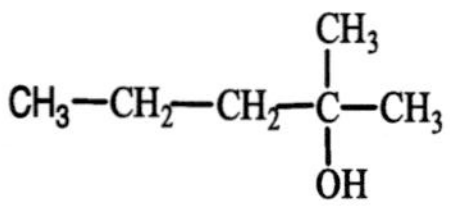	54.7	298	$n = 2$ 4.75 × 2	8.92	$\sum = 2.37$ 1.20	14.87
2-Methyl-3-pentanol	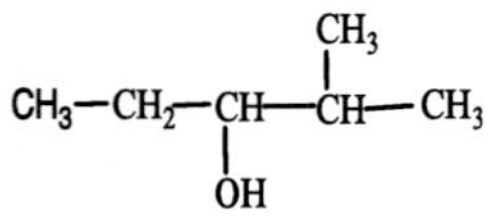	56.7	298	$n = 1$ 4.75	8.92	$\sum = 3.1$ 1.55	15.23
3-Methyl-3-pentanol	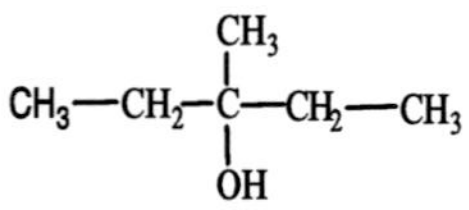	55.7	298	$n = 1$ 4.75	8.92	$\sum = 2.1$ 1.05	14.90

2,2-Dimethyl-1-butanol	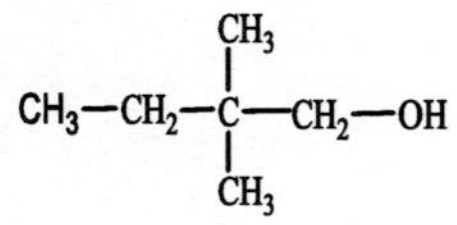	53.7	313	$n = 1$ 4.75	8.92	$\sum = 1.40$ 0.35	14.90
3,3-Dimethyl-1-butanol	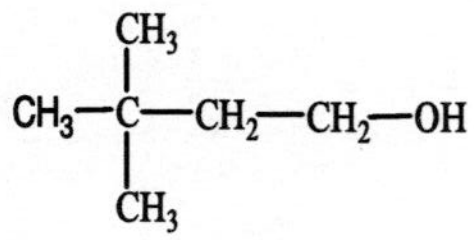	58.0	298	$n = 1$ 4.75	8.92	$\sum = 5.67$ 1.42	14.90
2,3-Dimethyl-2-butanol	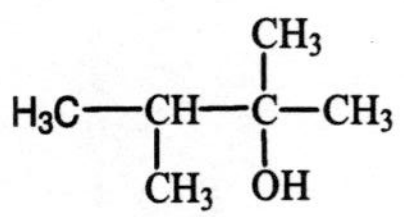	54.0	298	$n = 1$ 4.75	8.92	$\sum = 4.26$ 1.06	14.82
3,3-Dimethyl-2-butanol	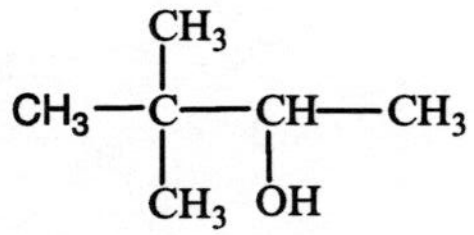	53.6	298	$n = 1$ 4.75	8.92	$\sum = 3.86$	14.82

this contribution is equal to the difference between the vaporization enthalpies, for example, of 1-hexanol and 2-methyl-1-hexanol (5.15)

$$2DH_3C \rightarrow H - CH_2 = \Delta_{vap}H^0(T)iso- = \Delta_{vap}H^0(T)1\text{-}ol \qquad (5.15)$$

where $\Delta_{vap}H^0(T)nA$ is the vaporization enthalpy of the alcohol of the normal structure with the same number of carbon atoms in the alkyl chain.

The energy of the hydrogen bond in the condensed state of the alcohol with a methyl group attached to a long chain is evaluated by the already discussed thermodynamic calculation underlain by the above founded concept that the energy of the specific intermolecular interaction is identical in symmetric ethers and alcohols. Therefore, we use for the alcohols under consideration the value of the specific interaction involving the propyl group equals 8.92 kJ mol^{-1}. Obviously, the energy contribution of the methylene groups into the vaporization enthalpy should also be taken into account. Consequently, the hydrogen bond energy is the difference between the vaporization enthalpy and the sum of all existing specific interactions.

$$\begin{aligned} DO - H \cdots O - H = (&\Delta_{vap}H^0(T)iso - 2D - O \\ &\rightarrow CH_3 - CH_2 - CH_2 - 2DisoH_3C \\ &\rightarrow H - CH_2 - D(CH_2)n)/2. \end{aligned} \qquad (5.16)$$

The same procedure is used in the thermodynamic calculation of energies of the specific intermolecular interactions of low stability and of the hydrogen bonds in the condensed state for alcohols with two methyl groups in the side chain. The difference between the vaporization enthalpy of the alcohol with the isostructural methyl groups and the alcohol of the normal structure, for instance, between 2,2-dimethyl-1-butanol and 1-butanol, corresponds to four methyl groups in the molecules of the nearest surrounding. Thus, the energy of the hydrogen bond in the alcohol with two isostructural methyl groups and a hydroxy group at the first carbon atom is calculated by the equation

$$\begin{aligned} DO - H \cdots O - H = (&\Delta_{vap}H^0(T)iso - 2D - O \\ &\rightarrow CH_3 - CH_2 - CH_2 - 4DisoH_3C \\ &\rightarrow H - CH_2 - D(CH_2)n)/2. \end{aligned} \qquad (5.17)$$

As we repeatedly stated before, the position of the hydroxy group at different carbon atoms of the alkyl chain essentially affects the capability of the methyl group in the shorter fragment of the chain to form the specific interactions. Therefore, at the presence in the alcohol molecule of an isostructural methyl group and a hydroxy group at their various reciprocal positions the versatile possibilities should be taken into account for the formation of the specific interactions both by the isostructural group and the terminal CH_3-group. Certain possibilities for the thermodynamic calculations of the hydrogen bond energies arise from the above found energies of the specific interactions of low stability formed by the methyl groups of alcohols with the same carbon skeleton and varying position of the hydroxy group in the chain, in particular, 1-pentanol, 2-pentanol, and 3-pentanol.

The analysis of the structure of molecules of 2-methyl-3-pentanol and 3-pentanol shows again that not only their structures are similar, but, moreover, the overall contribution of the energy of the specific interactions of low stability formed by the isostructural methyl groups into the vaporization enthalpy of 2-methyl-3-pentanol is equal to the difference between the vaporization enthalpy of this compound and of 3-pentanol.

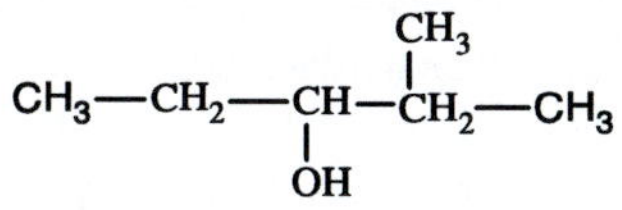

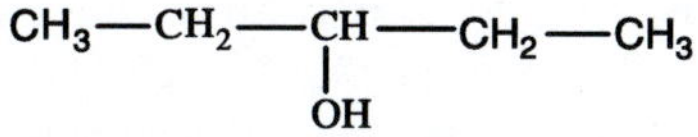

Hence from this value the energy of the specific interaction D*iso*$H_3C \rightarrow H–CH_2$ is obtained by the simple division by the number of such bonds. The next unknown value is the energy of the hydrogen bond. However, the above analysis showed that the energy of hydrogen bond of this class of alcohols in the condensed state is obtained from the equation

$$\begin{aligned} DO - H \cdots O - H = (\Delta_{vap}H^0(298)iso - 2D - O \\ \rightarrow CH_3 - CH_2 - CH_2 - k\,DisoH_3C \\ \rightarrow H - CH_2 - 2DCH_3 \rightarrow H - CH_2 - D(CH_2)n)/2, \quad (5.18) \end{aligned}$$

where $\Delta_{vap}H^0(T)$iso is the vaporization enthalpy of, e.g., 2-methyl-3-pentanol, and (k) is the number of specific interactions of the type D*iso*$H_3C \rightarrow H–CH_2$. The results of thermodynamic calculations (Table 5.11) demonstrate, firstly, the low energy of the specific intermolecular interactions formed by the isostructural methyl groups and, secondly, the insignificant difference in the energy of the hydrogen bonds in the liquid alcohols with one and two isostructural methyl groups with the alkyl chains of 1-pentanol, 2-pentanol, and 3-pentanol, 1-butanol and 2-butanol. The energies of the specific interactions of low stability tend to grow in the series

$$\begin{aligned} &2,2\text{ - Dimethyl - 1 - butanol }(0.35) < 3,3\text{ - Dimethyl - 2 - butanol }(0.96) \\ &\quad < 3\text{ - Methyl - 3 - pentanol }(1.05) < 2,3\text{ - Dimethyl - 2 - butanol }(1.06) \\ &\quad < 2\text{ - Methyl - 2 - pentanol }(1.20) \approx 2\text{ - Methyl - 1 - pentanol }(1.20) \\ &< 3,3\text{ - Dimethyl - 1 - butanol }(1.40) < 3\text{ - Methyl - 1 - pentanol }(2.40\,\text{kJ mol}^{-1}). \end{aligned}$$

This series corresponds to the energy variation in the specific interactions $DCH_3 \rightarrow H–CH_2$ in the range 0.35–2.40 kJ mol^{-1} with maximum values in

3,3-dimethyl-1-butanol (1.40) and 3-methyl-1-pentanol (2.40 kJ mol^{-1}). Thus, the location of the isostructural methyl group at the third carbon atom of the alkyl chain, most remote from the hydroxy group results in the maximum stability of this type specific interaction. It is therefore possible to conclude that the isostructural methyl group does not suffer from the influence of the reverse dative bond directed from the oxygen atom. Therefore, the energy of the specific interaction formed by the isostructural methyl group provides the value of the vaporization enthalpy equal to that of 1-hexanol (61.72 kJ mol^{-1}). A similar sequence in the variation of the vaporization enthalpy is observed in 2-methyl −1-hexanol (53.5) and 3-methyl-1-hexanol (57.4 kJ mol^{-1}). The reason of this trend is the participation of the isostructural CH_3-group in the shift of the electron density, and it is affected by the reverse dative bond directed from the oxygen atom. This occurs also under the condition that the methyl and hydroxy groups are attached to the same carbon atom. As a result the charge on the carbon atom of this methyl group changes, and this group forms less stable specific interactions of the type $D\mathit{iso}H_3C \rightarrow H–CH_2$. Therefore, the vaporization enthalpies of the alcohols under consideration can be of similar value. This conclusion follows from the comparison of the vaporization enthalpy of alcohols of the series:

$$\begin{aligned}&2\text{ - Hexanol }(58.3) > 2\text{ - Methyl - 2 - pentanol }(54.7)\\&\quad < 2\text{ - Methyl - 3 - pentanol }(56.7) > 3\text{ - Methyl - 3 - pentanol }(55.7)\\&\quad < 3\text{ - Hexanol }(58.6\,\text{kJ mol}^{-1}).\end{aligned}$$

Consequently, the vaporization enthalpy of 2-methyl-2-pentanol and 3-methyl-3-pentanol have the low values with respect to the other alcohols of this series. The somewhat higher value of the vaporization enthalpy of the 2-methyl-3-pentanol should be ascribed to the participation of the methyl group in the shift of the electron density and in its sensitivity to the reverse dative bond resulting in the enhanced stability of its specific interaction

$$\begin{aligned}&2\text{ - Methyl - 2 - pentanol }(1.20) < 2\text{ - Methyl - 3 - pentanol }(1.55)\\&\quad > 3\text{ - Methyl - 3 - pentanol }(1.05\,\text{kJ mol}^{-1}).\end{aligned}$$

Analogous regularity should appear also in the derivatives of 3-hexanol and therefore it should be regarded as a general rule.

It is not possible to follow the regularities in the energy variation of the analogous specific interactions in the derivatives of 1-octanol for their vaporization enthalpies has been measured at different temperatures in the range 365–383 K (Table 5.12). The growing number of carbon atoms in the alkyl chain of 2-octanol derivatives provides additional information in the already formulated order of variation in the vaporization enthalpy of alcohols since the growing number of compounds in the homologous series is reflected in the appearance of finer correlation between the structure and the thermodynamic properties.

Table 5.12 Vaporization enthalpy (kJ mol^{-1}) of methyl derivatives of heptanol and octanol

Alcohol	Structure	$\Delta_{vap}H^0(T)$ [21]	T (K)	Alcohol	$\Delta_{vap}H^0(T)$ [21]	T (K)
1-Hexanol	CH_3—$(CH_2)_3$—CH_2—CH_3—OH	55.2 ± 0.2	358	3-Pentanol	46.4	353
2-Methyl-1-hexanol	CH_3—$(CH_2)_3$—CH(CH_3)—CH_3—OH	53.5	390	2-Methyl-3-Hexanol	55.7	338
3-Methyl-1-hexanol	CH_3—$(CH_2)_3$—CH(CH_3)—CH_2—OH	57.4	399	3-Methyl-3-hexanol	53.6	338
1-Heptanol	H_3C—H_2C—$(CH_2)_3$—CH_2—CH_2–OH	62.6	351	2-Heptanol	54.4	366
2-Methyl-1-heptanol	CH_3—$(CH_2)_4$—CH(CH_3)—CH_2—OH	53.3	366	2-Methyl-2-heptanol	55.0	354
3-Methyl-1-heptanol	CH_3—$(CH_2)_2$—CH_2—CH(CH_3)—CH_2—CH_2—OH	53.4	375	3-Methyl-2-heptanol	48.0	357

(continued)

Table 5.12 (continued)

Alcohol	Structure	$\Delta_{vap}H^0(T)$ [21]	T (K)	Alcohol	$\Delta_{vap}H^0(T)$ [21]	T (K)
4-Methyl-1-heptanol	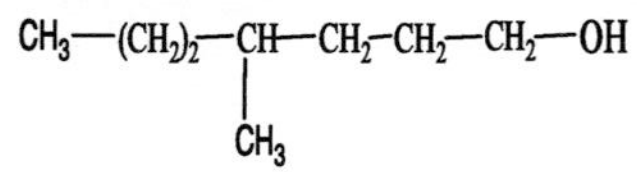	56.7	367	4-Methyl-2-heptanol	54.2	366
5-Methyl-1-heptanol	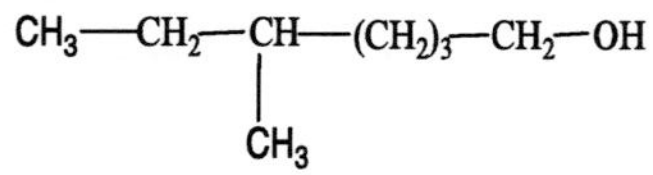	57.8	379	5-Methyl-2-heptanol	51.4	363
6-Methyl-1-heptanol	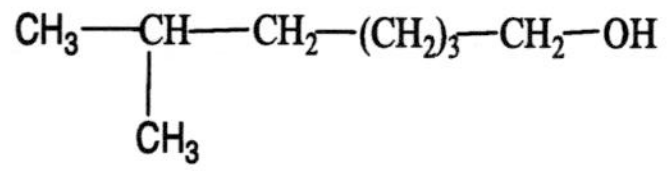	61.0	383	6-Methyl-2-heptanol	55.2	369
3-Heptanol	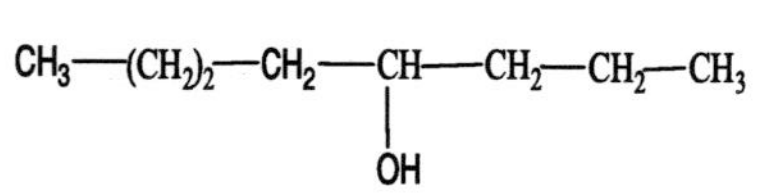	53.1	364	4-Methyl-3-heptanol	43.9	345
2-Methyl-3-heptanol	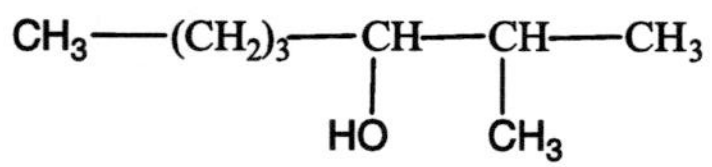	54.8	367	5-Methyl-3-heptanol	46.5	345
3-Methyl-3-heptanol	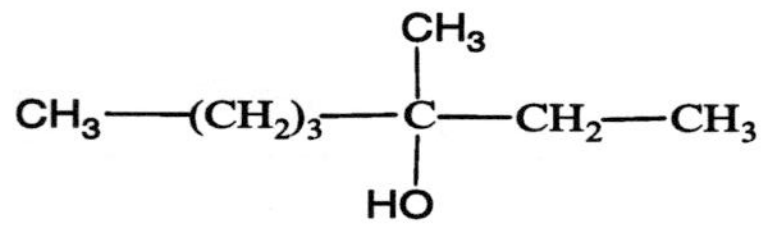	54.7	353	6-Methyl-3-heptanol	47.6	348

4-Octanol	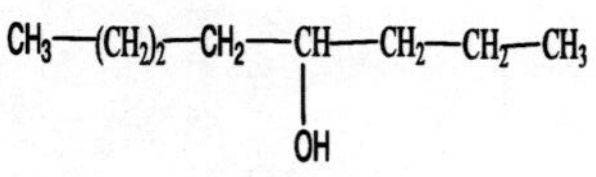	57.3	358	3-Methyl-4-heptanol	48.0	355
2-Methyl-4-heptanol	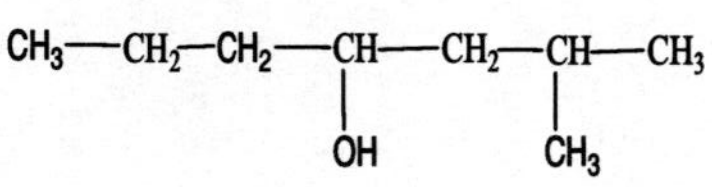	56.3	360	4-Methyl-4-heptanol	54.8	345

$$3\text{-Methyl-2-heptanol}\ (48.0,\ 356\,\mathrm{K}) < 5\text{-Methyl-2-heptanol}\ (51.4,\ 363\,\mathrm{K})$$
$$< 4\text{-Methyl-2-heptanol}\ (54.2, 366\ \mathrm{K})$$
$$< 2\text{-Methyl-2-heptanol}(55.6, 354\ \mathrm{K})$$
$$< 6\text{-Methyl-2-heptanol}\ (55.2\,\mathrm{kJ\,mol^{-1}}, 369\ \mathrm{K}),$$

$$2\text{-Methyl-2-heptanol}\ (55.6,\ 354\,\mathrm{K}) > 3\text{-Methyl-2-heptanol}\ (48.0,\ 356\,\mathrm{K})$$
$$< 4\text{-Methyl-2-heptanol}\ (54.2,\ 366\,\mathrm{K})$$
$$> 5\text{-Methyl-2-heptanol}\ (51.4,\ 363\,\mathrm{K})$$
$$< 6\text{-Methyl-2-heptanol}(55.2\,\mathrm{kJ\,mol^{-1}}, 369\,\mathrm{K}).$$

The comparison of the enthalpy characteristics of compounds in question reveals a very important feature. The structural and energy variations reflected in the vaporization enthalpies of 2-octanol derivatives are demonstrated by the sharp decrease in the enthalpy characteristics of related compounds: the vaporization enthalpy of 3-methyl-2-heptanol is by 7.0 kJ mol^{-1} lower than that of 2-methyl-2-heptanol. This drop in the enthalpy characteristic is so drastic on condition of the low stability of the specific intermolecular interactions that it indicates the decrease in the energy of the hydrogen bonds and the specific interactions. This follows also from the rough approximation in the thermodynamic calculation. Taking tentatively at 366 K the energy of the specific interaction D–O → CH_3–CH_2–CH_2 at 7.0 instead of 8.92 kJ mol^{-1}, the contribution into the vaporization enthalpy of four CH_2-groups of the 2-heptanol molecule at 19.0 kJ mol^{-1}, and the contribution of two specific interactions of low stability at 2.0 kJ mol^{-1}, then the energy of the hydrogen bond would be 8.0 kJ mol^{-1}. Although the calculation is rough, still the obtained value of the hydrogen bond energy shows the significant influence of the isostructural methyl group of the 3-methyl-2-heptanol on the stabilization of the intramolecular bonds resulting in the destabilization of the hydrogen bonds and the specific intermolecular interactions D–O → CH_3–CH_2–CH_2.

The detailed analysis of the vaporization enthalpy of 3-octanol derivatives also reveals the significant role of the isostructural methyl groups. Like the relations in the derivatives of 2-octanol, the position of the hydroxy group at the third carbon atom of the alkyl chain has the decisive and very important significance for the formation of the hydrogen bonds and various specific interactions (Fig. 5.12).

The location of the isostructural methyl group at the fourth and of hydroxy group at the third carbon atoms is accompanied with a drastic decrease in the enthalpy characteristic, and thus the vaporization enthalpy of this compound is by 10.2 kJ mol^{-1} lower than that of 2-methyl-3-heptanol and 3-methyl-3-heptanol.

$$2\text{-Methyl-3-heptanol}(54.8, 353\mathrm{K}) = 3\text{-Methyl-3-heptanol}(54.7, 353\mathrm{K})$$
$$> 4\text{-Methyl-3-heptanol}(43.6, 353\mathrm{K})$$
$$< 5\text{-Methyl-3-heptanol}(46.5, 354\mathrm{K})$$
$$< 6\text{-Methyl-3-heptanol}(47.6\mathrm{kJ\,mol^{-1}}, 348\mathrm{K}).$$

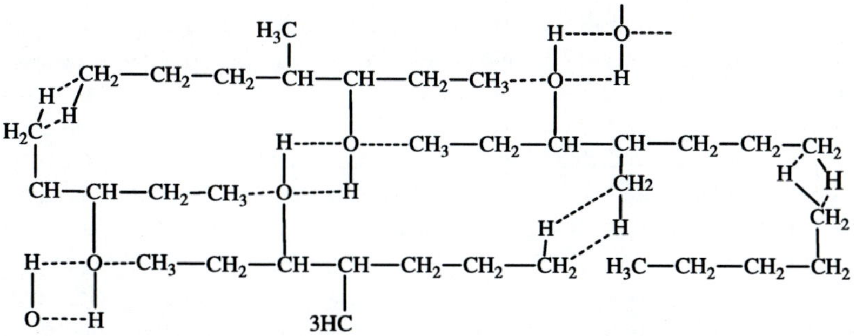

Fig. 5.12 Schematic picture of the liquid structure of 4-methyl-3-heptanol

The isostructural methyl group at the fifth and sixth carbon atoms provides slightly lesser destabilizing effect (Table 5.12). The comparison of the enthalpy characteristics of the compounds from this homologous series shows a considerable influence of the isostructural methyl group at the fourth carbon atom and of the hydroxy group at the third carbon atom on the shift of the electron density in the molecule of 4-methyl-3-heptanol which ensures the stabilization of the intramolecular bonds and destabilization of the hydrogen bonds and specific interactions. Thus, the location of the isostructural methyl group at the fourth carbon atom and of the hydroxy group at the third carbon atom in the alkyl chain of the alcohol possesses the key influence in the redistribution of the electron density in the molecule, in the change of the atomic charges, and consequently in the enthalpy characteristics of vaporization and sublimation.

Considering this conclusion from the viewpoint of the correlation between the energies of the intramolecular bonds and of the specific interactions we can state that the dependence of the vaporization enthalpy on the location of the isostructural methyl group in the alkyl chain of the members of the homologous series of 1-heptanol (Fig. 5.12) illustrates the stabilization of the latter in going to 3-methyl-1-heptanol and further to 4-methyl-1-heptanol and 5-methyl-1-heptanol. However, the similar dependence of the enthalpy characteristic in the homologous series of 2-heptanol containing side methyl groups has an opposite direction with a minimum at the vaporization enthalpy of 3-methyl-2-heptanol corresponding to the destabilization of the same types of the intermolecular interactions already at the location of the methyl group at the third carbon atom. As seen from Table 5.12, the influence of the isostructural methyl group and the hydroxy group appears even sharper at the location of these moieties at the fourth and third carbon atoms, respectively. Only later the trend appears toward some stabilization of the same types of specific interactions in 5-methyl-3-heptanol and 6-methyl-3-heptanol. Still the enthalpy characteristic of these compounds remains lower than in the 3-methyl-2-heptanol indicating the different energy of intramolecular interactions in the considered molecules of these homologous series. The data presented in Table 5.12 demonstrate that the dependence of the enthalpy characteristic on the position of the methyl and hydroxy group in the alkyl chain in the compounds of the homologous series of 3-heptanol is more pronounced that in the derivatives of 2-heptanol. It shows the stronger effect of intramolecular bonds stabilization in 4-methyl-3-heptanol.

Apparently, the possible influence of the location of the isostructural methyl and hydroxy groups at the fourth and third carbon atoms of the chain, respectively, on the second chain fragment is pronounced since the vaporization enthalpy contains the contribution from both alkyl groups situated on both sides of the hydroxy group in the molecule.

Taking into account that the shorter alkyl fragment forming the specific interaction of low stability is methyl, ethyl, or propyl group we should state that the main contribution into the vaporization enthalpy originates from the longer alkyl chain. From the energies of the specific interactions presented in Table 5.12, we estimate this contribution at about 40 kJ mol^{-1}. Such a low energy contribution of a fragment with five carbon atoms shows that the influence of the

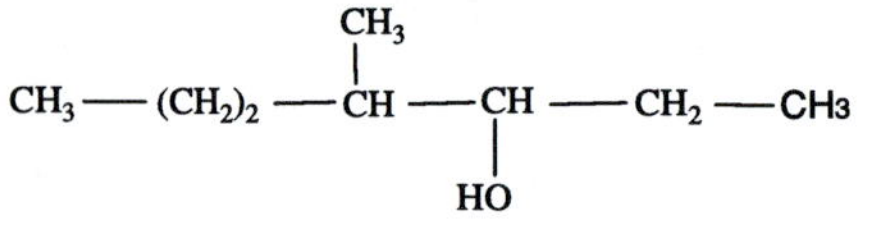

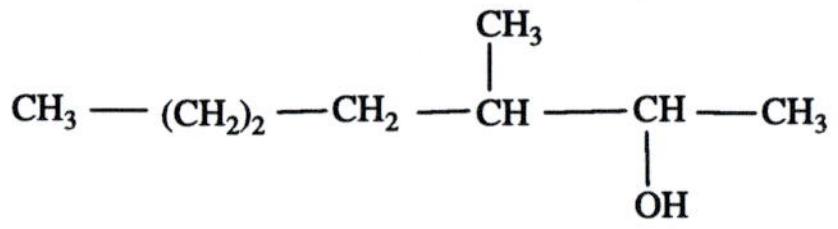

isostructural methyl and hydroxy groups at the fourth and third carbon atoms of the chain extends through the total chain. The obtained conclusion can be supplemented by the data on the effect of two and three isostructural methyl groups in the 3-pentanol derivatives. The data of Table 5.12 show that the decrease in the number of carbon atoms in the main chain of the heptanol derivatives from seven to five does not practically destabilize the hydrogen bond. Therefore, we have a possibility to estimate the energy of the specific interactions of low stability in 2.2-dimethyl-3-pentanol and 2,3-dimethyl-3-pentanol formed by the isostructural methyl groups amounting to 0.45 ($\sum = 1.80$) and 0.90 ($\sum = 3.60$) kJ mol^{-1}, respectively. In its turn the increase in the number of isostructural methyl groups and their location at the second and third carbon atoms in the 2,2,3-trimethyl-3-pentanol (Table 5.13) results in a decrease in the vaporization enthalpy by 2.3 kJ mol^{-1} making the vaporization enthalpy lower than that of 3-pentanol. This means that the isostructural methyl group at the third carbon atom destabilizes the existing specific intermolecular interactions. At the same time in compounds 2,2,4-trimethyl-3-pentanol and 2,4,4-trimethyl-3-pentanol the interactions of the same type formed by the isostructural methyl groups located at the second and fourth carbon atoms get stabilized (Table 5.13).

The replacement of a methyl group by an ethyl in 3-ethyl-3-pentanol results in the formation of a more stable specific interaction whose energy calculated with the use of experimentally established vaporization enthalpy from various sources amounts to 1.05 and 2.8 kJ mol^{-1}. As already shown the overall values of the energies of these interactions are equal to the difference between the vaporization enthalpies of 3-ethyl-3-pentanol and 3-pentanol, 2,2-diethyl-3-pentanol and 3-pentanol or, in particular, of 2,2,3-trimethyl-3-pentanol and 3-pentanol. However in the 3-ethyl-3-heptanol with seven carbon atoms in the main alkyl chain and

Table 5.13 Vaporization enthalpy and energies of specific interactions (kJ mol^{-1}) formed by side methyl and ethyl groups in liquid alcohols

Alcohols	Structure	$\Delta_{vap}H^0(T)$	T (K)	$DH_3C \rightarrow H\text{–}CH_2$
2,2-Dimethyl-3-pentanol	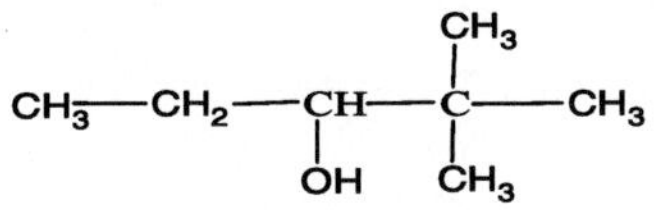	51.4	338	$\sum = 1.8$ 0.45
2,3-Dimethyl-3-pentanol	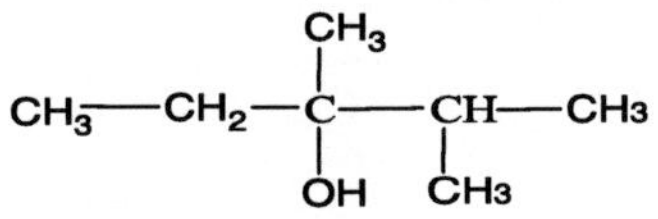	53.2	333	$\sum = 3.6$ 0.90
3-Pentanol	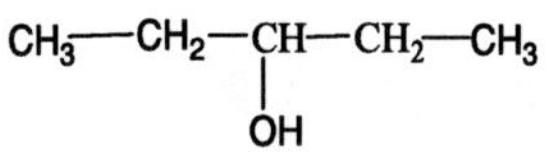	49.6	332	–
2,2,3-Trimethyl-3-pentanol	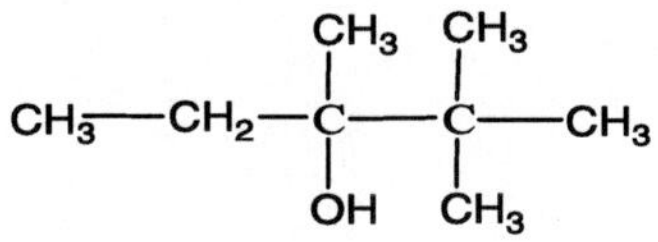	47.3	333	–
2,2,4-Trimethyl-3-pentanol	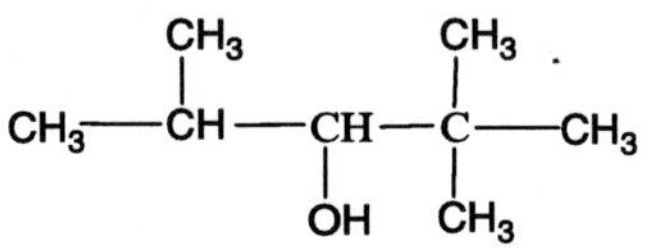	57.1	343	$\sum = 7.5$ 1.23
2,4,4-Trimethyl-3-pentanol	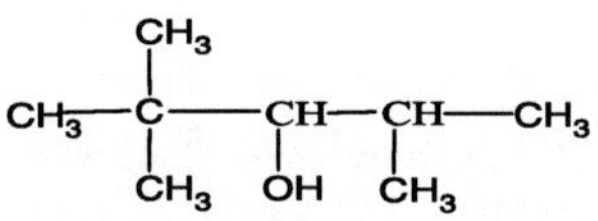	60.6	328	$\sum = 11.0$ 1.83
3-Heptanol	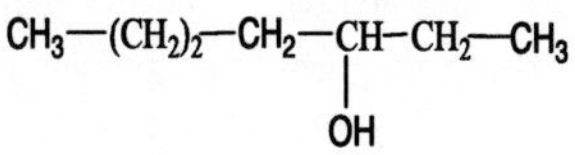	62.9	348	–
3-Ethyl-3-heptanol	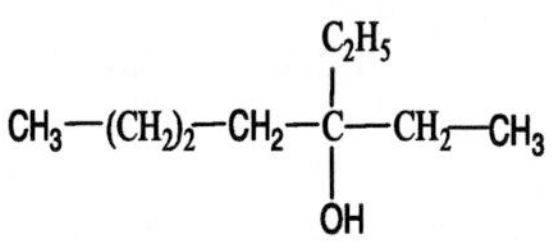	49.2	345	–

(continued)

Table 5.13 (continued)

Alcohols	Structure	$\Delta_{vap}H^0(T)$	T (K)	$DH_3C \rightarrow H\text{–}CH_2$
3-Ethyl-3-pentanol	$CH_3—CH_2—C(C_2H_5)(OH)—CH_2—CH_3$	51.3 55.2	332 323	$\Sigma = 2.2, 1.1$ $\Sigma = 5.6, 2.8$
2-Ethyl-1-hexanol	$CH_3—(CH_2)_3—CH(C_2H_5)—CH_2—OH$	60.2	362	$\Sigma = 5.0$ 2.5
1-Hexanol	$CH_3—(CH_2)_3—CH_2—CH_2—OH$	55.2 ± 0.2	358	–
2-Ethyl-1-pentanol	$CH_3—(CH_2)_2—CH(C_2H_5)—CH_2—OH$	60.2	347	$\Sigma = 6.4$ 3.2
1-Pentanol	$CH_3—(CH_2)_2—CH_2—CH_3—OH$	53.0 ± 0.2	343	–
3-Ethyl-1-butanol	$CH_3—CH(C_2H_5)—CH_2—CH_2—OH$	60.3 ± 0.2	298	$\Sigma = 7.0$ 3.5
1-Butanol	$CH_3—CH_2—CH_2—CH_2—OH$	53.3	298	–

consequently with four carbon atoms in its large fragment, the vaporization enthalpy is lower by 12.4 kJ mol^{-1} compared to the enthalpy characteristic of the 3-heptanol. This means that ethyl and hydroxy groups located both at the third carbon atom affect the electron density distribution in the molecule and significantly destabilize the hydrogen bonds and specific intermolecular interaction stabilizing simultaneously the intramolecular interactions. The observed higher vaporization enthalpy of 2-ethyl-1-hexanol and 2-ethyl-1-pentanol than that of 1-hexanol and 1-pentanol, respectively (Table 5.13) demonstrates the high stability of specific interaction formed by the ethyl group attached to the second carbon atom, and of the hydroxy group at the first carbon atom. This is evidently due to the sharply reduced influence of the ethyl group on the distribution of the electron density in the molecule. The energy of this interaction in 3-ethyl-1-butanol at the standard temperature (3.50 kJ mol^{-1}) is nearly twice lower than the energy of the specific interaction existing in the liquid diethyl ether (6.78 kJ mol^{-1}).

The conclusion on the weakening of the bonding of molecules and the stabilization of intramolecular bonds by the isostructural methyl groups in the positions four and three of the main alkyl chain and the positions three and two of hydroxy groups in the molecules of 4-methyl-3-heptanol and 3-methyl-2-heptanol, respectively, is an amazing phenomenon fundamentally important for organic chemistry and biochemistry.

The lack of published thermodynamic data for the other alcohols and the outstanding importance of the mentioned phenomenon in chemistry stimulated the thermodynamic studies of alcohols with two and three analogous fragments. The results of these studies and this phenomenon will be considered in the later paragraphs.

5.4 Energies of Specific Interactions of Liquid Polyhydric Alcohols

Thermodynamic methods are widely used in the investigation of natural and polymeric compounds [50]. At the same time their application due to several reasons is still restricted. Perhaps the main reason is the attachment to the sp^3-hybridization model. The successful thermodynamic analysis of relatively simple organic and organoelemental compounds and moreover of polyhydric alcohols is probable only at the rejection of the sp^3-hybridization model and at the employment of the concept of the specific intermolecular interactions formed by the essentially unshared $2s^2$ electron pair of the carbon atom and at the accounting for all bond vacancies of the molecule [10]. The indispensable condition of the thermodynamic analysis is the tendency of the bond vacancy to form the most stable interactions assuring to the system the energy minimum. The increase in the number of hydroxy groups in the molecule of the polyhydric alcohol results both in growing number of the hydrogen bond and a proportional increase in the number of specific intermolecular interactions. Therefore, the thermodynamic analysis of polyhydric alcohols is interesting both for the evaluation of the energies of the hydrogen bonds and for critical consideration of the obtained energy effects and for further development of the general understanding of the structures bound by the hydrogen bonds.

The presence of two hydroxy groups of the same polarity in the molecule of dihydric alcohols results in their high association with the formation of hydrogen bonds. The molecules of ethylene glycol exist mainly in the gauche form and are capable of homoassociation without the rupture of intramolecular H-bonds for in these configurations one hydroxy group can play the role of donor, and the other, of acceptor. The second cause of the ethylene glycol association is connected with the possibility of the rupture of the intramolecular oxygen bridge followed by further interaction of hydroxy groups like it occurs in aliphatic alcohols.

In the IR spectrum of the ethylene glycol solution in CCl_4 at any concentration a definite doublet is present with the maxima at 3,638, 3,607 cm^{-1} corresponding to the hydroxy groups of free molecules possessing an intramolecular hydrogen bond

H–O···H–O [51]. The position and pattern of the bands is conserved at dilution, only the intensity changes. The spectrum contains a weak band with a maximum in the region 3,500–3,200 cm^{-1} corresponding to the intermolecular association existing even at a very low concentration. This allows a conclusion that the energy of the hydrogen bonds is more stable than in water. The comparison of the Raman spectra with the Raman spectra of water with accounting for various conformations of the ethylene glycol was understood in [52] as the formation in more uniform net of hydrogen bonds in the liquid ethylene glycol than in water. The formation of ethylene glycol dimers was suggested [1]. The spectrum of pure ethylene glycol contains a symmetric strong absorption band at 3,350 cm^{-1} whose parameters are practically identical to the corresponding bands in the spectra of liquid monohydric alcohols. The lack of absorption bands in the high-frequency part of the spectrum was interpreted in [1] as the cleavage of intramolecular hydrogen bonds at the high concentration of the ethylene glycol resulting in association of the molecules similarly to aliphatic alcohols with the formation of the cooperative hydrogen bonds OH···OH···OH. Two hydroxy groups interact with the hydroxy groups of two different molecules; this provides the structure with a higher stability. The linear structure of the ethylene glycol molecule possesses six bond vacancies located at its ends with hydroxy groups. The likeness of the structure of molecules of the tetramethylene glycol and ethylene glycol results in the similarity of their properties.

The shift of the electron density from the middle methylene groups to the oxygen atoms of the hydroxy groups, in particular, in the structure of liquid and solid tetramethylene glycol H–O–CH_2–CH_2–CH_2–CH_2–O–H ensures the stabilization of the bonds formed. All the bond vacancies should be filled through the formation of stable hydrogen bonds by the hydroxy groups and of specific interactions by the essentially unshared $2s^2$ electron pairs of carbon atoms of the CH_2-groups with the electron pair of the oxygen atom. This results in the reciprocal replacement of the alcohol molecules of the nearest environment. The lone electron pairs of the oxygen atoms of dihydric alcohols provide a possibility of the formation of a planar structure. However, the coordination of four molecules in the nearest environment results in involving into the interaction of CH_2-groups securing the formation of a spatial screw structure of the liquid state. The participation of all four bond vacancies of each of the two hydroxy groups ensures the formation of four strong hydrogen bonds H–O···H–O and four specific intermolecular interactions with the carbon atoms of the methylene groups of the contiguous chains H–O → CH_2 via the dative mechanism (Fig. 5.13). Each molecule is bound in the chain with four

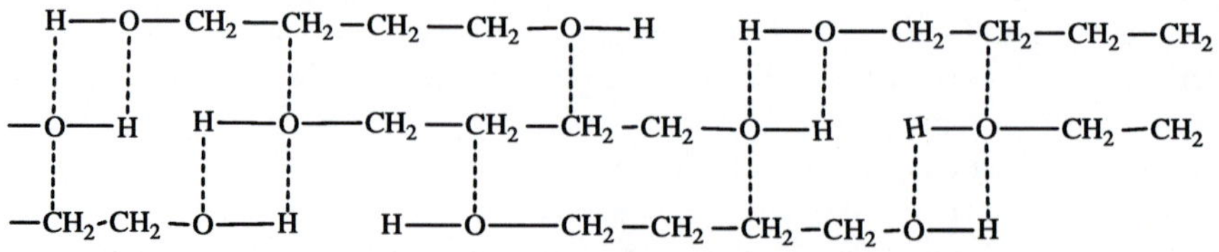

Fig. 5.13 Schematic picture of the liquid structure of dihydric 1,4-butanediol

hydrogen bonds and four specific intermolecular interactions. In this structure with four coordinated molecules and displaced chains, the linking of the latter occurs through the oxygen atoms of the hydroxy groups located mutually at the interatomic distance close to the interatomic distance in the hydroxy group and carbon atoms of the CH_2-groups most removed from the oxygen atom.

In the molecules of the dihydric alcohols of the normal structure, both oxygen atoms of the hydroxy groups form the reverse dative bond. Its influence also is finished at the third methylene group of the chain, and consequently only the middle carbon atom of the alkyl chain in the 1,7-heptanediol is free of the effect of the dative bond.

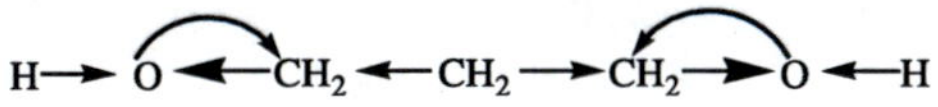

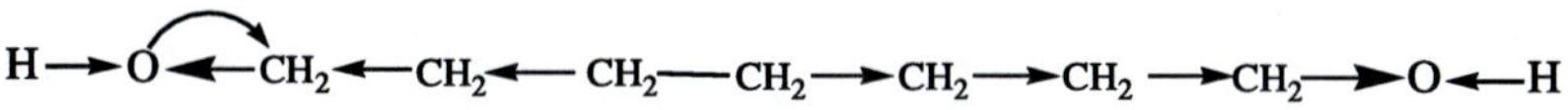

These CH_2-groups of molecules take part also in linking the chains by the essentially unshared $2s^2$ electron pairs of carbon atoms forming with positively charged hydrogen atoms of the analogous groups in the nearest surrounding interactions $H_2C \rightarrow H\text{–}CH$ of low stability. The contribution of the CH_2-group into the enthalpy characteristic equal to the overall value of the energies of interactions $H_2C \rightarrow H\text{–}CH$ is estimated from the difference in the vaporization enthalpies of two compounds of the homologous series where this group is not affected by the influence of the reverse dative bond. The energy of the contribution of these interactions into the vaporization enthalpy of the alcohols of homologous series 1-heptanol–1-octanol and 1,7-heptanediol–1,8-octanediol amounts to 4.75 kJ mol^{-1}.

This distribution of structural and energy parameters in the molecule is reflected in the intermolecular interactions in dihydric alcohols and underlies the behavior of the dependence of the vaporization enthalpy $\Delta_{vap}H^0(298) = f(\text{Cn})$ from the number of carbon atom in the carbon–carbon chain of the alcohols presented in Fig. 5.14. The position of the values of the enthalpy characteristics of the dihydric alcohols of the normal structures on two straight lines crossing at the vaporization enthalpy of 1,5–pentanediol (Fig. 5.14) essentially differs from the analogous dependence for monohydric alcohols, 1-propanol homologs from the series methanol–1-decanol (Fig. 5.3). Two straight lines describing the dependence $\Delta_{vap}H^0(298) = f(\text{Cn})$ and crossing at the enthalpy characteristic of 1,5-pentanediol (Fig. 5.14), and similar dependence of the 2-propanol homologs (Fig. 5.3) described by two crossing straight lines with the crossing point at the vaporization enthalpy of 2-butanol demonstrate the inadequacy of the variation character of the vaporization enthalpy of the initial members in the homologous series of dihydric alcohols of the normal structure and of 2-propanol. The character of the dependence $\Delta_{vap}H^0(298) = f(\text{Cn})$ of the initial members of the homologous series of 1-propanol shows that in the latter exists the compensating effect between the dissociation energy of the dimeric molecules and the decreasing effect of the reverse dative bond on the stabilization of specific interactions and hydrogen bonds (Fig. 5.3). At the same time this effect

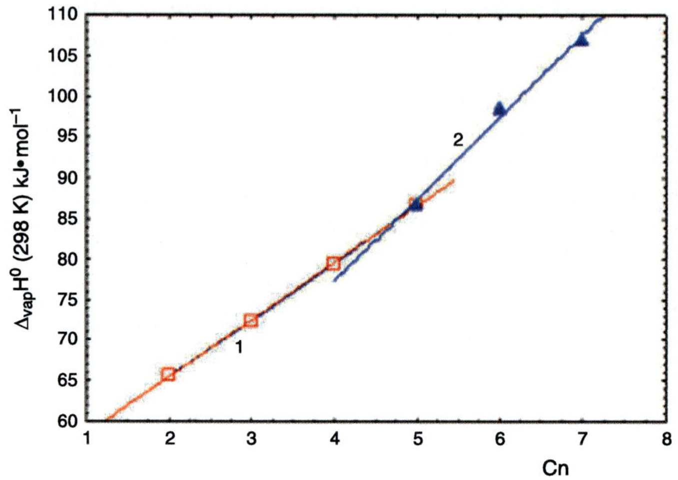

Fig. 5.14 Dependence of vaporization enthalpy of dihydric alcohols versus the number of carbon atoms in alkyl chain (1) Methanediol–Ethylene glycol–1,5-pentanediol, (2) 1,5-pentanediol–1, 7-heptanediol

is absent in the homologs of 2-butanol (Fig. 5.3) and its absence is still more obvious in the dihydric alcohols due to the cleavage of the large number of stable hydrogen bonds (Fig. 5.14). Yet in alcohols homologous to 2-butanol and glycols of the normal structure the effect of the reverse dative bond characteristically finishes on the 2-butanol and 1,5-pentanediol, respectively, namely, at the third carbon atom from the oxygen of the hydroxy group. It is possible to conclude that the dimerization practically finishes in these two homologous series at the alcohols containing three carbon atoms in the alkyl chain from the hydroxy group. The presence in the vapor of 2-methyl-1-propanol of 11.6 and 6.9% dimers, respectively, at 320.45 K and 348.95 K suggests the lesser presence of dimers in the vapor of 1-butanol. Ethylene glycol exists in dimers even in the dilute water solutions that show the high stability and dissociation energy of the hydrogen bonds of the ethylene glycol compared to the hydrogen bond energy in liquid water (10.99 kJ mol^{-1}).

The dimeric structure of dihydric alcohols can be represented by two structures distinguished by the number of hydrogen bonds. In a linear structure two hydrogen bonds can form. Therefore, the going over into vapor will require a cleavage of two hydrogen bonds and four specific interactions DH–O → CH_2–$(CH_2)n$. The decrease in the number of coordinated molecules in going into vapor from the condensed state is accompanied with the stabilization of the hydrogen bonds with decrease in their number from four to two.

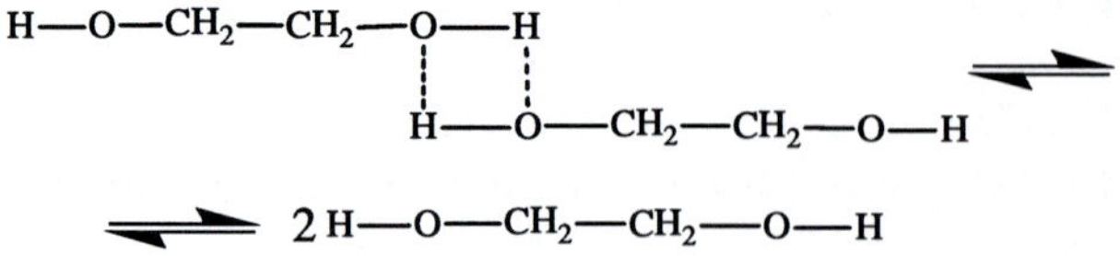

Nonetheless, the energy of each remaining H-bond in the dimer should be more stable that that existing in the liquid or solid state. The cyclic structure of a dimeric molecule of the dihydric alcohol is formed by four hydrogen bonds. The process of vaporization is connected with the rupture of the specific interactions D–O → CH_2–$(CH_2)n$ that are less stable than hydrogen bonds, and requires the transformation of a chain structure into cyclic where strains appear disregarding the mobile bending structure of the molecule of the dihydric alcohol.

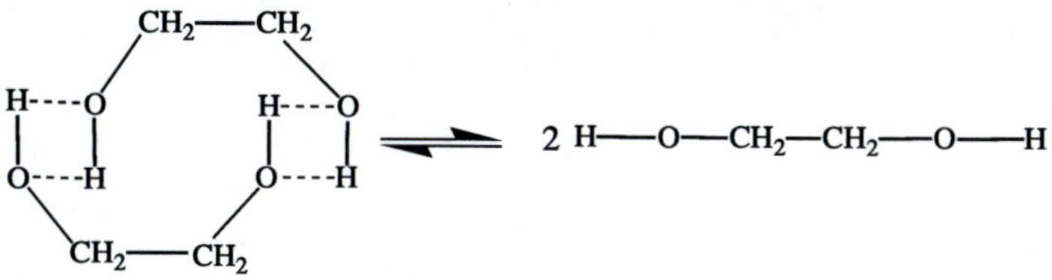

Fundamentally, these dimers may be present in vapor to a low degree for the rupture of only specific interactions during vaporization of the cyclic dimer should correspond to the vaporization enthalpy lower than that experimentally found in the monohydric alcohols.

The initial members of the dihydric alcohols should be characterized by the reduced stability of the hydrogen bonds and specific interactions from the alcohols with larger number of carbon atoms in the skeleton due to the varying negative charge on the oxygen of the hydroxy groups.

The dependence of the vaporization of the dihydric alcohols presented in Fig. 5.14 makes it possible to use extrapolation for estimating vaporization enthalpy of methanediol, heptanediol, and octanediol with the accuracy of the experimental measurements by the calorimetric method (Table 5.14).

To avoid the error in the estimation of the energy of the hydrogen bond, the contribution into the vaporization enthalpy should be accounted for of the CH_2-groups unaffected by the reverse dative bond and not involved into the stable specific intermolecular interactions –O → CH_2.

As was already shown, the energy of the intermolecular specific interaction H–O → CH_3–$(CH_2)n$ depending on the alkyl fragment is taken equal to the energy of the analogous specific interaction in the liquid dimethyl, diethyl, or dipropyl ether, respectively, and this ensures the creation of interconnected system of self-consistent energies of specific interactions with the pentacoordinate carbon atom and the energies of the hydrogen bonds. The dihydric alcohols of symmetric structure, starting with 1,3-propanediol, contain the growing number of methylene groups involved in the shift of the electron density from the central CH_2-group, e.g., of the 1,3-propanediol, to oxygen atoms and the reverse dative bond. The sum of the energies of the hydrogen bonds and the specific interactions is equal to the vaporization enthalpy; therefore, the energy of the hydrogen bond is calculated by (5.18a).

$$DH - O \cdots H - O = (\Delta_{vap}H^0(298\,K) - 4D - O \rightarrow CH_2)/4. \qquad (5.18a)$$

Table 5.14 Energies of hydrogen bonds and specific interactions (kJ mol^{-1}) of liquid polyhydric alcohols

Alcohol	Formula	$\Delta_{vap}H^0(T)$ [21]	T (K)	D–O → CH_3–$(CH_2)_n$	D*iso*CH_3	DCH_2	DH–O···H–O
Methanediol		56.8[a]	298	–	–	–	–
Ethylene glycol	$C_2H_6O_2$	65.6 ± 0.3	298	5.63 × 4	–	–	10.80
1,3-Propanediol	$C_3H_8O_2$	72.4 ± 0.3	298	5.63 × 4	–	–	11.3
1,4-Butanediol	$C_4H_{10}O_2$	79.3	298	6.78 × 4	–	–	13.05
1,5-Pentanediol	$C_5H_{10}O_2$	86.8	298	6.78 × 2 8.92 × 2	–	–	13.75
1,6-Hexanediol	$C_6H_{14}O_2$	98.5	298	8.92 × 4	–	–	15.7
1,7-Heptanediol	$C_7H_{16}O_2$	101.6[a]	298	8.92 × 4	–	4.75	15.3
1,8-Octanediol	$C_8H_{18}O_2$	108.3[a]	298	8.92 × 4	–	4.75 × 2	15.8
1,2-Propanediol	$C_3H_8O_2$	60.0	390	5.63 × 2 6.78 × 2	–	–	8.8
1,2-Butanediol	$C_4H_{10}O_2$	71.6	298	5.63 × 2 8.92 × 2	–	–	10.6
1,3-Butanediol	$C_4H_{10}O_2$	74.5	298	6.78 × 4	–	–	11.85
2,3-Butanediol	$C_4H_{10}O_2$	62.5, (58.4)	363 (332)	6.78 × 4	–	–	8.9
Glycerol	$C_3H_8O_3$	91.7	298	5.63 × 6	–	–	12.3
1,2,3-Butanetriol	$C_4H_{10}O_3$	68.1	390	5.63 × 4 6.78 × 2	–	–	5.4
2,3,4-Pentanetriol	$C_5H_{12}O_3$	78.9	443	5.63 × 2 6.78 × 4	–	–	6.8
2-Methyl-1,3-butanediol	$C_5H_{12}O_2$	62.4	414	6.78 × 4	$\sum = 2.70$ 1.35	–	8.8
3-Methyl-1,3-butanediol	$C_5H_{12}O_2$	60.3	361	6.78 × 4	$\sum = 0.60$ 0.30	–	8.2
1,3-Butanediol	$C_4H_{10}O_2$	59.7	398	6.78 × 4	–	–	8.2
2,2-Dimethyl-1,3-propanediol		79.4	415	6.78 × 2 5.63 × 2	$\sum = 5.40$ 1.35 $\sum = 1.20$	–	12.2
1,3-Propanediol	$C_3H_8O_2$	63.3	408	6.78 × 2 5.63 × 2	–	–	9.4

[a]Estimated by extrapolation

At the large number of CH_2-groups in the 1,7-heptanediol molecule, the methylene group in position four not involved in the formation of the stable specific interactions contributes additional energy into the enthalpy of vaporization or sublimation of the compound. The contribution of these groups grows with their increasing number in the molecule of the dihydric alcohol of the normal structure. Taking into account the stable specific interactions DH–O → CH_2–$(CH_2)n$ formed by the CH_2–$(CH_2)n$ fragment of the carbon chain and the energy contribution from the interactions of low stability H_2C → H–CH formed by the CH_2-groups the energy of the hydrogen bonds in the dihydric symmetric alcohols should be calculated by (5.18b).

$$\mathrm{DH-O\cdots H-O} = (\Delta_{\mathrm{vap}}H^0(298\,\mathrm{K}) - 4 - D - \mathrm{O} \rightarrow \mathrm{CH_2} - n\mathrm{DH_2C} \rightarrow \mathrm{H-C-H})/4. \tag{5.18b}$$

Thermodynamic characteristics of the vaporization processes are known for a limited number of polyhydric alcohols. The results of performed thermodynamic calculations (Table 5.14) for symmetric dihydric alcohols demonstrate the regular increase in the energy of the hydrogen bonds in the series ethylene glycol (ethanediol)–1,6-hexanediol:

$$\begin{aligned}&\text{Ethylene glycol }(10.8) < 1{,}3\text{ - Propanediol }(11.3) < 1{,}4\text{ - Butanediol }(13.05)\\&\quad < 1{,}5\text{ - Pentanediol }(13.75) < 1{,}6\text{ - Hexanediol }(15.70)\\&\quad \approx 1{,}7\text{ - Heptanediol }(15.30) \approx 1{,}8\text{ - Octanediol}(15.80\,\mathrm{kJ\,mol^{-1}}).\end{aligned}$$

This trend corresponds, on the one hand, to the decrease in the influence of the reverse dative bond on the charge on the carbon atoms in the alkyl chain with their removal from the oxygen atom, and on the other hand, to the decrease in the dimerization degree in the series ethylene glycol–1,6-hexanediol. The constant value within the error limits of the hydrogen bond energy in the series 1,6-hexanediol–1,8-octanediol shows that these dihydric alcohols practically have no dimers in the vapor phase.

The energy of the hydrogen bond in liquid alcohols of this homologous series reaches the maximum value and afterwards remains constant in the alcohols with the growing number of carbon atoms in the alkyl chain whose vaporization enthalpy increases with the increase in the number of CH_2-groups and is described by the dependence $\Delta_{vap}H^0(298) = f(\mathrm{Cn})$. The insignificant difference between the calculated energies of the hydrogen bonds in the 1,4-butandiol and 1,3-propanediol (0.6 kJ mol^{-1}) suggests that these compounds hardly differ in the degree of dimerization in vapor. Note that ethylene glycol formed dimers in dilute solutions in CCl_4 which proves indirectly the stability of the dimers of this compound [51].

The unsymmetrical located hydroxy groups in the molecules of 1,2-propanediol, 1,2-butanediol, 1,3-butanediol form in the liquid state a network structure (Fig. 5.15), cross-linked by the stable hydrogen bonds and specific interactions –O → $CH_2(CH_2)n$.

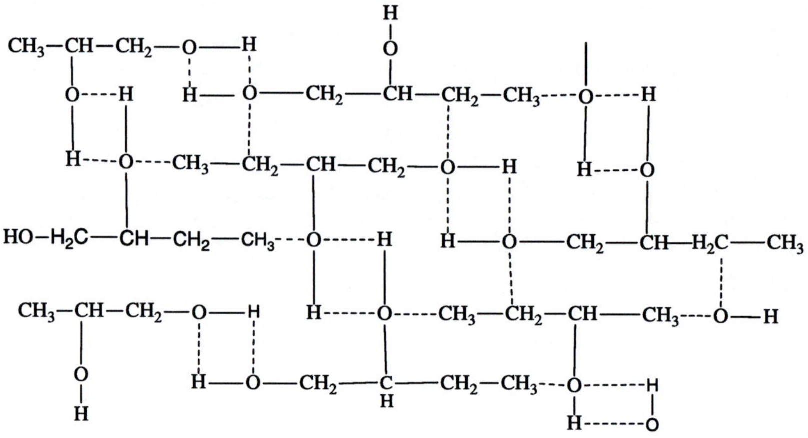

Fig. 5.15 Schematic picture of the liquid structure of dihydric alcohol 1,2-butanediol

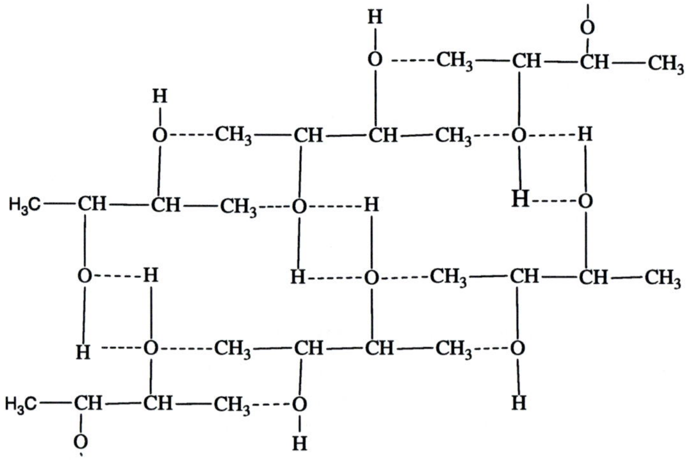

Fig. 5.16 Schematic picture of the liquid structure of dihydric alcohol 2.3-butanediol with unsymmetrical located hydroxy groups

The molecules of unsymmetrical dihydric alcohols with eight bond vacancies like 1,2-propanediol form four hydrogen bonds and four specific interactions distinguished by the length of the carbon chain depending on the location of the hydroxy group. In the liquid, 1,2-propanediol forms two specific interactions H–O → CH_3 (5.63 × 2) and two H–O → CH_2–CH_2 (6.78 × 2 kJ mol^{-1}). The molecule of 1,2-butanediol forms two interactions H–O → CH_3 (5.63 × 2) and two with a chain containing three carbon atoms H–O → $CH_2(CH_2)_2$ (8.92 × 2) (Fig. 5.14). The molecules of 1,3-butanediol and 2,3-butanediol form each four specific interactions of the same type D–O → CH_2–CH_2 with the energy 6.78 kJ mol^{-1} (Fig. 5.16). The results of calculations along (5.18a) are presented in Table 5.14.

The analysis of the found values of the hydrogen bonds (Table 5.14) leads to the following conclusions:

Firstly, the small difference in the hydrogen bond energies in liquid 1,2-Butanediol (10.6) ≈ Ethylene glycol (10.8) and 1,3-Propanediol (11.3) < 2,3-Butanediol (11.9 kJ mol^{-1}) shows the stability of these bonds and suggests that the dimerization of compounds in question in vapor is similar.

Secondly, the hydrogen bond energies in the dihydric alcohols with unsymmetrical located hydroxy groups fit to a series:

$$1{,}2\text{ - Butanediol }(10.6) < 2{,}3\text{ - Butanediol }(11.9) < 1{,}4\text{ - Butanediol }(13.1\,\text{kJ}\,\text{mol}^{-1}).$$

This series shows the stabilization of hydrogen bonds when the hydroxy groups are attached to the terminal carbon atoms of the alkyl chain favoring the shift of the electron density to oxygen atoms, supplying to them an enhanced negative charge and the donor properties.

Among the liquid trihydric alcohols where the enthalpy characteristics of vaporization have been measured, although at quite different temperature, glycerol, 1,2,3-butanetriol, and 2,3,4-pentanetriol (Table 5.14) should be mentioned. The three hydroxy groups ensure the existence of these compounds in liquid and crystalline state as a three-dimensional network. The experimentally obtained diffraction curves of glycerol intensity correspond to its lower association than that of ethylene glycol and the trend to the formation of the three-dimensional network of hydrogen bonds. The three hydroxy groups of glycerol form six hydrogen bonds. The bond vacancies belonging to the carbon atoms of CH_2-groups with their essentially unshared $2s^2$ electron pair form with oxygen atoms stable specific interaction $H–O \rightarrow CH_2$ with the energy equal to 5.63 kJ mol^{-1}. The three remaining bond vacancies of the oxygen atoms form H-bonds of low stability H–O···H–CH with the hydrogen atoms in the *trans*-position of the CH_2-groups whose energy may be estimated as equal to the analogous interaction in liquid methanol (0.60 kJ mol^{-1}) (Chap. 2). The performed calculations (Table 5.14) show that the energy of the hydrogen bond in liquid glycerol amounts to 12.25 kJ mol^{-1}, a higher stability than in the liquid 1,3-propanediol (11.3 kJ mol^{-1}). Considering this difference in the energy of the hydrogen bonds it is clear that the vaporization of glycerol is connected with the greater loss in the number of coordinated molecules than in vaporization of 1,3-propanediol.

The distinction in the structure of trihydric alcohols 1,2,3-butanetriol and 2,3,4-pentanetriol consists in larger number of carbon atoms in their skeleton. Therefore in the liquid, 1,2,3-butanetriol forms two specific interactions $–O \rightarrow CH_3$, two $H–O \rightarrow CH_2$, and two $–O \rightarrow CH_2–CH_2$. In the liquid, 2,3,4-pentanetriol forms two specific interactions $–O \rightarrow CH_2$, and four $–O \rightarrow CH_3–CH_2$ analogous to those existing in liquid methyl and ethyl ether. The performed calculation procedure of the energy of the hydrogen bonds by (5.18a) (Table 5.14) resulted in relatively low values for 1,2,3-butanetriol (5.4 kJ mol^{-1}, 443 K) and 2,3,4-pentanetriol

(6.8 kJ mol^{-1}, 443 K) compared to the value in liquid glycerol. Evidently, this low value mostly originates from the features of the interaction and from the unaccounted temperature dependence of the thermodynamic characteristics. Therefore, the obtained energies of the hydrogen bonds in 1,2,3-butanetriol and 2,3,4-pentanetriol are underestimated. At the same time the different value of the energy of the hydrogen bonds in trihydric alcohols

$$\text{1,2,3-Butanetriol}\ (5.4\,\text{kJ}\,\text{mol}^{-1}, 390\,\text{K}) < \text{2,3,4-Pentanetriol}\ (6.8\,\text{kJ}\,\text{mol}^{-1}, 443\,\text{K})$$

illustrated the dependence of the hydrogen bond energies in polyhydric alcohols on the location of the hydroxy groups in the carbon skeleton, the enhanced negative charge on oxygen and its donor properties at the second and the fourth carbon atoms of the chain in the 2,3,4-pentanetriol.

The dimerization in vapor and the partial components in the vapor–liquid equilibrium of the alcohols should be essentially affected by the isostructural methyl groups and evidently by the increased temperature. The presence of these obvious mutually complementary factors influencing the dimerization process requires the experimental study of this process in various members of dihydric alcohols and their methyl derivatives. The published information concerning this problem is lacking, and the available enthalpy characteristics of vaporization are often established at incommensurable temperatures. The lucky exceptions are 2-methyl-1,3 butanediol (414 K) and 3-methyl-1,3-butanediol (361 K). The vaporization enthalpies of these compounds are larger than that of 1,3-butanediol (Table 5.14). The difference between the vaporization enthalpy of the methyl-substituted alcohols and 1,3-butanediol is the sum of the energies of interactions formed by the isostructural methyl groups whose number doubles at the interaction of two molecules of the nearest surrounding.

$$2DCH_3 = \Delta_{\text{vap}}H^0(T)\ \text{m, b.} - \Delta_{\text{vap}}H^0(T)\text{b.} \qquad (5.21)$$

Therefore, the energy of the specific interaction of low stability formed by the essentially unshared $2s^2$ electron pair of the carbon atom of the isostructural methyl group in the liquid 2-methyl-1,3-butanediol at 414 K equals 1.35 kJ mol^{-1}, and in 3-methyl-1,3-butanediol, 0.30 kJ mol^{-1}, respectively. Note that these values are somewhat underestimated for the vaporization enthalpy of 1,3-butanediol was measured at lower temperature, 398 K.

However, these data show that the isostructural methyl group located in the molecule of 2-methyl-1,3-butanediol at the second carbon atom forms the interaction with higher stability than the methyl group in the 3-methyl-1,3-butanediol. The results of performed calculations of the hydrogen bond energies in methyl-substituted dihydric alcohols (Table 5.14) taking into account the temperatures of the experimental measurement of the vaporization enthalpy show the trend of their stabilization at the location of the isostructural methyl group at the second atom in the molecule of the 1,3-butanediol.

$$3\text{-Methyl-}1,3\text{-butanediol }(8.2,\ 361\,\text{K}) < 1,3\text{-Butanediol }(8.2,\ 398\,\text{K}) < 2\text{-Methyl-}1,3\text{-butanediol }(8.8\,\text{kJ mol}^{-1},\ 414\,\text{K}).$$

Whereas 1,3-propanediol has a high degree of dimerization in vapor and low vaporization enthalpy value [72.4 (298 K) and 63.3 (408 K)] the enthalpy characteristic of 2,2-dimethyl-1,3-propanediol has a higher magnitude at 415 K (79.4 kJ mol^{-1}). Consequently, at the small energy contribution of the methyl groups into the vaporization enthalpy the ability to dimerize in vapor sharply decreases in the alcohols with the isostructural methyl groups. However, the existing information is insufficient for the estimation of the degree of dimerization in this compound. Using the experimentally measured vaporization enthalpy value of the diethyl ether in the temperature range 366–432 K (26.6 kJ mol^{-1}) and taking the contribution of the isostructural methyl group equal to that in liquid 2-methyl-1,3-butanediol (Table 5.14), the energy of the hydrogen bond in 2,2-dimethyl-1,3-propanediol would be 13.25 kJ mol^{-1}. Taking the contribution of the isostructural methyl group into the vaporization enthalpy analogously to that in liquid 3-methyl-1,3-butanediol, the energy of the hydrogen bond would be 14.30 kJ mol^{-1}.

The data obtained show that the energy of the hydrogen bond in the liquid 2, 2-dimethyl-1,3-propanediol at 415 K is more stable than in liquid 1,3-propanediol at 298 K. This is possible only in the case when the dimers dissociate and the molecules go over into the vapor mainly as monomers [53], and their partial pressure grows with temperature. In this connection, the value of the vaporization enthalpy grows and the obtained energy of the hydrogen bond is near the real value.

5.5 Energies of Hydrogen Bonds and Specific Interactions in Cyclic Alcohols

From the values of the vertical ionizing potentials and bond energies of 1s carbon atom (E_{CB} (C1s)) in benzene derivatives with OR-groups as substituents it was concluded on the summary acceptor nature of the OH-group [6]. The analysis of the thermodynamic properties of alicyclic alcohols and the alcohols possessing a benzene ring makes it possible to quantitatively estimate this property. One of the most interesting problems arising in the study of the specific intermolecular interactions between complex molecules whose numerical values are determined for a limited range of compounds is the energy of the hydrogen bonds. From the vaporization enthalpy of cyclopentane and cyclohexane (Table 5.15) it follows that the sixth methylene group of the latter compound contributes 2.66 kJ mol^{-1} into the overall enthalpy characteristic. The replacement of a hydrogen in these molecules by OH group is accompanied with the formation of two hydrogen bonds that results in a significant increase in the vaporization enthalpy of the corresponding alcohols, and the difference in their enthalpy characteristics grows to 4.60 kJ mol^{-1}.

Table 5.15 Energies of hydrogen bonds and specific interactions (kJ mol^{-1}) in liquid cyclic alcohols

Compound	Formula	$\Delta_{vap}H^0(298)$ [21]	D–O → CH_2	DH–C → HC	DH–O···H–O
Cyclohexanol	$C_6H_{11}OH$	62.0 ± 0.3	5.63	0.60	10.40
Cyclohexane	C_6H_{12}	29.96	5.00	–	–
Cyclopentanol	C_5H_9OH	57.40	5.63	0.60	9.40
Cyclopentane	C_5H_{10}	27.30	5.47	–	–
Benzyl alcohol	$C_6H_5CH_2OH$	50.50	0.58x2	0.60	7.75
Benzene	C_6H_6	33.82	5.72	0.60	–

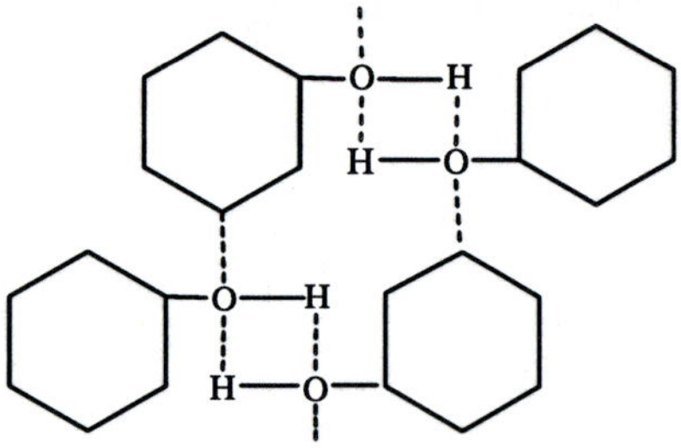

Fig. 5.17 Schematic picture of the liquid structure of cyclohexanol

This means that the energies of the hydrogen bonds existing in the solid and liquid states of these alcohols are also insignificantly different.

The assumed structure of liquid cyclohexanol can be schematically depicted with showing the bond vacancies of the molecule of this compound (Fig. 5.17). Two bond vacancies are formed by two unshared electron pairs of the oxygen, and the third one corresponds to the hydrogen of the hydroxy group. The principle of the occupancy of all bond vacancies in the formed structure of the alcohols under consideration is fulfilled by the formation of two hydrogen bonds and two specific intermolecular interactions between the oxygen atom and the carbon atom of the CH_2-group of the contacting hydrocarbon rings. This interaction between two species possesses a donor–acceptor character and is a specific intermolecular interaction. It forms at the expense of the essentially unshared $2s^2$ electron pair of the carbon atom of the ring and the unshared electron pair of the oxygen atom.

The cyclohexanol ring like the cyclohexane ring has in the nearest surrounding in the condensed state two analogous rings of the contacting molecules. Therefore, the vaporization of cyclohexane like that of the cyclohexanol ring occurs with energy expenditure equal the vaporization enthalpy (29.96 kJ mol^{-1}) minus the energy of the interaction formed by the hydrogen which has been replaced by the hydroxy group. The interaction of two molecules of the nearest surrounding requires double contribution of the substituted hydrogen atoms. This thermodynamic characteristic includes also the energies of the universal interactions. The vaporization of cyclohexanol involves the cleavage of two hydrogen bonds and two specific intermolecular interactions –O → CH_2 (Fig. 5.17). Similar to the alcohols of the normal structure the energy of the latter interactions should be taken equal to the energies of the specific interactions in the liquid (5.63) and solid (6.60 kJ mol^{-1})

methyl ether, respectively, and the energy of the formed interaction DH–C → HC of the hydrogen atom with the carbon atom of the contacting molecule (0.60 kJ mol^{-1}). Analogous interactions exist in the liquid cyclopentanol ($\Delta_{vap}H^0$(298)cyc.c.). Thus, the energy of the hydrogen bond is estimated from (5.22)

$$\mathrm{D-H-O\cdots H-O} = (\Delta_{\mathrm{vap}}H^0(298)\mathrm{alc.} - \Delta_{\mathrm{vap}}H^0(298)\ \mathrm{cyc.c} - 2\mathrm{D-O} \\ \rightarrow \mathrm{CH_2} + 2\mathrm{DH-C} \rightarrow \mathrm{H})/2. \qquad (5.22)$$

The results of calculations are presented in Table 5.14. The sublimation enthalpy of solid cyclopentanol is unknown, and the value reported for the cyclohexanol 60.7 (285 K) kJ mol^{-1} [42, 43] is essentially below the enthalpy characteristic of vaporization.

The molecules of benzyl alcohol contain four bond vacancies (Fig. 5.18) alongside the remaining bond vacancies of the benzene ring. Two bond vacancies of the hydroxy group form the hydrogen bonds, and the two unshared electron pairs of two oxygen atoms in the two molecules of the nearest surrounding form H-bonds of low stability –O···H–C with the positively charged hydrogen of the benzene ring.

This type of hydrogen bond is formed along the donor–acceptor mechanism. It is also a specific interaction of low energy equal to 0.60 kJ mol^{-1}. This type of hydrogen bond should be similar in energy to the specific interaction H_3C → H–C forming between two species along the donor–acceptor mechanism. The energies of this type of interaction are analogous to the interaction in liquid methanol equal to 0.60 kJ mol^{-1} Two bond vacancies of CH_2-groups of contacting molecules form the specific intermolecular interactions between a carbon atom and a positively charged hydrogen atom H_2C → H–C of the benzene ring also along the donor–acceptor mechanism with the participation of the essentially unshared $2s^2$ electron pair of the carbon atom acting as an electron donor. The energy of this interaction is also 0.60 kJ mol^{-1}.

$$\mathrm{D-H-O\cdots H-O} = (\Delta_{\mathrm{vap}}H^0(298)\mathrm{alc.} - (\Delta_{\mathrm{vap}}H^0(298)\mathrm{ben.} - 2\mathrm{DH-C} \\ \rightarrow \mathrm{H-C}) - 2\mathrm{DH-O\cdots H-C} - 2\mathrm{DH_2C} \\ \rightarrow \mathrm{H-C})/2. \qquad (5.23)$$

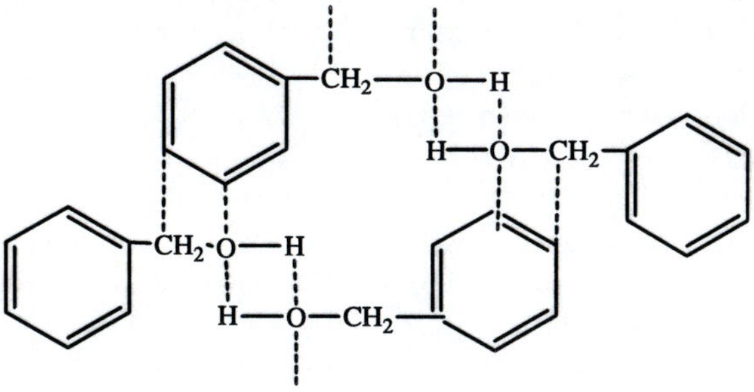

Fig. 5.18 Schematic picture of the liquid structure of benzyl alcohol

The performed thermodynamic calculations by (5.23) (Table 5.14) show that the two hydrogen bonds existing in the liquid benzyl alcohol have the energy below that of the hydrogen bonds in the cyclohexanol and cyclopentanol.

$$\text{Benzylalcohol} < \text{Cyclopentanol} < \text{Cyclohexanol}.$$

The cause of this phenomenon consists in a weak shift of the electron density from the strained benzene ring to the CH_2-group and consequently the oxygen atom possesses an insufficient electron excess due to the electron shift mainly from a hydrogen atom. Therefore, the negative charge on the oxygen atom in the benzyl alcohol is essentially smaller than in cyclopentanol and cyclohexanol. Just the overall charges on the oxygen atoms in these alcohols govern the energies of the arising hydrogen bonds and the sequence of their stability.

The low values of the energy of the hydrogen bonds in the alcohols under consideration are well consistent with the above-cited quantum-chemical calculations indicating the acceptor properties of the OH-group in the benzyl alcohol that are fundamentally correct [Boв]. The regular series of decreasing energy of the hydrogen bonds follows from the higher negative charge on the oxygen atom in the water molecule [6].

$$H_2O\,(10.99) > \text{Cyclohexanol}\,(10.40) > \text{Cyclopentanol}\,(9.40)$$
$$> \text{Benzylalcohol}\,(7.75\,\text{kJ mol}^{-1}).$$

The structure and bonding are interconnected properties of a compound and do not exist separately, therefore it looks promising to apply an analogous approach to the evaluation of the energies of specific interactions and hydrogen bonds in the other cyclic compounds.

The location of hydroxy groups at the carbon atoms of the benzene ring governs the variations in the enthalpy of vaporization and sublimation of compounds. These enthalpy data include the energies of hydrogen bonds and specific interactions in benzene derivatives [49] and hence make it possible to evaluate the energies of the hydrogen bonds and the specific interactions of more complex compounds.

The data on the energies of the hydrogen bonds compiled in Table 5.14 indicate that the conjugated stable benzene system provides its hydrogen atom with very restricted possibilities to participate in the intermolecular interactions in contrast to hydrogen atoms in the rings of cyclohexane and cyclopentane. Therefore, the enthalpies of vaporization (33.82) and sublimation (44.4 kJ mol^{-1}) of benzene correspond to the interaction of the benzene ring as a whole. Therefore for all benzene derivatives the energy contribution of all hydrogen atoms substituted by hydroxy or other groups (0.60 kJ mol^{-1}) should be taken into account. Accounting for the interaction of two contacting molecules the contribution of two substituted hydrogen atoms equals 1.20 kJ mol^{-1}. This means that the contribution of the partly substituted benzene ring into the vaporization enthalpy of phenol or alkylbenzene would amount to ($\Delta_{vap}H^0(298)$bn. -1.20 kJ mol^{-1}) or in the general case for the process of vaporization and sublimation ($\Delta_{vap}H^0(298)$ben. –K·DH–C $\rightarrow$ H) where

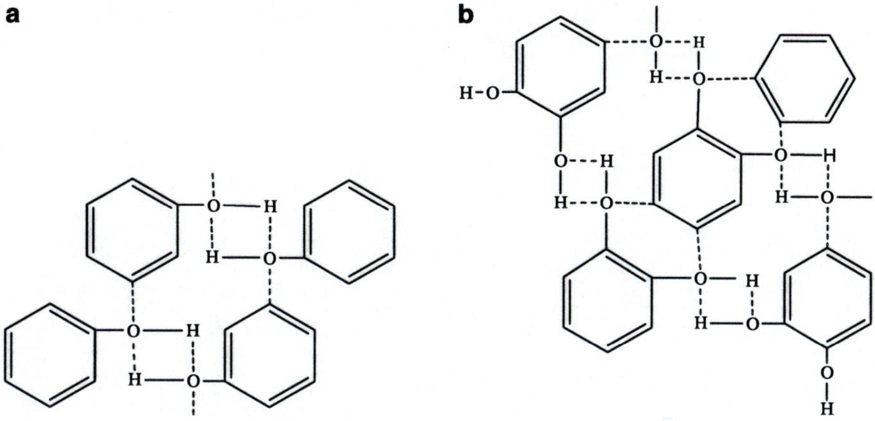

Fig. 5.19 Schematic picture of the liquid structure of phenol (**a**) and 1,2-dihydroxybenzene (catechol) (**b**)

"*K*" is the number of substituted hydrogens in the benzene ring. Consequently, the difference between the vaporization enthalpy of benzene derivatives $\Delta_{vap}H^0(298)$ der.bn. and the value ($\Delta_{vap}H^0$(298)bn. –K·D DH–C → H–C) comprises the overall energy value of the hydrogen bonds formed by the hydroxy groups DH–O···H–O and of the H-bonds of low stability DH–O···H–C (Fig. 5.19)

$$n\mathrm{DH}-\mathrm{O}\cdots\mathrm{H}-\mathrm{O}+n\mathrm{D}-O\cdots H-C$$
$$=(\Delta_{vap}H^0(298)\mathrm{der.bn.}-(\Delta_{vap}H^0(298)\mathrm{bn}-\kappa\mathrm{D\ DH}-\mathrm{C}\rightarrow\mathrm{H}-\mathrm{C}).\quad(5.24)$$

The factor "*n*" corresponds to the number of hydroxy groups substituting the hydrogen atoms in the benzene ring and forming each two hydrogen bonds. It should be taken into consideration that at the substitution of a hydrogen atom by any functional group with a single bond vacancy, in contrast to the hydroxy group, the number of the formed specific interactions in the liquid and crystalline state always doubles.

In (5.24) the energy of the hydrogen bond is unknown, and we determine it using (5.25)

$$n\mathrm{DH}-\mathrm{O}\cdots\mathrm{H}-\mathrm{O}=(\Delta_{vap}H^0(298)-(\Delta_{vap}H^0(298)\mathrm{bn}-\kappa\mathrm{DH}-\mathrm{C}$$
$$\rightarrow H)-n\mathrm{D}-O\cdots\mathrm{H}-\mathrm{C}).\quad(5.25)$$

The increased number of hydroxy groups in the molecules of 1,2-dihydroxybenzene and 1,2,3-trihydroxybenzene to two and three furnishes them four and six hydrogen bonds H–O···H–O and the same number of H-bonds of low stability of the type H–O···H–C, respectively (Fig. 5.19).

In the performed thermodynamic calculations, the weak H-bonds were taken into account. The established energies of the hydrogen bonds (Table 5.16) allow

Table 5.16 Energies of hydrogen bonds and specific intermolecular interactions (kJ mol^{-1}) of liquid hydroxybenzene derivatives

Compound	Formula	$\Delta_{evap}H^0(T)$			D*H–C → H–C/DH–O···H–C		∑DO···H–C	DH–O···H–O	
		Vap [47]	*T* (K)	Sub [48]	Liquid	Solid	Liquid/solid	Liquid	Solid
Benzene	C_6H_6	33.8	298	44.4	0.6	0.6	–	–	–
Phenol	C_6H_6O	58.8	298	69.7	0.60 × 2	0.60 × 2	23.8/24.7	11.9	12.4
1,2-Dhydroxybenzene	$C_6H_6O_2$	61.2	393	87.3	1.20 0.60 × 4	1.20 0.60 × 4	26.2/41.7	6.6	10.4
1,3-Dihydroxybenzene	$C_6H_6O_2$	74.3	407	87.5	1.20 0.60 × 4	1.20 0.60 × 4	39.3/41.9	9.8	10.5
1,4-Dihydroxybenzene	$C_6H_6O_2$	70.5	462	94.1	1.20 0.60 × 4	1.20 0.60 × 4	35.5/48.5	8.9	12.1
1,2,3-Trihydroxybenzene	$C_6H_6O_3$	72.4	440	116.9	1.80 0.60 × 6	1.80 0.60 × 6	25.6/70.5	4.3	11.8
1,2,4-Trihydroxybenzene	$C_6H_6O_3$			119.8		1.80 0.60 × 6	73.6	–	12.7
1,3,5-Trihydroxybenzene	$C_6H_6O_3$			131.7		1.80 0.60 × 6	85.5	–	14.2

D* is the number of substituted hydrogen atoms in the benzene ring and their contribution into the enthalpy characteristic some conclusion. The hydrogen bonds' energy in the liquid and solid phenol is nearly equal within the accuracy of the measurement of the thermodynamic values and the assumptions taken for the calculations of the energy of specific intermolecular interaction. In view of the character of the temperature dependence of the vaporization enthalpy, the established energies of the hydrogen bonds in the liquid dihydroxybenzenes at 393, 407, and 462 K, we should expect little difference in the energies of the hydrogen bonds in the liquid and solid 1,2-dihydroxybenzene, 1,3-dihydroxybenzene, and 1,4-dihydroxybenzene. The low value of the hydrogen bond energy in the liquid (4.3) and solid state (12.1 kJ mol^{-1}) of 1,2,3-trihydroxybenzene is also due to the difference in the temperature of measuring the vaporization (440 K) and sublimation enthalpy (298 K).

In all the dihydroxybenzenes, the energies of the hydrogen bonds have low values compared with that in liquid phenol. In the general case, they are described by the series:

$$\text{Dihydroxybenzene} < \text{Phenol} < \text{Trihydroxybenzene}.$$

The hydrogen bond energies of the alcohols of these homologous series are described by regular series of bond stability depending on the location of the hydroxy groups in their molecules with characteristic sequence of each series. Yet some general trends are also present.

Considering the temperature dependence of the thermodynamic properties the stabilization of the hydrogen bonds in the dihydroxybenzenes should be represented by a regular series:

$$1,2\text{-Dihydroxybenzene}\ (6.6,\ 393\,\text{K}) < 1,3\text{-Dihydroxybenzene}\ (9.8,\ 407\,\text{K}),$$
$$1,4\text{-Dihydroxybenzene}\ (8.9\,\text{kJ mol}^{-1},\ 484\,\text{K}).$$

The hydrogen bond energies of crystalline dihydroxybenzenes are described by the clearly expressed regularity of variations

$$1,2\text{-Dihydroxybenzene}\ (10.4) \approx 1,3\text{-Dihydroxybenzene}\ (10.5) <$$
$$1,4\text{-Dihydroxybenzene}\ (12.1\,\text{kJ mol}^{-1}).$$

Both trends illustrate the maximum effect on the stability of the hydrogen bonds of the most mutually remote hydroxy groups accompanied with the larger shift of the electron density ensuring the increased negative charges on the oxygen atoms and the stabilization of the hydrogen bonds.

The hydrogen bonds of trihydroxybenzenes are stabilized in the series corresponding to the location of the hydroxy groups in the benzene ring.

$$1,2,3\text{-Trihydroxybenzene}\ (11.8) < 1,2,4\text{-Trihydroxybenzene}\ (12.7) <$$
$$1,3,5\text{-Trihydroxybenzene}\ (14.2\,\text{kJ mol}^{-1}).$$

The maximum energy of the hydrogen bonds is attained at the symmetric position of the hydroxy groups in the benzene ring in 1,3,5-trihydroxybenzene.

It means that at this location of the hydroxy groups the most favorable distributions of the negative charges on the oxygen atoms and effect of the reverse dative bonds on carbon atoms are achieved.

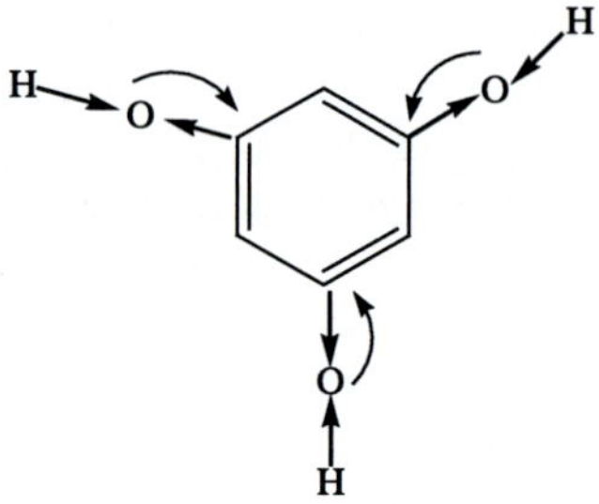

In alkylphenols, the methyl or ethyl groups form specific interactions with an unshared electron pair of oxygen –O → CH_3 or –O → CH_3CH_2 with the ethyl group linking the molecules together with the hydrogen bonds into a network structure (Fig. 5.20).

Equal number (two) of hydrogen bonds and specific intermolecular interactions forms in these structures. The dimethylphenols form additionally two specific interactions of low stability (H_3C → HC) through the interaction of methyl groups with CH-groups of benzene. Consequently, in the structures of liquid and solid trimethylphenols the number of the specific interactions of low stability (H_3C → HC) increases to four.

The results of the performed thermodynamic calculations (Table 5.17) show the low energy of hydrogen bonds in 3-methylphenol, essentially lower than in di- and trimethylphenols. This is caused by weak shift of the electron density from the carbon atom of this substituent to the benzene ring. In keeping with the data

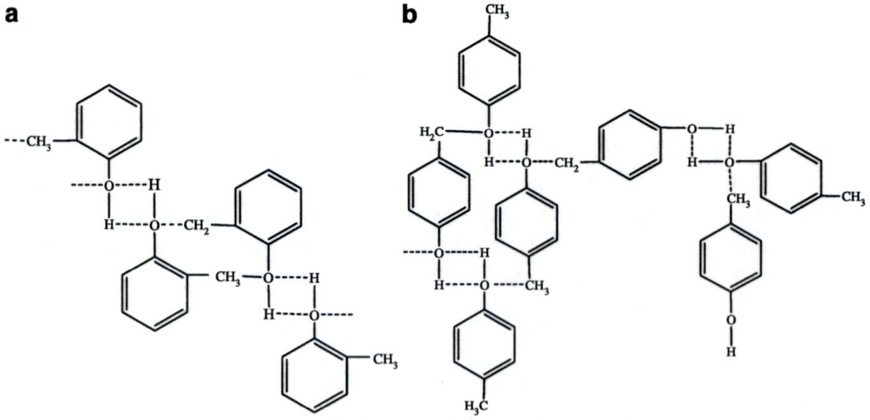

Fig. 5.20 Schematic picture of the network structure of condensed 2-methylphenol (**a**) and 4-methylphenol (**b**)

Table 5.17 Energies of hydrogen bonds and specific interactions (kJ mol^{-1}) of solid methyl-substituted phenols

Compound	Formula	$\Delta_{sub}H^0(T)$ [48]	T (K)	DH–C → H–C/DH–O → CH_3	∑D–O···H–O	DH–O···H–O
2-Dethylphenol	C_7H_8O	76.0 ± 0.8	288	0.6/6.6 × 2	18.3	8.7
3-Methylphenol	C_7H_8O	61.1 ± 0.7	Liq.	0.6/6.6 × 2	4.8	2.4
4-Methylphenol	C_7H_8O	73.9 ± 1.5	298	0.6/6.6 × 2	16.4	8.2
4-Ethylphenol	$C_8H_{10}O$	80.3 ± 0.5	278	0.6/8.83 × 2	18.9	9.5
2,3-Dimethylphenol	$C_8H_{10}O$	84.0	273–320	1.2/6.6 × 4	14.8	7.4
2,5-Dimethylphenol	$C_8H_{10}O$	85.0 ± 0.25	282–321	1.2/6.6 × 4	15.8	7.9
2,6-Dimethylphenol	$C_8H_{10}O$	75.6 ± 0.1	277–317	1.2/6.6 × 4	5.8	2.9
3,4-Dimethylphenol	$C_8H_{10}O$	85.7 ± 0.1	282–321	1.2/6.6 × 4	18.0	9.0
3,5-Dimethylphenol	$C_8H_{10}O$	82.8 ± 0.1	282–323	1.2/6.6 × 4	14.1	7.0
2,3,6-Trimethylphenol	$C_9H_{12}O$	86.7 ± 0.6	298	1.8/6.6 × 6	4.5	2.3
2,4,6-Trimethylphenol	$C_9H_{12}O$	82.8 ± 0.3	298	1.8/6.6 × 6	–	4.1
4-Bethyl-1,2-dihydroxybenzene	$C_7H_8O_2$	94.9 ± 1.0	298	0.6/6.6 × 2 D–O···HC = 1.2	35.8	9.0
2-Methyl-1,2-dihydroxybenzene	$C_7H_8O_2$	92.7 ± 1.4	351	0.6/6.6 × 2 D–O···HC = 1.2	38.6	9.6

presented in Table 5.17, the energy of the hydrogen bond is regularly stabilized in solid singly substituted methylphenols in the series

$$2\text{-Methylphenol } (9.1) > 4\text{-Methylphenol } (8.2) > 3\text{ - Methylphenol } (2.4\,\text{kJ mol}^{-1}).$$

Consequently, in the same sequence the donor properties of the essentially unshared $2s^2$ electron pair of the carbon atom of the methyl are diminished. The location of the methyl group next to the hydroxy group ensures the higher shift of the electron density to the oxygen atom virtually excluding the influence of the rigid system of the benzene ring, which is not possible in the 3-methylphenol. The location of the ethyl group in the position 4 in the 4-ethylphenol stabilizes the hydrogen bond (9.5) compared to 4-methylphenol (8.2 kJ mol^{-1}). The energies of the hydrogen bonds in dimethylphenols fit to the series

$$\begin{gathered}3{,}4\text{ - Dimethylphenol } (9.2) > 2{,}5\text{ - Dimethylphenol } (7.9) > \\ 2{,}3\text{ - Dimethylphenol } (7.4) > 3{,}5\text{ - Dimethylphenol } (7.0) > \\ 2{,}6\text{ - Dimethylphenol } (2.9\,\text{kJ mol}^{-1}) > 2{,}4\text{ - Dimethylphenol,}\end{gathered}$$

where the highest stability of the hydrogen bond is reached at the maximum removal of the methyl groups from the hydroxy in the positions 3,4 and at their maximum vicinity to the hydroxy group in the positions 2 and 5.

Trimethyl-substituted phenols are characterized by the extremely low energy of the hydrogen bond (Table 5.17).

The characteristic feature of the singly substituted methyl derivatives of dihydroxybenzenes is the stabilization of the hydrogen bond in the following series

$$2\text{ - Methyl - }1{,}2\text{ - dihydroxybenzene } (9.6) > 4\text{ - Dihydroxybenzene } (9.0\,\text{kJ mol}^{-1}).$$

It should be noted that the only published experimentally measured sublimation enthalpy of phenol (44.4 kJ mol^{-1}) reported without mentioning the accuracy of the experiment did not provide the necessary reliability for the estimation of the energy of the hydrogen bond in the solid phenol derivatives. Still the results obtained made it possible to estimate their values in the general system with the use of reliable enthalpy characteristics of all compounds considered in this chapter and to establish the regularities of the changes in the stability of these bonds.

5.6 Energies of Hydrogen Bonds and Specific Interactions in Solid Alcohols

The alcohols of the normal structure melt without polymorphous transformations (Table 5.18) due to the formation of the most stable dimeric forms underlying the structure of the unit cell of the solid state with the strong hydrogen bonds between

Table 5.18 Phase transition enthalpy (kJ mol^{-1}), energies of hydrogen bonds and specific interactions (kJ mol^{-1}) in solid alcohols (298 K)

Alcohol	$\Delta_{melt}H^0(T)$ [21]	$\Delta_{sub}H^0(T)$ [21]	$\Delta_{vap}H^0(T)$ [21]	D–O → $CH_3(CH_2)_n$	DH–O···O–H	$\sum DCH_2$
Methanol	4.85	43.37	39.25	6.59	15.10	–
Monodeuter-methanol	3.66	42.91		6.59	14.86	–
Ethanol	5.58	47.84	42.60	8.83	15.10	–
1-Propanol	5.37	52.85	48.30	11.61	14.80	–
1-Butanol	9.372	61.70	52.33	11.61	16.35	5.80
1-Pentanol	10.56	67.56	56.94	11.61	16.35	11.60
1-Hexanol	15.1	76.82	61.72	11.61	16.75	20.00
1-Heptanol	18.2	85.01	66.81	11.61	16.80	28.20
2-Propanol	5.41	50.73	45.34	8.83	15.10	$DH_2CH \rightarrow HCH_2$
						1.40×2
2-Butanol	6.00[a], 5.93	55.71 (61.71)	49.74	11.61	15.10	1.15×2
2-Methyl –1-propanol	6.32	57.40	50.79	11.61	15.10	2.00×2
2-Methyl –2-propanol	0.82[a], 6.70	54.34	46.81	8.83	15.10	1.65×4
2-Methyl –2-butanol	1.96[a], 4.46	55.62 (53.39)	49.20	11.61	15.10	0.55×4

[a]Polymorphous transformation

the molecules of the nearest surrounding. In their turn the dimers form between themselves the less stable specific intermolecular interactions involving the essentially unshared $2s^2$ electron pair of the carbon atom of the terminal methyl group of the carbon–carbon chain and the unshared electron pair of the oxygen of the hydroxy group of the alcohol.

Among the alcohols the polymorphous transformations occur in 2-butanol, 2-methyl-2-butanol, and 2-methyl-2-propanol. Two latter compounds are prone to the formation of stable specific interactions involving the terminal methyl groups. Two isostructural methyl groups of 2-methyl-2-propanol (*tert*-butyl alcohol) like the *tert*-butyl methyl ether are able to form less stable specific interaction. The high enthalpy of the polymorphous transition of 2-butanol (6.00 kJ mol^{-1}) is probably due to the participation of the second terminal methyl group. The enthalpy of this polymorphous transition is considerably greater than those in 2-methyl-2 propanol with an isostructural methyl group (0.82 kJ mol^{-1}) and 2-methyl-2 butanol with an isostructural methyl group at the carbon atom bearing the OH-group (1.96 kJ mol^{-1}).

The sublimation enthalpies measured by the calorimetric method were obtained in kJ/mol for alcohols methanol–1-propanol with the precision up to the second decimal digit and for alcohols 1-octanol–1-eicosanol, to the first decimal digit. As seen from the data presented in Fig. 5.21 at the number of carbon atoms in the alkyl chain of the alcohol growing up to 1-eicosanol the enthalpy characteristic increased five times as compared with the sublimation enthalpy of methanol. As already stated, the variation of the plot of the dependence $\Delta_{vap}H^0(298) = f(\text{Cn})$ in the initial part reflects the decrease in the influence of the reverse dative bond on the terminal methyl groups in the series methanol–1-propanol till the chain grows to three carbon atoms. The effect of the decrease in the influence of the dative bond in the crystalline alcohols is so large that unlike the liquid state of these alcohols (Fig. 5.21) it is not compensated by the evaporation of dimeric molecules.

The position of the point corresponding to the sublimation of 1-butanol on the crossing of two straight lines $\Delta_{sub}H^0(298) = f(\text{Cn})$ is the consequence of the significant change in the alcohols' ability to form dimers in vapor. This means

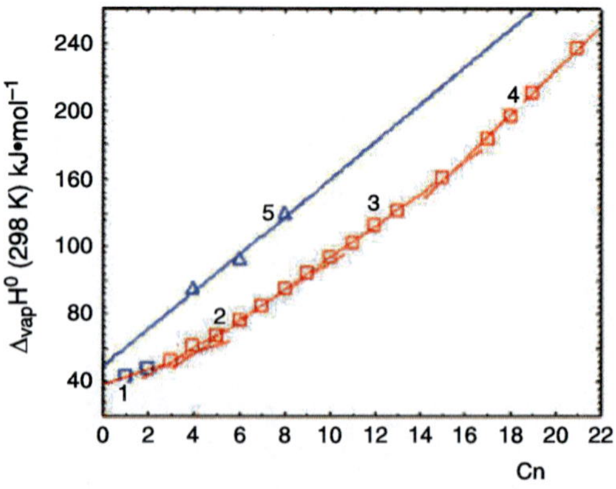

Fig. 5.21 Dependence of sublimation enthalpy of compounds versus the number of carbon atoms in alkyl chain: (1–4) 1-Akanols, (1) 1-Methanol–1-Propanol, (2) 1-Butanol–1-Octanol, (3) 1-Octanol–1-Hexadecanol and (4) 1-Hexadecanol–1-Eicosanol: (5) 1,4-Butandiol–1,8-Octandiol

that in the homologous series of 1-alkanols the alcohols having over four carbon atoms in the chain are virtually incapable to dimerize in vapor, Therefore, their going over into vapor does not considerably affect the enthalpy characteristic. Hence the subsequent increase in this characteristic is practically connected with the growing number of the methylene groups in the alkyl chain and their adequate contribution into the sublimation enthalpy in the series 1-butanol–1-octanol. The gradually increasing effect of the crystal field with the growing number of the methylene groups begins to appear since 1-octanol and the subsequent growing influence of the crystal field results in the description of the sublimation enthalpy of alcohols 1-octanol–1-hexadecanol and further of 1-hexadecanol–1-eicosanol with separate straight lines. A very attractive analogy appears in the nature of the dependence of the enthalpy characteristics on the number of carbon atoms in the alkyl chain of alcohols (Figs. 5.21 and 5.22) and ethers (Chap. 3) with their crossing in the point corresponding to the compound with three carbon atoms in the molecule and one hydroxy group. This is a general rule for the crystalline and liquid states of these compounds. This rule is also valid for symmetric and unsymmetrical ethers and ketones (Chaps. 3 and 4). In this connection, the vaporization enthalpies of 1-alkanols are an exception for in the first compounds of this homologous series the effect of the reverse dative bond is completely compensated by the dimerization in vapor.

This dependence describing the correlation between the structure and energy is extremely interesting for obtaining the enthalpy characteristics of unstudied compounds by interpolation and extrapolation. In the considered homologous series, the congener factors are the nature of the specific interactions originating from the involvement of the pentacoordinate carbon atom with the essentially unshared $2s^2$ electron pair and the unshared electron pair of the oxygen and on the other hand the common character of the hydrogen bonds. It is seen from the data presented

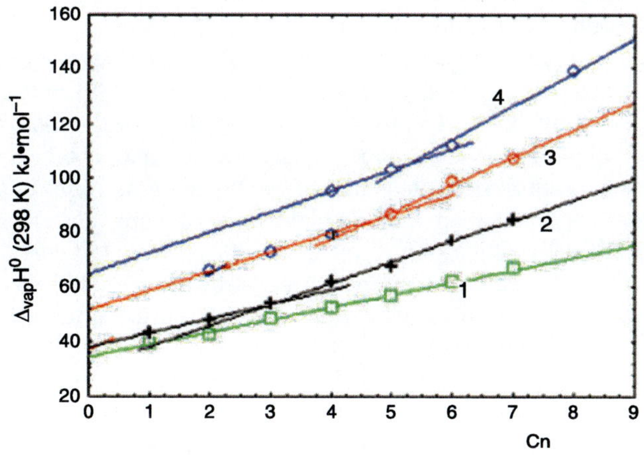

Fig. 5.22 Dependence of vaporization (1, 3) and sublimation (2,4) enthalpy of compounds versus the number of carbon atoms in alkyl chain: (1, 2) 1-alkanols, (3,4) symmetric dihydric alcohols

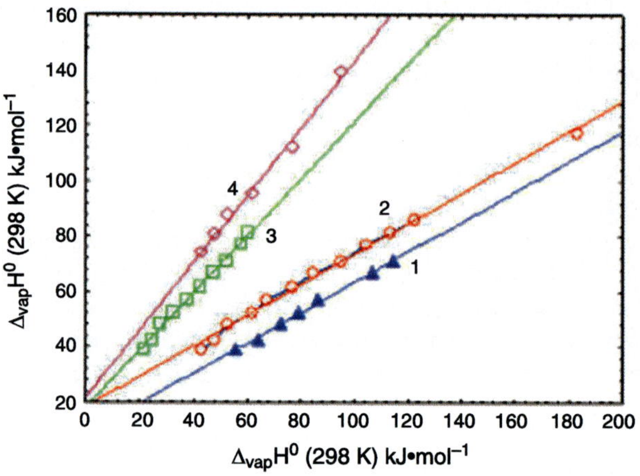

Fig. 5.23 Dependence of vaporization enthalpy of 1-alkanols versus vaporization enthalpy of dihydric symmetrical alcohols (1), sublimation enthalpy of 1-alkanols (2) versus vaporization enthalpy of methyl ethers (3), and sublimation enthalpy of symmetric dihydric alcohols versus sublimation enthalpy of 1-alkanols (4)

in Fig. 5.23 that the correlation between the vaporization enthalpy of 1-alkanols and symmetric dihydric alcohols (1), vaporization of 1-alkanols and sublimation of solid 1-alkanols (2), vaporization of 1-alkanols and vaporization of unsymmetrical methyl ethers CH_3–O–R (3), sublimation of dihydric symmetric alcohols and sublimation of 1-alkanols (4) are described by straight lines.

The description of the enthalpy characteristics in homologous series by relationships $\Delta H^0(T) = f(\mathrm{Cn})$ and of the enthalpy characteristics of compounds from different homologous series of alcohols with a large variation of the number of carbon atoms in the alkyl chain ensures the reliability of the values of the enthalpy characteristics obtained by interpolation and extrapolation.

The correlations depicted in Fig. 5.24 between the vaporization and sublimation enthalpy of 1-alkanols (1) and the vaporization enthalpy of 1-alkanols and the sublimation enthalpy of 2-alkanols (2) illustrate their description with linear dependences including the vaporization enthalpy of eicosanol obtained by extrapolation whereas the vaporization enthalpies of two homologous series of monohydric alcohols have been obtained experimentally. Analogous correlations of the enthalpy characteristics have been used to obtain the enthalpy characteristics of unstudied alcohols. The alcohols of the normal structure melt at high temperature with a large enthalpy due to the deep transformation in the structure of the solid state in the course of melting (Table 5.18).

The likeness of the specific intermolecular interactions in ethers and alcohols governed by the participation in the interaction of the terminal methyl group possessing the essentially unshared $2s^2$(c) electron pair on the carbon atom and of the unshared electron pair of the oxygen makes it possible to use the principle of the additive contribution of each type of interaction into the sublimation enthalpy of

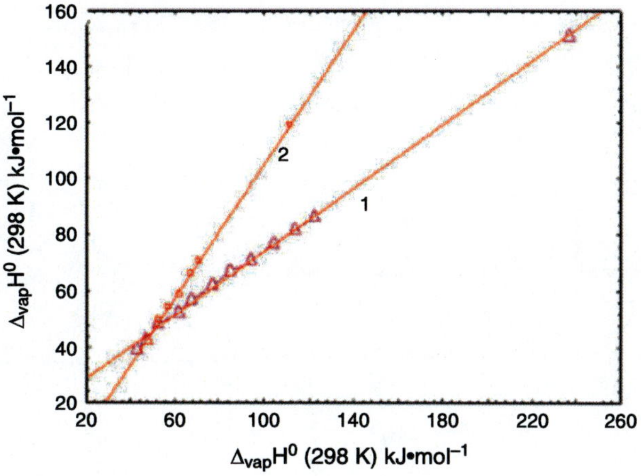

Fig. 5.24 Correlations between vaporization enthalpy and sublimation enthalpy of 1-alkanols (1) and vaporization enthalpy of 1-alkanols and sublimation enthalpy of 2-alkanols (2)

alcohols for the evaluation of the energy of the hydrogen bonds in solid methanol, ethanol, and 1-propanol. These alcohols possess the key importance in the thermodynamic calculations for the influence of the reverse dative bond is finished on the propyl chain, and on the butyl chain of 1-butanol begins the effect of the CH_2-group whose contribution to the sublimation or vaporization enthalpy originating from its involvement into the intermolecular interactions should be taken into account in the thermodynamic calculations. Therefore, the evaluation of the energies of the hydrogen bonds in the first three alcohols of the homologous series provides a possibility to estimate the energy of the interactions formed by the isostructural methyl and CH_2 groups.

In certain cases, the thermodynamic calculations of the hydrogen bond energy are performed with some error if the temperature dependence of the sublimation enthalpy is not accounted for and the thermodynamic characteristics have been measured not at the standard temperature. However, these calculations permit evaluation of quantitative values and establishment of regularities in their variations. This in its turn favors the development of the method of the thermodynamic analysis which is especially important since the ways have been lacking to obtaining the valid hydrogen bond energies. Further investigations can be used for refining these data.

The energies contributed by the CH_2-groups into the sublimation enthalpy are considerably higher than those for the vaporization enthalpy. The energy of the hydrogen bond in solid alcohols, the same as when they are in the liquid state, in the case of a branched molecule where the hydroxy group is attached to a chain of three carbon atoms, should remain constant starting from butyl alcohol. It is therefore possible to estimate the energy contributed by the isostructural methyl groups and to reveal the effect of the location of these groups in the alkyl chain on their contribution into the sublimation and vaporization enthalpy.

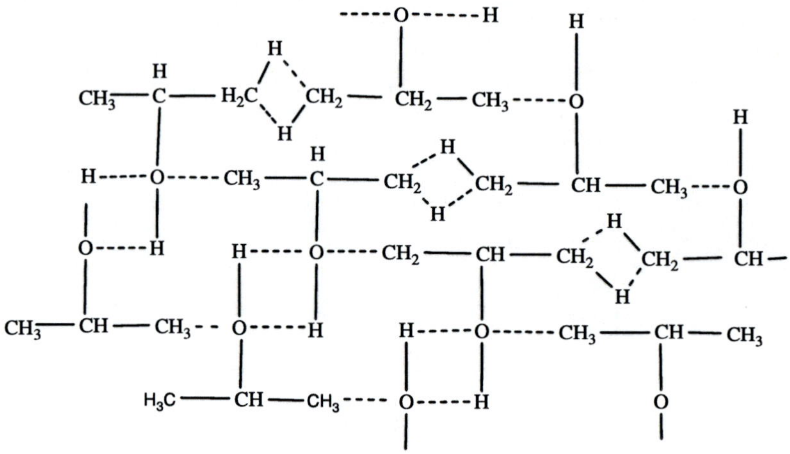

Fig. 5.25 Schematic picture of the condensed structure of 2-propanol

The lacking values of the energies of the specific intermolecular interactions formed by the terminal methyl groups of alkyl chains we take as equal to the energy of analogous interactions existing in solid ethers (Table 5.18). The molecule of 2-propanol compared to that of 1-propanol has an additional methyl group (Fig. 5.25) with a free bond vacancy of the essentially unshared $2s^2$ electron pair capable of interacting with the positively charged hydrogen atom of the contacting CH_3-group. Hence the propyl fragment of 1-propanol in the 2-propanol diminishes to ethyl, like that in ethanol. This means that the formed hydrogen bond and the specific interaction in 2-propanol are of the same energy as in solid ethanol. Applying the additivity principle we estimated the energy of the interaction $H_3C \rightarrow H\text{–}CH_2$ formed by the isostructural methyl group (Table 5.18).

The results obtained show that the change of the location of the hydroxy group in the alkyl chain of the alcohol and the elongation of the chain essentially affect the energy of the specific

	2-propanol	2-butanol
$\Delta_{sub}H^0(T)$	50.73	55.71 kJ mol^{-1}
	CH_3—CH(OH)—CH_3	CH_3—CH_2—CH(OH)—CH_3
$DH_3C \rightarrow H\text{–}CH_2$	1.40	1.15 kJ mol^{-1}

interaction of low stability formed by the isostructural methyl group.

The strongest decrease in the energy of this specific interaction and thus the decrease in its

	2-methyl-1-propanol	2-methyl-2-propanol	2-methyl-2-butanol
$\Delta_{sub}H^0(298)$	57.40	54.34	55.62 kJ mol^{-1}
	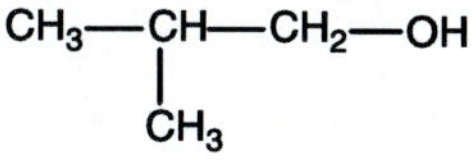	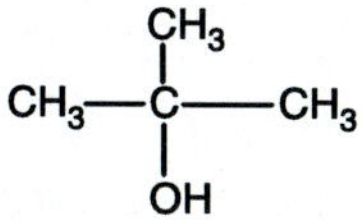	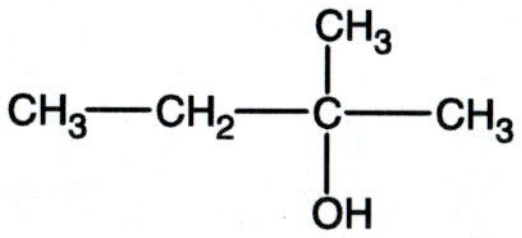
$D iso CH_3$	1.65	1.00	0.55 kJ mol^{-1}

contribution into the enthalpy characteristic occurs at the location of the isostructural methyl group at the same carbon atom like the hydroxy group, and at the elongation of the fragment forming the stable specific interaction.

The dependence of the sublimation enthalpy on the number of carbon atoms in the alkyl chain for the alcohols of the normal structure is shown in Fig. 5.21. This dependence for the sublimation process is of steeper character originating from the additional stabilization by the crystal field of the hydrogen bonds and the specific intermolecular interactions formed by the essentially unshared $2s^2$ electron pair of the carbon atom. The analogous trend in the dependence of vaporization and sublimation enthalpy should be conserved in the dihydric alcohols of the normal structure. It is seen even on a small number of the enthalpy characteristics of alcohols (Fig. 5.21). Evidently, the estimation of the values of the vaporization and sublimation enthalpy of a series of unstudied dihydric alcohols is less reliable (Table 5.19) yet it allows calculation of the hydrogen bond energies without accounting for dimerization in vapor.

High values of the difference between the vaporization and sublimation enthalpy of normal alcohols established at 298.15 K include not only the enthalpy characteristic of melting but for certain alcohols also the enthalpy of polymorphous transitions. The latter were experimentally measured only for a limited number

Table 5.19 Vaporization and sublimation enthalpy (kJ mol^{-1}) of monohydric 1-alkanols of normal structure (298.15 K)

Alcohol	Formula	$\Delta_{vap}H^0$ (*T*)	$\Delta_{sub}H^0$ (*T*) [48]
1-Octanol	$C_8H_{18}O$	70.98 [21]	94.7
1-Nonanol	$C_9H_{20}O$	76.86 [21]	104.3
1-Decanol	$C_{10}H_{22}O$	81.6 [47]	113.7
1-Undecanol	$C_{11}H_{24}O$	86.4*	122.7
1-Dodecanol	$C_{12}H_{26}O$	92.4*	132.3
1-Tridecanol	$C_{13}H_{28}O$	97.0*	141.7
1-Tetradecanol	$C_{14}H_{30}O$	101.7	150.4*
1-Pentadecanol	$C_{15}H_{32}O$	106.1*	160.6
1-Hexadecanol	$C_{16}H_{34}O$	112.0	171.6*
1-Heptadecanol	$C_{17}H_{36}O$	117.1*	183.2
1-Octadecanol	$C_{18}H_{38}O$	–	196.7
1-Nonadecanol	$C_{19}H_{40}O$	–	210.2
1-Eicosanol	$C_{20}H_{42}O$	–	223.4*
1-Henicosanol	$C_{21}H_{44}O$	–	237.0

*Established by extrapolation

Table 5.20 Phase transitions enthalpy (kJ mol^{-1}) of monohydric alcohols

Type of transition	T (K)	$\Delta_{pol}H^0(T)$	Type transformation	T (K)	$\Delta_{pol}H^0$ (T)
1-Tridecanol			1-Tetradecanol		
K(β) → liq	304.63	45.12	K(β) → liq	311.00	49.51
K(γ) → liq	304.83	41.4	K(γ) → liq	311.21	47.0
K(α) → liq	303.50	23.3	K(α) → liq	310.79	25.1
K(β) → K(γ)	301.6	3.6	K(β) → K(γ)	306	1/8
K(β) → K(α)	305.8	22.1	K(β) → K(α)	311.2	23.8
K(α) → liq	306.6	18.7	K(γ) → K(α)	311.6	23.0
1-Pentadecanol			1-Hexadecanol		
K(β) → liq	316.61	54.72	K(α) → liq	322.2	58.38
K(α) → liq	316.92	30.3	K(α) → liq	322.34	33.6
K(β) → K(α)	316.20	23.6	K(γ) → K(α)	322.2	23.7

of dihydric alcohols of the normal structure. The values of enthalpies of polymorphous transitions and melting of alcohols compiled in Table 5.20 show the high magnitude of these characteristics.

In particular, the transition from the $K(\beta)$ polymorphous form into $K(\alpha)$ in 1-tridecanol and 1-tetradecanol requires 22.1 and 23.8 kJ mol^{-1}, respectively, whereas their melting enthalpies amount to 23.1 and 25.1 kJ mol^{-1}. Thus, the enthalpy of the polymorphous transition is less by about 1 kJ mol^{-1} than the melting enthalpy. Consequently, the enthalpies of phase transition $K(\beta) \rightarrow$ liq of the alcohols in question equal 45.12 and 49.51 kJ mol^{-1} and is only nearly twofold less than their vaporization enthalpy (97 and 102 kJ mol^{-1}). This means that in solid dihydric alcohols the stabilization of the crystal field is significant, and consequently the stabilization occurs of the hydrogen bonds and of the specific interactions involving the CH_2-groups. The essential difference in their energies results in the alcohols of the series 1-tridecanol–1-hexadecanol in three polymorphous transitions having sharply dissimilar enthalpy characteristics. In this connection, the thermodynamic calculation of the sublimation enthalpy is strongly complicated by the lack of information of the polymorphous transitions of the alcohols.

Similar enthalpies of the overall polymorphous transitions and of the melting in the series 1-tridecanol–1-hexadecanol amounting to about 25 kJ mol^{-1} correspond, on the one hand, to the considerable disorder of the structure at the polymorphous transitions, and on the other hand, to the disintegration of the crystal structure in the course of melting; the enthalpy and entropy values of these processes are related to the number and energy of the ruptured bonds [49].

The two hydroxy groups in the molecules of dihydric alcohols ensure their high association in the liquid state due to the doubled number of the hydrogen bonds and the specific interactions whose stabilization in the crystal field results in still higher values of their sublimation enthalpy compared to that in the monohydric alcohols (Table 5.21).

As a result the conditions arise in the crystalline state for the formation of polymorphous forms and the stabilization of the hydrogen bonds and the specific interactions involving the fifth bond vacancy of the carbon atom and the hydrogen

Table 5.21 Vaporization and sublimation enthalpy (5.5 kJ mol^{-1}), energies of hydrogen bonds and specific intermolecular interactions (kJ mol^{-1}) in symmetric dihydric alcohols of normal structure (298.15 K)

Alcohol	Formula	$\Delta_{vap}H^0(T)$ Vap [47]	$\Delta_{sub}H^0(T)$ Sub [48]	DH–O → CH_2 Liquid	DH–O → CH_2 Solid	DH–O···H–O Liquid	DH–O···H–O Solid
Methanediol	CH_4O_2	56[a]	74.0[a]	5.63/ DO···HC = 5.5	6.60/ DO···HC = 6.50	8.2	11.8
1,2-Ethanediol	$C_2H_6O_2$	64[a]	81.0[a]	5.63 × 4	6.60 × 4	10.3	12.65
1,3-Propanediol	$C_3H_8O_2$	72.4	88.0[a]	5.63 × 2 6.78 × 2	6.60 × 2 8.83 × 2	11.9	14.3
1,4-Butanediol	$C_4H_{10}O_2$	79.3	95.3	6.78 × 4	8.83 × 4	12.05	15.0
1,5-Pentanediol	$C_5H_{12}O_2$	86.8	103.0[a]	6.78 × 2 8.83 × 2	8.83 × 2 11.60	13.9	15.5
1,6-Hexanediol	$C_6H_{14}O_2$	96.5[a]	112.0	8.83 × 4	11.60 × 4	15.3	16.4
1,7-Heptanediol	$C_7H_{16}O_2$	107	126.0[a]	8.83 × 4 10.5	11.60 × 4 13.8	15.3	16.5
1,8-Octanediol	$C_8H_{18}O_2$	117[a]	139.3	8.83 × 4 10.5x2	11.60 × 4 13.8x2	15.2	16.5
1,9-Nonanediol	$C_9H_{20}O_2$	124[a]	153[a]	8.83 × 4 10.5 × 2	11.60 × 4 13.8 × 4	15.3	16.5

[a]Established by interpolation

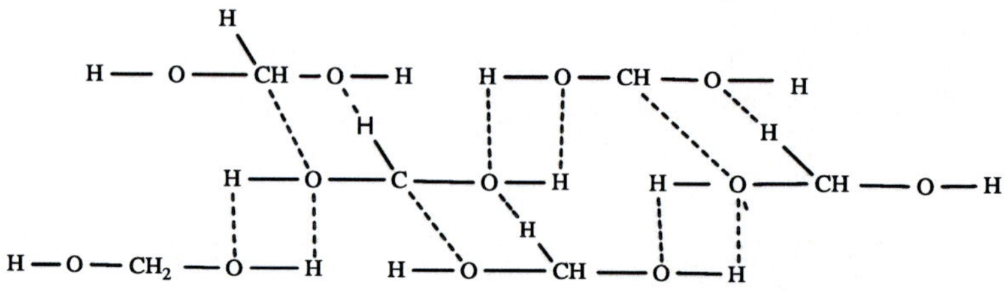

Fig. 5.26 Schematic picture of the network structure of liquid formaldehyde hydrate (methanediol)

atoms of the CH_2-groups. The structure of the liquid and the crystalline state in the nearest surrounding of a dihydric alcohol should be identical. The thermodynamic calculation of the energy of the hydrogen bonds in the dihydric alcohols has been already presented in Sect. 5.4, still the structures of formaldehyde hydrate (methanediol) and ethanediol require an additional consideration. Taking into consideration the rule of the occupancy of all bond vacancies in the dihydric methanediol we should account for the bond formed by the essentially unshared $2s^2$ electron pair of the carbon atom H–O → CH_2 and by the oxygen atom H–O···H–CH with the hydrogen atom located in the *trans*-position of the CH_2-group (Fig. 5.26).

The energy of the specific interaction of the carbon atom of the CH_2-group with the lone electron pair of the hydroxy group oxygen we take as equal to the energy of the analogous specific interaction DH–O → CH_2 in the liquid dimethyl ether. The energy of the hydrogen bond involving the hydrogen atom in the *trans*-position of the CH_2-group with the electron pairs still not brought into the interaction on two oxygen atoms H–O···H–CH we take equal to the energy of the hydrogen bond of the liquid formaldehyde (5.5 kJ mol^{-1}) based on the resemblance of the structures and the specificity of their interactions in the liquid state. In the liquid and crystalline ethanediol also formed four hydrogen bonds and four specific interactions H–O···CH_2 whose energies do not differ from the energies of analogous specific interactions in the corresponding states of the dimethyl ether.

The results of the performed thermodynamic calculations (Table 5.21) illustrate the regular increase in the energy of the hydrogen bonds in the liquid and solid dihydric alcohols in the series methanediol–1,7-heptanediol and the stabilization of the hydrogen bond in the crystals.

The obtained values of the energy of the hydrogen bonds in the saturated alcohols of normal and branched structures with different length of the alkyl chain, in the alcohols containing the benzene and alicyclic structures allow some general conclusions:

Firstly, the energies of the hydrogen bonds in the 1-alkanols of the normal structure grow from methanol to 1-propanol due to the weakening of the influence of the reverse dative bond existing between the oxygen atom of the hydroxy group and the contiguous carbon atom of the molecule. The energy of the hydrogen bond in the branched alcohols is stabilized at the increase to three of the carbon atoms of the alkyl chain. The isostructural methyl groups take part in the formation of the specific interactions of low stability.

Secondly, in the alcohols with over three carbon atoms in the chain the hydrogen bond energies remain practically constant;

Thirdly, the specific intermolecular interactions in the alcohols form with the participation of the essentially unshared $2s^2$ electron pair of the carbon atom of the terminal methyl group of the chain occurring in the pentacoordinate state. The energies of the specific intermolecular interaction stabilize with the number of carbon atoms in the alkyl chain growing to three and further remain practically constant.

Fourthly, the increasing number of hydroxy groups in alcohols results in decreasing energy (destabilization) of an isolated hydrogen bond.

Fifthly, the replacement of the alkyl chain by a benzene ring results in the destabilization of the hydrogen bond, and the replacement of the alkyl chain by an alicyclic ring stabilizes the hydrogen bond increasing its energy in cyclohexanol to the energy existing in 1-propanol and 1-butanol.

Sixthly, the energy of the hydrogen bond in alcohols grows (is stabilized) in the series: benzyl alcohol < tetramethylene glycol < methanol < cyclopentanol < ethanol < 1-propanol ≤ 1-butanol–1-dodecanol ≤ cyclohexanol.

5.7 Energies of Hydrogen Bonds and Specific Interactions in Unsaturated Alcohols

The thermodynamic properties of vaporization process have been measured for a limited number of unsaturated alcohols. The use of these data for the thermodynamic analysis is complicated by the fact that the majority of the enthalpy characteristics have not been measured at the standard temperature (Table 5.22). In the liquid phase, the normal unsaturated monohydric and dihydric alcohols form chain structures linked into a network by the specific intermolecular interactions. The difference consists in the capability of the liquid unsaturated alcohols to form the specific intermolecular interaction through the terminal CH_2- or CH-group, and that should show up in the energy of the specific interactions, of the hydrogen bonds, and finally in the enthalpy characteristics of the vaporization. The alcohols with the hydroxy groups localized not on the ends of the alkyl chain form the network structure of liquid state where the isostructural methyl groups also take part in the cross-linking. The resemblance of the structures of the molecules of saturated and unsaturated alcohols shows that the difference in the vaporization enthalpies of the 3-buten-1-ol (Table 5.22) and 1-butanol (1.40 kJ mol^{-1}) is due to the effect of the double bond. However, the experimental value of the enthalpy characteristic was published without indication of the measurement accuracy and at the temperature higher the standard one. The vaporization enthalpy of 3-buten-1-ol originates from the sum of two hydrogen bonds, two specific interactions $-O \rightarrow CH_2=CH-CH_2$, and from the contribution of the CH_2-group. The hydrogen bond energy in the liquid 3-buten-1-ol can be obtained from the difference between the vaporization enthalpy of this compound and the energies of two specific interactions along (5.4).

Table 5.22 Energies (5.5 kJ mol^{-1}) of hydrogen bonds and specific interactions in unsaturated alcohols

Alcohols	Structure	$\Delta_{vap}H^0(T)$	T (K)	$DCH_2/DisoC_3H_5$	D–O → CH_2=CH–CH_2	DH–O···H–O
3-Buten-1-ol	CH_2=CH-$(CH_2)_2$-OH	50.9	313	4.70	8.66 × 2	14.45[a]
1-Butanol	CH_3-$(CH_2)_3$-OH	52.39	298	4.70	8.66 × 2	15.00 ± 0.05
5-Hexen-1-ol	H_2C=HC–$(CH_2)_3$–CH_2–OH	60.2 ± 0.1	298	4.70 × 3	8.66 × 2	14.40
1-Hexanol	H_3C–H_2C–$(CH_2)_3$–CH_2–OH	61.72	298	4.70 × 3	8.92 × 2	14.90
1-Penten-3-ol	CH_3–CH_2–CH(OH)–CH=CH_2	49.9 ± 0.1	313	$DisoC_3H_5$ = 3.25	8.66 × 2	14.40
3-Pentanol	CH_3–CH_2–CH(OH)–CH_2–CH_3	53.6	298	$DisoC_3H_7$ = 6.05	8.92 × 2	14.85
2-Buten-1,4-diol	HO-CH_2-CH=CH-CH_2-OH	69.0	433	–	6.78 × 4	8.6[a] 12.5[b]
1,4-Butanediol	HO-CH_2-CH-CH-CH_2-OH	79.3	298	–	6.78 × 4	13.05

2-Methyl-2-propen-1-ol	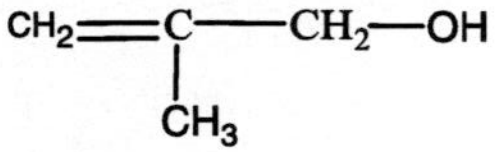	51.9	298	D*iso*CH_3 = 5.80	8.66 × 2	14.40
3-Methyl-2-buten-1-ol	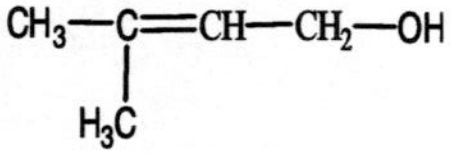	48.7	360	∑D*iso*CH_3 = 4.10[c]	8.66 × 2[a]	13.7
2-Methyl-3-buten-2-ol	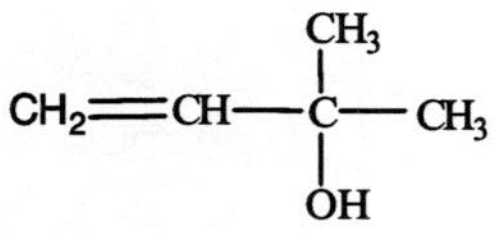	43.1 ± 0.1	331	∑D*iso*CH_3 = 4.10[c]	8.66 × 2[a]	10.85
2-Methyl-3-butyn-2-ol	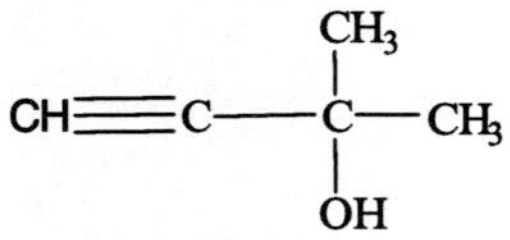	49.5	294–380	∑DisoCH_3 = 4.10[c]	8.50 × 2	14.20
6-Methyl-5-hepten-2-ol	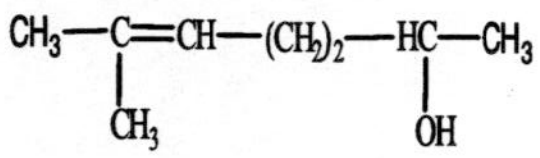	57.0	329	4.7 × 3 ∑D*iso*CH_3 = 4.10[c]	8.66 × 2[a]	10.75
6-Methyl-3-hepten-2-ol	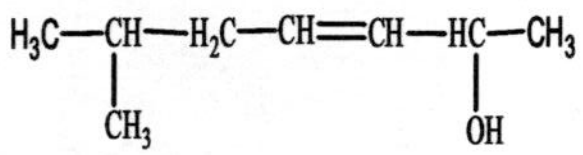	59.7	329	4.7 × 3 ∑D*iso*CH_3 = 4.10[c]	8.66 × 2[a]	12.10

[a]The temperature dependence of the energy of the specific interaction was disregarded[b]Estimated from the hydrogen bond energies of 3-buten-1-ol–1-Butanol and 5-hexen-1-ol–1-hexanol[c]Estimated from the data of the saturated alcohols

The data of Table 3.4 show that the diethyl ether forms the specific interaction (6.78 kJ mol^{-1}) more stable by 0.26 kJ mol^{-1} than the interaction in the divinyl ether (6.52 kJ mol^{-1}). Analogous difference should exist between the energies of the specific interactions formed by a saturated three-carbon chain and analogous chain containing a double bond. This conclusion makes it possible to evaluate the energies of the specific interactions in the liquid 5-hexen-1-ol and the other unsaturated alcohols. The performed calculations (Table 5.22) illustrate the constant value of the interaction energy and its lower stability compared with the specific interactions of the saturated alcohols. An interesting compound is the 1-penten-3-ol whose structure resembles that of symmetric ketones and ethers by the even distribution of the electron density providing an equal possibility both to the saturated and unsaturated fragments of the carbon skeleton to form the specific interactions. However, only one of these chains forms the specific interaction with the oxygen atom $–O \rightarrow CH_3–CH_2–CH_2$, the second one practically acts like an isostructural propene group. The same role plays the propyl group of the liquid 3-pentanol (Table 5.22).

In the dihydric 2-butene-1,4-diol like in the monohydric alcohol, the stability of the hydrogen bond is also lower than in the 1,4-butanediol. Therefore, the value of the vaporization enthalpy measured at 433 K should essentially increase at the standard temperature and be close to the enthalpy characteristic of the 1,4-butanediol. Consequently, by analogy with monohydric alcohols, the energy of the hydrogen bond of the liquid 1,4-butenediol should be lower in the value than in the liquid 1,4-butanediol, and it may be estimated at close to 12.5 kJ mol^{-1}.

The obtained value of the energy of the hydrogen bond in the unsaturated alcohols makes it possible to estimate the energy contribution of the isostructural methyl group into the enthalpy characteristic of 2-methyl-2-propen-1-ol. The energy of the interaction formed by the isostructural methyl group is calculated by (5.26) taking into account the energies of the hydrogen bonds and the specific interactions

$$\begin{aligned} \mathrm{D}iso\mathrm{CH_3} = (&\Delta_{\mathrm{vap}}H^0(298)iso - 2\mathrm{DH-O\cdots H-O} - 2\mathrm{D-O} \\ &\rightarrow \mathrm{CH_2 = CH - CH_2}). \end{aligned} \tag{5.26}$$

The energy of this interaction (5.80 kJ mol^{-1}) is over two times greater than that of the analogous interaction in the liquid 2-methyl-1-propanol (2.50 kJ mol^{-1}).

The hydrogen bond energy of the liquid 3-methyl-2-buten-1-ol at 360 K may be estimated if the contribution of two methyl groups would be taken equal to analogous groups in 2-methyl-2-propanol (5.26). The obtained value is somewhat underestimated due to disregarded temperature dependence of the energy of specific interaction (Table 5.22).

The replacement of hydrogen at the second carbon atom in 3-buten-2-ol and 3 butyn-2-ol provides for each obtained compound, 2-methyl-3-buten-2-ol and 2-methyl-3-butyn-2-ol two isostructural methyl groups. Therefore, the calculation of the hydrogen bond energy is performed by (5.27).

$$2\mathrm{DH-O\cdots H-O} = (\Delta_{\mathrm{vap}}H^0(298)iso - 2\mathrm{D-O} \\ \rightarrow \mathrm{CH_2 = CH - CH_2} - 2\ \mathrm{D}iso\mathrm{CH_3})/2. \quad (5.27)$$

Making the calculations with the defined assumptions we obtained the underestimated energy of the hydrogen bond in 2-methyl-3-buten-2-ol due to disregarded temperature dependence of the energy of specific interaction (Table 5.22). In the liquid, 2-methyl-3-butyn-2-ol the specific interaction D–O → $CH_2{\equiv}CH–CH_2$ is formed by the three-carbon chain containing a triple bond, and this should affect the shift of the electron density in the molecule and the decrease in the charges on the oxygen atom and the terminal carbon atom. Consequently, the energy of the hydrogen bond should be destabilized compared with that in the alkyl chains with the double and ordinary bonds. Therefore, the calculation using the energy of the specific interaction D–O → $CH_2{=}CH–CH_2$ would give an underestimated value of the hydrogen bond energy. At the same time using the energy of the hydrogen bond for the liquid unsaturated compounds at 14.40 kJ mol^{-1} we obtain an underestimated value of the energy of the specific interaction D–O → $CH_2{\equiv}CH–CH_2$. The most realistic values of the energies of the specific interaction and the hydrogen bond are the mean values of both performed calculations (Table 5.22).

Unsaturated alcohols 6-methyl-5-hepten-2-ol and 6-methyl-3-hepten-2-ol illustrate the stabilization of the hydrogen bond and the specific interaction –O → $CH_2{=}CH_2–CH_2$ by 2.7 kJ mol^{-1} because of the approach of the double bond to the carbon atom attached to the hydroxy group. The results of calculations of the energies of the mentioned interactions are compiled in Table 5.22.

The comparison of the data of this table shows that the energies of hydrogen bonds and specific interactions of saturated and unsaturated alcohols are described by a regular series of their decreasing values:

Firstly, the energies of the hydrogen bonds of all alcohols with the number of carbon atoms in the chain three and more fit to the series

$$\text{1-Butanol } (15.00) > 3\text{ - Buten - 1 - ol } (14.40) > 3\text{ - Butyn - 2 - ol } \\ (14.20\,\mathrm{kJ\,mol^{-1}}).$$

Secondly, the decrease of the energy of the specific interactions involving the pentacoordinate carbon in the series

$$\mathrm{D-O \rightarrow CH_3 - CH - CH_2}\ (8.92) > \mathrm{D-O \rightarrow CH_2 = CH - CH_2}\ (8.66) > \mathrm{D-O} \\ \rightarrow \mathrm{CH \equiv C - CH_2}\ (8.50\,\mathrm{kJ\,mol^{-1}})$$

reflects the weakening of the bonding between the molecules at replacing the ordinary bond by a double and triple one in the skeleton of the molecules of alcohols.

Thirdly, the energies of the hydrogen bonds in liquid monohydric alcohols (14.80) are more stable compared to symmetric dihydric alcohols (14.00 kJ mol) due to the decreased negative charge on the oxygen atom. The energy of the hydrogen bond in the unsaturated and saturated symmetric dihydric alcohols is described by an inequality 2-butene-1,4-diol (13.05) > 1,4-butanediol (12.5 kJ mol^{-1}).

5.8 Thermodynamics of Vaporization of Acetylene Alcohols and Isophitol. Energies of Specific Interactions

The published information on thermodynamic investigation of alcohols from isophitol series with double and triple bonds [3,7-dimethylocten-6-yn-1-ol-3 (dihydrolinalool), 3,7,11-trimethyldodecyn-1-ol-3 (acetylene alcohol C_{15}), 3,7,10, 15-trimethylhexadecyn-1-ol-3 (acetylene alcohol C_{20}), and 2,6,10,14-tetramethyl-hexadecen-1-ol-3 (isophitol)] is absent. The energies of the hydrogen bonds were unknown and the specific intermolecular interactions were not discussed. The thermodynamic characteristics of the other unsaturated alcohols were not measured apparently due to the difficulties presented by their purification to the extent ensuring the possibility of the thermodynamic investigation and of obtaining reliable thermodynamic properties of compounds suitable for the thermodynamic analysis.

The second complex problem in the thermodynamic investigation of these compounds is the necessity to have a reliable test method for controlling the state of the compound under study in the process of the investigation. It is important to ensure the absence in the system of side processes whose occurrence introduces additional errors into the measurement of the thermodynamic properties. The control method should provide a possibility to take an account of the effect of the side reactions on the main process. The availability of information on the stability of the substances and the constant composition of the initial compounds permit a valid choice of the investigation procedures for solving the mentioned problems aiming at the establishment of the reliable thermodynamic properties of compounds comparable to analogous properties of the other compounds.

In order to confirm the high reliability of the performed study and the precision of the obtained experimental data we regarded as necessary to go into details of the procedure and to report on certain experimental results.

The investigations of vapor–liquid equilibrium were carried out by the static tensiometric method with a membrane zero-gauge along a developed procedure approbated on a large number of compounds belonging to versatile classes. The procedure ensures obtaining precision experimental results [53]. The vapor pressure was measured at fixed temperature in the course of heating and cooling of the system with the manometer MChR-3 with an accuracy ±0.1 mm Hg. The temperature in the liquid thermostat was maintained with the accuracy ±0.05 K. The investigations were carried out in a wide range of temperatures and pressures.

The vaporization enthalpy of the initial compounds was measured by a calorimeter calibrated by the vaporization enthalpy of methanol and benzene in the temperature range 298.15–343.15 and 298.15–353.25 K, respectively. The experimental findings for the vaporization enthalpy of methanol and benzene at the limiting temperature of the mentioned ranges are well consistent with the most valid published data: for methanol at 298.15 K 37.31 ± 0.32 and 37.40 [55], 37.28 [55], at 343.12 K 34.71 ± 0.48 and 34.85 [55] kJ mol^{-1}; for benzene at 298.15 K 33.80 ± 0.27 and 33.82 [55], 33.84 [56], at 353.25 K 30.70 ± 0.38, 30.75 [55], 30.73 kJ mol^{-1} [55, 57]

Acetylene alcohols possess a large number of isomers. For instance, the acetylene alcohol C_{20}, 3,7,10,15-tetramethylhexadecyn-1-ol-3 is practically inseparable from its isomer 3,7,11,15-tetramethylhexadecyn-1-ol-3. Therefore, the isolation of these compounds in the pure state is a complex experimental task. The content of the main substance in the initial compound was 75–80%. In the course of the primary fractional distillation, the main fraction was separated and subjected to rectification.

The primary purification of the acetylene alcohol C_{15} was carried out by repeated vacuum distillation at the residual pressure 0.066 kPa. The fraction boiling at the temperatures 361.15–363.95 K was collected. The analysis performed in the course of the rectification by means of GLC revealed the presence of difficultly divisible non-volatile impurities in an amount of 0.9%. The purification of the acetylene alcohol C_{20} was carried out by repeated vacuum distillation at the residual pressure 0.053 kPa. The fraction boiling at the temperatures 400.35–403.35 K was collected. The GLC analysis revealed the presence of nonvolatile impurities in an amount of 7%.

The repeated control of the impurity content by GLC resulted in its minimum presence. The degree of the purity of the acetylene alcohols C_{15} (3,7,11-trimethyldodecyn-1-ol-3) and C_{20} (3,7,10,15-tetramethylhexadecyn-1-ol-3) was tested by the molecular refraction. The gas removal was performed by repeated (no less than ten times) freezing cycles in a vacuum. The purified and free of gas acetylene alcohols were stored in the dark and cold in sealed ampules.

The procedure of charging the substances into the membrane ampule was somewhat complicated since the acetylene alcohols C_{15} and C_{20} were oily fluids, and the quantitative distillation of the weighed sample required preliminary heating. The dihydrolinalool (3,7-dimethylocten-6-yn-1-ol-3) was easily distilled into the membrane chamber of the zero-gauge and was sufficiently heat resistant.

The isophitol was a viscous oily fluid, therefore the process of gas removal was carried with it for 6–8 h. Its distillation into ampules also was time consuming.

The measured pressures of the saturated and unsaturated vapor of the acetylene alcohols are depicted in Fig. 5.27. The transition to the region of unsaturated vapor is clearly seen and on the line of the gas expansion the Gay–Lussac law is valid.

The calculated values of the molecular masses of the acetylene alcohols C_{15} and C_{20} (Table 5.23) are within the experimental error identical to the values calculated from the empirical formulas.

This result shows that the studied compounds are present in the vapor as monomeric molecules. At temperatures exceeding 614 K the acetylene alcohol C_{20} was found to suffer irreversible decomposition, and the gain in the vapor pressure attained 333.3 Pa in 2 h. The linear dependence $\log P = f(1/T)$ in all

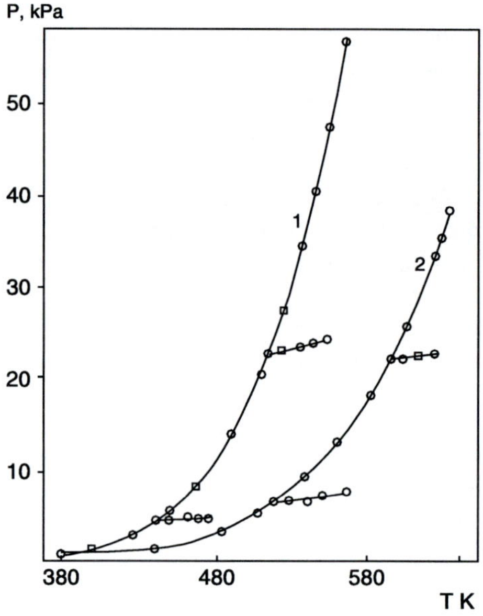

Fig. 5.27 Dependence of the pressure of saturated and unsaturated vapor of acetylene alcohol C_{15} (1) and acetylene alcohol C_{20} (2) versus temperature (T)

Table 5.23 Mean molecular mass of acetylenes alcohol C_{15}, C_{20} and isophitol at temperature of transition of saturated into unsaturated vapor

Compound	Formula	Run no.	M calc.	M exp.
Acetylene alcohol C_{15}	$C_{15}H_{28}O$	1	224.42	215.9 ± 1.2
		2		215.9 ± 1.0
Acetylene alcohol C_{20}	$C_{20}H_{38}O$	1	294.52	291.8 ± 1.4
		2		291.1 ± 2.1
Isophitol	$C_{20}H_{40}O$	1	295.34	295.7
Dihydrolinalool	$C_{10}H_{16}O$	1	152.23	150.8
		2		153.7

studied range of temperatures indicates the absence of both volatile and nonvolatile impurities. Apparently, only the isomeric forms of compounds are presently characterized by practically the same boiling point. The P–T dependences were approximated by Clapeyron–Clausius and Antoine equations [54]. The performed measurements of the saturated vapor pressure underlie the reliability of the calculated thermodynamic characteristics of vaporization of the compounds under consideration. The study of the temperature dependence of the pressures of saturated and unsaturated vapor for isophitol and dihydrolinalool was also carried out in a wide temperature and pressure range. Inasmuch as the experimentally measured molecular masses are the same within the error limit as those calculated from the empirical formulas, these compounds are present in vapor as monomer molecules (Table 5.23). The clearly seen transition from the saturated into unsaturated vapor is due to the high purity of the studied objects and to the absence of the side processes

Table 5.24 Vaporization enthalpy (5.5 kJ mol^{-1}) and entropy (J mol^{-1} K^{-1}) of acetylene alcohols C_{15} and C_{20}, isophitol and dihydrolinalool

Compound	Temperature interval, K	$\Delta_{vap}H^0$ (*T*)	$\Delta_{vap}S^0$ (*T*)
Acetylene alcohol C_{15}	381–572	43.4 ± 0.28	91.1 ± 1.1
Acetylene alcohol C_{20}	401–609	44.00 ± 0.32	88.9 ± 1.0
Isophitol	411–521	44.67 ± 0.30	98. ± 1.0
Dihydrolinalool	350–445	50.72 ± 0.30	107.8 ± 1.3

Table 5.25 Calorimetric data of vaporization enthalpy (kJ mol^{-1}) of acetylene alcohols C_{15} and C_{20}

Compound	Temperature			
	413.25	423.35	442.95	464.35
Acetylene alcohol C_{15}	44.93 ± 0.18	43.93 ± 0.14	43.76 ± 0.15	43.61 ± 0.17
Acetylene alcohol C_{20}	44.39 ± 0.13	44.32 ± 0.18	44.22 ± 0.16	44.10 ± 0.18

in the investigated temperature range. The initial data on the saturated vapor are described by the Antoine equation [54].

The calculated values of the enthalpy and entropy of vaporization are presented in Table 5.24.

The results of the calorimetric measurement of the vaporization enthalpy of acetylene alcohols C_{15} and C_{20} are given in Table 5.25. They are well consistent with the vaporization enthalpy of the same compounds determined from the tensimetric compounds at the medium temperatures 475 and 500 K for alcohols C_{15} and C_{20}, respectively. The performed calorimetric measurements of the vaporization enthalpy of dihydrolinalool at various temperatures showed that in the range 415–444 K the intensity of vaporization was high. In this case at the insignificant pressure drop in the calorimetric cell, the efficiency of the registration was not sufficient and the error grew. On the whole the results of the direct measurement of the vaporization enthalpy are well consistent with the data obtained by tensimetry [54].

The hydrogen bond energies are stabilized in the series of the 1-alkenols of normal structure with the growing number of carbon atoms in the molecule from methanol to 1-butanol due to the decrease in the influence of the reverse dative bond between the oxygen atom and adjacent carbon atom of the alkyl chain (Table 5.4). The subsequent increase in the number of carbon atoms in the chain does not affect the stabilization of the hydrogen bond, and the growth of the vaporization enthalpy is provided by the contribution of the interaction of the growing number of methylene groups with the carbon atoms of the contacting molecules.

This feature of saturated and unsaturated alcohols was not observed in the vaporization enthalpies of the acetylene alcohols C_{15} and C_{20} (Table 5.26) where the corresponding values are close to each other (C_{15}, 43.43 and C_{20}, 44.00 kJ mol^{-1}) and consequently are independent of the number of the carbon atoms in the chain. This experimental fact is important not only for the thermodynamic analysis of the correspondence between the structure and energy, for the problems of the hydrogen

Table 5.26 Vaporization enthalpy (5.5 kJ mol^{-1}) of dihydrolinalool, acetylene alcohols C_{15} and C_{20}, and isophitol

Compound	Structure	$\Delta_{vap}H^0(T)$	$\Delta_{vap}S^0(T)$
Dihydro-linalool	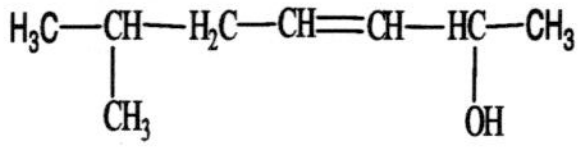 3,7-dimethy-6-octen-1–yn-3-ol	50.78 ± 0.30	107.2 ± 1.3
Acetylene alcohol C_{15}	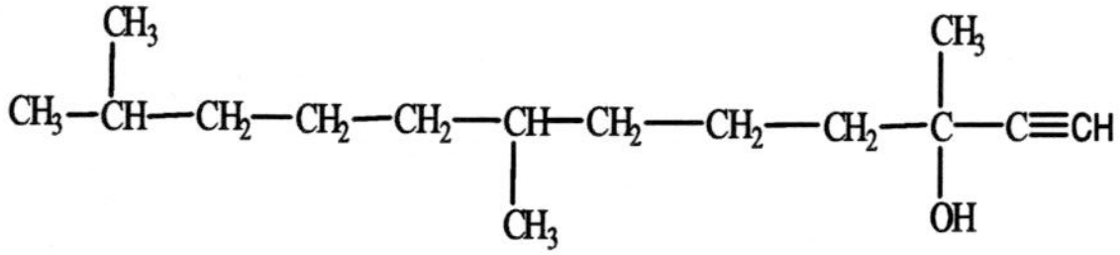 3,7,11-trimethyldodecyn-1-ol-3	43.43 ± 0.28	91.1 ± 1.1
Acetylene alcohol C_{20}	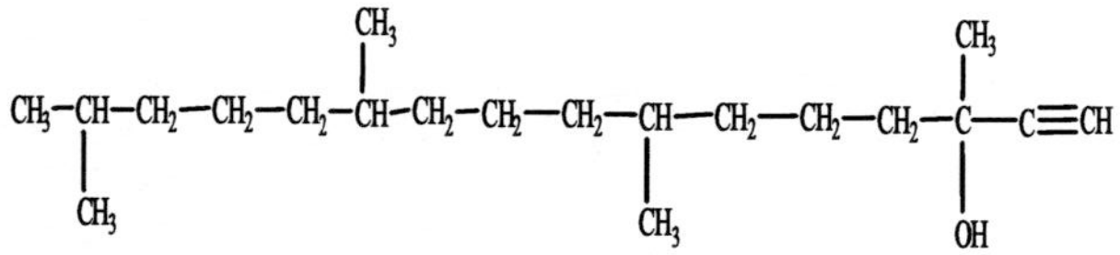 3,7,11,15 – tetramethylhexadecyn-1-ol-3	44.00 ± 0.32	88.2 ± 1.0
Isophitol C_{20}	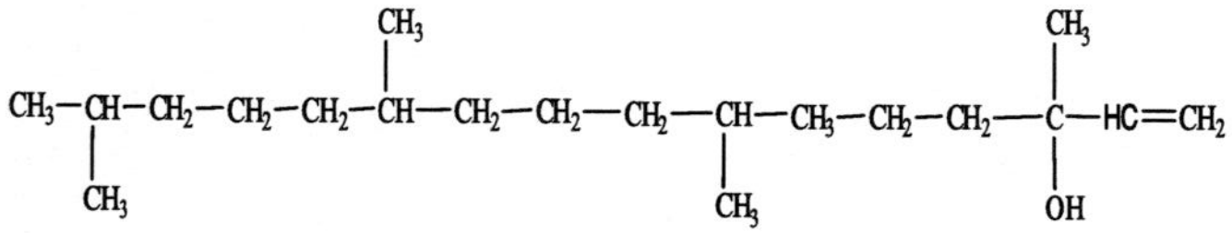 2,6,11,15 – tetramethylhexadecen-1-ol-3	44.67 ± 0.30	98.4 ± 1.0

bonds and specific intermolecular interactions. It is especially significant for the understanding of the features of the interactions in the polyfunctional compounds, peptides, and proteins. The difference in the vaporization enthalpy of the acetylene alcohols C_{15} and C_{20} (0.57 kJ mol^{-1}) corresponds to the energy of the specific interaction $H_3C \rightarrow H–CH_2$, for the former acetylene alcohol has one methyl group less. It allows an assumption that the isostructural methyl groups in the acetylene alcohols form the intermolecular interaction with the same energy as in the branched saturated alcohols. Therefore, the three methyl groups in the acetylene alcohol C_{15} and four methyl groups in the alcohol C_{20} contribute to the vaporization enthalpy 1.70 and 2.20 kJ mol^{-1}, respectively. The remaining energies 41.73 and 41.80 kJ mol^{-1} are contributed to the vaporization enthalpy of liquid alcohols C_{15} and C_{20}, respectively, by the existing specific intermolecular interactions involving the pentacoordinated carbon atom of the terminal methyl group, by the hydrogen bonds and the four-carbon fragments of the carbon–carbon chain between the side methyl groups. The said energies are only by 0.80 kJ mol^{-1} lower than the vaporization enthalpy of ethanol (42.80 kJ mol^{-1}).

The constant energy of the specific intermolecular interactions in the acetylene alcohols is the consequence of the following facts:

Firstly, the constant effect of the triple bond $–C \equiv C–$ in the fragment

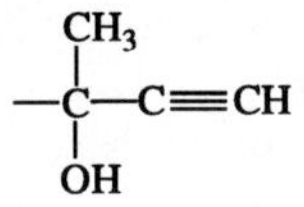

of the acetylene alcohols.

Secondly, the vaporization enthalpy including the energy of the total number of the specific interactions of liquid acetylene alcohols does not depend on the number of the carbon atoms in the chain and originates from the number of four-carbon fragments and the terminal methyl group.

Thirdly, the terminal methyl group is separated by the carbon atom bearing the isostructural methyl group from a linear fragment of three methylene groups,

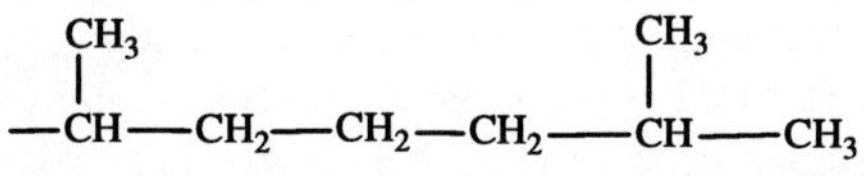

and therefore the terminal methyl group forms with the oxygen of the hydroxy group specific interaction analogous to that existing in the liquid dimethyl ether.

Just the even distribution in the chain of the isostructural methyl groups prevents the contribution of the CH_2-groups of the fragments into the vaporization enthalpy of the acetylene alcohols. It means that the hydrogen and carbon atoms of the methylene groups of the fragments virtually do not take part in the interaction with the analogous groups of the other chains. Consequently, the effect of the isostructural methyl groups is observed on the charges on the hydrogen and carbon atoms of the CH_2-groups, on the displacement of the electron density in the alkyl chain of the fragment; this fact distinguishes these groups of similar groups in the alkenols.

Therefore, these isostructural methyl groups govern the low value of the vaporization enthalpy of the acetylene alcohols and the alcohols having a double bond at the beginning of the chain.

Let us consider now the effect of the isostructural methyl groups on the vaporization enthalpy on branched pentanols. As a result of the shift of the methyl group of the 3-methyl-1-butanol to the central atom of the chain in 2-methyl-1-butanol the vaporization enthalpy decreases. A more significant effect on the vaporization enthalpy and on the energy of the hydrogen bond is caused by the shift of the hydroxy group to the central carbon atom in 2-pentanol and 3-pentanol (Table 5.9).

3-Methyl–1-butanol

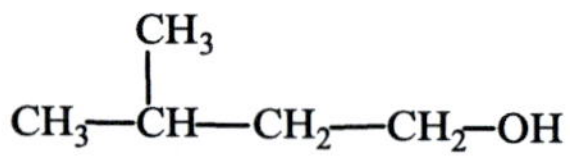

54.3 kJ mol^{-1}

2-Methyl –2-butanol

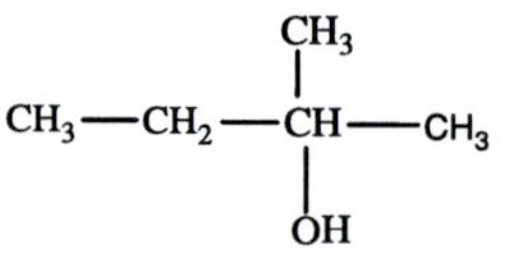

49.2 kJ mol^{-1}

2-Methyl-3-butyn-2-ol

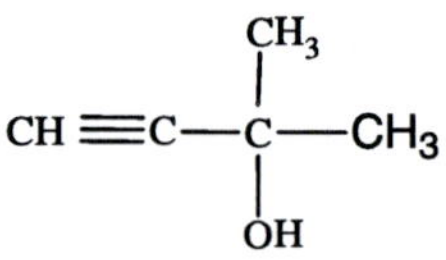

49.5 kJ mol^{-1}

The largest decrease in the vaporization enthalpy occurs at the simultaneous location of the methyl and the hydroxy groups at the same carbon atom, like in 2-methyl-3-butyn-2-ol and 2-methyl-2-butanol, namely, their effect on the distribution of the electron density increases and is revealed in the diminished donor properties of the carbon atoms in the methyl groups.

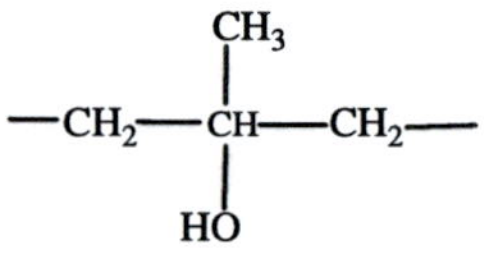

Just in this event the vaporization enthalpy of 2-methyl-2-butanol (49.2 kJ mol^{-1}) approaches the vaporization enthalpy of 2-methyl-3-butyn-2-ol. It means that the isostructural methyl group makes the molecule a rigid system and thus destabilizes the specific intermolecular interactions of alcohols. The maximum destabilizing effect is observed when both isostructural methyl and hydroxy groups are attached to the second carbon atom. The vaporization enthalpy of 2-methyl-2-butanol exceeds by 5.20 kJ mol^{-1} the enthalpy characteristic of the acetylene alcohol C_{20} evidently due to the isostructural methyl groups at both ends of the fragment taking into account the weak influence of the double and triple bonds on the shift of the electron density. In this connection, isophitol with a double bond at the end of the chain –C–CH=CH_2 is especially interesting. The increase in the vaporization enthalpy by 0.67 kJ mol^{-1} compared to the acetylene alcohol C_{20} should be ascribed to the general stabilization of the molecule and evidently it should be reflected in the stabilization of the hydrogen bond. The molecules of dihydrolinalool, the acetylene alcohols C_{15} and C_{20} contain equal number of free bond vacancies, and their structure in liquid state is schematically represented in Fig. 5.28.

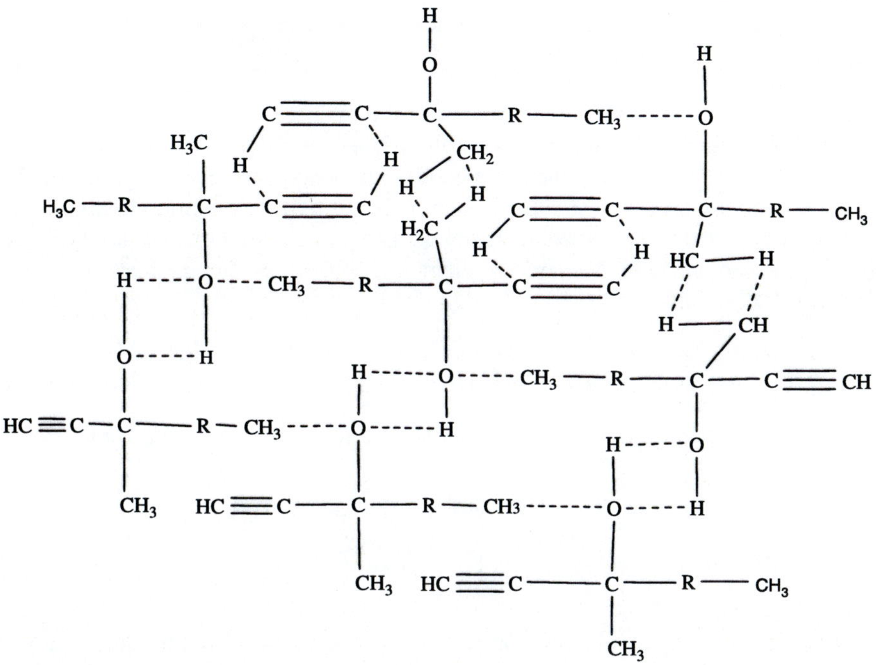

Fig. 5.28 Schematic picture of the structure of liquid acetylene alcohols

The vaporization enthalpies of isophitol, acetylene alcohol C_{20}, saturated and unsaturated ketones C_{18} lie in the range 43.19–44.67 kJ mol^{-1} and differ by 1.5 kJ mol^{-1}. The presence in the molecules of these compounds of three identical fragments shows that these fragments make the same contribution into the vaporization enthalpy. This fact makes it possible to evaluate the contribution of each fragment from the difference between the vaporization enthalpies of the unsaturated ketones (Chap. 4) amounting to 6.15 kJ mol^{-1}.

The energy of the specific interaction forming between the terminal methyl group and the oxygen atom of the hydroxy group H–O → CH_3 (Fig. 5.28) may be taken equal to the energy of the analogous interaction in dimethyl ether for isophitol, acetylene alcohol, and dihydrolinalool. Taking into account that the contribution of the isostructural methyl groups linked to each fragment is included into the value 6.15 kJ mol^{-1} the only unaccounted isostructural methyl group is that attached to the same carbon as the hydroxy group. It was already shown that the interaction formed by the isostructural methyl group contributes to the vaporization enthalpy 0.60 kJ mol^{-1}. A similar contribution originates from the specific interaction formed by the hydrogen and a carbon of the group –C ≡ CH. As a result we get an equation containing all the contributions of the interactions formed in liquid dihydrolinalool and alcohols C_{15} and C_{20}

$$\Delta_{\mathrm{vap}}H^0(T) = 2\mathrm{DH-O\cdots H-O} + 2\mathrm{D-O \rightarrow CH_3} + 2\mathrm{D-C \rightarrow H-C-} + \mathrm{DH_3-C \rightarrow H-CH_2} + n \cdot 6.15 \qquad (5.28)$$

where "n" is the number of the fragments in the molecule.

Thus, we are able to estimate the energies of the hydrogen bonds in the acetylene alcohols. The replacement of the triple bond of the acetylene alcohols by the double bond in isophitol ensures a small stabilization of the hydrogen bond. The vaporization enthalpy of this compound includes the energy of three types of the intermolecular interaction, and one among them is identical to the interaction of the terminal CH_2-group

$$\Delta_{\mathrm{vap}}H^0(T) = 2\mathrm{DH-O\cdots H-O} + 2\mathrm{D-O \rightarrow CH_3} + n \cdot \mathrm{DH_3C \rightarrow H-CH_2} + 6.15 \times 3. \qquad (5.29)$$

The replacement of the triple bond of the acetylene alcohols by the double bond in isophitol

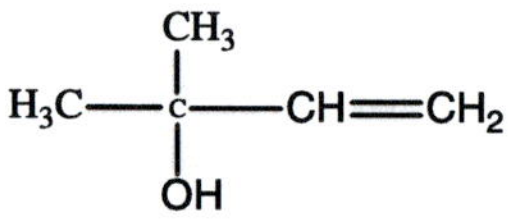

results in relative decrease in the shift of the electron density from the carbon atom of the CH_2-group. Therefore, the hydrogen bond energies of the acetylene alcohol C_{20} and the isophitol should not strongly differ. The results of the performed calculations of the hydrogen bond energies are compiled in Table 5.27.

As seen from the table, the hydrogen bond energies in dihydrolinalool and acetylene alcohols C_{15} and C_{20} decrease with the growing number of fragments from one to three in the alcohol C_{20}. Consequently, the increased stability of the hydrogen bond in the liquid dihydrolinalool as compared to the analogous bonds in the liquid saturated alcohols (15.00 kJ mol^{-1})

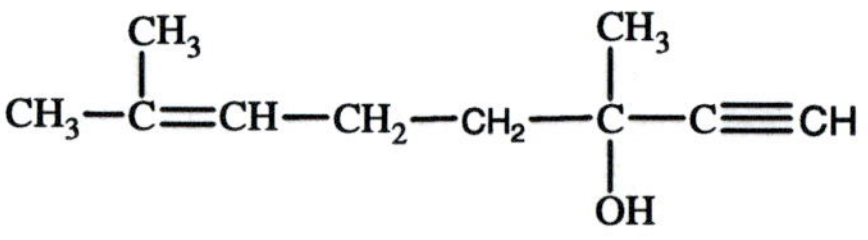

is due to the single fragment in the molecule of the dihydrolinalool ensuring the shift of the electron density and the increase in the negative charge on the oxygen atom and the positive charge on the carbon of the terminal methyl group.

Table 5.27 Energies (5.5 kJ mol^{-1}) of hydrogen bonds and specific interactions of acetylene alcohols and isophitol

Compound	$\Delta_{\mathrm{vap}}H^0(T)$	D–O → CH_3	DH_3–C → H–CH_2	DH–C → H–C–	DH–O···H–O
Dihydrolinalool	50.78 ± 0.30	5.63	0.60	0.60	15.9
Acetylene alcohol C_{15}	43.43 ± 0.28	5.63	0.60	0.60	9.60
Acetylene alcohol C_{20}	44.00 ± 0.32	5.63	0.60	0.60	6.55
Isophitol	44.67 ± 0.30	5.63	0.60	DHC → H–CH 0.60	6.90

5.9 Extrastabilizing Effect of Isostructural Methyl Groups

The important result of investigation of the organic compounds thermodynamics is the establishment for the first time the fact of extremely low value of the thermodynamic characteristics of the vaporization of saturated and unsaturated ketones C_{13} and C_{18} and acetylene alcohols C_{15} and C_{20}. The small difference between the vaporization enthalpy of saturated and unsaturated ketones and acetylene alcohol C_{20} and isophithol with equal number of carbon atoms in the chain is also important.

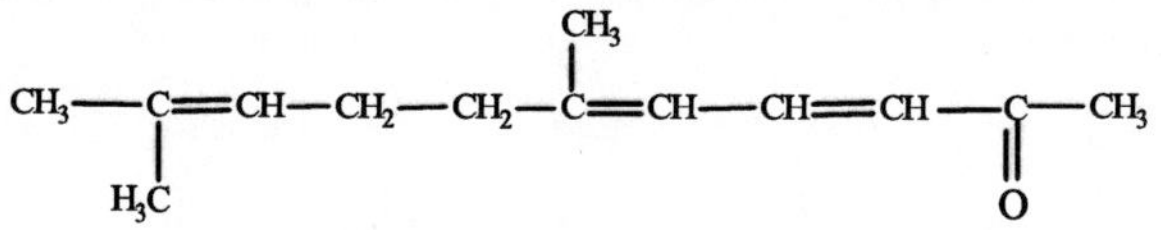

6,10-dimethyl-3,5,9-undecatrien-2-one

Pseudoionone $\Delta_{vap}H^0(T) = 36.84$ kJ mol^{-1} $\Delta_{vap}S^0(T) = 83.9 \pm 1.0$ J mol^{-1} T^{-1}

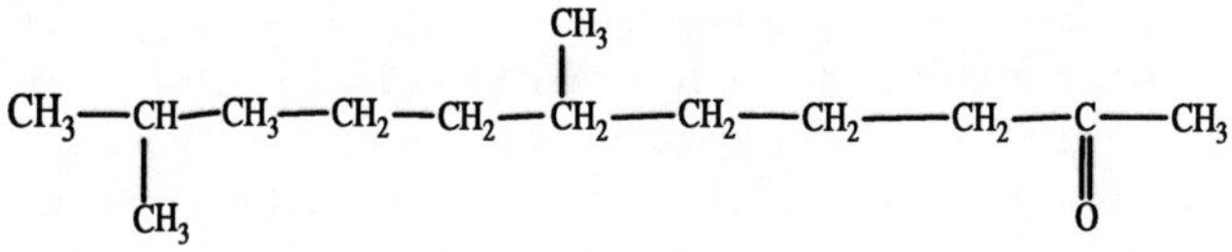

6,10-dimethyl-2-undecanone

Hexahydropseudoionone $\Delta_{vap}H^0(T) = 37.93$ kJ mol^{-1} $\Delta_{vap}S^0(T) = 83.9 \pm 1.0$ J mol^{-1} T^{-1}

Revealing the physicochemical essence of this phenomenon even in a general way provides an understanding of a number of problems in the intramolecular and intermolecular interactions and their relation to the position of the isostructural methyl groups and multiple bonds in the skeleton of the molecule.

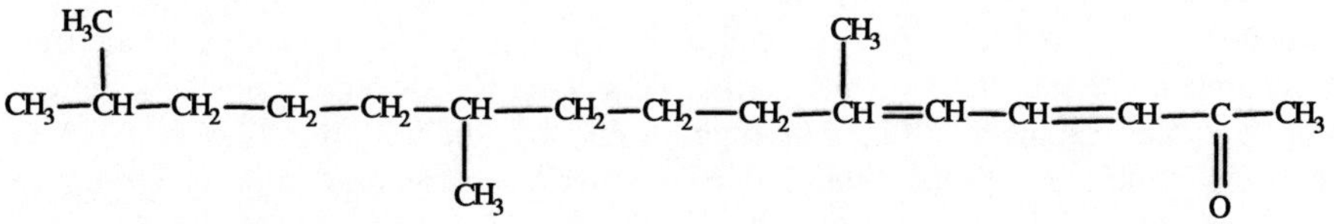

Unsaturated ketone C_{18} $\Delta_{vap}H^0(T) = 43.19$ kJ mol^{-1}

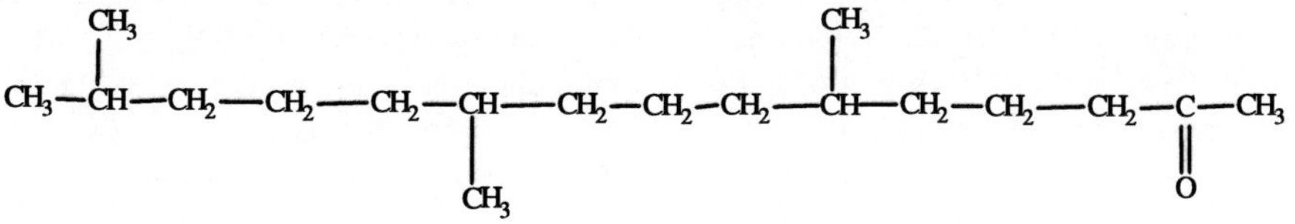

Saturated ketone C_{18} $\Delta_{vap}H^0(T) = 44.36$ kJ mol^{-1}

We should first of all understand the priority of the influence between the double bond –C=C– and the isostructural methyl group. The given data show that the three double bonds –C=C– in the pseudoionone located in two fragments reduce its vaporization enthalpy by 1.09 kJ mol^{-1} compared to that of the saturated hexahydropseudoionone. At the same time two double bonds in the unsaturated ketone C_{18} located in the first fragment after the carbonyl group also lead to the decrease in the vaporization enthalpy by virtually the same value (1.17 kJ mol^{-1}). If the three double bonds of the pseudoionone located sequentially in each of two fragments

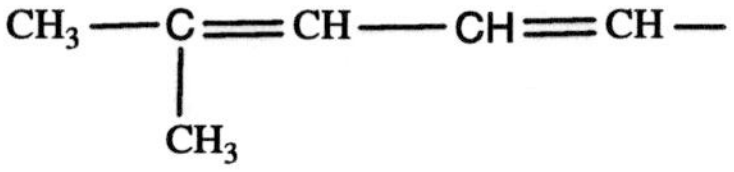

affect the shift of the electron density, then the removed from the carbonyl oxygen free of the double bonds CH_2-groups of molecules of the hexahydropseudoionone should participate in the intermolecular interactions $CH_2 \rightarrow H$ (0.60 kJ mol^{-1}). Hence three and two double bonds –C=C– in the pseudoionone and the unsaturated ketone C_{18} (6,10,14-trimethylpentadecadien-3,5-one-2) affect the vaporization enthalpy to the same extent reducing it by 1.10 and 1.17 kJ mol^{-1} compared with the same characteristic of the hexahydropseudoionone and saturated ketone C_{18}, respectively. Therefore, the role of the three and two double bonds in the unsaturated ketones C_{13} and C_{18} is strictly determined and is limited to the redistribution of the electron density in the molecules of the pseudoionone and 6,10,14-trimethylpentadecadien-3,5-one-2. This role becomes apparent in the energies of the specific intermolecular interactions existing in their liquid state and in the vaporization enthalpy. The nearly equal (1.10 and 1.17 kJ mol^{-1}) but not an essential decrease in the vaporization enthalpy of these compounds shows the weak influence of two and three double bonds in the carbon–carbon chain on the distribution of the electron density. Therefore, we are obliged to state that the main, priority, influence factor resulting in sharp decrease in the vaporization enthalpy of saturated and unsaturated ketones C_{13} and C_{18} and acetylene alcohols C_{15} and C_{20} is the presence in their structure of two and three isostructural methyl groups dividing the alkyl chain into fragments containing four carbon atoms. This conclusion we are going to illustrate with definite examples. The vaporization enthalpy of the saturated ketones C_{13} and C_{18} containing, respectively, two and three isostructural methyl groups is incomparably lower than those of the saturated unsymmetrical ketones of normal structures. Even more impressive example is the 2-tridodecanone containing in the chain 12 carbon atoms and possessing the vaporization enthalpy of 76.7 kJ mol^{-1}, exceeding by 39.9 kJ mol^{-1} the vaporization enthalpy of the hexahydropseudoionone containing 13 carbon atoms in the molecule. The vaporization enthalpy of 2-octadecanone exceeds by 59.0 kJ mol^{-1} the vaporization enthalpy of its isomer 6,10,14-trimethylpentadecadien-3,5-one-2. These data allow a conclusion that the low values of the vaporization enthalpy of the pseudoionone and hexahydropseudoionone, the saturated and unsaturated ketones C_{18} originate from two and three isostructural methyl groups, respectively, uniformly located in the structure of the molecule dividing the chain into fragments of four carbon atoms. An important conclusion follows that the isostructural methyl group creates in the fragment of four carbon atoms a rigid system of the electron density distribution resulting in the reduced capability of its carbon and hydrogen atoms to take part in the interactions, namely, diminishing their charges and donor–acceptor properties. This effect significantly exceeds the stabilizing influence of one or two double bonds present in the considered fragments. Thus, we have a reason for a conclusion that one, two, and even three double bonds in an unsaturated ketone or alcohol exceedingly weakly affect the shift of the electron density in the respective molecule. This statement was confirmed by the thermodynamic analysis of the data for ketones (Chap. 4) and alcohols (Chap. 5). The different behavior is observed only in alcohols and ketones with a small number of carbon atoms where the effect of the double bond is larger than in the unsaturated ketone C_{18} with the isostructural methyl groups bordering the fragments of chain containing four carbon atoms. This extrastabilizing effect of the isostructural methyl group is equally pronounced in the absence or the presence in the fragment of one or two double bonds. This means that the –CH= and $–CH_2–$ groups in these fragments in the presence of the isostructural methyl group equally participate in the intermolecular interactions. However, their contribution to the vaporization enthalpy is incomparably smaller than the contribution of $CH_2–$ and –CH= groups in the saturated and unsaturated ketones with the short normal chain.

As already mentioned, the difference in the vaporization enthalpy of pseudoionone (C_{13}) and unsaturated ketone C_{18} amounts to 6.35 kJ mol^{-1}, of hexapseudoionone (C_{13}) and saturated ketone C_{18}, to 6.43 kJ mol^{-1}. Just this energy is contributed to the vaporization enthalpy by the four-carbon fragment containing one, two double bonds or all ordinary bonds independent of the number of the fragments. This conclusion allowed the use of the extrapolation method for the estimation of the vaporization enthalpy of yet unstudied unsaturated ketones C_{23} and C_{28} equal 49.54 and 55.89 kJ mol^{-1} and of saturated ketones C_{23} and C_{28} amounting to 50.79 and 57.20 kJ mol^{-1}, respectively. This sequence can be extended in reasonable measure. The vaporization entropies of these compounds may be estimated also based on the vaporization entropy difference of ketones C_{18} and C_{13} contributed by the fragment to the entropy characteristic of unsaturated (4.4 J mol^{-1} K^{-1}) and saturated (3.9 kJ mol^{-1}) ketones. The obtained values of the vaporization entropy characteristics $\Delta_{vap}S^0(T)$ for the unsaturated ketones C_{23} and C_{28} are 91.4 J mol^{-1} K^{-1} and 95.6 J mol^{-1} K^{-1}, for the saturated ketones C_{23} and C_{28}, 91.7 and 95.6 J mol^{-1} K^{-1}, respectively.

The extrastabilizing effect of the isostructural methyl group can be estimated from the vaporization enthalpies of saturated ketone C_{13} with isostructural methyl groups and of saturated ketone C_{13} of the normal structure, and also from the vaporization enthalpy of saturated ketones C_{18} with isostructural methyl groups and with a normal structure. Taking into account the number of four-carbon fragments in each saturated ketone C_{13} and C_{18} we obtain an equation for estimating the extrastabilizing effect of the isostructural methyl group

$$\text{D ee.c.} = (\Delta_{vap}H^0(298)\ \kappa n - \Delta_{vap}H^0(298)\ \kappa.iso)/\kappa \qquad (5.30)$$

where "κ" is the number of fragments. The established extrastabilizing effect for two different pairs of saturated ketones C_{13} and C_{18} equals 19.4 and 19.3 kJ·mol^{-1} at two and three fragments, respectively. The values of this effect in unsaturated ketones C_{13} and C_{18} equal 20.0 and 19.7 kJ mol^{-1}. This equation is also valid for the estimation of the overall numerical value of the extrastabilizing effect involving the double bond and the isostructural methyl group in the fragment of the chain of the unsaturated alcohols with double bond and in the acetylene alcohols. Note that the fragments under consideration in the acetylene alcohol C_{15} produce the extra-stabilizing effect and decrease the vaporization enthalpy to 43.43 kJ mol^{-1}, and in the alcohol C_{20}, to 44.00 kJ mol^{-1}. This value exceeds the vaporization enthalpy of the unsaturated ketone C_{13} (pseudoionone) by 6.60 kJ mol^{-1} and the vaporization enthalpy of the saturated ketone C_{13} (hexahydropseudoionone), by 5.49 kJ mol^{-1}.

The latter value indicates that the extrastabilizing effect of the isostructural methyl group in the acetylene alcohol is less by this magnitude.

It is important to mention here that the acetylene alcohols with one and two four-carbon fragments, 3,7-dimethyl-6-octaen-1-ynol-3 and acetylene alcohol C_{15} form the hydrogen bonds strongly differing by energy, 15.9 and 9.60 kJ·mol^{-1}, respectively. Therefore for the thermodynamic analysis acetylene alcohol C_{20} and isophitol should be used, for their hydrogen bond energies do not significantly differ (6.55 and 6.90 kJ mol^{-1}, respectively). Just the considerable difference in the energies of the hydrogen bonds of the acetylene alcohols C_{15} and C_{20} resulted in a slight difference (0.55 kJ mol^{-1}) in their vaporization enthalpy at two and three fragments in the structure of their molecules.

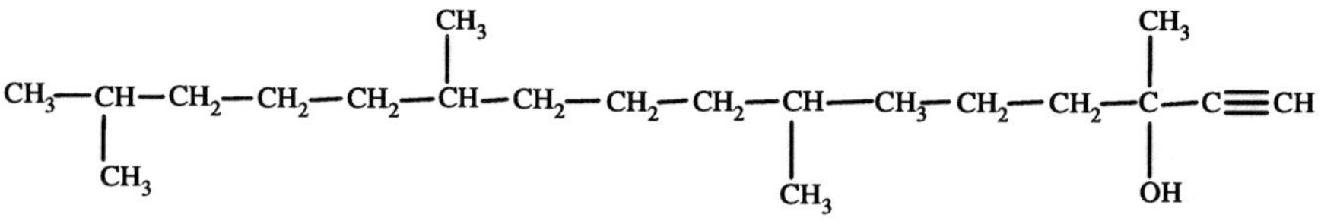

acetylene alcohol C_{20} 3,7,11,15-trimethylhexadecyn-1-ol-3 $\Delta_{vap}H^0(T) = 44.00$ kJ mol^{-1}

The important feature of the acetylene alcohol C_{20} and isophitol (C_{20}) is the small difference in the energies of their hydrogen bonds and vaporization enthalpy (0.67 kJ mol^{-1}). Their molecules contain three four-carbon fragments and a triple or a double bond.

Therefore, we are able to make a conclusion that the replacement of a triple bond in the acetylene alcohol C_{20} by a double bond (isophitol) increases the vaporization enthalpy by 0.67 kJ mol^{-1}.

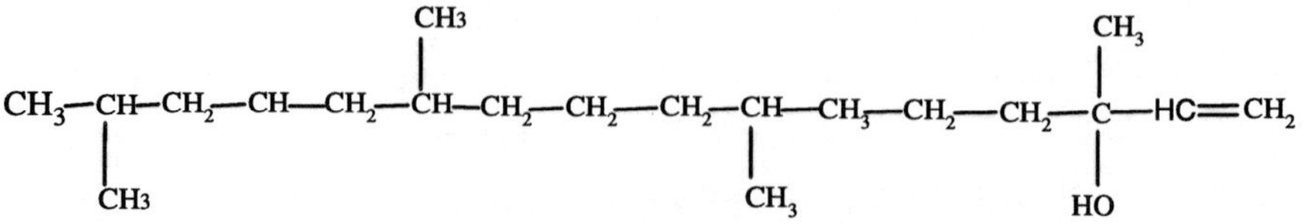

3,7,11,15-tetramethylhexadecyn –1-ol-3 (isophitol) $\Delta_{vap}H^0(T)$ 44.67 kJ mol^{-1}

These conclusions permit the estimation of the vaporization enthalpy of yet unstudied alcohols from acetylene and isophitol series with the accuracy corresponding to the performed experiments. We employed this obvious conclusion and estimated the vaporization enthalpy of the acetylene alcohol C_{25} and isophitol analog C_{25} at 44.53 and 45.34 kJ mol^{-1}, respectively.

The data necessary for the calculation of the extrastabilization effect from the vaporization enthalpy we extracted from the vaporization enthalpy of the saturated alcohols with the same number of carbon atoms in the molecule as that of the unsaturated alcohols. The said vaporization enthalpy was estimated at 97.0 and 122.1 kJ mol^{-1} by extrapolation of the dependence $\Delta_{vap}H^0(298) = f(\mathrm{Cn})$ obtained from the experimental thermodynamic characteristics (Sect. 5.4). The data obtained show that the effect of the extrastabilization by the isostructural methyl groups in the acetylene alcohol C_{20} and isophitol amounts to 26.0 and 25.8 kJ mol^{-1}, respectively. Here the digit after the decimal point is meaningless for it is within the error of the performed thermodynamic calculations. The important feature is the high value of the extrastabilization effect of the isostructural methyl groups in the unsaturated alcohols with a triple or double bond in the vicinal position to the carbon atom linked to the hydroxy group.

For the investigation of the possible existence of the extrastabilization effect in the alcohols with

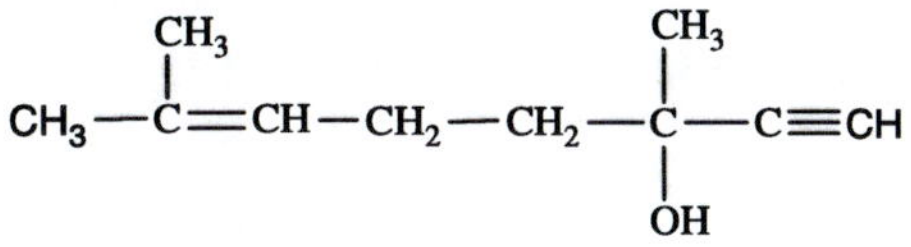

3,7-dumethyloctaen-6-in –1-ol-3
$\Delta_{vap}H^0(T) = 50.78$ kJ mol^{-1}

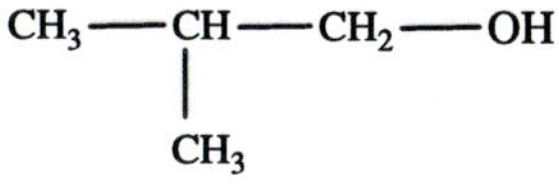

2-methyl-1-propanol (isobutyl alcohol)
$\Delta_{vap}H^0(T) = 50.79$ kJ mol^{-1}

shorter carbon–carbon chain the 3,7-dimethylocta-6-en-1-yn-3-ol, the acetylene alcohol with an additional double bond contained in the fragment with four carbon atoms, may be of interest.

With the characteristic for the studied acetylene alcohols fragment of four carbon atoms and one isostructural methyl group the vaporization enthalpy of 3,7-dimethylocta-6-en-1-yn-3-ol coincides with that of the isobutyl alcohol. However, these values are less than the vaporization enthalpy of 1-pentanol (56.95) by 5.15 kJ mol^{-1}.

Taking into consideration that the vaporization enthalpy is connected with the number and the energy of the specific interaction in the liquid compounds [12] we should state that the energy of the hydrogen bonds and the specific intermolecular interactions should be more stable in 1-pentanol. The above cited energy of the hydrogen bond in the liquid dihydrolinalool exceeding by 1.5 kJ mol^{-1} the energy of similar bonds in the unsaturated alcohols with the number of carbon atoms over three (14.40) shows the appearance of the stabilizing effect. Thus, the acetylene alcohol with a single four-carbon fragment between two isostructural methyl groups also is characterized by the extrastabilization effect although it is not so pronounced and consequently appears with a lesser energy contribution. This effect is still less pronounced in the isobutyl alcohol.

All branched ketones also possess low vaporization enthalpy compared with that of the compounds with the normal structure, and this also is due to the extrastabilization effect.

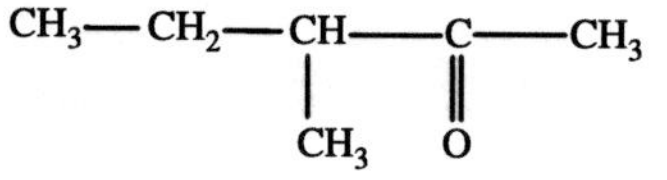

3-methyl –2-pentanone 37.18

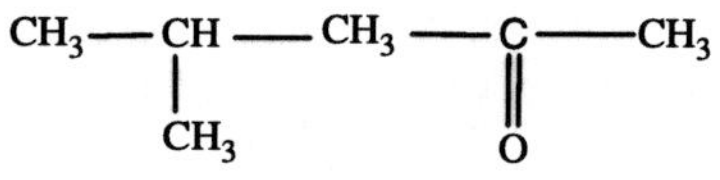

4-methyl-2-pentanone (isobutyl methyl ketone) 37.69 kJ mol

This effect becomes more and more pronounced with the growing number of carbon atoms. This effect is clearly reflected in the value of the vaporization enthalpy of 2-hexanone (42.22), exceeding approximately by 5 kJ mol^{-1} the enthalpy characteristics of isomeric 3-methyl-2-pentanone and 4-methyl-2-pentanone. However, the effect of the isostructural methyl group is not felt in the positions 3 and 4.

References

1. Nefedov VI (1984) Infra red – spectroscopy of chemical compounds. Chemistry, Moscow
2. Tucker EE, Lippert E (1976) High resolution NMR studies of hydrogen bonding. In: Schusters P et al (eds) The hydrogen bond: recent development in theory and experiments, vol 2. North Holland, Amsterdam, pp 791–831
3. Sokolov ND (1972) Hydrogen Boyd. Russ J Mendeleev Chem Soc 17:299–308
4. Pereliygin IS (1984) IR spectroscopy investigation of intermolecular interactions in nonaqueous solutions. Russ J Mendeleev Chem Soc 29:24–29
5. Krestov GA, Berezin BD (1999) Basic law of chemistry. Science, Moscow
6. Nefedov VI, Vovna VI (1989) Electronic structure of organic and element organic compounds. Nauka, Moscow
7. Baev AK (1996) Five-coordinated carbon atom at intermolecular interactions. Russ J Coord Chem 22:399–402
8. Baev AK (2001) Pentacoordinated carbon atom in intermolecular interactions in solutions of alkyl compounds of non-transition elements. In: Abstract of sixth international symposium on carbanion chemistry, Marburg, Germany, p 75
9. Baev AK, Korolkov DV (2002) Phenomenon of dative bonds in element organic compounds with alkyl ligand. In: Abstract XVth FECHEM conference on organometallic chemistry. University of Zurich, Zurich, p 350
10. Baev AK (2003) Novel approaches to the structure of functional solvents and element organic compounds with saturated hydrocarbon ligands and their solvates in binary systems. In: Book of Abstract of 27th international conference on solution chemistry, Debrecen, Hungary
11. Baev AK (2004) Insolence of model sp^3-hybridization of carbon atom and chemistry of non aqueous solutions. In: IX international conference on the problem of salvation and complex formation in solutions. Plyos, Russia, pp 62–64

12. Baev AK (2004) Lack of co-ordination thermodynamic properties of element organic compounds with sp^3-hybridization of carbon atom and its insolence. In: Russian scientific symposium on thermo chemistry and calorimetric. Niznie Novgorod, Russia, pp 40–41
13. Baev AK (2002) Thermodynamics of solutions non-electrolytes with five- coordinated carbon atom at specific intermolecular interactions. In: Abstract XIV international conference on chemical thermodynamic, St. Petersburg, p 135
14. Vitalev SE, Baev AK (1993) Thermodynamic vaporization and association of trimethylaluminum. Russ J Izv Vischich Ucheb Zaved Chem Chem Technol 36:36–40
15. Baev AK, Vitalev SE (1989) Thermodynamic vaporization and association of triethylaluminium. Vesti Akad Navuk Belarusii 2:58–63
16. Baev AK, Shishko MA (2002) Thermal behavior of trialkylaluminiums. Russ J Appl Chem 74:26–29
17. Narten AH, Habenschusse A (1984) Hydrodgen bonding in liquid methanol and ethanol determined by x-ray diffraction. J Chem Phys 80:3387
18. Tanaka Y, Ohtamo N, Atakawa K (1984) The structure of liquid alcohols by neutron diffraction. 1. Molecules of methyl alcohols. Bull Chem Soc Jpn 57:644
19. Newton MD, Boer FP, Lipscomb WN (1966) Molecular orbital for organic systems parameterized from SCF – model calculations. J Am Chem Soc 88:2367–2384
20. Jorgensen WL, Ibrahim M (1982) Quantum and statistical mechanical studies of liquids. 20. Pressure dependence of hydrogen bonding in liquid methanol. Theoretical study of liquids. J Am Chem Soc 104:373
21. Vasiliev IA, Petrov VM (1984) Thermodynamic properties of oxygen organic compounds. Chemistry, Leningrad
22. Baev AK (2001) Thermodynamic evaporation of alkyl compounds of elements second-six groups. Russ J Izv Vischich Ucheb Zaved Chem Chem Technol 44:3–13
23. Magaril PZ (1970) Mechanism and kinetics homogeny thermal of transformation hydrocarbon. Chemistry, Moscow
24. Hovard NV, Legen AC, Regol A, Wallwork FC (1989) Ground-state relational of t-butyl isocyanide and an r_s-geometry of its heavy-atom skeleton. J Mol Struct 197:181–191
25. Euiacek F, Kratochvil M, Hobza P (1999) Adenine ... 2.4 – Difluoroluene (Modified Base) par: potential and free – energy surfaces non – empirical and empirical potential studies. Chem Phys Lett 312:393–398
26. Yuchnovich GV, Tarakanova EG (1998) Hydrogen bond C-H···O in liquid methanol. J Mol Struct 447:257
27. Mandado M, Musquera RA, Graria AM (2002) On the effect of electron correlation and conformational changes on the distortion of the change distribution in alkyl chain. J Chem Phys Lett 335(5–6):529–537
28. Ohtaki H, Shin-ichi I (1994) Complexation in nonaqueous solvents. Chemistry of nonaqueous solutions. Current progress. VCH, Germany
29. Ivanov EV, Puchovskie YuP, Abrosimov VK (2004) About formation of weak hydrogen bonds CH···O in liquid methanol (Abstract). In: IX International conference on the problem of solvation and complex formation in solutions. Plyos, Russia, p 345
30. Krupin SV, Puhovskoi VA (2004) Investigation of new configuration hydrogen bonding dimer of methanol by method of molecular orbital and functional density. In: IX international conference on the problem of salvation and complex formation in solutions. Plyos, Russia, p 136
31. Nour DAA, Rogowski F (1966) Further consideration about the structure of n-propyl alcohols by electron diffraction. Z Naturforsch 21b:996–1002
32. Biddiscombe BD, Collerson RR, Handley R (1963) Cleaning and vapor pressure of propane and butane alcohols. J Chem Soc 3:1954–1957
33. Ambrose D, Townsend R (1963) Thermodynamic properties of organic oxygen compounds. Part IX: The critical properties and vapors pressure above five atmospheres of six aliphatic alcohols. J Chem Soc 6:3614–3625
34. Wadso J (1966) Thermochemical properties of diacetomide, N-butyldiacetomide and phenyldiacetomide. Acta Chem Scand 20:544–552

35. Polak J, Benson GG (1971) Enthalpies of vaporization of some aliphatic alcohols. J Chem Thermodyn 3:235–242
36. Ambrose D, Broderick BE, Townsend R (1974) The critical temperatures and pressures of thirty organic compounds. J Appl Chem Biotechnol 24:359–372
37. Hales JL, Cox JD, Lees ED (1963) Measurements conductivity of vapor and vaporization enthalpy of isopropanols. Trans Faraday Soc 59:1544–1554
38. Rid P, Prausnic Dis, Shervud T (1977) The properties of gas and liquids. McGraw Hill, New York
39. Chickos JS, Acree WE (2003) Total phase change entropies and enthalpies. Update on their estimation and application fluorocarbon – hydrocarbon molecules. J Phys Chem Rev Data 32:699–721
40. Counsell JP, Lees EB, Martin JF (1971) Thermodynamic properties of organic oxygen compounds. Part XXVII (±) – butan-2-ol. J Chem Soc A 4:661–664
41. Sachek AI, Peshenko AD, Markovnik VS et al (1982) Temperature dependence of vapor pressure butyls alcohols. Thermodynamics of organic compounds. Gorkei 11:94–98
42. Ambrose D, Sprake CHS (1980) Equations for the correlation and estimation of vapor pressures. J Chem Thermodyn Data 2:631–645
43. Baev AA, Baev AK (2011) Thermodynamics of Vaporization isostructural Propanol and Butanol. Russ J Phys Chem (in press)
44. Ambrose D, Sprake CHS (1970) Thermodynamic properties of organic oxygen compounds. XXV. Vapor pressure and normal billing temperatures of aliphatic alcohols. J Chem Thermodyn Data 2:631–645
45. Boublic T, Frie V, Hala E (1973) The vapor pressures of pure substances. Academic, Amsterdam
46. Butler JAF, Ramchandant CN, Thomson DW (1936) The solubility of non-electrolytes. Part I. The free energy of hydration of some aliphatic alcohols. J Chem Soc 280–286
47. Chickos JS, Acree WE Jr (2002) Enthalpies of vaporization of organic and organometallic compounds, 1880–2002. J Phys Chem Rev Data 31:519–879
48. Chickos JS, Acree WE Jr (2003) Enthalpies of sublimation of organic and organometallic compounds, 1910–2001. J Phys Chem Rev Data 32:537–698
49. Baev AK (1972) Complex formation ability halogenides of second – six groups periodical system. Gen Appl Chem 5:35–51
50. Baev AK (2006) Solution chemistry and technology liquid phase materials. In: Zacharov AG (ed) Achievements and perspectives. Ivanovo, Russia, p 315
51. Solomonov BN, Novikov VB, Varfolomeev MA, Faizullin DA, Klimovskie AE (2004) Enthalpies of specific Solvation and Non-electrolytes in associated solvent. Accounting of reorganization of solvent (Abstracts). In: IX International conference on the problem of salvation and complex formation in solutions. Plyos, Russia
52. Chumaevskei NA, Rodnikova MN, Ippolitov EG, Tripolskaia TA, Kayumova DB, Sirotkin DA (2004) Some peculiarities of structure compounds with hydrogen bonds on the results of Raman spectroscopy: H_2O, H_2O_2 $HOCH_2CH_2OH$ (Book of Abstracts). In: IX International conference on the problems of salvation and complex formation in solutions. Plyos, Russia, p 46
53. Baev AK (1987) Chemistry of gas-heterogenic systems of element organic compounds. Science and Technology, Minsk, Belarus
54. Baev AA, Baev AK (2011) Thermodynamic saturated and unsaturated ketones C_{13} and C_{18} and energies of specific interactions. Russ J Phys Chem 85:736–744
55. Majer V, Svoboda V (1985) Enthalpies of vaporization of organic compounds: a critical review and data complication. Blackwell Scientific, Oxford, p 300
56. Svoboda V, Vesely F, Holub R, Rick J (1973) Enthalpy data of liquids. I. The dependence of heats of vaporization of methanol, propanol, butanol, cyclohexene and benzene on temperature. Collect Czech Chem Commun 38:3539–3543
57. Wadso J (1966) Calorimetric for a determination of enthalpy vaporization small quantity of compounds at 25 C^0. Acta Chem Scand 20:544–552

Chapter 6
Specific Intermolecular Interactions and H-bonds in Aldehydes and Esters

6.1 Specific Interactions and H-bonds in Aldehydes

6.1.1 Saturated Aldehydes

The problems of specific interactions under discussion are of necessity connected with using reliable data on the thermodynamic properties of vaporization of compounds; therefore, let us analyze the available enthalpy characteristics of aldehydes and select the most precise experimental data. The vaporization enthalpy of aldehydes of normal structure [1, 2] is presented in Fig. 6.1 as a functional dependence $\Delta_{vap}H^0(298) = f(C_n)$ described by two straight lines crossing at the vaporization enthalpy of butanal (C_4). The first line corresponds to the regular variation in the vaporization enthalpy in the series formaldehyde–butanal due to the decrease in the effect of the reverse dative bond between the carbonyl oxygen atom and the contiguous carbon atom with growing number of carbon atoms in the alkyl chain up till four. The second linear dependence describes the enthalpy characteristics in the series pentanal–hexadecanal and corresponds to the growing energy contribution to the vaporization enthalpy of aldehydes from the increasing number of the CH_2 groups.

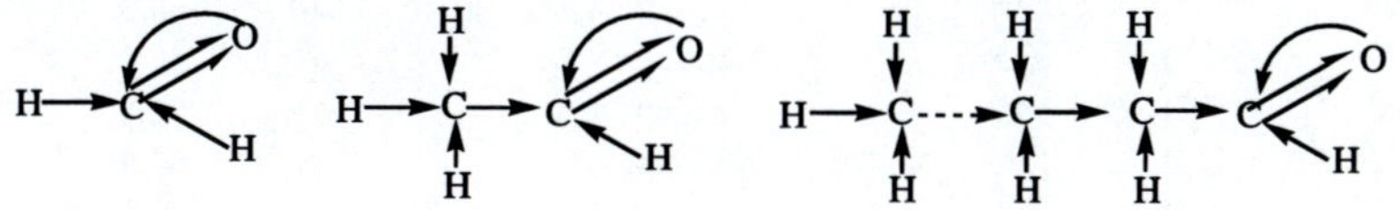

The crossing of the straight lines near the vaporization enthalpy of butanal is quite natural for it shows the end of the influence of the reverse dative bond on the terminal methyl group at four carbon atoms in the alkyl fragment. Thus, the angle formed by the crossing lines (Fig. 6.1) gives a qualitative notion of the stability of the specific interactions formed in the liquid aldehydes and of their energies in pentanal–hexadecanal.

A.K. Baev, *Specific Intermolecular Interactions of Organic Compounds*,
DOI 10.1007/978-3-642-21622-0_6, © Springer-Verlag Berlin Heidelberg 2012

The dependence depicted in Fig. 6.1 provides a possibility to refine the existing values of the vaporization enthalpy and also to use extrapolation and interpolation for estimating the enthalpy characteristics of yet unstudied compounds (Table 6.1).

A more detailed picture is seen on comparing the character of the variation in the vaporization enthalpy of the first members of compound classes' ketones, ethers, and aldehydes. It follows from the dependences shown in Fig. 6.2 that the vaporization enthalpies synchronously changing in the first members of ketones and ethers series adequately reflect the effect of the reverse dative bond on the energy of the specific interactions $=O \rightarrow CH_3–(CH_2)_n$ and $–O \rightarrow CH_3–(CH_2)_n$, respectively. However, the energies of these interactions are governed by different charges on the atoms of the carbonyl and ether oxygen.

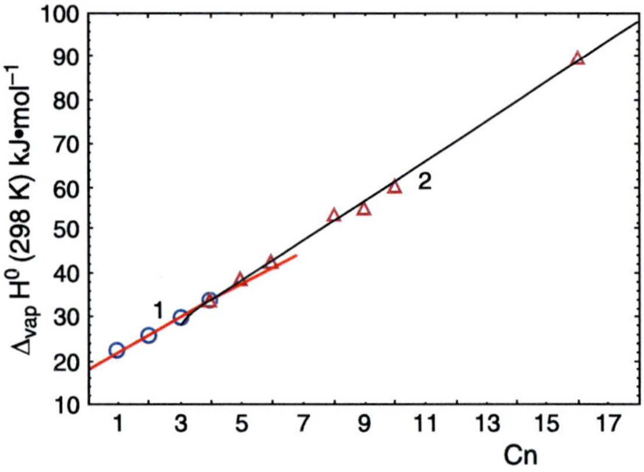

Fig. 6.1 Dependence of vaporization enthalpy of aldehydes on the number of carbon atoms in the alkyl chain

Table 6.1 Vaporization enthalpy of aldehydes (kJ mol^{-1}) at 298 K

Compound	Formula	$\Delta_{vap}H^0(298)$ [2]	Compound	Formula	$\Delta_{vap}H^0(298)$ [2]
Formaldehyde	CH_2O	24.2 22.0[a]	3-Methyl hexanal	$C_7H_{14}O$	42.3
Acetaldehyde	C_2H_4O	25.71	3,4-Dimethyl pentanal	$C_7H_{14}O$	42.4
Propanal	C_3H_6O	29.82	Formaldehyde[a]	CH_2O	22.0
Butanal	C_4H_8O	33.84	Heptanal[a]	$C_7H_{14}O$	46.8
Pentanal	$C_5H_{10}O$	38.3	Octanal[a]	$C_8H_{16}O$	51.6
Hexanal	$C_6H_{12}O$	42.3	Nonanal[a]	$C_9H_{18}O$	54.2
Octanal	$C_8H_{16}O$	53.8	Decanal[a]	$C_{10}H_{20}O$	61.0
Nonanal	$C_9H_{18}O$	55.2	Undecanal[a]	$C_{11}H_{22}O$	65.6
Decanal	$C_{10}H_{20}O$	60.4 ± 0.1	Tridecanal[a]	$C_{13}H_{26}O$	74.8
Undecanal	$C_{11}H_{22}O$	69.3	Tetradecanal[a]	$C_{14}H_{28}O$	74.6
Dodecanal	$C_{12}H_{24}O$	70.2	Pentadecanal[a]	$C_{15}H_{30}O$	84.2
Hexadecanal	$C_{16}H_{32}O$	89.7	Heptadecanal[a]	$C_{17}H_{34}O$	93.6
3-Hexadecanal	$C_{16}H_{32}O$	89.9	Octadecanal[a]	$C_{18}H_{36}O$	98.6
4-Hexadecanal	$C_{16}H_{32}O$	88.7			

[a]Determined by interpolation and extrapolation

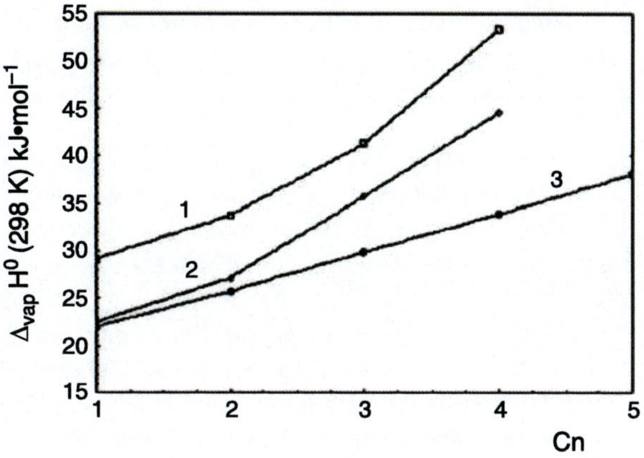

Fig. 6.2 Dependence of vaporization enthalpy on the number of carbon atoms in the alkyl chain (1) symmetric ketones, (2) symmetrical ethers, (3) aldehydes

Therewith a distinguishing feature appears in the shift in the electron density along the alkyl chain to the oxygen atom in the symmetric and unsymmetrical compounds with growing number of carbon atoms in the chain. The characteristic distinction of the aldehydes from ketones and ethers consists in the presence of a single alkyl chain attached to the atom bearing the oxygen, and therefore, the shift in the electron density from a single alkyl fragment provides for the decreased charge on the oxygen atom of the aldehyde and decreased donor properties.

As a consequence, at the presence of two alkyl chains in ketones and ethers, the effect of the reverse dative bond is more pronounced and is clearly seen on the dependence $\Delta_{vap}H^0(298) = f(C_n)$ in Fig. 6.2. Interestingly, in the ethers and the aldehydes, the vaporization enthalpies do not significantly differ, on the one hand, in the formaldehyde and the acetaldehyde, and, on the other hand, in dimethyl and diethyl ethers. The respective differences amount to 0.54 and 1.43 kJ mol^{-1}. Taking into account that two molecules are involved in the interaction, the energy per one molecule is 0.27 and 0.71 kJ mol^{-1}, or to one specific interaction in ethers and ketones corresponds to a value on the border of the error of the experiment in formaldehyde and somewhat exceeding this error in dimethyl ether. In this state, the energy of the specific interactions in the dimethyl ether and the energy of the H-bond in the formaldehyde are not very different, $D-O \rightarrow CH_3 - > D=O\cdots H-CH-$. At the same time, the larger value of the vaporization enthalpy of the diethyl ether compared with the enthalpy characteristic of acetaldehyde shows the high energy contribution at the replacement of hydrogen in the formaldehyde by the CH_3 group. Therefore, the energies of the specific interactions of the diethyl ether and the acetaldehyde are equal within the accuracy of the experimental measurements of the vaporization enthalpy by the calorimetric method $D-O \rightarrow CH_3CH_2- = D=O \rightarrow CH_3 - CH -$, $D=O\cdots H-C-CH_3$. Consequently, in the energies of two types of the intermolecular interactions in liquid aldehyde, the following inequality is valid:

$$D=O \rightarrow CH_3 - CH - > D=O\cdots H-C-CH_3.$$

Quantum-chemical calculations showed that the hyperconjugation and the inductive effect in going from H_2CO to CH_3CHO and further to $(CH_3)_2CO$ resulted in the destabilization of the π-orbital by 1.26 and 1.9 eV. These values exceed the change both in their vertical ionization potentials by 0.67 and 1.17 eV and in E(O1s) by 1.04 and 1.60 eV [3]. Consequently, the charge on the oxygen of the acetaldehyde is intermediate in these compounds. Therefore, the decreased negative charge on the oxygen atom of the acetaldehyde compared to that of ketone $(CH_3)CO$ corresponds to the same sequence in the stability of their specific intermolecular interactions. It means also that the specific interactions existing in the liquid formaldehyde have a reduced stability compared to those formed in the liquid acetaldehyde and liquid dimethyl ether. This conclusion is confirmed by the vaporization enthalpies of ketones and aldehydes [2] (Fig. 6.2), since the types of the specific interactions and their energies govern the value of the enthalpy characteristics [4–8]. Similar to ethers, ketones, and alcohols, the aldehyde molecules possess four bond vacancies (Fig. 6.3). Their characteristic feature consists in the formation by the H–C fragment 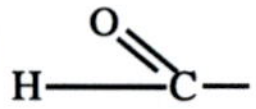 of a H-bond of low stability. In this respect, the aldehydes are especially interesting for they provide a possibility to estimate the energy of this donor–acceptor interaction $= \mathrm{O}\cdots\mathrm{H}-\mathrm{C}-(\mathrm{CH}_2)_n\mathrm{CH}_3$ occurring at the carbonyl oxygen atom of the aldehydes. In the formaldehyde, two hydrogen atoms and the carbonyl oxygen atom form with the molecules of the nearest surrounding in liquid and solid state four equivalent H-bonds $D = \mathrm{O}\cdots\mathrm{H}-\mathrm{CH}-$ whose energy is estimated from the vaporization enthalpy $\Delta_{vap}H^0(298)/4$. The aldehyde molecules with two and more carbon atoms in the alkyl fragment form two specific interactions and two hydrogen bonds. Taking into account the equal energy of the specific interactions in the liquid acetaldehyde and diethyl ether, the energy of the hydrogen bond is calculated by the equation

$$D = \mathrm{O}\cdots\mathrm{H}-\mathrm{C}-\mathrm{CH}_3 = \frac{\Delta_{vap}H^0(298) - 2D - \mathrm{O} \rightarrow \mathrm{CH}_3 - \mathrm{CH}-}{2}. \tag{6.1}$$

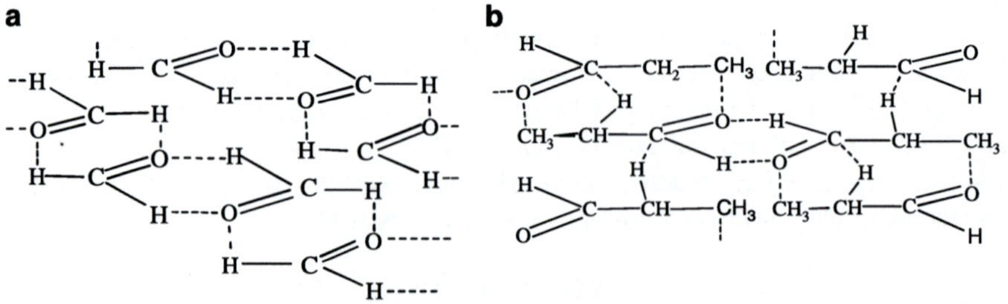

Fig. 6.3 Schematic picture of the liquid structure of formaldehyde (**a**) and propanal (**b**)

Table 6.2 Energies of hydrogen bonds and specific interactions (kJ mol^{-1}) in liquid saturated aldehydes (298 K)

Compound	Formula	$\Delta_{vap}H^0(T)$ [2]	D=O → CH_3–$(CH_2)_n$–C–	D = O···H – C –	ΣDCH$_2$
Formaldehyde	HCHO	22.0	–	5.50	–
Acetaldehyde	CH_3CHO	25.71	$n = 0$ 6.78	6.10	–
Propanal	CH_3CH_2CHO	29.82	$n = 0$ 8.92	6.00	–
Butanal	$CH_3(CH_2)_2CHO$	33.84	$n = 1$ 8.92	6.10	3.8
Pentanal	$CH_3(CH_2)_3CHO$	38.0	$n = 2$ 8.92	6.20	7.6
Hexanal	$CH_3(CH_2)_4CHO$	42.3	$n = 3$ 8.92	5.80	11.4

The results of the performed calculations (Table 6.2) show the decreased stability of the H-bonds in the liquid formaldehyde (5.5) and acetaldehyde (6.1 kJ mol^{-1}) in agreement with the assumptions based on the quantum-chemical calculations and the character of the dependences of the vaporization enthalpy of ketones, ethers, and aldehydes.

The error introduced by the insignificant values of the nonspecific interactions into the energy of this type of hydrogen bond does not exceed the error in the experimentally measured enthalpy characteristics. The energy of this type of H-bond in the aldehyde with three carbon atoms in the alkyl chain is estimated based on the principle of the additive contribution of two specific interactions D=O → CH_3–CH_2– to be equal to the energy of the interactions in the dipropyl ether.

$$\mathrm{D{=}O\cdots H-C-CH_2-CH_3} = \frac{\Delta_{vap}H^0(298) - 2\mathrm{D{=}O \rightarrow CH_3-CH_2-C-H-}}{2}. \tag{6.2}$$

For the aldehydes containing a large number of methylene groups, the contribution of the latter into the enthalpy characteristic should be taken into account. This value is obtained from the difference in the vaporization enthalpies of two remote members of this series (3.8 kJ mol^{-1})

$$\begin{aligned}&\mathrm{D = O\cdots H - C - CH_2 - CH_3}\\&\quad = \frac{\Delta_{vap}H^0(298) - 2\mathrm{D = O \rightarrow CH_3 - CH_2 - C - H -} + n\cdot \mathrm{DCH_2}}{2}.\end{aligned} \tag{6.3}$$

The results of the calculations (Table 6.2) show the constant energy of the hydrogen bonds in the series acetaldehyde–hexanal notwithstanding the length of the chain.

This rule is one of the main facts used for the estimation of the vaporization enthalpy of unstudied aldehydes with a long alkyl chain. The second key value is the contribution of the methylene group to the vaporization enthalpy at the number of carbon atoms in the alkyl chain exceeding three, namely, when the terminal methyl group is no more affected by the reverse dative bond.

The current information on the thermodynamic properties of solid aldehydes is scanty. However, even this information permits the estimation of the hydrogen bond energies and leads to interesting conclusions. The published data on the melting enthalpy of propanal and butanal (Table 6.3) are approximately two and four times less than that on the vaporization and sublimation enthalpy, respectively. These values indicate the significant transformations of the crystal structure in the course of melting [5]. The absence of polymorphous transformations in the aldehydes is due apparently to the little difference between the energies of the hydrogen bond and the specific intermolecular interactions formed by the pentacoordinate carbon atom involving the essentially unshared $2s^2$-electron pair.

The above-substantiated possibility to apply the energies of the specific interactions existing in ethers for the estimation of the hydrogen bond energies in the liquid aldehydes permits the use of this assumption in calculating the energies of the corresponding interactions in solid aldehydes. The validity of the application of the additivity principle of the energies of the specific interactions in the ethers for the estimation of the hydrogen bond energies in the aldehydes was confirmed by the mentioned quantum-chemical calculations, which stated that the hyperconjugation and the inductive effect in going from H_2CO to CH_3CHO and further to $(CH_3)_2CO$ resulted in the destabilization of the π-orbital stronger than the vertical ionization potential. Therefore, based on the results of the quantum-chemical calculations, we can assume the adequacy of the energies of the specific intermolecular interactions in aldehydes and ethers. In this connection, the energies of the specific interactions in solid ethers compiled in Table 3.3 serve as the basis of estimation of the energies of the hydrogen bonds in the aldehydes. The analogy of the type of specific interaction in the dipropyl ether and propanal and the same value of the CH_2 group contribution to the sublimation enthalpy of butanal provided a possibility to estimate the hydrogen bond energies in these aldehydes (Table 6.3).

The energies of the hydrogen bonds in the solid aldehydes obtained with the use of the energy of the specific interactions of symmetric ketones are underestimated. Hence, the presence of a carbonyl group in aldehyde and ketone molecules cannot be a basis for the estimation of the hydrogen bond energy in the aldehydes. The obtained hydrogen bond energies of solid aldehydes illustrate their stabilization in the series formaldehyde–butanal, with attaining the maximum value in the latter.

The additional confirmation of the validity of the performed estimations of the enthalpy characteristics of the unstudied saturated aldehydes follows from the correlations between the vaporization enthalpies of compounds of two homologous series belonging to the same or even different classes. The correlation between the vaporization enthalpy of saturated aldehydes and ethers, unsaturated ethers and 1-alkanols, and 1-alkanols and 2-alkanols shown in Fig. 6.4 is shown by

Table 6.3 Energies (kJ mol^{-1}) of hydrogen bonds and specific interactions in solid aldehydes

Aldehyde	$\Delta_{melt}H^0(T)$ [2]	$\Delta_{sub}H^0(T)$ [2]	From solid ethers			From solid ketones		
			$D{=}O \rightarrow CH_3(CH_2)_2$	$D = O \cdots H - C -$	DCH_2	$D{=}O \rightarrow CH_3(CH_2)_2$	$D = O \cdots H - C -$	DCH_2
Formaldehyde	–	24.2	–	6.05	–	–	–	–
Propanal	8.59	38.41	11.61	7.60	–	11.91	7.30	–
Butanal	11.11	44.95	11.61	7.95	5.80	11.91	7.95	5.20

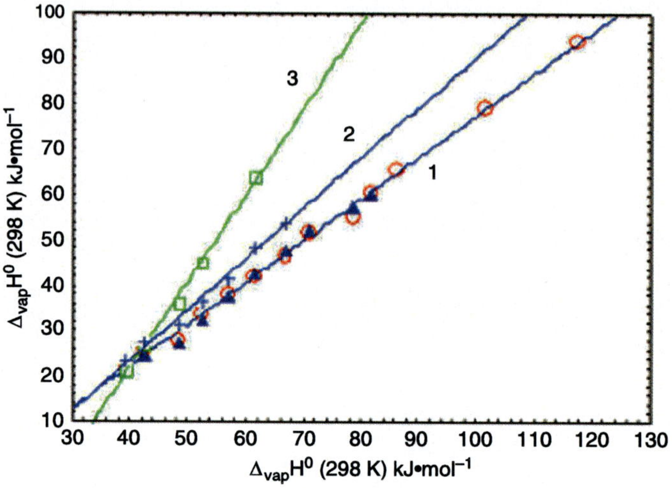

Fig. 6.4 Correlation of vaporization enthalpy of saturated aldehydes with that of unsymmetrical methyl ethers (1), vinyl R ethers and 1-alkanols (2), 2-alkanols and 1-alkanols (3)

straight lines within the accuracy of the experimental measurements of the vaporization enthalpy.

In turn, these dependences make it possible to refine the performed estimation of the values of the enthalpy characteristics.

6.1.2 *Unsaturated Aldehydes*

The vaporization enthalpies of unsaturated aldehydes at the standard temperature are known for crotonaldehyde, decenal, dodecenal, tetradecenal, and hexadecenal. For aldehydes of the series decenal–hexadecenal, the published enthalpy characteristics [6] belong to the molecules with the position of the double bond at different carbon atoms of the chain and to the presence of *E* and *Z* forms. The following general rules govern the enthalpy characteristics:

First, the vaporization enthalpy of two configuration isomers corresponds to the condition $\Delta_{vap}H^0(298)\ E < \Delta_{vap}H^0(298)\ Z$, with the maximum difference observed in 5-decenal (0.7), 6-dodecenal (2.0), 7-tetradecenal (1.0), and 8-hexadecenal (1.0 kJ mol^{-1}).

Second, the values of the vaporization enthalpy increase from the named carbon atom of the chain (C_2 and C_3, respectively) to the aldehyde group with the difference in *Z* and *E* forms not exceeding 0.1–0.6 kJ mol^{-1}. Note that for the majority of aldehydes in the homologous series decenal, dodecenal, tetradecenal, and hexadecenal, the vaporization enthalpies were measured with the best accuracy of the method of determination of this parameter.

The dependences of vaporization enthalpy on the number of carbon atoms in the chain presented in Fig. 6.5 for unsaturated aldehydes of the series 2-decenal–2-hexadecenal and 3-decenal–3-hexadecenal fit well to straight lines. These dependences permit the use of interpolation and extrapolation for estimating vaporization enthalpies of unstudied aldehydes of various series with the accuracy of the experimental measurement (Table 6.4).

The obtained values of the vaporization enthalpy of aldehydes make it possible to refine the essentially overestimated experimental enthalpy characteristic of nonenal (56.1 kJ mol^{-1}), obtained at 379 K and reported without indicating the error of the measurement [9]. In the molecules with fragments containing 2, 3, and even 4 carbon atoms, the gradually decreasing influence of the reverse dative bond on the energy of the specific interactions is observed. Afterward it is replaced by the contribution of the CH_2 groups to the vaporization enthalpy spreading up to octenal–nonenal and further strengthening in the successive members of the homologous series. Therefore, the dependence of the vaporization enthalpy on the number of the carbon atoms may be described by several straight lines whose number is

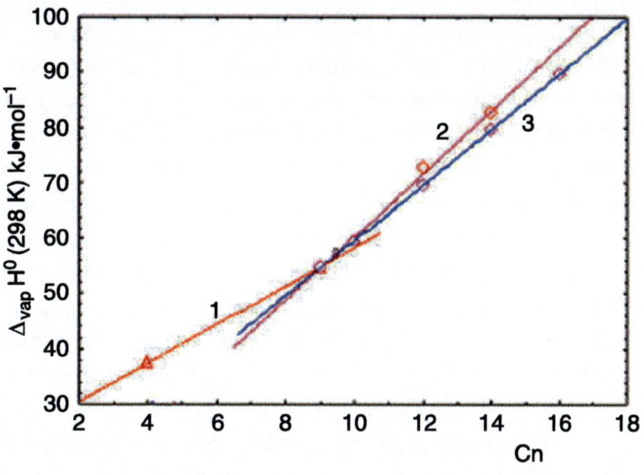

Fig. 6.5 Dependence of vaporization enthalpy on the number of carbon atoms in the alkyl chain (1) unsaturated aldehydes with the double bond at the C^2 and C^3, (2) aldehydes with the double bond at the C^2, (3) aldehydes with the double bond at the C^3

Table 6.4 Vaporization enthalpies (kJ mol) of unsaturated aldehydes (298 K)

Compound	Formula	$\Delta_{vap}H^0(T)$	Compound	Formula	$\Delta_{vap}H^0(T)$
2-Butenal[a]	C_4H_6O	37.3[a]	2-Undecenal	$C_{11}H_{20}O$	64.5
2-Pentenal	C_5H_8O	41.0	3-Dodecenal[a]	$C_{12}H_{22}O$	69.6[a]
2-Hexenal	$C_6H_{10}O$	44.4	3-Tridecenal	$C_{13}H_{24}O$	74.8
2-Heptenal	$C_7H_{12}O$	47.9	3-Tetradecenal[a]	$C_{14}H_{26}O$	79.9[a]
2-Octenal	$C_8H_{14}O$	51.3	3-Pentadecenal	$C_{15}H_{28}O$	84.7
2(3)-Nonenal[a]	$C_9H_{16}O$	56.1[a] 54.6	3-Hexadecenal[a]	$C_{16}H_{30}O$	89.7[a]
2(3)-Decenal[a]	$C_{10}H_{18}O$	59.9[a]	3-Heptadecenal	$C_{17}H_{32}O$	94.5

[a]Experimental data [6]

connected with the errors in the experimental data. Note that the vaporization enthalpy of the saturated aldehydes with long alkyl chains is described with a single straight line within the accuracy of the experimental data (Fig. 6.1).

It was previously shown that the contribution of the CH_2 group of the alkyl chain to the vaporization enthalpy of saturated ethers and ketones differs insignificantly from the analogous contribution of the CH group in their unsaturated analogs. Therefore, we take the contribution of the single CH group of crotonaldehyde as equal to that of methylene group in butanal. Taking into account insignificant difference in the energy of the specific interactions of the saturated and unsaturated ethers, ketones, and alcohols and neglecting the possible insignificant difference between the energy of the specific interactions in the saturated and unsaturated aldehydes, assuming it to be equal to the analogous bond in the liquid saturated ethers, we estimated the hydrogen bond energy $D = O \cdots H - C - CH = CH - CH_3$ in the liquid crotonaldehyde at 7.9 kJ mol^{-1}. This hydrogen bond energy coincides with the energy of the same type of bond calculated from the estimated vaporization enthalpy of pentenal. Therefore, the validity of the estimated vaporization enthalpy of unsaturated aldehydes is confirmed by the performed calculation of the energy of the hydrogen bonds (Table 6.5). The vaporization enthalpy of propenal is published [6] at two nonstandard temperatures (Table 6.5). The vaporization enthalpy of propenal reduced to the standard temperature (31.1 kJ mol^{-1}) was used for estimation of the hydrogen bond energy (Table 6.5). Its lower value than that of the hydrogen bond in the butanal is due to the weaker shift of the electron density to the oxygen atom along the chain from the terminal methylene group of the propenal. Consequently, the increased number of carbon atoms in the butenal and the presence of the terminal methyl group essentially strengthen the shift in the electron density to the oxygen atom and increase its negative charge, thus stabilizing the hydrogen bond. The identical hydrogen bond energy in butenal and pentenal (Table 6.5) indicates its equal contribution to the enthalpy characteristics in all unsaturated aldehydes with the growing number of carbon atoms in the chain, regardless of the position of the double bond in this chain.

The values obtained show that the contribution of the ethyl group exceeds twice that of the methyl group. The contribution of the isostructural methyl group in 2-methyl-2-butenal (1.9 kJ mol^{-1}) was obtained from the difference in the vaporization enthalpies of this compound and butenal, and the contribution of the isostructural ethyl group (4.0 kJ mol^{-1}) was calculated from the difference of the enthalpy characteristics of 2-ethyl-2-hexenal and 2-hexenal (Table 6.5). These values show practically double contribution of the isostructural ethyl group compared to that of the methyl group. The combined contribution of the isostructural methyl and ethyl groups is estimated from the difference of the vaporization enthalpies of 2-ethyl-4-methyl-2-pentenal and 2-pentenal equal 5.7 kJ mol^{-1}. Taking the contribution of the isostructural methyl group at 1.9 kJ mol^{-1}, we get the contribution of the ethyl group at 3.8 kJ mol^{-1}. This means that the contribution of the isostructural methyl group to the vaporization enthalpy of 2-ethyl-4-methyl-2-pentenal does not depend on its location in the molecule of 2-pentenal and 2-butenal. Analogously from the difference in the enthalpy characteristics of

Table 6.5 Energies (kJ mol^{-1}) of specific interactions and H-bonds in unsaturated aldehydes (298 K)

Compound	Formula	Structure	$\Delta_{vap}H^0(T)$ [9]	D=O → CH_2=CH–C–H/ D=O → CH_3–$(CH)_n$–C–	D = O···H – C	ΣDCH₂
Propenal	C_3H_4O	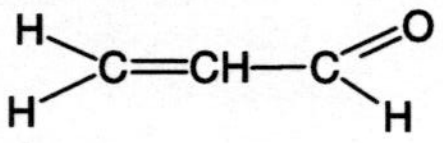	32.9 (265 K) 30.4 (319 K) 31.1[a]	8.92	6.60	–
Crotonaldehyde	C_4H_6O	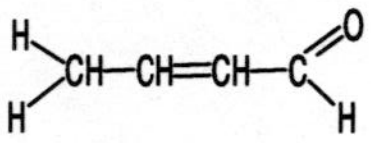	37.3 ± 0.4	$n = 1$ 8.92	7.90	3.6
2-Pentenal	C_5H_8O	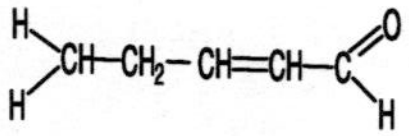	41.0[a]	$n = 2$ 8.92	7.80	7.2
2-Methyl-2-butenal	C_5H_8O	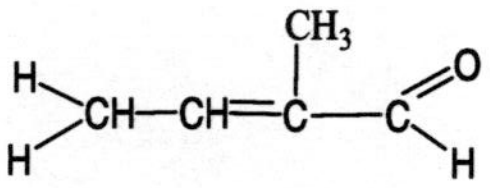	39.2	$n = 1$ 8.92	7.90	$\Sigma DCH_3 = 1.9$
2-Ethyl-4-methyl-2-pentenal	$C_8H_{14}O$	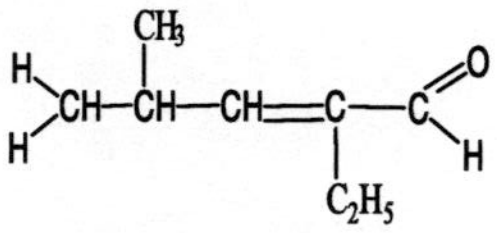	46.7 (320 K)	$n = 2$ 8.92	7.90	$\Sigma DCH_3 = 1.9$ $\Sigma DC_2H_5 = 3.6$

(continued)

Table 6.5 (continued)

Compound	Formula	Structure	$\Delta_{vap}H^0(T)$ [9]	D=O → CH_2=CH–C–H/ D=O → CH_3–$(CH)_n$–C–	D = O···H – C	ΣDCH$_2$
2-Ethyl-2-hexenal	$C_8H_{16}O$	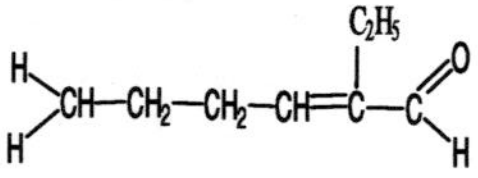	48.4 (341 K)	$n = 3$ 8.92	7.90	$\Sigma DC_2H_5 = 3.9$
2–Ethylacrolein	C_5H_8O	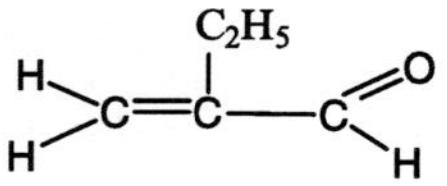	36.8 ± 0.4	8.92	6.60 (7.7)[b]	$\Sigma DC_2H_5 = 5.7$ (3.6)[b]

[a]Determined by interpolation
[b]Method of additive contribution

2-ethylacrolein and propenal is obtained the contribution of the side ethyl group in the 2-ethylacrolein amounting to 5.7 kJ mol^{-1}. This originates from the involvement of the side ethyl group in the shift of the electron density in the 2-ethylacrolein molecule, resulting in the growth of the negative charge on the oxygen atom. Therefore, the formed hydrogen bond is more stable than that in the propenal. The energy value of this interaction may be estimated taking the energy contribution of the side ethyl group equal to that in the 2-ethyl-2-hexenal (3.6 kJ mol^{-1}). The obtained energy value of the hydrogen bond (7.7 kJ mol^{-1}) illustrates the validity of the assumption in the involvement of the side ethyl group in the electron density shift in the 2-ethylacrolein molecule. It should be noted that considering certain arbitrariness of the taken assumptions, it is necessary to carry out reasonable combination of the used approaches in order to obtain the most reliable information concerning the effect of various factors on the energy of the hydrogen bond in the unsaturated aldehydes.

It was shown above that the thermodynamic properties of organic compounds from various homologous series correlate among themselves, and it is also valid for the vaporization enthalpy of the unsaturated aldehydes (Fig. 6.6).

The initial part of the plot corresponds to the weakening of the influence of the reverse dative bond on the energy of the specific interactions terminating at propenal in the correlation plots with the vaporization enthalpies of ethers and alcohols and at butenal in the correlation with the enthalpy characteristic of ketones. This dependence is due to the differentiating influence of the oxygen atom with two ordinary bonds of ether and one of alcohol, on the one hand, and of the carbonyl oxygen of ketone, on the other hand.

The description of the vaporization enthalpy by a single straight line or two crossing lines at the value of the vaporization enthalpy of nonenal reflects the

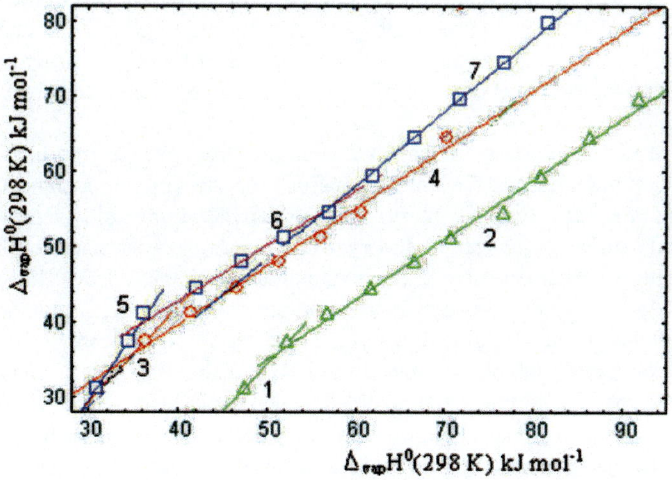

Fig. 6.6 Correlation of vaporization enthalpy of unsaturated 3-alkenals with the vaporization enthalpy of 1-alkanols (1, 2), 2-ketones (3, 4), and unsymmetrical ethers R–C(–O)–C_2H_5 (5, 6, 7)

difference in the contribution of methylene groups in pentenal–nonenal and nonenal–heptadecenal (Fig. 6.6). Interestingly, the unsaturated aldehydes are characterized by the enhanced stability of the hydrogen bonds and growing contribution of methylene groups in the vaporization enthalpy in compounds of the series nonenal–hexadecenal, regardless of the place of the double bond in the alkyl chain (Fig. 6.6). The comparison of the hydrogen bond energy in propenal, butenal, and liquid 2-furaldehyde (furfural) (7.38 kJ mol^{-1}) (Chap. 4)

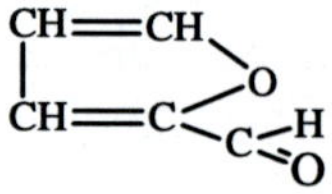

reveals the enhanced stability of the H-bond formed by the oxygen atom of the aldehyde group. This is apparently due to the prevailing shift in the electron density from the two fragments with double bonds to the endocyclic oxygen atom rather than to the aldehyde group.

6.2 Energies of Specific Interactions and H-bonds in Esters

6.2.1 Esters of Formic, Acetic, and Propionic Acids

Esters being derivatives of carboxylic acids and alcohols combine in the structure the effects of the acid and alcohol fragments. Therewith the distribution of the electron density is essentially governed by the carbonyl oxygen. Yet this class of compound attracts limited attention. It is maybe connected, first, with the difficulty to base on them the development of general theoretic applications, and second, because in the most numerous studies of alcohols of normal structure and of carboxylic acids, the clear concept of types of hydrogen bonds and their energies has been lacking.

The experimental and theoretical studies on the structure, electron density distribution in the molecules, and the thermodynamic properties of compounds of the three mentioned classes are complementary. But their value is significantly increased when they are connected by a formal theory using the molecular parameters for the understanding of the macroscopic phenomenological characteristics. In this sense, the thermodynamic properties of vaporization are especially important for they are related to the associative interactions in the condensed state of compounds.

The esters make it possible to reveal the effect of the ester group oxygen on the energy of the specific intermolecular interaction, and the influence of the alkyl group of the alcohol residue on the energy of the previously unknown type of the hydrogen bond formed by the hydrogen atom of the acid residue of the formic acid $= O \cdots H - C$.

As shown in the previous paragraph of this chapter, the description of the vaporization enthalpy of the related compounds by the dependence $\Delta_{vap}H^0(298) = f(C_n)$

provides a possibility to establish the most valid values of the specific interactions and to estimate the enthalpy characteristics of unstudied compounds with an accuracy of the calorimetric measurements. These correlations also permit acquiring information on the effect of the reverse dative bond between the oxygen atom and the contiguous carbon atom on the energy of the specific interactions of the terminal methyl groups and its spreading on the carbon atoms of the alkyl chain. The dependences $\Delta_{vap}H^0(298) = f(C_n)$ presented in Fig. 6.7 for methyl, ethyl, propyl, and butyl esters of the formic and acetic acids reproduce the dependences shown on the aldehydes, esters, ketones, and alcohols. The pattern of this dependence as two straight lines crossing at the enthalpy characteristic of a compound

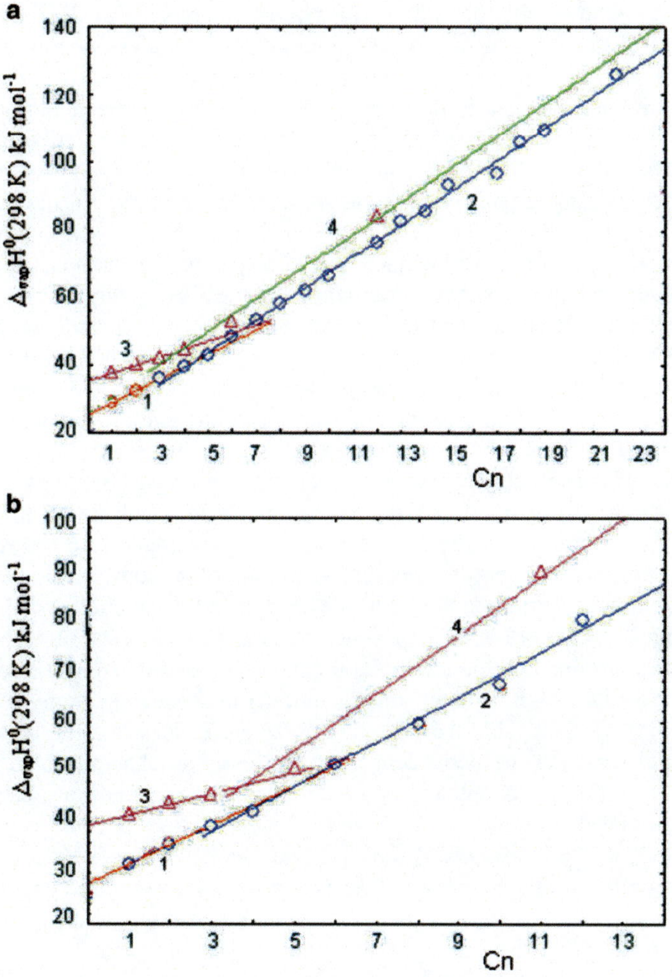

Fig. 6.7 Vaporization enthalpy versus the number of carbon atoms in the alkyl chain of esters **(a)** (1, 2) methyl esters, (3, 4) propyl esters; **(b)** (1, 2) ethyl esters, (3, 4) butyl esters

with three carbon atoms in the alkyl group of the acid residue reflects the similarity of the influence of the reverse dative bond in all series of esters.

This result provides a possibility to extend the dependence to the esters formed from other alcohols and carboxylic acids, where the number of experimental data is limited. The increase in the vaporization enthalpy in the first three compounds C_1–C_3 in each homologous series of esters corresponds to the weakening of the influence of the dative bond on the terminal methyl groups forming the specific interactions involving the pentacoordinate carbon atom. Further growth of the enthalpy characteristics reflects the contribution of the CH_2 groups originating from the participation of the atoms of these groups in the intermolecular interactions linking the chains. The similar character of the vaporization enthalpy dependence on the number of carbon atoms of esters and ethers (Fig. 6.8) is due to the likeness of the types of the specific intermolecular interactions in these class compounds.

Therefore, the dependence of the vaporization enthalpy on the number of carbon atoms in the chain in various homologous series of esters and the similarity of these dependences to the character of the analogous dependences in all considered classes of oxygen-containing organic compounds are a general property illustrating, on the one hand, the diminishing influence of the reverse dative bond on the intermolecular interactions in first three–four members of the respective series, and, on the other hand, the demonstration of the contribution of the CH_2 group to the enthalpy characteristics in all subsequent compounds illustrates the general sense of the reverse dative bond.

Evidently, the structural and energy parameters of the specific intermolecular interactions governed by the shift in the electron density and the distribution of the charges on the atoms in the molecules of esters determine the correlation dependences of the enthalpy characteristics and the stability of the forming specific interactions. These statements are the key points in the analysis of the obtained dependences, in their use in the interpolation and extrapolation for estimating the lacking experimental thermodynamic properties of compounds, and in the evaluation of the reliability and refinement of the energies of these characteristics in the event of significant errors in the experimental data. The vaporization enthalpies of various esters obtained by the interpolation and extrapolation from the correlation between the vaporization enthalpy and the number of the carbon atoms in the alkyl chain are presented in Table 6.6. The data were corrected using the most precise experimental data and the dependences of the enthalpy characteristics of esters (Figs. 6.7–6.9). This list of enthalpy characteristics of compounds may be extended in the case of need.

The ester molecule consists of the acid and alcohol residues. It is capable of forming three types of interactions. One type forms involving the oxygen of the carbonyl group and the hydrogen atom of the terminal methyl group. The other two types are formed by the interaction of the oxygen from the alcohol residue possessing two unshared electron pairs and the methyl groups of the alkyl of the alcohol residue and the acid fragment, respectively. Therefore, the thermodynamic calculations can be performed with the use of the energies of the specific

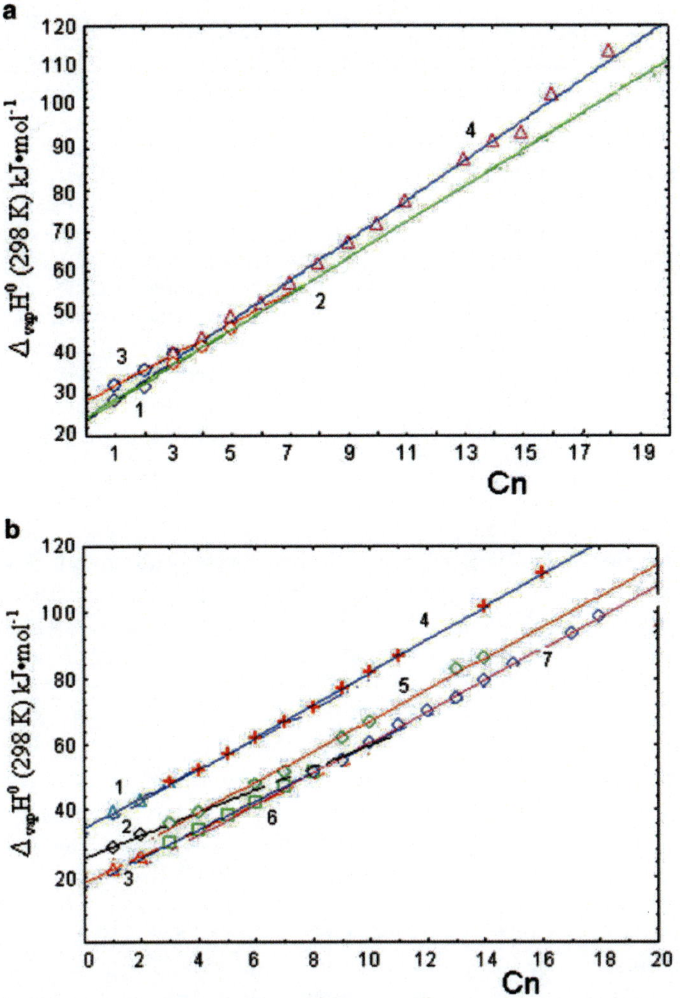

Fig. 6.8 (**a**) Dependence of vaporization enthalpy on the number of carbon atoms in the alkyl chain of esters: (1, 2) R* formiate, (3, 4) R* acetate (R* alcohol fragment). (**b**) Vaporization enthalpy versus the number of carbon atom in the alkyl chain of alcohols (1, 4), methyl esters (2, 5), aldehydes (3, 6, 7); (1, 2, 3) compounds with 1–3 carbon atoms, (4, 5) 3–16 carbon atoms in the chain, (6) 3–8 carbon atoms in the chain, (7) 8–18 carbon atoms in the chain

intermolecular interactions existing in the ethers in order to understand the role of the acid residue of the complex composition and to estimate the energies of the formed interactions. The higher vaporization enthalpy of esters compared to that of ethers is due to the contribution of the most stable specific interactions formed by the bond vacancies of the acid residue. Two oxygen atoms of ester possess four bond vacancies, and the carbon atoms of two terminal methyl groups of the alkyl

Table 6.6 Estimated values of vaporization enthalpy of esters (kJ mol^{-1}) at 298 K

Compound	Formula	$\Delta_{vap}H^0(T)$	Compound	Formula	$\Delta_{vap}H^0(T)$
Methyl pentanoate	$C_6H_{12}O_2$	42.9	Propyl pentanoate	$C_8H_{16}O_2$	48.8
Methyl undecanoate	$C_{12}H_{24}O_2$	71.6	Propyl hexanoate	$C_9H_{18}O_2$	54.3
Methyl hexadecanoate	$C_{17}H_{34}O_2$	95.8	Propyl heptanoate	$C_{10}H_{20}O_2$	59.1
Methyl eicosanoate	$C_{21}H_{42}O_2$	111.3	Propyl octanoate	$C_{11}H_{22}O_2$	64.3
Methyl heneicosanoate	$C_{22}H_{46}O_2$	115.7	Propyl nonanoate	$C_{12}H_{24}O_2$	69.3
Methyl docosanoate	$C_{23}H_{44}O_2$	124.5	Propyl decanoate	$C_{13}H_{26}O_2$	74.6
Ethyl heptanoate	$C_9H_{18}O_2$	55.2	Propyl undecanoate	$C_{14}H_{28}O_2$	79.6
Ethyl nonanoate	$C_{11}H_{22}O_2$	63.6	Propyl tridecanoate	$C_{16}H_{32}O_2$	90.6
Ethyl undecanoate	$C_{13}H_{26}O_2$	74.1	Propyl tetradecanoate	$C_{17}H_{34}O_2$	95.2
Pentyl acetate	$C_7H_{13}O_2$	46.8	Heptadecyl acetate	$C_{19}H_{38}O_2$	108.2
Butyl hexanoate	$C_{10}H_{20}O_2$	56.1	Butyl decanoate	$C_{14}H_{28}O_2$	78.3
Butyl heptanoate	$C_{11}H_{22}O_2$	61.6	Butyl undecanoate	$C_{15}H_{30}O_2$	83.8
Butyl octanoate	$C_{12}H_{24}O_2$	67.3	Butyl tridecanoate	$C_{17}H_{34}O_2$	94.7
Butyl nonanoate	$C_{13}H_{26}O_2$	72.6			

chains of the acid and alcohol residues provide two bond vacancies owing to the essentially unshared 2s^2-electron pairs of the pentacoordinate carbon. However, the possibilities of the carbonyl oxygen atom are limited due to the decreased frequency of the C=O vibrations (by 17%) [1]. Taking into account the tendency of the system to attain the energy minimum, the bond vacancies of this oxygen atom play the secondary role and form the bonds of lower energy involving the hydrogen atoms of the methylene or methyl group located in the *trans*-position with respect to the oxygen of the alcohol residue [10]. In contrast, the more active bond vacancies of the oxygen atom from the alcohol residue with an enhanced negative charge form stable specific interactions with the carbon atoms of the terminal methyl groups of the alkyl chains of both alcohol and acid residues. As has been already mentioned, these interactions form between two atoms from two molecules and are strictly localized on definite reacting species. It is, therefore, possible to describe the structure of the liquid ester by the schematic picture of Fig. 6.10.

Notwithstanding the general similarity of the structure of ester molecules, a significant difference exists in the features of the intermolecular interactions of formiates, on the one hand, and the esters of the carboxylic acids with the larger number of carbon atoms in the chain, on the other hand.

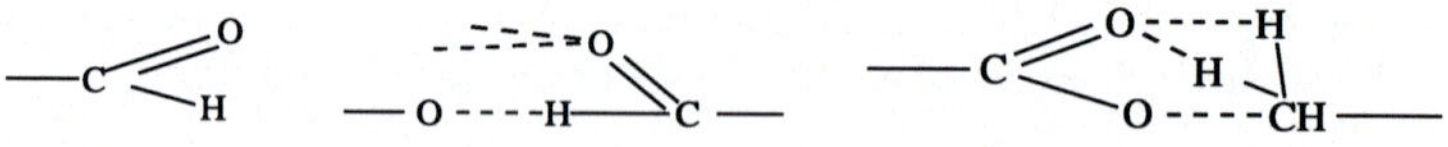

The fundamentally important feature for the formic acid esters is the presence of a hydrogen atom attached to the carbonyl carbon atom capable of donor–acceptor interactions with the oxygen atom of the alcohol residue with the formation of a stable hydrogen bond. It is, therefore, important to obtain information on the energy of this type of hydrogen bond with the oxygen atom of the alcohol residue and to elucidate the influence of the alkyl chain of the alcohol residue on the stability of

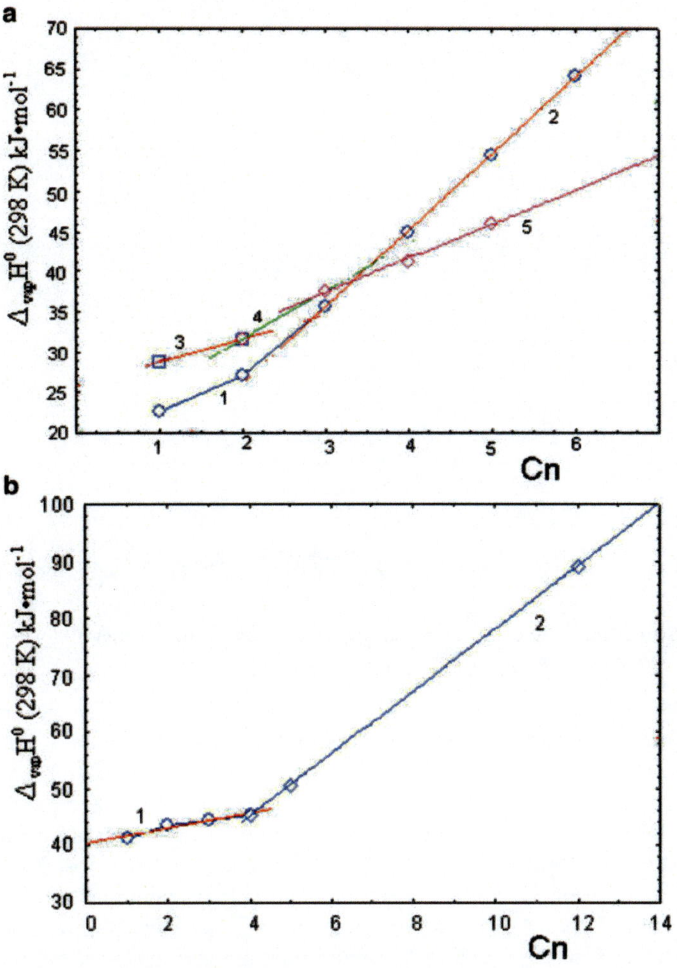

Fig. 6.9 Dependence of vaporization enthalpy on the number of carbon atoms in the alkyl chain (**a**) symmetric ethers (1, 2) and esters of formic acid (3, 4, 5); (**b**) butyl esters

this bond. In this connection, the results should be applied of the above estimation of the energy of the hydrogen bond of low stability formed by the carbonyl oxygen atom with the hydrogen of the methylene or methyl group located in the *trans*-position to the alcohol oxygen. The estimated value is close to 0.60 kJ mol^{-1}. The second bond vacancy of this oxygen atom interacts with the second hydrogen of the same group (Fig. 6.10). Two such oxygen atoms of the two molecules of the nearest surrounding contribute to the vaporization enthalpy with the energy equal to 2.40 kJ mol^{-1}. The terminal methyl group of the alkyl chain of the alcohol residue forms a specific interaction whose energy is equal or at least close to the energy of the analogous interactions in liquid ethers. This is confirmed by the performed

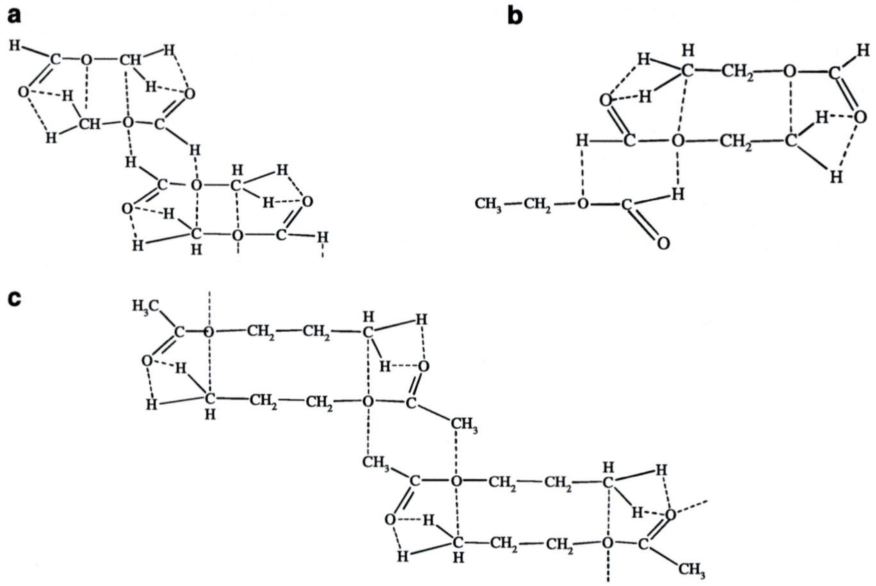

Fig. 6.10 Schematic picture of the liquid structure of esters (a) methyl formiate, (b) ethyl formiate, (c) propyl acetate

thermodynamic calculations for compounds from other classes of organic substances. One of the factors leading to the enhanced vaporization enthalpy of the formiates (Fig. 6.10) is two interactions formed by the oxygen atom with the terminal methyl group of the alcohol residue. The schematic picture of the liquid structure of methyl formiate shown in Fig. 6.10 depicts four hydrogen bonds of low stability (0.60 kJ mol^{-1}) $=\mathrm{O}\cdots\mathrm{H}-\mathrm{CH}_2-$, two hydrogen bonds formed by the hydrogen atom linked to the carbonyl carbon with oxygen atom of the alcohol residue $-\mathrm{O}\cdots\mathrm{H}-\mathrm{C}-$, and two specific interactions $-\mathrm{O}\rightarrow\mathrm{CH}_3-$ with the energy equal to that existing in the liquid dimethyl ether. Therefore, considering that the vaporization enthalpy is equal to the sum of the energies of the existing hydrogen bonds and the specific interactions involving the pentacoordinate carbon, the energy of the hydrogen bond should be estimated from (6.4)

$$\mathrm{D}-\mathrm{O}\cdots\mathrm{H}-\mathrm{C}- = \frac{\Delta_{\mathrm{vap}}H^0(T) - 2\mathrm{D}-\mathrm{O}\rightarrow\mathrm{CH}_3 - 4\mathrm{D}=\mathrm{O}\cdots\mathrm{H}-\mathrm{CH}_2}{2}. \tag{6.4}$$

In the calculations of the energies of this type of hydrogen bond in the liquid esters of the formiate homologous series, the change in the energy of the specific interaction formed by the alkyl chain of the alcohol residue when the number of carbon atoms grows to three is taken into account.

$$D-O\cdots H-C-=\frac{\Delta_{vap}H^0(T)-2D-O\rightarrow CH_3-(CH_2)_n-4D=O\cdots H-CH_2}{2}. \tag{6.5}$$

Also, the energy contribution from the growing number of CH_2 groups of the alkyl chain is taken into consideration.

$$D-O\cdots H-C- \\ =\frac{\Delta_{vap}H^0(T)-2D-O\rightarrow CH_3-(CH_2)_n-4D=O\cdots H-CH_2-n\cdot DCH_2}{2}. \tag{6.6}$$

The calculations utilize the energies of the specific interactions existing in the ethers (Chap. 3, Table 3.1), and the interactions of the CH_2 groups are accounted for at the number of carbon atoms in the alkyl chain exceeding three.

It follows from Table 6.7 that the energy of the hydrogen bond $D-O\cdots H-C-$ in the liquid methyl formiate equals 7.50 kJ mol^{-1}, and it is more stable than the energy of the specific interactions $-O\rightarrow CH_3-$ between the methyl group and the oxygen of the methanol residue. The energy of this type of hydrogen bond grows with the growing number of carbon atoms in the chain of the alcohol residue, and more rightly, with the weakening of the influence of the reverse dative bond in this chain:

Methyl formiate (7.50)<Ethyl formiate (7.80)<Propyl formiate (8.60 kJ mol^{-1}).

The esters of other carboxylic acids also form in the liquid state the chain network structure (Fig. 6.10) that is cross-linked with the hydrogen bonds of low stability and with the specific interactions formed by the terminal methyl groups of the acid and alcohol residues. The energy of specific interactions $-O\rightarrow CH_3-(CH_2)_k-C-$ formed by the alkyl chain of the acid residue is estimated from (6.7)

$$D-O\rightarrow CH_3-(CH_2)_k-C- \\ =\frac{\Delta_{vap}H^0(T)-2D-O\rightarrow CH_3-(CH_2)_n-4D=O\cdots H-CH_2-n\cdot DCH_2}{2}. \tag{6.7}$$

In the acetates, two interactions are formed by the methyl groups of the acetyl

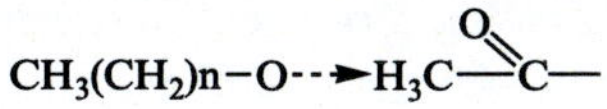

and two remaining interactions $DCH_3-(CH_2)_nO\rightarrow CH_3-(CH_2)_nO$ involve the terminal methyl groups of the alkyl chain of the alcohol residue (6.8)

$$\mathrm{D-O \rightarrow CH_3-C-}$$
$$= \frac{\Delta_{\mathrm{vap}}H^0(T) - 2\mathrm{D-O \rightarrow CH_3-(CH_2)}_n - 4\mathrm{D=O \cdots H-CH_2} - n \cdot \mathrm{DCH_2}}{2}. \tag{6.8}$$

The energy of the specific interactions $2\mathrm{D{-}O} \rightarrow CH_3{-}(CH_2)_n{-}$ at the appropriate number of the methylene groups is identical to that existing in the liquid methyl ether or in ethyl or propyl fragment, respectively. The performed calculations of the energy of this type of specific interaction (Table 6.7) in the liquid acetates illustrate its stabilization with increasing alcohol chain.

$$\text{Methyl acetate (9.30)} < \text{Ethyl acetate (9.90)} = \text{Propyl acetate (9.80 kJ mol}^{-1}).$$

In propionates, the energy of the specific interactions formed by the alkyl chain of the acid residue remains constant within the accuracy limits of the experimental measurements of the vaporization enthalpy

$$\text{Methyl propionate (11.00)} = \text{Ethyl butyrate (10.90)} = \text{Propyl propionate (10.95)}$$
$$= \text{Isopentyl propionate (11.00 kJ mol}^{-1})$$

except for ethyl propionate where the measurement error has not been indicated.

According to the data of Table 6.7, the specific interactions formed by the propionate fragment are more stable than those of the acetates.

The results of the performed calculations of the energy of the specific interactions confirm the conclusions from the analysis of the dependences of the vaporization enthalpies of esters on the number of carbon atoms in the respective alkyl chains. The data obtained illustrate the weakening of the influence of the reverse dative bond with growing number of carbon atoms in the alkyl chains of both the alcohol and acid residues in going from the methyl to a propyl fragment. This regular series corresponds to the intramolecular redistribution of the electron density governed by the variation in the number of carbon atoms in going from the methyl to the propyl fragment. However, the constancy of the energy of the specific interactions formed by the acid residue in the propionate indicates that at the propyl chain in the acid residue, the maximum stabilization of this specific interaction is attained in the esters. This conclusion is confirmed by the energy of the specific interaction in the liquid ethyl butyrate. Here, attention should be directed to the enhanced stability of the hydrogen bonds in the formiates and to the specific interaction originating from the enhanced negative charge on the alcohol oxygen due to the neighboring carbonyl oxygen [4], as shown by the quantum-chemical calculations and the photoelectron spectra of acids. These data also indicate the prevailing stabilization of the specific interaction formed by the alcohol residue. The stabilization of the specific interaction grows from 5.63 even to the energy of 9.30 in the methyl acetate and from 8.92 to 9.90 kJ mol^{-1} in the

Table 6.7 Energies of hydrogen bonds and specific interactions (kJ mol^{-1}) in liquid esters

Compound	Structure	$\Delta_{vap}H^0(T)$ [2, 9]	D–O → CH_3–$(CH_2)_n$–	D = O···H – CH_2	D – O···H – C
Methyl formiate	HC(=O)OCH_3	28.67	5.63 $n = 0$	∑D = 2.40	7.50
Ethyl formiate	HC(=O)OCH_2CH_3	31.6 ± 0.1	6.78 $n = 1$	∑D = 2.40	7.80
Propyl formiate	HC(=O)O$(CH_2)_2CH_3$	37.49	8.92 $n = 2$	∑D = 2.40	8.60
Butyl formiate	HC(=O)O$(CH_2)_3CH_3$	41.25	8.92 $n = 3$	∑D = 2.40 3.70	8.60
Pentyl formiate	HC(=O)O$(CH_2)_4CH_3$	46.0	8.92 $n = 4$	∑D = 2.40 4.70	8.60
			D–O → $CH_3(CH_2)_n$–		D–O → CH_3–C–
Methyl acetate	CH_3C(=O)OCH_3	32.30	5.63 $n = 0$	–	9.30
Ethyl acetate	CH_3C(=O)O$(CH_2)CH_3$	35.62	6.78 $n = 1$	0.60 × 4	9.90
Propyl acetate	CH_3C(=O)$OCH_3(CH_2)_2$	39.83	8.92 $n = 2$	0.60 × 4	9.80
Butyl acetate	CH_3C(=O)$OCH_3(CH_2)_3$	43.60	8.92 $n = 3$	0.60 × 4 4.5	9.90
Pentyl acetate	CH_3C(=O)$OCH_3(CH_2)_4$	46.8	8.92 $n = 4$	0.60 × 4 4.6	9.90
Hexyl acetate	CH_3C(=O)$OCH_3(CH_2)_5$	51.1	8.92 $n = 5$	0.60 × 4 11.0:3	9.90
			D–O → CH_3–$(CH_2)_n$–		D–O → CH_3–CH_2–C–
Methyl propionate	CH_3CH_2C(=O)OCH_3	35.85	5.63 $n = 0$	0.60 × 4	11.0
Ethyl propionate	CH_3CH_2C(=O)OCH_2CH_3	39.26	6.78 $n = 1$	0.60 × 4	11.50
Ethyl butyrate	$CH_3(CH_2)_2$C(=O)O$(CH_2)CH_3$	42.0	6.78 $n = 1$	0.60 × 4 4.1	10.90
Propyl propionate	CH_3CH_2C(=O)–O–$(CH_2)_2CH_3$	42.14	8.92 $n = 2$	0.60 × 4	10.95
Isobutyl propionate	$C_7H_{14}O_2$	44.9 (286)*	8.92 $n = 2$	0.60 × 4 D*iso*CH_3 = 1.8	11.40*(286 K)
Isopentyl propionate	$C_8H_{16}O_2$	44.1	8.92 $n = 3$	0.60 × 4 D*iso*CH_3 = 1.8	11.00
Dimethoxymethane	$CH_2(OCH_3)_2$ – O···H – CH– = 0.95	26.41 –		D–O → CH_3–5.63	

propyl acetate, and also to 11.00 in methyl propionate and 10.95 kJ mol^{-1} in propyl propionate.

The previously disregarded possibility of hydrogen bond formation in aldehydes and esters was revealed in this study and as a result, the stability series of these bonds were determined:

$$\text{Formaldehyde (5.50)} < \text{Acetaldehyde (6.10)} < \text{Propanal (6.00)} < \text{Butanal} = \text{Pentanal (6.10 kJ mol}^{-1}\text{)};$$

unsaturated aldehydes:

$$\text{Propenal (6.60)} < \text{Crotonaldehyde (7.90 kJ mol}^{-1}\text{)};$$

esters:

$$\text{Methyl formiate (6.60)} < \text{Ethyl formiate (7.90)} < \text{Propyl formiate (8.69)} < \text{Butyl formiate (8.60 kJ mol}^{-1}\text{)}.$$

Important information on the interaction of molecules involving the hydrogen atom was obtained of dimethoxymethane (dimethylformal). Two oxygen atoms of this compound form by their unshared electron pairs four specific interactions with the molecules of the nearest surrounding in the liquid and crystalline state; in these interactions, carbon atoms of methyl groups are involved, and in the other two interactions, the hydrogen atoms of the methylene group take part. Thus, each molecule of the dimethoxymethane forms four specific interactions $-O \rightarrow CH_3-$ with the energy similar to that in liquid dimethyl ether, and four hydrogen bonds of low stability $-O\cdots H-CH$ whose energy should be estimated from the difference between the vaporization enthalpy and the sum of the energies of four specific interactions:

$$D-O\cdots H-CH = \frac{\Delta_{vap}H_0(298) - 4D-O \rightarrow CH_3-}{4}. \quad (6.9)$$

The obtained energy value for this hydrogen bond (1.00 kJ mol^{-1}) is somewhat larger than the value used above in the performed calculations (0.60 kJ mol^{-1}). The valuable meaning of the obtained energies of hydrogen bond of low stability consists in the fact that the performed calculations of the energy of the specific interactions and the hydrogen bonds in the aldehydes and esters correctly describe the nature of the interactions and create a self-consistent system of energy values of the discussed specific interactions formed by the pentacoordinate carbon atom and of various types of hydrogen bonds.

6.2.2 *Methyl, Ethyl, Propyl, and Butyl Esters*

The obtained conclusions can be supplemented with the analysis of the effect of acid residues on the energy of the specific interactions formed by the alcohol residues in the methyl, ethyl, and propyl esters. In this connection, the refined values of enthalpy characteristics of vaporization of esters (Table 6.7), the energies of the specific interactions formed by the alkyl chain of the acid residue, and the energy of the H-bonds of low stability $D = O \cdots H - CH_2 -$ should be taken into account.

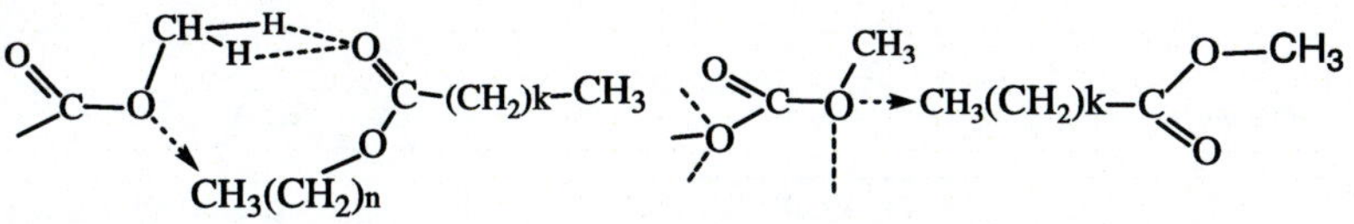

The results of the performed thermodynamic calculations of the energy of the specific interactions in liquid esters formed by methyl, ethyl, and propyl alcohol residues are given in Tables 6.8 and 6.9. The data in Tables 6.8 and 6.9 show that the specific interaction $DO = C - O \rightarrow CH_3 - (CH_2)_n - C(=O)$ in liquid methyl esters is of higher energy than the analogous interactions in liquid ethyl and propyl esters by 1.8 and 2.1 kJ mol^{-1}, respectively. Reduced stability was found in methyl and ethyl acetates (Table 6.8).

The energy of the specific interactions formed by propyl esters with longer acid residue remains constant in the total series of methyl and ethyl esters. The values of the energy of the specific interactions in liquid propyl esters presented in Table 6.9 illustrate the small fluctuations in the energies of the interactions and their stabilization by 2.0 kJ mol^{-1} compared with those in liquid propanol (8.92 kJ mol^{-1}).

The butyl esters are characterized by a little difference in the vaporization enthalpy in the series butyl formiate–butyl butyrate (Fig. 6.9b) originating from, first, the high stability of the hydrogen bond formed by the acid residue of the formic acid; second, the growing stabilization of the specific interaction formed by the alcohol residue with the increasing number of the carbon atoms; and third, the material termination of the influence of the reverse dative bond on the energy of the specific interactions in butyl propionate.

The results of the performed thermodynamic calculations of the energy parameters of interactions confirm these conclusions (Table 6.10). The maximum energy value of the specific interactions $D–O \rightarrow CH_3–(CH_2)_n–C–$ in liquid butyl propionate and the same value of this energy in all subsequent esters of this homologous series provided a possibility to estimate the contribution of the CH_2 group of the fourth carbon atom of the butyl fragment to the vaporization enthalpy of this compound.

The comparison of the energy of the specific interactions $D–O \rightarrow CH_3–(CH_2)_n–$ in esters formed by the alcohol residue shows their stabilization in the series of

Table 6.8 Energies (kJ mol^{-1}) of hydrogen bonds and specific interactions of liquid methyl and ethyl esters (298 K)

Compound	$\Delta_{vap}H^0(T)$ [2, 4]	D–O → CH_3–	D = O···H – CH_2 –	D–O → CH_3–$(CH_2)_n$–C–
Methyl acetate	32.3 ± 0.1	5.63 $n = 0$	0.60 × 4	9.30 $n = 0$
Methyl propionate	35.6 ± 0.4	5.63	0.60 × 4	11.00 $n = 1$
Methyl butyrate	39.3 ± 0.2	5.63	0.60 × 4 $\Sigma DCH_2 \times 1 = 3.6$	11.00 $n = 2$
Methyl pentanoate	42.9	5.63	0.60 × 4 $\Sigma DCH_2 \times 2 = 7.2$	11.00 $n = 3$
Methyl hexanoate	48.0 ± 0.1	5.63	0.60 × 4 $\Sigma DCH_2 \times 3 = 12.0$ 12.3	11.10 $n = 4$
Methyl heptanoate	51.6 ± 0.5	5.63	0.60 × 4 $\Sigma DCH_2 \times 4 = 16.0$	11.00 $n = 5$
		D–O → CH_3–CH_2–		D–O → CH_3–$(CH_2)_n$–C–
Ethyl acetate	35.3	6.78	0.60 × 4	9.70
Ethyl propionate	39.5 ± 0.1	6.78	0.60 × 4	11.75
Ethyl butyrate	42.0 ± 01	6.78	0.60 × 4 $\Sigma DCH_2 \times 1 = 2.6$	11.70 $n = 2$
Ethyl pentanoate	46.7	6.78	0.60 × 4 $\Sigma DCH_2 \times 2 = 7.6$ 7.3	11.70 $n = 3$
Ethyl hexanoate	51.5 ± 0.1	6.78	$\Sigma DCH_2 \times 3 = 11.90$	11.70 $n = 4$

Table 6.9 Energies (kJ mol^{-1}) of specific interactions of propyl esters (298 K)

Compound	$\Delta_{vap}H^0(T)$ [2, 4]	D–O → $CH_3(CH_2)_n$–	DC = O···H – CH_2 –	D–O → CH_3–$(CH_2)_n$–C–
Propyl acetate	38.2	8.92 $n = 0$	0.60 × 4	9.00
Propyl propionate	42.1	8.92 $n = 1$	0.60 × 4	10.90
Propyl butyrate	44.3	8.92 $n = 2$	0.60 × 4 $\Sigma DCH_2 \times 1 = 2.2$	10.90
Propyl pentanoate	49.2	8.92 $n = 3$	0.60 × 4 $\Sigma DCH_2 \times 2 = 7.0$	10.90
Propyl hexanoate	54.3	8.92 $n = 4$	0.60 × 4 $\Sigma DCH_2 \times 3 = 10.2$	10.90
Propyl heptanoate	59.4	8.92 $n = 5$	0.60 × 4 $\Sigma DCH_2 \times 3 = 16.4$	11.35

compounds belonging to the same homologous series with the growing number of carbon atoms in the alcohol alkyl chain.

$$\text{Methyl acetate } (9.30) < \text{Ethyl acetate } (9.70) < \text{Propyl acetate } (10.70) < \text{Butyl acetate } (11.00 \text{kJ mol}^{-1}).$$

Table 6.10 Energies (kJ mol^{-1}) of specific interactions of butyl esters (298 K)

Compound	$\Delta_{vap}H^0(T)$ [2, 4]	D–O → CH_3–$(CH_2)_3$–	D = O···H – CH_2	D–O → CH_3–$(CH_2)_n$–C–
Butyl formiate	41.25	8.92 $DCH_2 = 3.6$	2.40	D – O···H – C – 8.7
Butyl acetate	43.68	8.92	2.40	11.1
Butyl propionate	44.7	8.92	2.40 $DCH_2 = 2.5$	11.00
Butyl butyrate	45.4	8.92	2.40 $DCH_2 = 4.1$	11.00
Butyl pentanoate	50.4 ± 0.8	8.92	2.40 $DCH_2 = 8.1$	11.10
Butyl dodecanoate	89.2	8.92	2.40 $DCH_2 = 47.6$	11.00

In the homologous series of the propionates, the energies of the specific interactions are constant and have higher stability than that of acetates.

6.2.3 Esters of Dicarboxylic Acids

The performed analysis of structures and energies in the esters of monocarboxylic acids permits the consideration of esters originating from dicarboxylic acids. The dependence of the vaporization enthalpy of the saturated esters of dicarboxylic acids on the number of methylene groups between the ester groups presented in Fig. 6.11 indicates the notable decrease in the influence of the reverse dative bond in the series dimethyl oxalate–dimethyl glutarate, as shown by the growing vaporization enthalpy of the respective esters. The overall energy of these interactions equals the difference between the vaporization enthalpies of these compounds. The subsequent increase in the enthalpy characteristic is due to the energy contribution of the CH_2 groups and obviously to the growing number of specific interactions of low stability formed by these groups. Thus, the influence of the reverse dative bond ends in the presence of five carbon atoms in the molecule of dimethyl glutarate including the carbonyl carbon atoms. The increase in the number of carbonyl groups in the esters of dicarboxylic acids is necessarily reflected in the intermolecular bonding of the molecules of liquid dimethyl oxalate due to the enhanced positive charge on the carbon atom of the carbonyl group and the enhanced negative charge on the oxygen of the alcohol residue, due to the strong shift in the electron density under the effect of the oxygen atoms of these groups.

In liquid and crystalline states of the ethers, a chain network structure forms whose bonding originates from the stable specific interactions involving the oxygen atoms of the alcohol residue and carbon atoms of the terminal methyl groups or the carbon atoms of the carbonyl groups (Fig. 6.12). The characteristic feature of esters consists in the formation of four specific interactions D–O → C = O in the liquid

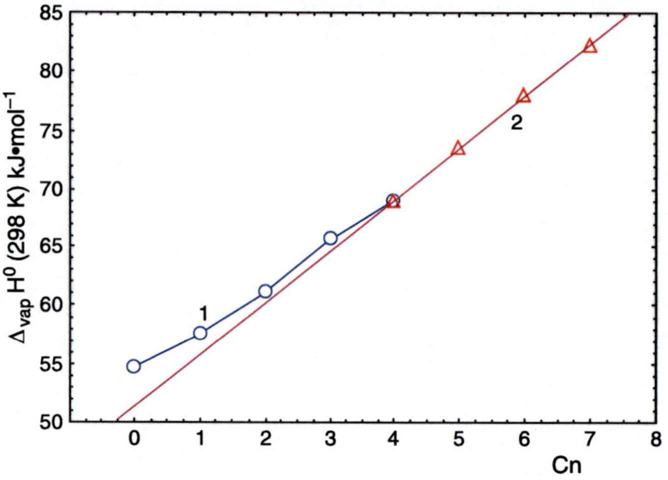

Fig. 6.11 Dependence of vaporization enthalpy on the number of carbon atoms in the alkyl chain of esters of dicarboxylic saturated acids (1) decrease in the effect of the reverse dative bond; (2) increasing contribution of the growing number of methylene groups

dimethyl oxalate and dimethyl malonate by the carbon atoms of the ester groups with the oxygen atoms of the alcohol residue from the molecule of the nearest surrounding (Fig. 6.12). The methyl groups of the alcohol residues form four specific interactions D–O → CH_3– with the analogous oxygen atoms of the molecules of the nearest surrounding: the energy of these interactions is identical to that in the liquid dimethyl ether. The carbonyl oxygen atom in its turn forms eight hydrogen bonds of low stability $=O\cdots H-CH$.

Thus, we obtain the unknown value of the energy of the specific interaction from the difference between the vaporization enthalpy and the sum of the energies of the hydrogen bonds of low stability and the interactions D–O → CH_3–.

$$D-O\rightarrow C=O=\frac{\Delta_{vap}H^0(T)-4D-O\rightarrow CH_3-8D=O\rightarrow H-CH_2-}{4}. \tag{6.10a}$$

The larger number of methylene groups in the molecules of dimethyl succinate and dimethyl glutarate and the corresponding extension of the chain between two ester groups favor the weakening of the effect of the reverse dative bond in the acid residue and consequently, the formation of more stable interactions by the carbon atoms of this residue. Taking into account that the shift in electron density in the chain of acid residue due to the presence of two oxygen atoms is stronger than that in the alcohol residue, the carbon atom of the carbonyl group acquires a larger positive charge. Therefore, the carboxylic fragment of ester forms a specific interaction of the stability superior than that involving the terminal methyl group of the alcohol residue.

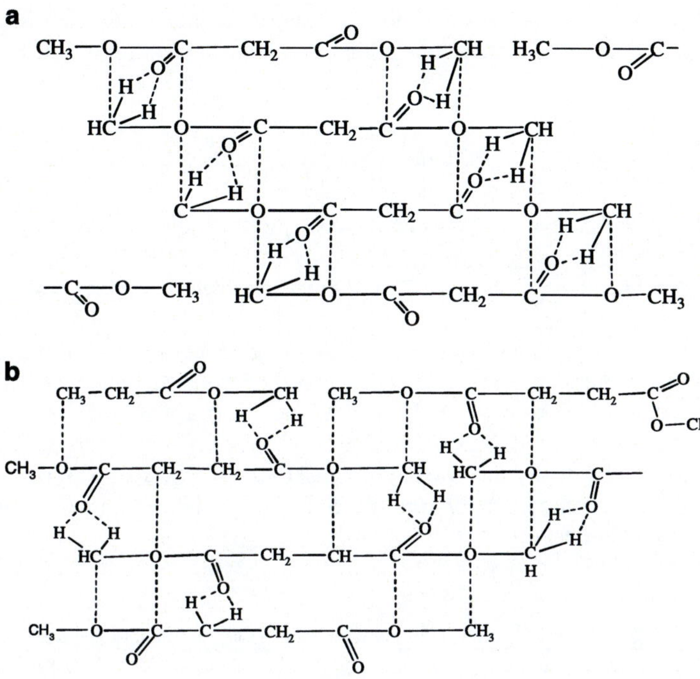

Fig. 6.12 Schematic picture of the liquid structure of esters dimethyl malonate (**a**) and dimethyl succinate (**b**)

The energy value of the specific interaction formed by the alcohol residue equals the energy of the analogous interaction of the corresponding symmetric ether, and the known energy of the hydrogen bond of low stability makes it possible to calculate the energy of the interaction D–O → $(CH_2)_n$–C– from (6.10b)

$$\mathrm{DO} = \mathrm{C}-\mathrm{O}\rightarrow(\mathrm{CH}_2)_n-\mathrm{C}- \\ = \frac{\Delta_{\mathrm{vap}}H^0(T)-4\mathrm{D}-\mathrm{O}\rightarrow\mathrm{CH}_3-8\mathrm{D}-\mathrm{O}\rightarrow\mathrm{CH}_3-}{4}. \tag{6.10b}$$

The results of the performed calculations (Table 6.11) illustrate the above-established rule of the stabilization of specific interactions in going from dimethyl malonate to diethyl malonate due to the increasing alkyl chain in the alcohol fragment, and further stabilization in going to dimethyl glutarate. The stabilization of the interaction is completed at the dimethyl glutarate having five carbon atoms in the acid residue, one methylene group less than that in the monohydric alcohol fragments. In all subsequent esters of this homologous series, the energy of these types of specific interactions is constant. The distinguishing feature of the esters of dicarboxylic acids is the reduced stability of the specific interaction compared with the stability of analogous interactions in the esters of monocarboxylic acids:

$$\text{Methyl acetate } (9.30) > \text{Dimethyl malonate } (8.20\,\text{kJ mol}^{-1}\,\text{DCH}_2),$$
$$\text{Methyl propionate } (11.00\,\text{kJ mol}^{-1}) > \text{Dimethyl glutarate } (9.60\,\text{kJ mol}^{-1}),$$

and further in the compounds with more carbon atoms in the acid residue:

$$\text{Methyl pentanoate } (11.00\,\text{kJ mol}^{-1}) > \text{Dimethyl suberate } (9.60\,\text{kJ mol}^{-1}).$$

The replacement of dimethyl ester by diethyl ester does not result in the stabilization of the specific interactions under consideration: dimethyl malonate (8.20) = diethyl malonate (8.20 kJ mol^{-1}). The calculation performed as an example for estimating the energy of the specific interactions in esters with a large number of carbon atoms in the chain and with an isostructural methyl group in the alcohol residue at the nonstandard temperature illustrates low reliability of the experimentally measured characteristic.

Table 6.11 Energies (kJ mol^{-1}) of specific interactions of esters of dicarboxylic acids (298 K)

Compound	Formula	$\Delta_{vap}H^0(T)$ [11]	DCH_2	D–O → CH_3–	D–O → $(CH_2)_n$–C
Dimethyl oxalate	$C_4H_8O_4$	54.7 ± 0.5	$n = 0$	5.63	6.85 $n = 0$
Dimethyl malonate	$C_5H_{10}O_4$	57.5 ± 0.3	$n = 1$	5.63	8.20
Dimethyl succinate	$C_6H_{12}O_4$	61.0 ± 0.3	$n = 2$	5.63	8.40
Dimethyl glutarate	$C_7H_{14}O_4$	65.7 ± 0.4	$n = 3$	5.63	9.60
Dimethyl adipate	$C_8H_{16}O_4$	69.0 ± 0.2	$n = 4$ $DCH_2 = 3.7$	5.63	9.60
Dimethyl pimelate	$C_9H_{18}O_4$	73.5 ± 0.4	$n = 5$ $2DCH_2 = 8.1$	5.63	9.60
Dimethyl suberate	$C_{10}H_{20}O_4$	78.1 ± 0.2	$n = 6$ $3DCH_2 = 12.4$	5.63	9.60
Dimethyl azelate	$C_{11}H_{22}O_4$	82.3 ± 0.4	$n = 7$ $4DCH_2 = 16.8$	5.63	9.60
Dimethyl subecate	$C_{12}H_{24}O_4$	84.4 ± 0.5 (86.7)	$6DCH_2 = 21.2$	5.63	9.60
Dimethyl undecandioate	$C_{13}H_{26}O_4$	(88.8)	$7DCH_2 = 25.6$	5.63	9.60
Diethyl malonate	$C_7H_{12}O_4$	64.7 ± 0.2 [9]	-	6.78 × 4	8.2
Diisopropyl oxalate	$C_8H_{14}O_4$	57.6 [9]	341 K $n = 0$	6.78 × 4 $DisoCH_3 = 3.6$	5.5
Dipropyl oxalate	$C_8H_{14}O_4$	57.8 [9]	341 K $n = 0$	8.92 × 4	4.3
Diethyl glutarate	$C_7H_{14}O_4$	55.7 [9]	353 $n = 3$	6.78 × 4 $n = 1$	6.0

6.2.4 Esters with Isostructural Methyl Group

Thermodynamic properties of the vaporization of esters containing an isostructural methyl group are known for 11 compounds, but some data have been obtained at nonstandard conditions and without indication of the measurement error. Yet it is possible to elucidate the contribution of the side methyl group to the vaporization enthalpy and to perform the calculations of the energy of the specific interactions formed by acid residues of the formic and acetic acids. To this end, let us compare the enthalpy characteristics of compounds possessing the isostructural methyl group and their analogs of normal structure (Table 6.12).

A characteristic feature of esters with an isostructural methyl group in the alcohol residue is the vaporization enthalpy value less than the vaporization enthalpy of the ester with the same number of carbon atoms but of normal structure of the alcohol chain, and vaporization enthalpy value larger than that of the ester with normal structure and with the chain shorter by one carbon atom. This means that the formation of isopropyl acetate from ethyl acetate results in an increase in the enthalpy characteristic by 1.8 kJ mol^{-1}, and the replacement of the CH_2 group in propyl acetate by the isostructural methyl group reduces the vaporization enthalpy by 2.8 kJ mol^{-1}. Analogously, the subtraction of the vaporization enthalpy of propyl acetate from that of the isobutyl acetate gives the energy contribution of the isostructural methyl group equal to 1.6 kJ mol^{-1}.

The introduction of an isostructural methyl group into butyl acetate with the formation of isopentyl acetate results in the increase of the enthalpy characteristic by 5.8 kJ mol^{-1}. This value is too large compared to the magnitudes of this contribution found in liquid ethers, ketones, and alcohols. Therefore, the vaporization enthalpy reported without indicating the error of the measurements is overestimated and should not be used in calculations.

The schematic picture of the structure of liquid isopentyl acetate presented in Fig. 6.13 illustrates the interaction of the isostructural methyl groups with the formation of a specific interaction $DH_3C \rightarrow H\text{–}CH_2$ of low stability with the energy of 1.8 kJ mol^{-1} and with the transfer of the electron density from the essentially unshared $2s^2$ electron pair of the carbon atom.

The structure of the molecule of isopropyl acetate possessing an isostructural methyl group and a linear chain of two carbon atoms, like in ethyl acetate, makes it possible to use the value of the energy of the specific interaction $2D\text{–}O \rightarrow CH_3\text{–}C\text{–}$ measured for the liquid ethyl acetate in the calculation of the contribution of the isostructural methyl group to the enthalpy characteristic of the compound with the same number of carbon atoms in the alcohol residue. This contribution is equal to the difference between the vaporization enthalpy of the ester and the sum of the energies of the specific interactions and the H-bonds of low stability of the ethyl acetate existing with the same energy in liquid isopropyl acetate:

Table 6.12 Energies (kJ mol^{-1}) of specific interactions of liquid esters with isostructural methyl group at 298 K

Compound	Formula	$\Delta_{vap}H^0(T)$ [2, 9]	D–O → CH_3–$(CH_2)_n$–	D = O···H – C	D–O → CH_3–C–	D*iso*CH_3
Ethyl acetate	$C_4H_8O_2$	35.2	6.78 $n = 0$	2.40	9.70	–
Isopropyl acetate	$C_5H_{10}O_2$	37.0	D*iso*CH_3 = 1.8 6.78 $n = 0$	2.40	9.70	1.8
Isobytyl acetate	$C_6H_{12}O_2$	39.8	D*iso*CH_3 = 1.6 6.78 $n = 1$	2.40	11.10	=1.6
Propyl acetate	$C_5H_{10}O_2$	38.2	6.78 $n = 1$	2.40	11.10	–
Butyl acetate	$C_5H_{10}O_2$	41.0 ± 0.5	6.78 $n = 2$	2.40 DCH_2 = 2.8	11.10	–
Isopentyl acetate	$C_6H_{12}O_2$	46.8 ± 0.1 (43.8)[a]	D*iso*CH_3 = 1.8 6.78 $n = 2$	2.40 DCH_2 = 4	(11.10)	1.8
Compound	Formula	$\Delta_{vap}H^0(T)$ [2, 9]	D–O → CH_3–$(CH_2)_n$–	D = O···H – C	D–O → $CH_3(CH_2)_n$–C–	D*iso*CH_3
Isopropyl isobutyrate	$C_7H_{14}O_2$	43.8 (43.4)[a] (272 K)	D*iso*CH_3 = 1.8 × 2 8.92	2.40	9.7	1.8 × 2
Isobytyl propionate	$C_7H_{14}O_2$	44.9 (43.8)[a] (286 K)	D*iso*CH_3 = 1.8 8.92	2.40	10.90	1.8
Isobutyl isobutyrate	$C_8H_{16}O_2$	44.5 ± 0.3	D*iso*CH_3 = 1.8 × 2 8.92	2.40	10.3 ± 0.3	1.8 × 2
Isobutyl butyrate	$C_8H_{16}O_2$	41.7 (45.6)[a]	D*iso*CH_3 = 1.8 8.92	2.40 DCH_2 = 2.2	(10.90)	1.8
Isopentyl propionate	$C_8H_{16}O_2$	44.1 (46.0)[a]	D*iso*CH_3 = 1.8	2.40 DCH_2 = 2.2	10.90	1.8
Propyl isovalerate	$C_8H_{16}O_2$	44.3	D*iso*CH_3 = 1.8 8.92	2.40 DCH_2 = 1.0	10.60	1.8
Isobutyl formiate	$C_5H_{10}O_2$	38.2 (258 K) (39.2)[a]	D*iso*CH_3 = 1.8 8.92	2.40	8.6	1.8
Isopentyl formiate	$C_6H_{12}O_2$	38.9 (267 K) (43.2)[a]	D*iso*CH_3 = 1.8 8.92	2.40 DCH_2 = 4	8.6	1.8

[a]Obtained on the base of additive contribution of the energy of specific interactions at 298 K

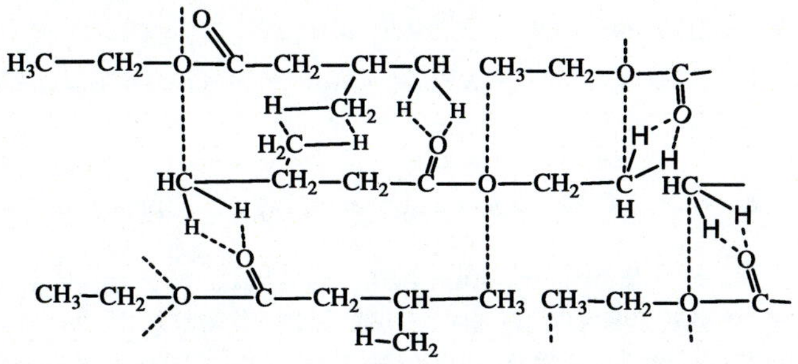

Fig. 6.13 Schematic picture of the liquid structure of isopentyl acetate

$$\mathrm{D}iso\mathrm{CH_3}-= \frac{\Delta_{\mathrm{vap}}H^0(T)-2\mathrm{D}-\mathrm{O}\rightarrow\mathrm{CH_3}-(\mathrm{CH_2})_n-2\mathrm{D}-\mathrm{O}\rightarrow\mathrm{CH_3}-\mathrm{C}-4\mathrm{D}{=}\mathrm{O}\cdots\mathrm{H}-\mathrm{CH_2}}{2}. \quad (6.11)$$

Similar calculation of the contribution of the isostructural methyl group to the enthalpy characteristic was carried out for isobutyl acetate (Table 6.12). The values of the energy of the specific interactions formed by the alcohol residue –O → CH_3–$(CH_2)_n$– are practically identical to those reported in Tables 6.9 and 6.10.

The published vaporization enthalpy of isopropyl isobutyrate was measured at 272 K; therefore, let us first estimate its vaporization enthalpy under standard conditions. To this end, we shall use the already established energies of the specific interactions (Tables 6.9 and 6.10) of the isopropyl alcohol residue containing a two-carbon straight chain and of the isobutyl acid residue containing a three-carbon straight chain, and also the energy contributions of two isostructural methyl groups. The calculated vaporization enthalpy of isopropyl isobutyrate (43.4) is by 0.4 kJ mol^{-1} below the experimentally measured value at 272 K; this discrepancy should be attributed to the temperature dependence.

In keeping with this, we may estimate the vaporization enthalpy of isopentyl acetate based on the principle of additive contribution of all existing specific interactions.

In this calculation, we should take into account the energy contributed to the vaporization enthalpy by the methylene groups at their number over three in the alkyl chain. The value of this energy contribution to the enthalpy characteristic is due to the termination of the influence of the reverse dative bond on the stability of the specific interaction. Therefore, the energy contribution of the CH_2 group is determined from the difference between the enthalpy characteristics of the compounds of the homologous series:

$$\Delta_{vap}H^0(T) = 2D - O \rightarrow CH_3 - (CH_2)_n - +2D - O \rightarrow CH_3 - (CH_2)_nC- \\ + 4D = O \cdots H - CH_2 + DisoCH_3 - +n \cdot DCH_2. \tag{6.12}$$

The obtained value of the vaporization enthalpy of isopentyl acetate is more realistic.

Using the value of the contribution of the isostructural methyl group and the energy of the specific interactions given in Tables 6.9 and 6.10, we estimated the vaporization enthalpies of isobutyl propionate, isobutyl butyrate, isopentyl propionate, isobutyl formiate, and isopentyl formiate, and also the energy of the specific interactions for isobutyl isobutyrate and propyl isovalerate (Table 6.12).

The results of the thermodynamic analysis evidence that disregarding the number of the isostructural methyl groups in the ester molecules and the number of carbon atoms in the alcohol and acid residues at the use of the principle of additive contribution of the existing interactions, the obtained values of the energy of these interactions and the vaporization enthalpy are reasonable. This allows a conclusion that the isostructural methyl group takes part in the redistribution of the electron density in the alkyl chain, and no steric effect is observed in the interactions of the molecules of the nearest surrounding.

6.2.5 *Solid Esters*

The specificity of the interactions in solid compounds, in particular, in esters, suffers the influence of the crystal field stabilizing all types of intermolecular interactions.

Sublimation enthalpies of esters were measured only for three compounds (Table 6.13) [2, 11]. Yet the limited number of thermodynamic data still leads to interesting conclusions. Using the analogy of energy variation in the interactions in other classes of organic compounds in going from the liquid to the solid state, we estimated the energy of the specific interaction of the carbonyl oxygen with the hydrogen atom of the terminal methyl group located in the *trans*-position at 0.80 kJ mol^{-1}.

In keeping with the previously confirmed statement, the energy of the specific interactions involving the bond vacancy of the oxygen atom and the carbon atom of the terminal methyl group of the alcohol residue $DC(=O){-}O{\cdots}H_2C{-}(CH_2)_n{-}O{-}C{\leqslant}O$ was to be taken equal to the energy of the specific interaction in the appropriate solid symmetric ether. Thus, (6.4) remains valid at replacement of the vaporization by sublimation enthalpy for the calculation of the energy of the specific interaction formed by the oxygen atom of the alcohol residue and the carbon atom of the terminal methyl group of the acid residue $DC(=O){-}O{\cdots}H_2C{-}(CH_2)_n{-}O{-}C{\leqslant}O$. The performed thermodynamic calculations (Table 6.11) illustrate the regular stabilization of the specific interaction –O → CH_3–$(CH_2)_n$–C– in the solid and liquid states

Table 6.13 Energies (kJ mol^{-1}) of specific interactions of solid esters at 298 K

Esters	$\Delta_{melt}H^0(T)$ [2]	$\Delta_{sub}H^0(T)$ [2]	D CH_2	D–O → CH_3–$(CH_2)_n$–	D–O → CH_3–$(CH_2)_n$C– Solid	Liquid
Ethyl acetate	10.48	46.10	–	8.83; $n = 1$ $D = O \cdots H - CH = 0.8 \times 4$	12.60 $n = 1$	9.80
Butyl acetate	14.59	58.19	5.6	11.61 $D = O \cdots H - CH = 0.8 \times 4$; $n = 2$	12.75 $n = 2$	11.10
Ethyl butyrate	10.92	52.97	5.6	8.83 $D = O \cdots H - CH = 3.2$; $n = 2$	13.25 $n = 2$	11.70

of esters at replacing the ethyl by the butyl chain in the alcohol residue in butyl acetate and at changing the acetate by butyrate fragment in esters ethyl acetate (12.60) < butyl acetate (13.25 kJ mol^{-1}).

6.2.6 Unsaturated Esters

The rules of variations in energies of specific interactions in esters revealed in Sect. 6.2 provide a possibility to perform the thermodynamic analysis and calculate the energy of the interactions in unsaturated esters. These calculations in the framework of the formal theory are necessary for the critical understanding of the obtained energy parameters and for the development of the theory of the nature of the specific intermolecular interactions, the discovery of previously unknown types of hydrogen bonds and evaluating their energy, favoring the advances in other fields of organic and physical chemistry. The molecular structure of saturated and unsaturated esters is similar, the number of bond vacancies involved in the formation of stable specific interactions is the same, and the unsaturated esters in the liquid state also form the network structure. It was shown before that the energy contributions of the methylene and methine groups of the saturated and unsaturated compounds, respectively, to the enthalpy characteristic are virtually equal, and this fact extends the possibility of the thermodynamic analysis. In this connection, the important information consists in the value of the energy of the specific interaction formed by the unsaturated acid residue through its alkyl chain. The enthalpy characteristics of the unsaturated esters measured at standard or close to standard conditions are limited to compounds listed in Table 6.14. The vaporization enthalpies of acrylates and methacrylates (Fig. 6.14) illustrate the dependences on the number of carbon atoms in the alcohol residue common to other homologous series of esters.

Notwithstanding the difference in the accuracy of the experimental measurements of the vaporization enthalpy at the standard temperature for three first members of these two homologous series and at higher temperature for butyl and pentyl esters, the dependence $\Delta_{vap}H^0(T) = f(C_n)$ feels the influence of the reverse dative bond on the

Table 6.14 Energies (kJ mol^{-1}) of specific interactions in liquid unsaturated esters

Compound	Formula	$\Delta_{vap}H^0(T)$ [2, 9]	T (K)	D–O → CH_3–$(CH_2)_n$–	DO → CH_2=CH–C–
Vinyl acetate	$C_4H_6O_2$	34.4	302	D–O → CH_3=C–=6.78	D–O → CH_3–C–=9.8
Methyl acrylate	$C_4H_6O_2$	34.2 38.2(36.2)[a]	331 244	5.63 $n = 0$	11.30
Ethyl acrylate	$C_5H_8O_2$	41.4 ± 0.6 36.8 ± 0.4	259 298	6.78 $n = 1$	10.40
Propyl acrylate	$C_6H_{10}O_2$	37.9 41.4[a]	298	8.92 $n = 2$	8.8 10.60
Butyl acrylate	$C_7H_{12}O_2$	47.3 ± 0.3 44.8[a]	298	8.92; $n = 3$; $DCH_2 = 4$	10.60
Pentyl acrylate	$C_8H_{14}O_2$	(46.0)[a]	–	8.92; $n = 4$; $DCH_2 = 4.6$	10.60
Isobutyl acrylate	$C_7H_{12}O_2$	43.0	305	8.92; $n = 2$; D*iso*CH_3 = 1.8	10.60
Methyl methacrylate	$C_5H_8O_2$	38.8 ± 0.1	298	5.63; $n = 0$; D*iso*CH_3 = 1.4	11.35
Ethyl methacrylate	$C_6H_{10}O_2$	38.3 (39.3)[a]	300	$n = 1$; 6.78; D*iso*CH_3 = 2.5	10.65
Propyl methacrylate	$C_7H_{12}O_2$	41.6	319	8.92; $n = 2$; D*iso*CH_3 = 2.1	(10.70)
Butyl methacrylate	$C_8H_{14}O_2$	49.0 45.50[a]	350 298	$n = 3$; 8.83; D*iso*CH_3 = 2.1; $DCH_2 = 1.9$	10.60
Pentyl methacrylate	$C_9H_{16}O_2$	47.6	354	8.83; $n = 4$; D*iso*CH_3 = 2.5; $2DCH_2 = 6.0$	10.6

[a]Obtained from the dependence $\Delta_{vap}H^0(T) = f(C_n)$ basing on the additivity principle

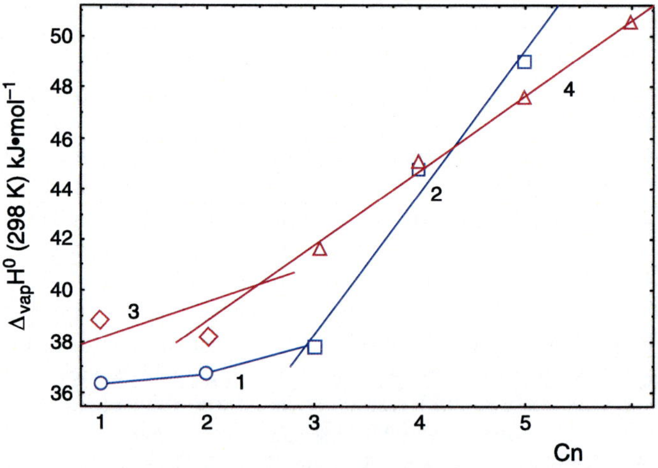

Fig. 6.14 Dependence of vaporization enthalpy of unsaturated esters on the number of carbon atoms in the alkyl chain of alcohol fragment: R-acrylates (1 and 2), R-methacrylates (3 and 4)

stabilization of the specific interactions terminated at the propyl esters. The vaporization enthalpy of methyl methacrylate measured with accuracy at ±0.1 kJ mol^{-1} and the dependence $\Delta_{vap}H^0(T) = f(C_n)$ make it possible to estimate the enthalpy characteristic of the ethyl methacrylate at 40.0 kJ mol^{-1}.

Taking into consideration the temperature dependence of the vaporization enthalpy and the existence of two measurements for the methyl acrylate at 331 and 244 K (Table 6.12), its value at the standard temperature should be estimated at 36.2 kJ mol^{-1}. The character of dependence of the vaporization enthalpy for acrylates and methacrylates on the length of the alkyl chain of the alcohol is irregular due to the experimental errors. For instance, the value of the vaporization enthalpy of the ethyl methacrylate is obviously underestimated [9]. Therefore, using the general character of the dependence $\Delta_{vap}H^0(T) = f(C_n)$, we corrected these values (Table 6.14). The similarity of the molecular structure and of the number of bond vacancies in the saturated and unsaturated esters permit the application of (6.7) and (6.11) for the calculation of the energy of the specific interactions formed by the acid residue D–O → CH_2=CH–C–.

The contribution of the isostructural methyl group to the vaporization enthalpy of the first members of esters in the homologous series of acrylates and methacrylates was estimated from the difference in the vaporization enthalpy of methacrylates and acrylates. In the thermodynamic calculation, the energy of the specific interactions D–O → CH_2=CH– of liquid vinyl acetate formed by the vinyl alcohol group was equaled to the energy of the similar interaction formed by the acetate group (6.78 kJ mol^{-1}). The energy of this specific interaction (Table 6.14) differs by 0.1 kJ mol^{-1} from that obtained for ethyl acetate (9.7 kJ mol^{-1}), and this difference is within the limits of the experimental error.

The energy of the specific interaction D–O → CH_2=CH–C– is estimated from the difference of the vaporization enthalpy and the overall energy of the specific interactions

$$\mathrm{D-O\rightarrow CH_2=CH-C-} = \frac{\Delta_{vap}H^0(T) - 2\mathrm{D-O\rightarrow CH_3-(CH_2)}_n - 4\mathrm{D=O\cdots H-CH_2-D}iso\mathrm{CH_3}}{2}. \tag{6.13}$$

The data presented in Table 6.14 illustrate equal value of the energy of the specific interactions formed by the acid residue of the acrylic and methacrylic acids with ethyl, propyl, and longer alkyls. The obtained energies of the considered specific interactions are lower by 0.40 kJ mol^{-1} compared to the energy of the interactions formed by the acid residue of the propionic acid in whose homologous series, the energies of these interactions are also constant within the accuracy of the measurement of the vaporization enthalpy (Table 6.10). This fact permits taking the energy of the specific interaction D–O → CH_2=CH–C– in the propyl acrylate at 10.60 kJ mol^{-1} and estimating the vaporization enthalpy of the propyl acrylate (Table 6.14). The data compiled in Tables 6.8–6.10 and 6.14 allow a conclusion that the constancy of the energies of the specific interactions of the esters in the homologous series of propionates and butyrates and of unsaturated acrylates and methacrylates is due to the completion of the influence of the reverse dative bond on the terminal methyl groups.

Table 6.15 Energies (kJ mol^{-1}) of specific interactions of liquid unsaturated esters of dicarboxylic acids

Compound	Formula	$\Delta_{vap}H^0(T)$ [2, 9]	T (K)	D–O → CH_3–CH_2–	D–O → CH_2=$(CH)_2$–C–
Diethyl fumarate	$C_8H_{12}O_4$	53.2	341	6.78 $D=O\cdots H-CH_2 = 1.20 \times 4$	5.30 $n = 1$
Diethyl maleate	$C_8H_{12}O_4$	55.2	345	6.78 $D=O\cdots H-CH_2 = 1.20 \times 4$	5.80 $n = 1$

The esters of unsaturated dicarboxylic acids form the same types and number of specific interactions as their saturated analogs. Therefore, the estimation of their energies remains the same, and by analogy to the esters of monocarboxylic unsaturated acids, we assume the energy of the interaction formed by the alkyl chain of the alcohol residue to be equal to the energy of the interaction existing in the symmetric ethers. The validity of this approach was illustrated by examples of esters of saturated monocarboxylic and dicarboxylic acids and unsaturated monocarboxylic acids. In the specificity of the interactions of esters of this series should also appear the influence of the decreasing number of the CH groups in the chain between the carbonyl groups. The key role in this effect belongs to the reverse dative bond depending on the number of carbon atoms in the alkyl chain of the acid residue.

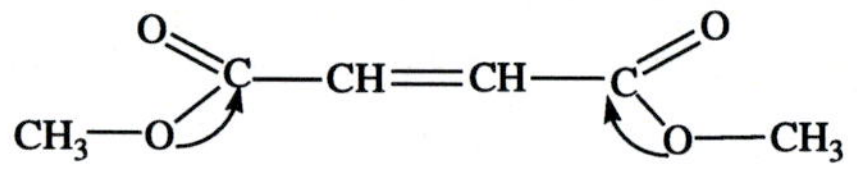

Diethyl fumarate

The calculations were performed without considering the temperature dependence of the energy of the specific interactions formed by the alcohol residue D–O → CH_3–C–, which led to the underestimation of the energy of the interactions D–O → CH_2–$(CH)_n$–C– involving the acid residue (Table 6.15).

Yet the obtained values of the energies of the specific interactions formed by the acid residue show a regular decrease in the stability of these interactions due to the reduced negative charge on the oxygen atom caused by a lesser shift in the electron density from the central carbon atoms of the chain connecting both ester groups to the carbon atoms of the carbonyl groups. The obtained values of the bond parameters illustrate the higher stability of the specific interactions formed by the *cis*-structure of the maleic acid residue compared to those formed by the *trans*-structure of the fumaric acid residue.

6.2.7 Cyclic Esters

The specific features of the interactions of aromatic compounds containing a single benzene ring were revealed in the preceding chapters. Still we return to this problem once more. The benzene molecule has a structure of a flat regular hexagon with a distance of 1.4 Å between the carbon atoms. The real lengths of the ordinary bonds (1.54 Å) and the double bonds (1.34 Å) are essentially different. The charges on the carbon atoms in benzene are different and they are alternating in the ring. Consequently, in the network structure of the liquid and crystalline states of benzene, the molecules are oriented in such a manner that the carbon atom with the reduced charge is located over the carbon atom of another molecule having the enhanced charge. The linking of rings to each other results from the partial shift in the electron density from the essentially unshared $2s^2$-electron pairs of a carbon atom to the hydrogen and carbon atoms of other contacting rings with the reduced negative charge. The energy contribution of these bonds to the enthalpy characteristic of benzene vaporization per each CH group amounts to 5.76 kJ mol^{-1}. The energy of the interaction of low stability H–C → H–C was estimated in the preceding chapters at 0.60 kJ mol^{-1}. The energy of this bond should be accounted for at the replacement of the hydrogen atom of the benzene ring by a methyl or any functional group. Therefore, the energy contribution of methyl group to the vaporization enthalpy can be obtained from the equation

$$\mathrm{D}iso\mathrm{CH_3} = \frac{\Delta_{\mathrm{vap}}H^0(T)iso - \Delta_{\mathrm{vap}}H^0(T)\mathrm{bz} - \mathrm{DC} \rightarrow \mathrm{H}}{2}. \tag{6.14}$$

Taking into account that in the liquid state each molecule interacts with its proper methyl group and analogous group of the molecule of the nearest surrounding, the obtained energy value (4.1 kJ mol^{-1}) corresponds to two such groups.

The calculation of the structure of the monomer methanol, bond and torsion angles, lengths of the intramolecular bonds, the electron density on the atoms, and spectral characteristics has shown [10, 12] that the hydrogen atoms in the methyl group are unequal. It was shown above that the hydrogen atom of the methyl group located in the *trans*-position with respect to the hydrogen of the hydroxy group possesses an enhanced bonding characteristic compared to the other two hydrogen atoms of the same methyl group. The energy of its interaction does not exceed the energy of the interaction formed by the CH group of benzene (0.60 kJ mol^{-1}). The charges on the hydrogen atoms in the methyl group differ insignificantly, and therefore, the energy of the interactions of two other hydrogen atoms of these groups may be taken at 0.60 kJ mol^{-1}. This fact provides a possibility to estimate the contribution to the vaporization enthalpy of toluene from the specific interaction formed by its methyl group by subtracting from the vaporization enthalpy of toluene that of benzene and of the contribution from the substituted hydrogen atom $\Delta_{vap}H^0(T)\mathrm{tol} - (\Delta_{vap}H^0(T) - 0.60)/2$ obtaining 2.35 kJ mol^{-1}. Analogously, the contribution of a methyl group in 1,2-dimethylbenzene is estimated by $\Delta_{vap}H^0(T)\mathrm{dmb} - (\Delta_{vap}H^0(T) - 1.20)/4$ at 2.48 kJ mol^{-1}.

The distribution of the structural and energy parameters of the intermolecular interactions in the liquid aldehydes and esters containing a rigid planar benzene ring

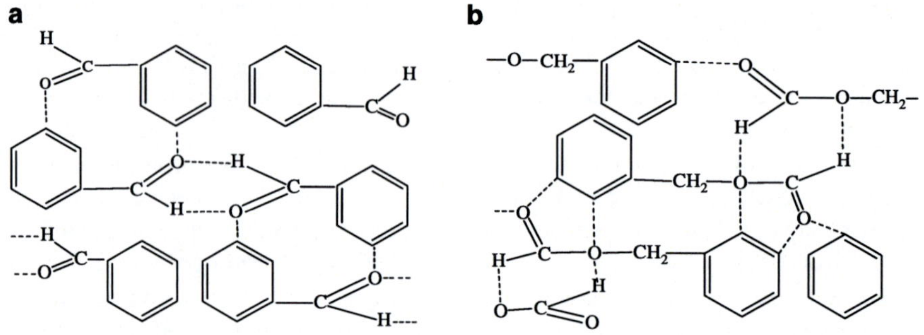

Fig. 6.15 Schematic picture of the fragment of the network liquid structure of benzaldehyde (**a**) and benzyl formiate (**b**)

should be represented by a schematic picture of a network structure where all the bond vacancies belonging to the benzene ring, the oxygen atom, and the hydrogen atom of the aldehyde fragment (Fig. 6.15), to the terminal methyl groups of alkyl chains of the alcohol and acid residues of the esters are occupied. In the structure of the liquid aldehyde with a benzene ring, the following interactions are existing: two hydrogen bonds formed by a hydrogen with the carbonyl oxygen and by the hydrogen of the same aldehyde group $\mathrm{D-O\cdots H-C-}$, and two interactions of low stability formed by the carbonyl oxygen atom with the CH group of the benzene ring D=O → CH. In liquid benzyl formiate and phenyl formiate, two analogous hydrogen bonds, four hydrogen bonds of low stability $\mathrm{D(H-C-)=O\cdots H-C}$ with the hydrogen atoms of the CH groups of the benzene ring, and two specific interactions of the oxygen atom of the alcohol residue with the CH group of the benzene ring D–C–O → CH– are formed. The rigid benzene ring weakly affects the specific interactions; therefore, the energy contribution of the benzene ring minus the substituted hydrogen atoms $\Delta_{\mathrm{vap}}H^0(298)\mathrm{bz} - n{\cdot}\mathrm{DH}$ is constant for all phenyl-containing compounds: benzyl alcohol, phenol, benzoic acid, and the other acids with the benzene ring. Taking the energy of the formed hydrogen bond to be equal to the energy of analogous bond in the liquid benzaldehyde, we may estimate the energy of the interactions between the oxygen atom of the alcohol residue and the CH group of the benzene ring. The energy value would indicate the type of the formed interaction. Therefore, we estimate the hydrogen bond energy in the liquid benzaldehyde and the energy of the specific interactions in the liquid benzyl formiate and phenyl formiate from (6.15) and (6.16), respectively:

$$\mathrm{D-O\cdots H-C-} = \frac{\Delta_{\mathrm{vap}}H^0(298)\mathrm{b.a.} - \left(\Delta_{\mathrm{vap}}H^0(298)\mathrm{bz} - n\cdot\mathrm{DH}\right) - 2\mathrm{D-O\rightarrow CH-}}{2}, \tag{6.15}$$

$$D-O\rightarrow CH- = \frac{\Delta_{vap}H^0(298) - (\Delta_{vap}H^0(298)bz - n\cdot DH) - 2D-O\rightarrow CH - 4D=O\cdots H-CH}{2}. \quad (6.16)$$

The results of the performed calculations of the energy of interactions (Table 6.16) permit a conclusion that their low values of about 0.6 kJ mol^{-1} are due to the formation of the hydrogen bonds of low stability between the oxygen atoms of the alcohol residue with the hydrogen atoms H–C of the benzene rings in formiates and benzaldehyde.

In the calculation of the energy of the specific interactions of benzene derivatives, the substitution differs by 0.1 kJ mol^{-1} from that obtained for ethyl acetate (9.7 kJ mol^{-1}), and this difference is within the limits of the experimental error.

The energy of the specific interaction D–O → CH_2=CH–C– is estimated from the difference of the measurement of the vaporization enthalpy, which makes it unnecessary to search for a more precise estimation of the energy of the interactions of low stability.

In the structure of the liquid phenyl acetate, benzyl acetate (Fig. 6.16), two stable specific interactions DC–O → CH_3–C(–O), two specific interactions with the same oxygen of the alcohol residue and the carbon atom of the benzene ring DC–O → CH<, and four hydrogen bonds of low stability formed by the carbonyl oxygen and the hydrogen of the benzene CH groups are present.

Thus, the energy of the specific interactions formed by the CH group of benzene and the oxygen atom of the alcohol residue is defined by the equation

$$D-O\rightarrow CH = \frac{\Delta_{vap}H^0(298) - \left(\Delta_{vap}H^0(298)b - n\cdot DH\right) - 2D-O\rightarrow CH_3-(CH_2)_n-C - 4D-O\cdots H-C}{2}. \quad (6.17)$$

In the calculations of the energy of the specific interactions in methyl 2-phenylpropionate and ethyl 2-phenylpropionate, the contribution to the vaporization enthalpy of a compound from the benzene ring with one substituted hydrogen was taken into account $\Delta_{vap}H^0(298)bz - n\cdot DH$:

$$D-O\rightarrow CH = \frac{\Delta_{vap}H^0(298) - \left(\Delta_{vap}H^0(298)bz - n\cdot DH\right) - 2D-O\rightarrow CH_3-(CH_2)_n-C - 4D=O\cdots H-}{2}. \quad (6.18)$$

The specific interactions formed by the methyl and ethyl groups in these liquid compounds were also accounted for in the calculations.

Table 6.16 Energies (kJ mol^{-1}) of hydrogen bonds of liquid cyclic hydrocarbons, benzaldehyde, and benzyl formiate (298 K)

Compound	Formula	$\Delta_{vap}H^0(T)$ [2, 9]	$(\Delta_{vap}H^0(T)$–DH$n)$ $D = O \cdots H - CH_2$	D=O → H–C–	$D = C - O \cdots H - C -$
Benzene	C_6H_6	33.8	33.2	–	–
Toluene	C_7H_8	37.9	37.12	–	–
1,2-Dimethy lbenzene	C_8H_{10}	42.4	41.27	–	–
Cyclohexane	C_6H_{12}	33.0 ± 0.1	32.4	–	–
Benzaldehyde	C_7H_6O	49.6	33.2 $D = O \cdots H - CH_2 = 1.2$	–	7.60
Benzyl formiate	$C_8H_6O_2$	51.6 (313 K)	33.2 $D = O \cdots H - CH_2 = 1.2$	0.90	7.60
Phenyl formiate	$C_7H_6O_2$	52.9 ± 0.6	33.2 $D = O \cdots H - CH_2 = 1.2$	1.65	7.60

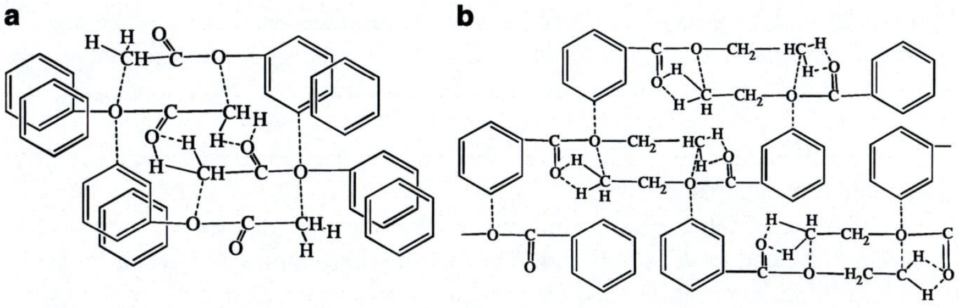

Fig. 6.16 Schematic picture of the network layer liquid structure of phenyl acetate (**a**) and benzyl acetate (**b**)

Table 6.17 Energies (kJ mol^{-1}) of specific interactions of liquid esters with benzene ring at 298 K (4DC–O → HC = 2.4 kJ mol^{-1})

Compound	Formula	$\Delta_{vap}H^0(T)$ [2, 9]	$\Delta_{vap}H^0(T)$ –DH*n*	D–O → CH_3 –$(CH_2)_n$–C–	D–O → CH bz
Phenyl acetate	$C_8H_8O_2$	53.6	33.2	6.78 $n = 0$	2.0
Benzyl acetate	$C_9H_{10}O_2$	55.5		6.78 $n = 0$	3.20
Benzyl propionate	$C_{10}H_{12}O_2$	59.0 (313 K)		8.92 $n = 1$	2.80
Methyl benzoate	$C_8H_8O_2$	55.6 ± 0.1	33.2	5.63	4.40
Ethyl benzoate	$C_9H_{10}O_2$	55.9	33.2	6.78	3.40
Isobutyl benzoate	$C_{11}H_{14}O_2$	58.1	D*iso*CH_3 = 1.8; 33.2	6.78	1.40
Benzyl benzoate	$C_{14}H_{12}O_2$	77.7 (313 K)	33.2 × 2	2.20	2.20
Methyl *o*-methylbenzoate	$C_9H_{10}O_2$	57.3 ± 0.2	37.4; D–O → H–C = 1.2	5.63	3.10
Methyl *m*-methylbenzoate	$C_9H_{10}O_2$	60.3 ± 0.1	37.4	5.63	4.60
Methyl 2-phenylpropionate	$C_{10}H_{12}O_2$	62.8 ± 0.8	37.4	8.92	2.70
Ethyl 2-phenylpropionate	$C_{11}H_{14}O_2$	64.0 ± 0.5	41.8; 41.6	8.92	2.90
Cyclohexyl acetate	$C_8H_{14}O_2$	51.7 ± 0.2	32.4	6.78	2.20
Cyclohexyl butyrate	$C_{10}H_{18}O_2$	60.0 ± 0.8	32.4	8.92 $DCH_2 = 3$	2.20

It follows from the results of calculations of the parameters of the specific interactions presented in Table 6.17 that the energy of the interactions involving the CH group of the benzene ring significantly exceeds the energy of the interaction formed by the hydrogen of this group (0.60 kJ mol^{-1}). The obtained values of the energy do not exceed the overall energy value of the interactions formed by the carbons and hydrogens of these groups of the benzene ring (5.78 kJ mol^{-1}) or the CH_2 groups of the cyclohexane (5.50 kJ mol^{-1}). This permits a conclusion on the validity of the performed calculations of the parameters of the interactions, and these energies should be attributed to the specific interactions formed by the carbon atoms of the benzene ring. In this connection, it is reasonable to carry out their thermodynamic analysis.

In phenyl esters, the smallest energy of the specific interaction D–O → CH illustrates the trend to the stabilization at the growing number of carbon atoms in the acid residue. The stabilization of this interaction occurs also at the replacement of the methyl group in methyl 2-phenylpropionate by the ethyl in going to ethyl 2-phenylpropionate;

The stabilization of the considered type of specific interactions is observed in going from phenyl to benzyl acetate, which should be ascribed to the influence of the methylene group in the benzyl acetate on the shift in the electron density to the benzene ring.

$$\text{Phenyl acetate } (2.00) < \text{Benzyl acetate } (3.20\,\text{kJ mol}^{-1}).$$

At the same time, the decreased stability of the same type of specific interaction in esters with the growing number of the carbon atoms in the acid residue originates from the enhanced difference between the charges of the carbon atom of the methyl group of the propyl chain in the acid residue and the oxygen of the carbonyl group compared with the same group in benzyl acetate.

$$\text{Benzyl acetate } (3.20) < \text{Benzyl propionate } (2.80\,\text{kJ mol}^{-1}).$$

As a result, the propionate fragment acquires more electron density from the oxygen atom of the alcohol residue and forms stronger interaction (8.92 kJ mol^{-1}) than the acetate fragment in the benzyl acetate. The decrease in the negative charge on the oxygen atom negatively affects its donor properties. This is the cause of the increase in the stability of the formed specific interaction with the CH group of benzene.

A similar trend is observed in benzoates with a growing number of carbon atoms in the alcohol residue.

$$\text{Methyl benzoate } (4.40) < \text{Ethyl benzoate } (3.40) < \text{Benzyl benzoate } (2.20) < \text{Isobutyl benzoate } (1.40\,\text{kJ mol}^{-1}).$$

Consequently, the growing number of carbon atoms in the chain of the alcohol residue, first, ensures the stabilization of the formed specific interaction D = C–O → CH_3–$(CH_2)_n$–C(=O) from 5.63 to 8.92 kJ mol^{-1} and decreases the negative charge and the donor properties of the oxygen atom, and second, supplies less electron density to the CH group of the rigid system of the benzene ring, and therefore, the formed specific interaction D=C–O → CHbz has a reduced stability. In this respect, benzyl benzoate is interesting where the rigid benzene rings in the alcohol and acid residues possess reduced acceptor and donor properties and thus form the same type of specific interactions with a lesser stability. Therefore, it is expectable that the most stable specific interaction of this type is formed in methyl benzoate (4.40 kJ mol^{-1}).

The stabilization of this type of specific interaction is essentially affected by the place of the ester group in the benzene ring

$$\text{Methyl}\,m\text{-methylbenzoate}\,(4.60) < \text{Methyl}\,o\text{-methylbenzoate}\,(3.10\,\text{kJ mol}^{-1}).$$

The *meta*-position of the ester group with respect to the methyl group in the ring in the methyl *m*-methylbenzoate stabilizes the interaction by 1.5 kJ mol^{-1} compared to the methyl *o*-methylbenzoate.

The replacement of the ethyl group in the ethyl 2-phenylpropionate by a methyl reflects only the trend in the variation of the stability of the specific interactions due to the insignificant difference in the charges on the terminal carbon atoms of these alkyl groups.

$$\text{Ethyl 2-phenylpropionate}\,(2.90) < \text{Methyl 2-phenylpropionate}\,(2.70\,\text{kJ mol}^{-1}).$$

When the alcohol function is performed by a cyclohexyl group, the stability of the specific interaction is lower than that of the specific interaction formed by the benzene ring in benzyl acetate

$$\text{Cyclohexyl acetate}\,(2.20) < \text{Benzyl acetate}\,(3.20\,\text{kJ mol}^{-1}).$$

The obtained conclusions can be supplemented by the analysis of the specific interactions in liquid methyl esters of benzenecarboxylic acids, whose enthalpy characteristics have been established by calorimetric investigations [8, 9, 11] and using also the reliable value for vaporization enthalpy of methyl benzoate reported by different authors. In the thermodynamic calculations, the twofold and threefold increase in the number of specific interactions in the liquid dimethyl phthalate and trimethyl trimellitate and the number of low stable hydrogen bonds from 4 to 12 with the energy contribution growth from 2.40 to 4.80 and 7.20 kJ mol^{-1}, respectively, was taken into account (Table 6.18).

The data obtained illustrate the regular decrease in the energy of the specific interaction D>C–O → CH< at the replacement of the methyl group by the ethyl group in the acid residue.

$$\text{Dimethyl phthalate}\,(4.55) > \text{Diethyl phthalate}\,(4.30\,\text{kJ mol}^{-1})$$

The essence of this phenomenon was revealed in this paragraph and it consists in the formation of the more stable interaction D–O → CH_3–CH_2– by the ethyl group of the diethyl phthalate compared with that formed by the methyl group of the

Table 6.18 Energies (kJ mol^{-1}) of specific interactions of liquid methyl esters of benzenecarboxylic acids at 298 K

Compound	Formula	$\Delta_{vap}H^0(T)$ [14–16]	$\Delta_{vap}H^0(T)$bz–DHn	D–O → CH_3–	DO → CH–
Methyl benzoate	$C_6H_5COOCH_3$	56.7 ± 0.3	33.2	5.63 × 2	5.40
Dimethyl phthalate	1,2-$C_6H_4(COOCH_3)_2$	78.0 ± 0.7	32.6	5.63 × 4	4.55
Trimethyl 1,2,4-trimellitate	1,2,4-$C_6H_4(COOCH_3)_3$	95.1 ± 0.4	46.1	5.63 × 6	1.25
Diethyl phthalate	$C_6H_4(COOC_2H_5)_2$	81.1 ± 0.8	32.6	6.78 × 4	4.30

dimethyl phthalate D>C–O → CH_3– due to the decrease in the donor properties of the oxygen atom. Therefore, the oxygen atom supplies the reduced electron density to the CH group of benzene.

The next important feature in the regular variation of the stability of the specific interaction is the influence of the number of carboxy groups substituting the hydrogen atoms in the CH groups of benzene.

$$\text{Methyl benzoate } (5.40) > \text{Dimethyl phthalate } (4.55) > \text{Trimethyl } 1,2,4\text{-trimellitate } (1.25\,\text{kJ mol}^{-1}).$$

References

1. Lebedev YuA, Miroschnichenko EA (1981) Thermochemistry of the vaporization of organic compounds. The vaporization and sublimation enthalpy and saturated pressure. Nauka, Moscow
2. Vasiliev IA, Petrov VM (1984) Thermodynamic properties of oxygen organic compounds. Chemistry, Leningrad
3. Nefedov VI, Vovna VI (1989) Electronic structure of organic and elementorganic compounds. Nauka, Moscow
4. Baev AK (1972) Complex formation ability halogenides of second – six groups periodical system. General and applied chemistry, vol 5. Vaisheshika Educating, Minsk, pp 35–51
5. Baev AK (2003) Novel approaches to the structure of functional solvents and elementorganic compounds. In: Abstract international conference on solution chemistry, Debrecen, Hungary, p 79
6. Baev AK (2004) Insolvence of model sp^3-hybridization of carbon atom and chemistry of non aqua solutions. In: IX international conference. The problem of solvation and complex formation in solutions, Plyos, Russia, pp 62–64
7. Baev AK (2004) Lack of co-ordination thermodynamic properties of elementorganic compounds with sp^3-hybridization of carbon atom and its insolence. Russian scientific symposium on thermo chemistry and calorimetry, Nizhny Novgorod, Russia, pp 40–41
8. Baev AK (1969) Phase condition and complex formation ability of halogenide metals. General and applied chemistry, vol 1. Vaisheshika Educating, Minsk, pp 207–218
9. Chickos JS, Acree WE Jr (2002) Enthalpies of vaporization of organic and organometallic compounds, 1880–2002. J Phys Chem Rev Data 31:519–879
10. Ivanov EV, Puchovskie YuP, Abrosimov VK (2004) About formation of weak hydrogen bonds CH···O in liquid methanol abstract. In: IX international conference. The problem of solvation and complex formation in solutions, Plyos, Russia, p 345
11. Chickos JS, Acree WE Jr (2003) Enthalpies of sublimation of organic and organometallic compounds, 1910–2001. J Phys Chem Rev Data 32:537–698
12. Krupin SV, Puhovskoi VA (2004) Investigation of new configuration hydrogen bonding dimer of methanol by method of molecular orbital and functional density. In: IX international conference. The problem of solvation and complex formation in solutions, Plyos, Russia, p 136
13. Maksimuk YuV, Kabo GJ, Kozyro FF et al (1998) Standard enthalpies of formation of some methyl esters of benzene carboxylic acids. J Chem Eng Data 29:293–298
14. Maksimuk YuV, Tarachanov PK (2000) Additivity of standard enthalpies formation of benzene carboxylic acids and their methyl esters at vapor phase. Vesty NAN Belarus Ser Chem Sci 1:48–52

15. Maksimuk YuV (2000) Thermodynamic properties of benzene carboxylic acids, their esters and some alcohols. Theses of PhD, Belarusian State University, Minsk, Belarus
16. Verevkin SP, Kozlova SA, Emelyanenko VN (2006) Vapor pressure and enthalpies of vaporization, and critical parameters of a series of linear aliphatic dimethyl esters of dicarboxylic acids. J Chem Eng Data 51:1898–1905

Chapter 7
Types and Energies of Specific Intermolecular Interactions in Carboxylic Acids

7.1 Specific Interactions in Saturated Monocarboxylic Acids

The quantum-chemical calculations and the analysis of relative intensity in the spectra of ionization with metastable atoms [1, 2] underlie the assumption that the concept of specificity of the σ- and π-orbitals of HCOOH follows from their correlation with the molecular orbitals of H_2CO_3 and H_2O. The orbital $7a(\gamma)(\eta_O{}^i)$ is formally nonbonding; however, in the first photoemission band, a long progression is present of the C=O vibrations with the frequency reduced by 17% [2, 3]. It was, therefore, suggested that the weakening of the C=O bond was caused by the elimination of the ionic component $C^+{=}O^-$. The acid properties of the hydroxy group of the formic acid are due to the low electron density on the oxygen compared to the molecule of methanol $HOCH_3$ originating from the σ- and π-acceptor properties of the C=O group [4]. The change in the vertical $(\eta_O{}^i)$ and (2π)-ionization potentials of the carbonyl group in the acetic acid and methyl formiate $HCOOCH_3$ from 0.6 and 0.5 eV to 0.5 and 1.0 eV, respectively, illustrate the prevailing localization of this molecular orbital [1]. The photoelectronic spectra of the acid dimers showed that the formation of two hydrogen bonds suffers a counter shift in the ionization energy scale [1, 5]. Consequently, the hydrogen bond energy of formic acid is lower than that of acetic acid. Alcohols, ketones, ethers, and carboxylic acids have identical numbers of free bond vacancies forming the hydrogen bonds and the specific intermolecular interactions. However, a distinguishing feature of the carboxylic acids is the presence of the carbonyl oxygen reducing the electron density on the oxygen atom of the hydroxy group and increasing the positive charge on the contiguous carbon atom. This fact ensures the enhanced acceptor properties of the carbon atom of the methyl group of the acid residue. This atom stabilizes the specific interaction that it forms. In keeping with the quantum-chemical calculation, the reduced electron density on the oxygen atom of the hydroxy group results in the lower stability of the hydrogen bonds. Evidently, the carbonyl oxygen forms considerably less strong specific interactions with the hydrogen atom of the CH_2 group or the hydrogen of the methyl group located in the

A.K. Baev, *Specific Intermolecular Interactions of Organic Compounds*,
DOI 10.1007/978-3-642-21622-0_7, © Springer-Verlag Berlin Heidelberg 2012

trans-position with respect to the hydroxy group. The energy of this type of intermolecular interactions was estimated in Chap. 3 at 0.60 and 1.00 kJ mol^{-1} for liquid and solid states, respectively. The most stable interactions in liquid carboxylic acids are the hydrogen bonds formed by the hydroxy groups of two molecules of the nearest surrounding. The interaction of the molecules proceeds by the rule "head to head" and "tail to tail" (Fig. 7.1). As a result, the molecules are linked into chains, on the one hand, by the stable hydrogen bonds, and, on the other hand, by the specific interactions. Each chain contains two rings, one ring is formed by the hydrogen bonds and another one by less strong specific intermolecular interactions H–O → CH_3–$(CH_2)_n$–O. The contacting chains are bound by the same specific intermolecular interactions.

In the liquid and solid formic acid, two hydrogen bonds of high stability arise, and two hydrogen bonds of low stability are formed by the oxygen atom of the carbonyl group and the hydrogen atom. The unshared electron pair of the oxygen atom of the hydroxy group forms a specific interaction with the carbonyl carbon atom H–O → C=O, with the energy of the same value as that of the same type of interactions in the dimethyl ether. The presence in the formic acid of a carboxy group and a hydrogen atom attached to the carbonyl carbon in the fragment

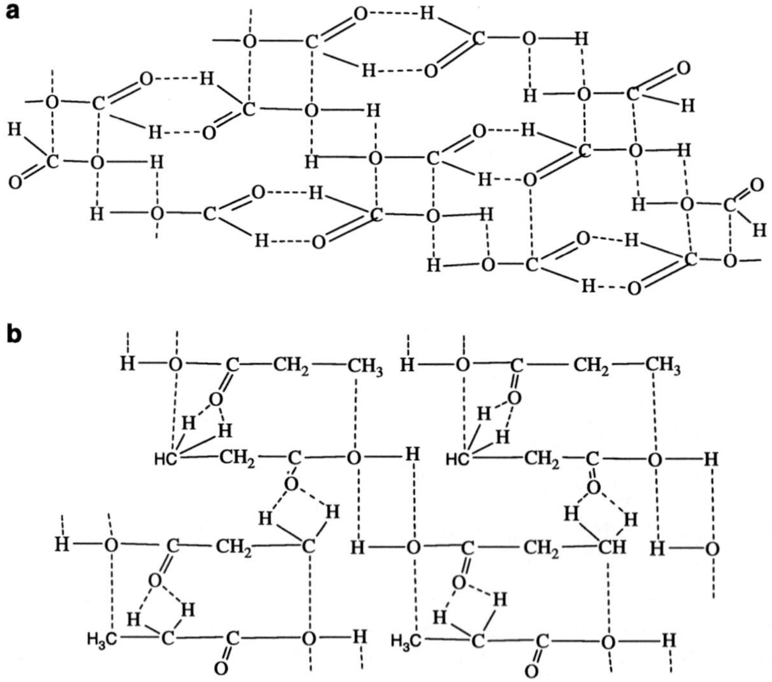

Fig. 7.1 Schematic picture of the liquid structure of carboxylic acids (**a**) formic acid, (**b**) propionic acid

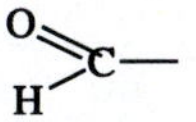 provides a possibility to form a hydrogen bond of low stability with the free bond vacancy of the carbonyl oxygen atom $= O \cdots H - C -$.

The acids possessing the terminal methyl group form less strong H-bonds between the oxygen atom of the carbonyl group and the hydrogen atom of the terminal group $= O \cdots H - CH_2$ situated inside the large cycle of the chain or between the analogous methyl groups and oxygen atoms of the adjacent layers, thus cross-linking them.

The hydroxy groups of two molecules in the nearest surrounding in their turn form two most stable hydrogen bonds $H - O \cdots H - O$ (Fig. 7.1a) that govern the vaporization of dimeric and more complex associates.

Each molecule forms also incomparably weaker specific interactions between the terminal methyl groups of the alkyl chain and the oxygen atoms of the hydroxy groups from the molecules of the nearest surrounding. These chains composed of dimeric forms linked by the stable hydrogen bonds underlie the liquid and solid states of all carboxylic acids. Hence, in the liquid carboxylic acids, a small number of carbon atoms in the alkyl group chains are present consisting of dimeric molecules connected mutually by the specific intermolecular interactions and by less stable H-bonds. Just these bonds of reduced strength underlie the minimal endothermicity at the going into vapor of dimeric molecules. Partial fractions of dimers and monomers in vapor depend on the energies of the hydrogen bonds and the specific intermolecular interactions, and from their relative value,

$$DH - O \cdots H - O > DH - O \rightarrow CH_3(CH_2)_n - C-, \tag{7.1}$$

$$DH - O \cdots H - O \approx DHO \rightarrow CH_3(CH_2)_n - C-, \tag{7.2}$$

$$DH - O \cdots H - O < DH - O \rightarrow CH_3(CH_2)_n - C - . \tag{7.3}$$

However, even the fulfillment of the last condition does not exclude the presence of dimers in vapor, since the second condition of their presence in the vapor depends on the temperature and overall and partial pressure of the monomer forms [6]. These conditions are valid for the association in vapor of all compounds, and they have been considered in detail in Chap. 5.

The definite indication of the association in the vapor of formic and acetic acids is the sharply decreased vaporization enthalpy compared to this characteristic in alcohols, ketones, ethers, and aldehydes (Fig. 7.2), which are volatile and contain in the vapor (save first members of the homologous series of alcohols) only monomeric molecules. Compared to the vaporization enthalpy of alcohols, the enthalpy characteristics of propionic, butyric, and valeric acids are also too low. An important feature of this dependence is the synchronous variation character of the vaporization enthalpy of the carboxylic acid and symmetric ethers that possess the enhanced enthalpy characteristics due to the presence of two alkyl chains in the ether molecules. A significant characteristic of this dependence consists in the sharp change in the effect of the reverse dative bond and in the degree of association in the

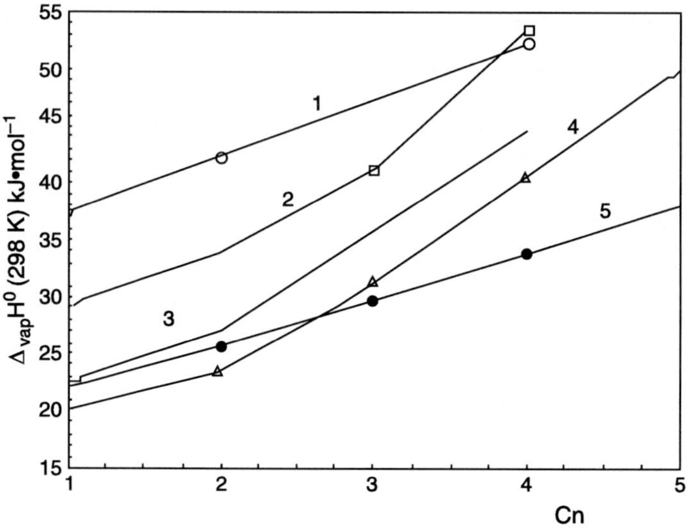

Fig. 7.2 Dependence of vaporization enthalpy [7] on the number of carbon atoms in the alkyl chain (1) saturated alcohols, (2) symmetric ketones, (3) symmetric ethers, (4) saturated monocarboxylic acids, (5) aldehydes

vapor in going from the formic acid to acetic and further to propionic acid. Consequently, the growth of the enthalpy characteristic in the series acetic acid–butyric acid is governed by the hydrogen bond and the specific interaction formed by the alkyl chain analogous to that in the other discussed compound classes.

The dimerization of the carboxylic acids is so common that the thermodynamic characteristics of these compounds reported in handbooks always correspond to dimers. Yet note that the presumed network structure of liquid and crystalline states of the carboxylic acids does not disagree with the existing notions on the network structure of organic compounds with the hydrogen bonds [8–10]. The fundamental difference consists in the fact that the schematic picture presented in Fig. 7.1 contains both all existing specific interactions of the hydrogen bonds and the interactions formed by the pentacoordinate carbon atom. Besides, regarding the polymorphous transitions and melting as chemical processes of dissociation, we should note that the structures of the crystal and the liquid insignificantly differ at melting [11–13]. The raising of the temperature results in the dissociation, with subsequent simplification of the liquid structure and the formation of hexa-, tetra-, and dimeric forms, namely, in the solvent are present in equilibrium several associates in concentrations corresponding to their thermodynamic characteristics. The analysis of associate structures frequently appears in publications without attempts of evaluation of the energy of the hydrogen bond, and the specific interactions involving the carbon atoms of the chains are totally disregarded [14]. Abnormally low vaporization enthalpy of the formic and the acetic acid is due to their vaporization prevailingly as dimers whose stabilization grows with simplification of the structure of the chain and the ring. It also shows that the energies of two types of hydrogen bonds in the formic acid $H-O\cdots H-O$ and $H-O\cdots H-C-$ also are essentially different.

$$DH-O\cdots H-O > DH-O\cdots H-C-.$$

The schematic picture of the liquid structure of the acid (Fig. 7.1) shows the formation in the nearest surrounding of two stable hydrogen bonds, two hydrogen bonds of low stability whose energy was formerly estimated at 0.60 kJ mol^{-1}, and three specific interactions DH–O $\rightarrow$ C=. Hence, we can write (7.4) as

$$DHO\cdots H-O = \frac{\Delta_{vap}H^0(T) - 2DH-O\cdots H-C-3DH-O\rightarrow C=}{2}. \quad (7.4)$$

Taking the energy of the stable hydrogen bond formed by the hydroxy groups of the molecules of the nearest surrounding as being identical in formaldehyde and formic acid and equal to 5.5 kJ mol^{-1}, since this value was established for the formaldehyde, we obtained the energy of the specific interaction in the liquid formic acids DH–O $\rightarrow$ C= equal 2.3 kJ mol^{-1} (Table 7.1). This value is too low compared to the energy of the interaction existing in the liquid dimethyl ether (5.63 kJ mol^{-1}), taking into consideration the obviously underestimated energy of the hydrogen bond.

Thus, this value contradicts the real energies of the hydrogen bonds in esters and alcohols and the relative energies in the discussed types of interactions. Although the thermodynamic calculations performed for the estimation of the energies of the hydrogen bonds are erroneous, they have a deep sense; for here on the definite quantitative data, the already drawn conclusion is substantiated. This makes reasonable the extention of the conclusion to the yet unstudied carboxylic acids with a small number of carbon atoms in the alkyl chain.

The energies of the stable hydrogen bonds of liquid acetic, propionic, and other carboxylic acids should be determined from the difference between the vaporization enthalpy of the acid and the sum of the energies of two existing types of interactions, in particular, of two interactions DH–O $\rightarrow$ CH_3–$(CH_2)_n$ and two H-bonds of low stability (7.5)

$$2DH-O\cdots H-O = \frac{\Delta_{vap}H^0(T) - 2DH-O\rightarrow CH_3(CH_2)_n - 2D=O\cdots H-CH-D(CH_2)_k}{2}. \quad (7.5)$$

In the calculation of the hydrogen bond energies in the liquid butyric acid and the acids with the greater number of carbon atoms, the contribution to the vaporization enthalpy of CH_2-groups is additionally taken into account. In the performed thermodynamic calculations, all the above-discussed approaches are taken into consideration that ensures obtaining valid and self-consistent energies of the hydrogen bond and the specific interactions in different carboxylic acids formed by the pentacoordinate carbon atom (Table 7.1). The most important result of these calculations is that the "hydrogen bond energies" of formic, acetic, and propionic acids possess lower energy than the energy of the specific interactions (Table 7.1). This permits a conclusion on the high stability of the hydrogen bonds and on their vaporization with the rapture of the specific interactions; the content of trimeric forms of molecules in vapor should be extremely low. This permits a conclusion that the obtained "energies of the hydrogen bonds" are not real due to the dimerization of the formic acid in vapor and to high degree of association in vapor of the propionic acid. The "hydrogen bond energy" of the valeric acid is more reasonable, but it is considerably lower than the energy of the hydrogen bond in pentanol. The obtained series of values corresponds to the transition from the complete dimerization in the vapor of the formic acid to a more developed dissociation in vapor of the valeric acid.

The vaporization enthalpy of propionic acid established [15] by the calorimetric method under the standard conditions (31.10 kJ mol^{-1}) is in agreement with the value presented in Table 7.1. Inasmuch as under the conditions of the calorimetric method it is impossible to estimate the partial pressures of the monomer and the dimer, the low value of the vaporization enthalpy suggests that prevailingly dimers go over into the vapor. A number of suggestions on the association of this compound in vapor is present in the literature, but no attempts has been performed to prove it. This problem can be solved by static methods with the measurements at equilibrium in the system, for instance, with the membrane zero-gage which permits the measurement of the molecular mass in vapor, partial pressures of dimeric and monomeric forms, and the thermodynamic characteristics of dimers' dissociation [6]. A more detailed picture of association in the vapor phase of the carboxylic acids was revealed by the thermodynamic studies performed by combining the torsion-effusion and effusion methods on the mass loss at higher temperature cited in [7, 14]. The results obtained substantiate the absence of dimers in the vapor of all carboxylic acids in the series hexanoic acid–octadecanoic acid. The vaporization enthalpies of these compounds are measured with the accuracy up to the second digit after the decimal point. However, the measurements were carried out at a wide temperature range, from 352.36 K for hexanoic acid to 443.13 K for octadecanoic acid (Table 7.2). Since at the increase in the temperature the enthalpy

Table 7.1 Energies (kJ mol^{-1}) of hydrogen bonds in liquid carboxylic acids (298 K)

Acids	Structure	$\Delta_{vap}H^0(T)$ [7]	DH–O → $CH_3(CH_2)_n$	D = O··· H – C	DH–O → CH_2	DH – O··· H – O
Formic	HCOOH	20.10	DH–O → C=O = 5.63	0.60 × 2	–	3.8
Acetic	CH_3COOH	23.36	6.78	0.60 × 4	–	4.9
Propionic	CH_3CH_2COOH	31.14	$n = 2$, 8.92	0.60 × 4	–	5.5
Butyric	$CH_3(CH_2)_2COOH$	40.45	$n = 3$, 8.92	0.60 × 4	–	10.1
Valeric	$CH_3(CH_2)_3COOH$	49.74	$n = 4$, 8.92	0.60 × 4	5.0	13.5

Table 7.2 Vaporization enthalpy (kJ mol^{-1}) of carboxylic acids [7]

Acids	Formula	T (K)	$\Delta(\Delta_{vap}H^0(T))$	$\Delta_{vap}H^0(T)$ [7]
Hexanoic	$C_6H_{12}O_2$	352.36	6.89	64.89 ± 0.02
Heptanoic	$C_7H_{14}O_2$	349.61	4.5	69.39 ± 0.04
Octanoic	$C_8H_{16}O_2$	370.18	2.95	72.30 ± 0.06
Nonanoic	$C_9H_{18}O_2$	374.26	3.96	76.26 ± 0.04
Decanoic	$C_{10}H_{20}O_2$	389.39	2.66	78.92 ± 0.04
Undecanoic	$C_{11}H_{22}O_2$	395.80	3.36	82.18 ± 0.03
Dodecanoic	$C_{12}H_{24}O_2$	406.24	2.77	84.95 ± 0.04
Tridecanoic	$C_{13}H_{26}O_2$	416.13	2.79	87.74 ± 0.06
Tetradecanoic	$C_{14}H_{28}O_2$	417.43	3.62	91.36 ± 0.09
Pentadecanoic	$C_{15}H_{30}O_2$	426.59	3.21	94.57 ± 0.07
Hexadecanoic	$C_{16}H_{32}O_2$	426.59	3.14	97.71 ± 0.09
Heptadecanoic	$C_{17}H_{34}O_2$	436.04	3.01	100.72 ± 0.06
Octadecanoic	$C_{18}H_{36}O_2$	443.15	2.09	102.81 ± 0.13

characteristic of vaporization decreases, the vaporization enthalpies presented in Table 7.2 tend to increase the deviation from the standard vaporization enthalpy. Although the accuracy of the experiment is virtually the same for all compounds, in the absolute value, they may be different by several kJ mol^{-1} at the boarders of the temperature dependence of the $\Delta_{vap}H^0(T)$ (352.36–445.14 K). Hence we can state that the error introduced into the hydrogen bond energy calculated with the use of the thermodynamic analysis for the total series of compounds under consideration should be maximum for butyric and valeric acids. Consequently, the deviation of the experimental temperature from the standard conditions necessarily shall result in a larger error in the thermodynamic calculation of the hydrogen bond energy. Yet within the limits of these errors, the hydrogen bond energies of the carboxylic acids would be in agreement with each other. The published value of vaporization enthalpy of butyric acid for vaporization of its monomer forms amounts to 58 ± 4 kJ mol^{-1} [14]. The thermodynamic calculation with the use of this value resulted in the hydrogen bond energy of $D\mathrm{H-O\cdots H-O}$ in butyric acid equal to 18.60 ± 4 kJ mol^{-1}, which is valid and corresponds to the real energy in case of vaporization of the monomer molecules.

An important feature of the carboxylic acids corresponding to the shift in the electron density in their alkyl chains is the different change in the vaporization enthalpy in going from the compound with odd number of carbon atoms to the compound with even number than in going from the compound with even number of carbon atoms to the compound with odd number of carbons (cf. Chap. 2).

Note that the reverse dative bond between the oxygen atom and the contacting carbon atom loses its influence at second–third carbon atoms of the methylene groups (Fig. 7.2) and consequently, it does not affect all successive groups. Therefore, in the thermodynamic calculations of the hydrogen bond energies, it is necessary to take into account the growing contribution to the enthalpy of vaporization and sublimation of the increasing number of the methylene groups in the alkyl chain. Yet before performing the thermodynamic calculations, it is necessary to establish the enthalpy characteristics of the vaporization when monomer molecules go over into vapor.

7.2 Energies of Specific Interactions in Monocarboxylic Acids

Although the interest in the hydrogen bonds of the carboxylic acid is considerable, yet their energies are not established. As shown by the preceding analysis, the high association on the first members of this homologous series hampered the solution of this problem. The polyfunctional character of the carboxylic acids additionally complicates this problem. Probably just these reasons cause the lack in the literature of attempts to find the way to the establishment of the energy of the hydrogen bonds in compounds in question. However, the real cause of the absence of such attempts in the current scientific literature lies in the adherence to the sp^3-hybridization model attractive, since it is simple and easy for understanding. We believe that the thermodynamic characteristics of vaporization and sublimation play the key role in the evaluation of the hydrogen bond energies in the solid and liquid carboxylic acids considering the new approaches formulated in Chap. 2 for the use of the enthalpy characteristics of vaporization.

The dependence of the vaporization enthalpy of 13 carboxylic acids: hexanoic–octadecanoic and butyric, propionic, and valeric [7] (Table 7.3), on the number of carbon atoms in their alkyl chains (Fig. 7.3) fits to three straight lines: propionic to decanoic, decanoic to haxadecanoic, and hexadecanoic to octadecanoic. All along, the vaporization enthalpy regularly grows. The experimentally measured melting enthalpies [7] and vaporization enthalpies made it possible to estimate the sublimation enthalpy in good agreement with that reported in [16].

An important feature of the dependences presented in Fig. 7.3 is the crossing of the plots of sublimation and vaporization at the point of zero carbon atoms indicating the similarity of the vapor composition in the sublimation and vaporization. Taking into account the general rule of decreasing the association degree in the vapor with the growing length of the alkyl chain [6], we should state that the dependence $\Delta_{vap}H^0(T) = f(C_n)$ corresponds to the process of going into the vapor of only monomer molecules of the carboxylic acids. Therefore, the important help in the interpretation of the thermodynamic properties of the carboxylic acids can be obtained from the relationships describing the variation of enthalpy of vaporization and sublimation of ethers and ketones that do not form dimers in vapor. At identical regular variation of the thermodynamic properties of the carboxylic acids and ethers and ketones depending on the number of carbon atoms in the alkyl chain, the thermodynamic analysis should account for the already established role of the reverse dative bond in the change in the thermodynamic characteristics of lower acids, from formic to acetic and propionic.

In this connection, the vaporization enthalpy of butyric acid [14] should be mentioned for it fits excellently to the plot $\Delta_{vap}H^0(T) = f(C_n)$. The coinciding of

Table 7.3 Energies (kJ mol^{-1}) of hydrogen bonds and specific interactions in liquid and solid carboxylic acids (298 K)

Acids	$\Delta_{m.p.}H^0(T)$ [7]	$\Delta_{avap}H^0(T)$ [7]		Solid $\sum D = O\cdots H-CH_2 = 1.00\times 4$			Liquid $\sum D = O\cdots H-CH_2 = 0.60\times 4$		
		Sub	Vap	$D\text{–}O \rightarrow CH_3(CH_2)_n$	DCH_2	$DH-O\cdots HO$	$DO \rightarrow CH_3(CH_2)_n$	DCH_2	$DH-O\cdots HO$
Formic	3.4	49.7	46.3	$DH-O\cdots HC$ 7.25	–	16.60	$DHO\cdots HC$ 5.62	–	16.90
Acetic	6.4	56.1	50.0	$n = 0$; 8.83	–	17.30	6.78	–	17.00
Propionic	8.4	62.9	54.9	$n = 0$; 11.60	–	17.35	8.92	–	17.35
Butyric	11.0	69.0	58.2 ± 0.2	$n = 1$; 11.60	4	18.90	$n = 1$; 8.92	3.0	17.50
Valeric	14.0	75.0	63.0 ± 0.5	$n = 2$; 11.60	10	18.75	$n = 2$; 8.92	7.0	17.30
Hexanoic	16.8	81.7	64.82	$n = 3$; 11.60	16	18.60	$n = 3$; 8.92	10.0	16.50
Heptanoic	2.04[a], 15.44	86.90	69.39	$n = 4$; 11.60	22.1	18.90	$n = 4$; 8.92	13.1	17.20
Octanoic	21.35	93.65	72.30	$n = 5$; 11.60	28.1	18.25	$n = 5$; 8.92	17	17.35
Nonanoic	5.385[a], 19.82	100.8	76.26	$n = 6$; 11.96	34.3	19.60	$n = 6$; 8.92	20.3	16.85
Decanoic	27.80	106.72	78.92	$n = 7$; 11.60	40.3	19.45	$n = 7$; 8.92	24.9	17.60
Undecanoic	8.15[a], 25.98	115.87	82.16	$n = 8$; 11.60	46.4	19.10	$n = 8$; 8.92	28.	16.95
Dodecanoic	36.3	121.25	84.95	$n = 9$; 11.96	53.8	20.10	$n = 9$; 8.92	31.0	16.85
Tridecanoic	8.73[a], 33.79	130.20	87.74	$n = 10$; 11.60	61.2	20.90	$n = 10$; 8.92	33.7	16.95
Tetradecanoic	45.10	136.46	91.36	$n = 11$; 11.60	68.5	20.35	$n = 11$; 8.92	37.4	16.85
Pentadecanoic	49.8	144.3	94.51	$n = 1,2$; 11.60	75.8	20.60	$n = 12$; 8.92	40.4	16.90
Hexadecanoic	53.7	151.4	97.71	$n = 13$; 11.60	83.1	20.55	$n = 13$; 8.92	43.7	16.85
Heptadecanoic	7.44[a], 51.34	159.6	100.72	$n = 14$; 11.60	90.4	21.00	$n = 14$; 8.92	47.0	16.75
Octadecanoic	61.8	164.62	102.81	$n = 15$; 11.60	98.2	19.60	$n = 15$; 8.92	49.1	16.75
Nonadecanoic	9.16 57.8	172.9	105.0	$n = 16$; 11.60	105.5	20.10	$n = 16$; 8.92	51.4	16.70
Eicosanoic	69.2	179.2	109.9	$n = 17$; 11.60	112.8	19.60	$n = 17$; 8.92	54.7	16.70

[a]Polymorphous transformation

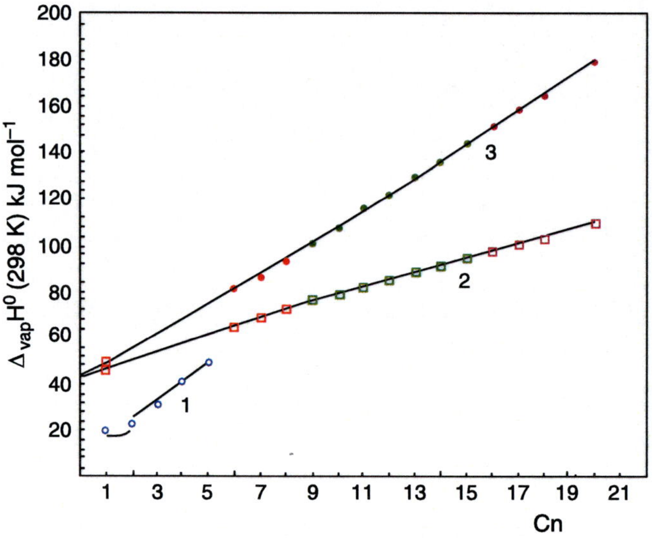

Fig. 7.3 Dependence of vaporization enthalpy [7, 14] of saturated monocarboxylic acids on the number of carbon atoms in the alkyl chain (1) vaporization with association in the vapor phase, (2) vaporization, and (3) sublimation of monomer molecules [7, 14]

the experimentally measured vaporization enthalpy with this dependence is caused by the presence in the vapor of only monomer molecules of the butyric acid in conformity to the above mentioned rule. On the same straight line are the points corresponding to the vaporization enthalpy of propionic and valeric acids experimentally measured at the temperature where the dimers are totally dissociated in vapor and recalculated to the standard temperature. The position of the vaporization enthalpy of the propionic acid measured at low temperature and recalculated to the standard conditions is by 23 kJ mol^{-1} lower than its position on the dependence $\Delta_{vap}H^0(T) = f(C_n)$ corresponding to the monomeric state of vapor. It shows the high degree of propionic acid association in the vapor state. The vaporization enthalpies of the formic and acetic acids are still less for they are totally dimerized in the vapor, and for the formic acid, it is possible to presume a presence in the vapor of higher associates. The low values of the vaporization enthalpy of dimeric molecules of the formic and acetic acids result from the high energy of the hydrogen bonds between the monomer molecules of the nearest surrounding (Fig. 7.1). These bonds form strong dimers in the crystals and consequently the dimers of these acids in vapor. Less stable specific interactions existing between the dimers in the liquid and crystalline states of the acids are easily broken in the course of melting (partially) and at vaporization (completely) [11–13]. The investigation of the association process and of the dissociation energies of the dimers compared with the energy of their going into vapor illustrates the erroneous notion that the bulky molecules require higher energy expenditure in the vaporization.

As was shown in Chap. 4, the localization of the enthalpies of vaporization and sublimation of 2-butanone and of vaporization of 2-propanone on the plots crossing with those corresponding to the vaporization of ketones with the larger amount of methylene groups was due to the effect of the reverse dative bond. The weakening of its effect on the energies of the hydrogen bonds and the specific intermolecular interactions achieves practically the maximum value in compounds with a chain of three carbon atoms. However, in the carboxylic acids, this influence of the reverse dative bond on the dependence of the vaporization enthalpy on the number of the carbon atoms in the chain is not observed because of the high and total dimerization in the vapor of the propionic and acetic acids, respectively.

Another feature of the carboxylic acids common to all saturated organic compounds is the growing influence of the methylene groups on the thermodynamic properties commencing with the butyl chain. Consequently, the answer to the question of difference in the hydrogen bond energies in the liquid and solid carboxylic acids, of the effect of the reverse dative bond and its regular stabilization in the series formic acid–butyric acids, and of the contribution of the CH_2-groups to the thermodynamic properties of vaporization and sublimation can be obtained from the results of the thermodynamic calculations.

The thermodynamic calculations of the energies of the hydrogen bonds and the specific intermolecular interactions formed by the terminal methyl groups with the participation of the essentially unshared $2s^2$(c)-electron pair of the carbon atom at the length of the alkyl chain exceeding three carbon atoms took into account the contribution of the CH_2-groups to the enthalpy of vaporization and sublimation. The decreasing contribution of the methylene groups at their growing number to the vaporization enthalpy of the carboxylic acids also should be taken into consideration (Fig. 7.3) as it is observed in the dependence of the vaporization enthalpy on the number of carbon atoms in the alkyl chain. This is the result of the decreased energy contribution to the vaporization enthalpy at the large number of CH_2-groups in the saturated monocarboxylic acids. In contrast, the sublimation enthalpies of the carboxylic acids are described by a more complex dependence on the number of carbon atoms in the alkyl chain that should be ascribed to the various stabilizing effects of the crystal field the most pronounced in the series nonanoic–hexadecanoic acids and further diminishing in the heptadecanoic–octadecanoic acids. By using the dependences presented in Fig. 7.3, the enthalpy characteristics of the formic and acetic acids at complete dissociation in the vapor were obtained by extrapolation, and also those of the unstudied nonadecanoic and eicosanoic acids (Table 7.3).

The energy contributions from various types of specific interactions in liquid and solid carboxylic acids are taken as equal to those into the enthalpy of sublimation and vaporization of ethers (Table 7.3). The established energies of the hydrogen bonds in formic, acetic, and propionic acids (Table 7.3) illustrate the regular increase in their stabilization with growing number of carbon atoms in the chain and weakening of the influence of the reverse dative bond in the same series. The attention is attracted by the enhanced hydrogen bond energies in the series propionic–octanoic acids, and the decrease in the energy to 16.90 kJ mol^{-1} further occurs.

Consequently, the influence of the specific interactions formed by the growing number of CH_2 groups somewhat decreases and attains certain constant value due to the small or practical absence of the difference in the structure of the liquid state, including associated molecules of various complexity. At the same time, the carboxylic acids having a definite crystal structure illustrate the stabilization of the hydrogen bonds with the growing number of methylene groups in the molecules of butyric–octanoic, nonanoic–dodecanoic, and tridecanoic–heptadecanoic acids. The subsequent decrease in the stability of the hydrogen bond in the octadecanoic–eicosanoic acids indicates that the maximum possible stabilization of the hydrogen bond occurs at 15 CH_2 groups in the molecule, and further begins the destabilizing effect. The energy of the hydrogen bonds in solid carboxylic acids is higher than that in solid alcohols and approaches 21 kJ mol^{-1} in the heptadecanoic acid. The regular series of the hydrogen bond stabilization in solid and liquid acids is well-consistent with the quantum-chemical calculations carried out in [1] and discussed above.

$$\text{Aldehydes} < H_2O < \text{Alcohols} < \text{Carboxylic acids.}$$

Therefore, it can be stated that the performed thermodynamic calculations provided reliable values of the hydrogen bond energies and substantiated the insignificant errors and consequently the validity of their values for the total series of saturated monocarboxylic acids. The energies of the hydrogen bonds in the series formic–butyric acids were established for the vaporization of monomer molecules. Consistent energies were obtained for the total series of saturated monocarboxylic acids. The established stabilization of the hydrogen bond in the series formic acid–butyric acid is analogous to the regular stabilization of the hydrogen bond in the series methanol–butanol.

7.3 Saturated Monocarboxylic Acids with Isostructural Methyl Group

Many properties of organic compounds with side alkyl groups which cannot be understood using the attractive model of the sp^3-hybridization are ascribed to the steric effect. However, the well-grounded rejection of this model and the developed conception of the new structural and thermodynamic description of the observed properties of the branched compounds led to the constructive conclusions. Thermodynamic analysis of the carboxylic acids makes it possible to reveal the main regularities of the effects of isostructural methyl and other alkyl groups on the thermodynamic properties of compounds and on the energies of the hydrogen bonds and the specific intermolecular interactions formed by these groups.

The monocarboxylic acids such as esters, alcohols, ketones, and ethers are characterized by the change in the enthalpy characteristics at the replacement of hydrogen in a CH_2 group of the alkyl chain by an isostructural methyl group, and in contrast to alcohols, this replacement causes an increase in the enthalpy of vaporization and sublimation (Table 7.4). This fact suggests that the isostructural methyl group takes lesser part in the distribution of the electron density than the CH_2 group of the chain. On the contrary, the replacement of a hydrogen in a CH_2 group by the isostructural methyl group is accompanied by a diminished contribution to the enthalpy characteristic, thus demonstrating its influence on the redistribution of the electron density in the molecule of the carboxylic acids and on the charges on the oxygen and on the carbon atom of the terminal methyl group of the alkyl chain.

The carbonyl oxygen atom with its characteristic negative charge interacts with the hydrogen atoms of the terminal methyl groups of the alkyl chain. The isostructural methyl groups transfer a part of the electron density from the essentially unshared $2s^2$-electron pair of the carbon atom to the analogous group of the molecule of the nearest surrounding. As a result, a specific interaction $H_3C \rightarrow H\text{–}CH_2$ of low stability formed involving the pentacoordinate carbon atom (Fig. 7.4). This interaction provides the linking of chains into a network and makes an additional contribution to the vaporization enthalpy of the carboxylic acids. Increasing the number of isostructural methyl groups to two in the molecule of trimethylacetic acid (Fig. 7.4) is accompanied with an inadequately large contribution to the vaporization enthalpy (Table 7.4). The increased vaporization enthalpies of the compounds with an isostructural methyl group considering the enthalpy characteristic of the corresponding carboxylic acids lacking the side methyl group provides a possibility to estimate the energy contribution of the isostructural methyl or ethyl groups. For instance, the difference between the vaporization enthalpies of 2-methylpropionic acid (55.5 kJ mol^{-1}) and propionic acid (54.9 kJ mol^{-1}), trimethylacetic acid (57.6 kJ mol^{-1}), and propionic acid (54.9 kJ mol^{-1}) gives the contribution of one and two such groups to the vaporization enthalpy, respectively. The structure of liquid 2-methylpropionic acid consists of a chain composed of rings bound by two hydrogen bonds $H-O\cdots H-O$ and two stable specific interactions $H\text{–}O \rightarrow CH_3\text{–}CH(CH_3)\text{–}C$ whose destruction during melting or in sublimation and vaporization governs the going over to vapor of dimers or monomers. The two other types of interactions, the H-bonds of low stability $=O\cdots H-CH_2$ formed by the carbonyl oxygen and the terminal methyl groups of the alkyl chains, and the interactions of the isostructural methyl groups provide the cross-linking of the chains (Fig. 7.4). Note that the obtained values of the energy contribution of the isostructural methyl groups correspond to two isostructural groups of the interacting molecules in the nearest surrounding in the structure of the liquid state. The energy contributions of isostructural methyl groups presented in Table 7.4 show the increase in this value at the position of two side groups in the molecules of trimethylacetic acid and 3,3-dimethylbutanoic acid at the second and third carbon atoms of the chain. Nonequality of the contribution from the first and the second methyl group amounting to 0.60 and 1.35 kJ mol^{-1} in the trimethylacetic acid, and 2.40 and 2.70 kJ mol^{-1} in the 3,3-dimethylbutanoic acid illustrates the role of the second isostructural methyl group in the redistribution of the electron density in the molecule. This is revealed in the stabilization of the specific interactions formed by the isostructural CH_3 groups due to the increased negative charge on the carbon

Table 7.4 Energies (kJ mol^{-1}) of hydrogen bonds and specific interactions of branched monocarboxylic acid

Acids	Formula	Structure	$\Delta_{vap}H^0(T)$ [7]	T (K)	Bond energies		
					DH – O··· H – O	D–O → $CH_3(CH_2)_n$–C	DH_3C → H–CH_2iso
$\sum D = O \cdots H - CH_2 = 0.60 \times 4\,kJ\,mol^{-1}$							
Propionic	$C_3H_6O_2$	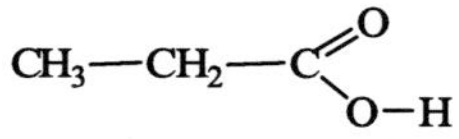	54.9	298	17.35	8.92	–
2-Methyl-propionic	$C_4H_8O_2$	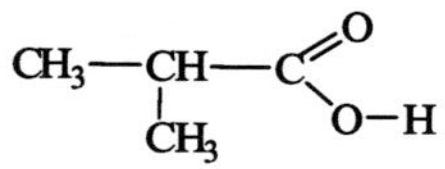	55.5 ± 0.3	298	17.35	8.92	0.60
Trimethylacetic	$C_5H_{10}O_2$	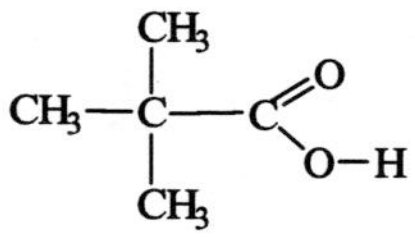	57.6 ± 0.2	320	17.35	8.92	1.35 × 2
Butyric	$C_4H_8O_2$	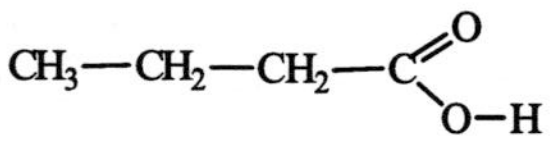	58.8 ± 0.7	298	17.35	8.92	DCH_2 = 3.90
3-Methylbutanoic	$C_5H_{10}O_2$	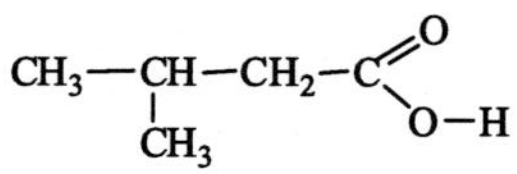	61.2 ± 0.7	298	17.35	8.92	2.40 DCH_2 = 3.90

3, 3-Dimethylbutanoic	$C_6H_{12}O_2$	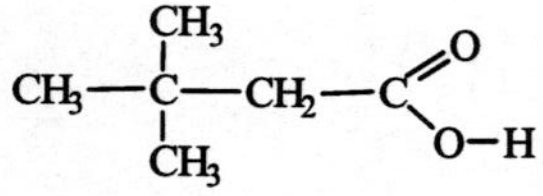	64.0 ± 0.4	304	17.35	8.92	2.70 × 2 $DCH_2 = 3.90$
Hexanoic	$C_6H_{12}O_2$	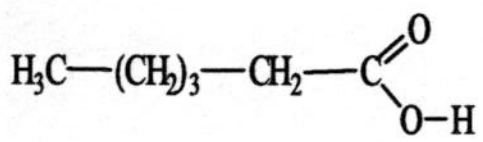	69.1	298	17.35	8.92	$DCH_2 = 16.40$
2-Ethylhexanoic	$C_8H_{12}O_2$	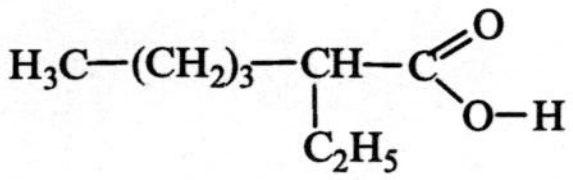	75.6 ± 0.8	298	–	8.92	$DC_2H_5iso = 6.5$ $DCH_2 = 3.90$

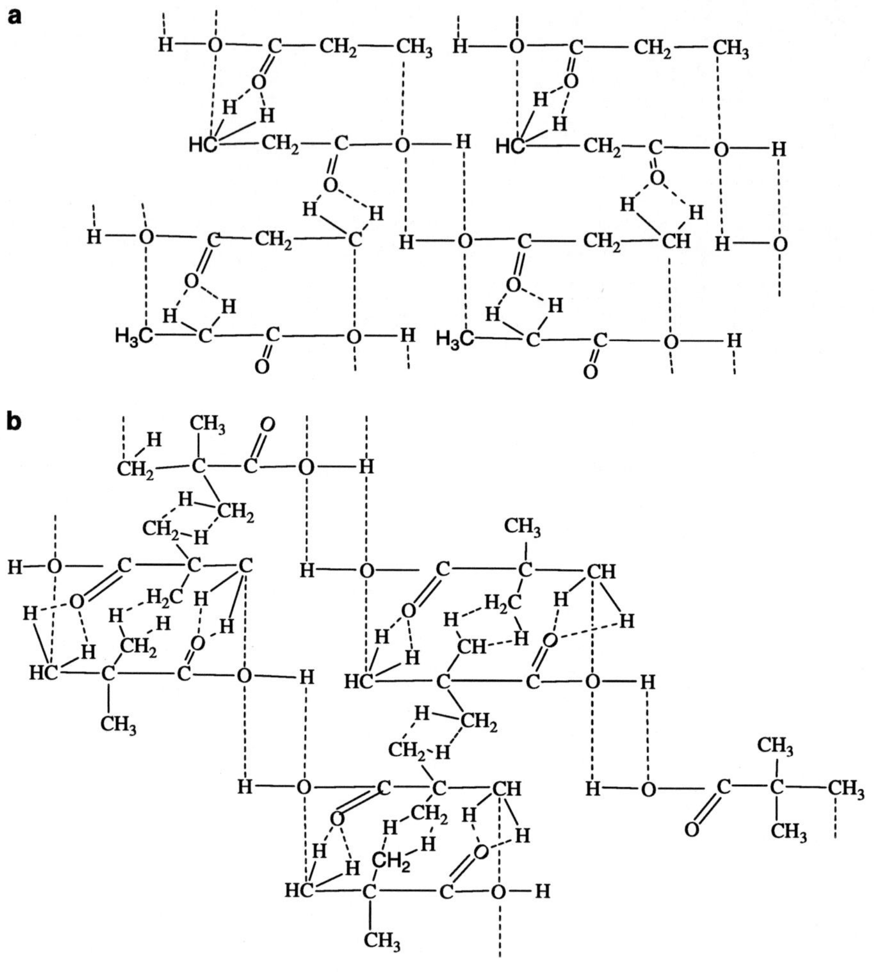

Fig. 7.4 Schematic picture of the liquid structure of 2-methylpropionic acid (**a**) and trimethylacetic acid (**b**)

atoms of these groups. The remoteness from the carboxy group at the location of the isostructural methyl groups in the molecules of 3-methylbutanoic and 3,3-dimethylbutanoic acids results in the stabilization of the specific interaction from 0.60 to 2.40 kJ mol^{-1} and from 1.35 to 2.70 kJ mol^{-1}, respectively. The side ethyl group in 2-ethylhexanoic acid forms a specific interaction with an energy 6.50 kJ mol^{-1}, close to the energy of the specific interaction in the liquid diethyl ether (6.78 kJ mol^{-1}).

Thus, the specific interactions of the isostructural groups are stabilized in going from the methyl to the ethyl group, and this stabilization should finish at the propyl group.

7.4 Energies of Hydrogen Bonds in Hydroxycarboxylic Acids

The presence in the hydroxyacetic acid of two hydroxy groups with very different properties suggests that the hydrogen bonds in the liquid state form by the interactions between similar acid and alcohol groups, respectively, and these hydrogen bonds make different energy contribution to the vaporization enthalpy. However, in this case, one bond vacancy of the oxygen atoms remains free and capable of forming specific interactions. Therefore, the structure where the oxygen atoms interact with the carbon atoms and the carbonyl oxygen atom interacts with the hydrogen atoms of the CH_2 group should acquire the enhanced energy stability. It follows from the schematic structure shown in Fig. 7.5 that the molecule of the hydroxyacetic acid forms 12 specific interactions, among them two are formed by the alcohol hydroxy group and two by the acid group. Two remaining vacancies of the oxygen atoms of the alcohol hydroxy group of the interacting molecules in the nearest surrounding form two specific interactions H–O → CH_2 with the carbon atoms of the methylene groups, and two oxygen atoms of the acid groups form, respectively, two specific interactions H–O → C=O with the carbon atoms of the carboxy groups. The carbonyl oxygen atoms of the interacting molecules in their turn form four H-bonds of low stability with the hydrogen atoms of CH_2 groups =O → H–CH with an energy of 0.60 kJ mol^{-1}.

The energy of the specific interaction formed by each carbon atom should be taken as equal to the energy of the interactions in the liquid dimethyl ether. However, for the two different hydrogen bonds, it is possible to obtain only the mean energy from (7.6).

$$\mathrm{H-O\cdots H-O} = \frac{\Delta_{\mathrm{vap}}H^0(T) - 2\mathrm{DH-O \rightarrow CH_2} - 2\mathrm{DH-O \rightarrow C{=}O} - 4\mathrm{D{=}O\cdots H-CH}}{4}. \quad (7.6)$$

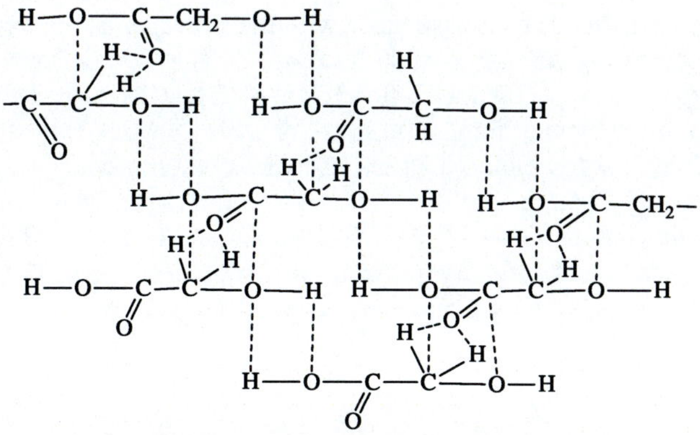

Fig. 7.5 Schematic picture of the liquid structure of hydroxyacetic acid

The obtained hydrogen bond energy corresponds to 363 K and is underestimated due to disregarding the temperature dependence of the energy of the specific interactions and the H-bonds of low stability (Table 7.5).

The molecules of 3-hydroxypropionic acid and 2-hydroxybutyric acid have a hydroxy group in the middle of the alkyl chain, and it essentially affects the shift in the electron density in the chain and the distribution of the charges on the atoms bearing the bond vacancies and forming specific interactions of various energies. These hydroxy groups are like analogous groups in the dihydric alcohols where the second hydroxy group is attached to the second or third carbon atom of the alkyl chain. It was shown before that in alcohols with analogous location of the hydroxy groups, for instance, in the 1,3-butanediol molecule, one terminal methyl group acts like an isostructural methyl group with the corresponding energy contribution to the enthalpy characteristic. The location of the second hydroxy group in the 1,3-pentanediol molecule at the third carbon atom provides the terminal ethyl fragment with the functions of an isostructural ethyl group which contributes to the vaporization enthalpy the energy value essentially reduced compared to the ethyl group of the diethyl ether and equal to the energy value of the analogous isostructural group. In case of the propyl terminal group, its contribution to the enthalpy characteristic increases to the energy value of the specific interaction in the liquid dipropyl ether. It was shown in Sect. 7.3 that the isostructural ethyl group in monocarboxylic acids contributes to the enthalpy characteristic by 6.5 kJ mol^{-1}, by 0.28 kJ mol^{-1} less than the energy of the specific interaction of the liquid diethyl ether. This fact is a direct confirmation of the effect of the reverse dative bond on the terminal methyl group of the propyl fragment partially operating like an isostructural group.

The hydroxy groups in the molecules of 3-hydroxypropionic acid and 2-hydroxybutyric acid complicate the structure of their liquid state that comprises a branched network (Fig. 7.6). The alcohol hydroxy group and the carboxy fragment form each two hydrogen bonds with different energies and of high stability, especially of the H-bond formed by the acid hydroxy group.

Similar to the already discussed carboxylic acids, the carbonyl oxygen atoms of the 3-hydroxypropionic acid and 2-hydroxybutyric acid form in the nearest surrounding four H-bonds of low stability with the hydrogen atoms of the methylene groups possessing an energy of 0.60 kJ mol^{-1}. The oxygen atom of the acid hydroxy group forms a specific interaction H–O → CH_3–$(CH_2)_n$– with the terminal methyl group of the alcohol fragment. Since its chain contains two carbon atoms, both in the 3-hydroxypropionic acid and 2-hydroxybutyric acid, we are able to estimate its energy at 6.5 kJ mol^{-1}. The oxygen atom of the acid hydroxy group forms the specific interactions H–O → CH–CH_2–C– and H–O → CH–C– in liquid 3-hydroxypropionic acid and 2-hydroxybutyric acid, respectively. Thus, the mean energy of the hydrogen bonds should be calculated from (7.7)

$$\mathrm{DH-O\cdots H-O} = \frac{\Delta_{\mathrm{vap}}H^0(T)\mathrm{hc.a.} - 2\mathrm{DH-O} \rightarrow \mathrm{CH_3-(CH_2)}_n - 2\mathrm{DH-O} \rightarrow \mathrm{CH-(CH_2)}_n\mathrm{-C} - 4\mathrm{D{=}O\cdots H-CH}}{4}. \tag{7.7}$$

Table 7.5 Energies (kJ mol^{-1}) of hydrogen bonds and specific interactions in monocarboxylic hydroxy acids

Acids	Formula	Structure	$\Delta_{vap}H^0(T)$ [7]	T (K)	Bond energies		
					DH – O···H – O	DHO → CH_3	DH – O···H – CH
Hydroxyacetic	$C_2H_4O_3$	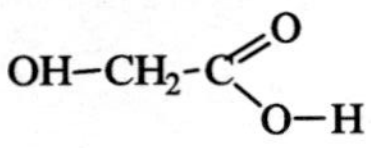	51.8	362	6.5	5.63	0.60 × 4
3-Hydroxypropionic	$C_5H_{10}O_3$	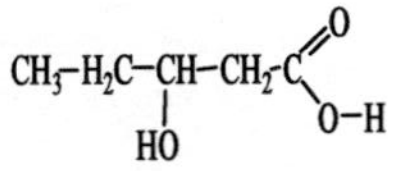	62.2	341	7.0	6.50 × 2 8.92 × 2	0.60 × 4
2-Hydroxybutyric	$C_4H_8O_3$	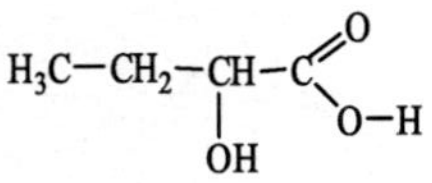	61.2	376	8.0	6.50 × 2 6.78 × 2	0.60 × 4

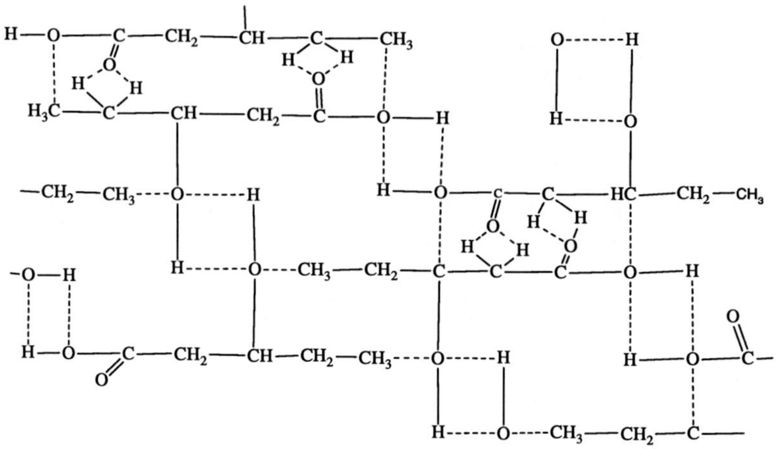

Fig. 7.6 Schematic picture of the liquid structure of 3-hydroxypropionic acid

Note that the CH group of these fragments is directly linked to the oxygen atom in the molecule and in the formed specific interaction; therefore, it should destabilize the hydrogen bond.

The results of the calculations of the average values of the hydrogen bond energies formed by the alcohol and acid hydroxy groups are underestimated since the applied energies of the specific interactions correspond to the standard conditions. Similar values of the vaporization enthalpies of 3-hydroxypropionic acid and 2-hydroxybutyric acid measured at different temperature, and the analogous types and number of specific interactions suggest a conclusion that the propyl fragment of the molecule contributes to the vaporization enthalpy essentially less energy than the employed value of 8.92 kJ mol^{-1}. This fact is a definite indication that the propyl fragment operates as an isostructural group in the molecule of the 3-hydroxypropionic acid. It should be noted in this connection that taking into account the temperature dependence, the obtained mean value of the hydrogen bond energy in the liquid 2-hydroxybutyric acid should be increased by 3–4 kJ mol^{-1}, and its standard value should be 11–12 kJ mol^{-1}. Hence, the performed thermodynamic analysis for the hydroxycarboxylic acids and the obtained values of the hydrogen bond energies made it possible to obtain a definite notion of the energies of the specific interactions and hydrogen bonds.

7.5 Energies of Specific Interactions in Methoxy- and Ethoxycarboxylic Acids

The ethoxycarboxylic acids containing an oxygen atom in the chain possess higher values of vaporization enthalpy than those of the carboxylic acids with the same number of carbon atoms in the chain; for instance, compare butyric acid

(58.8 kJ mol^{-1}) and ethoxyacetic acid (69.1 kJ mol^{-1}). This fact originates from the formation of two additional specific interactions by the unshared electron pairs of the oxygen atom of the methoxy and ethoxy groups in 2-methoxyacetic and ethoxyacetic acids, respectively (Table 7.6).

The molecules of 3-methoxypropionic acid and 2-methoxypropionic acid have each ten bond vacancies (Fig. 7.7) that form two stable hydrogen bonds $H-O\cdots H-O$, four H-bonds of low stability $=O\cdots H-C$, and two specific interactions formed by the methoxy group ($-O \rightarrow CH_3$–O). Acid fragments form two stable interactions –O $\rightarrow$ CH_2C–.

Hence the hydrogen bonds energies should be estimated from the difference between the vaporization enthalpy of compounds and the sum of the energies of all other specific interactions and the H-bonds of low stability along (7.8)

$$D{H-O\cdots H-O=}\,\frac{\Delta_{vap}H^0(T)\text{mtc.a.}-2D{-O\rightarrow CH_2-CH_2}-2D{-O\rightarrow CH_2C}-2D{-O\rightarrow CH_3-(CH_2)_n}-4D{=O\cdots H-CH_2}}{2}. \tag{7.8}$$

The calculated hydrogen bond energy of the 2-methoxyacetic acid was established at 348 K and, therefore, is underestimated since all values of other specific interactions were estimated at the standard temperature. The hydrogen bond energy of the ethoxyacetic acid is valid, and its low value (12.50 kJ mol^{-1}) compared with the corresponding energy of acetic acid (17.60 kJ mol^{-1}) is due to the effect of the oxygen atom of the ethoxy group reducing the shift in the electron density to the oxygen of the carboxy group.

7.6 Energies (kJ mol^{-1}) of Specific Interactions in Peroxycarboxylic Acids

Carboxylic acids containing a peroxy group possess reduced vaporization enthalpy compared with compounds having a common carboxy group (Table 7.7). The appearance of the peroxy oxygen leads to an increase in the number of bond vacancies and to a significant change in the electron density distribution in the molecules of the peroxy acids. The increase by two bond vacancies in the peroxy acid compared with that in a usual acid, and a simultaneous decrease in the vaporization enthalpy of these compounds compared to that in the carboxylic acids indicate the stabilization first of all of the most stable hydrogen bonds.

Let us consider the definite values of the vaporization enthalpy. The difference in the vaporization enthalpy of the acetic acid (50.0 kJ mol^{-1}) and peroxyacetic acid (44.2 kJ mol^{-1}) amounts to 5.8 kJ mol^{-1}, of the propionic acid (54.9 kJ mol^{-1}) and peroxypropionic acid (43.2 kJ mol^{-1}) to 11.7 kJ mol^{-1}, and of the butyric acid (58.8 kJ mol^{-1}) and peroxybutyric acid (45.1 kJ mol^{-1}) to 13.7 kJ mol^{-1}. However, the difference in the corresponding characteristics of the decanoic acid (118.8 kJ mol^{-1})

Table 7.6 Energies (kJ mol^{-1}) of hydrogen bonds and specific interactions in methoxy- and ethoxycarboxylic acids

Acids	Structure	$\Delta_{vap}H^0(T)$ [15]	T (K)	Bond energies			
				DH – O···H – O	D–O → CH_3/ D–O → CH_3–CH_2–	D=O → H–CH_2	D–O → CH_2–C–
2-Methoxyacetic	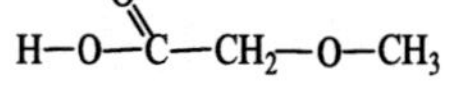	54.2	348	12.50	5.63	0.60 × 4	6.78
Ethoxyacetic	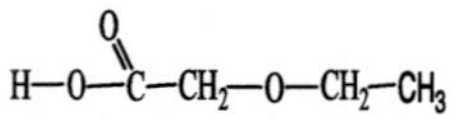	69.1	298	14.80	6.78	0.60 × 4	6.78

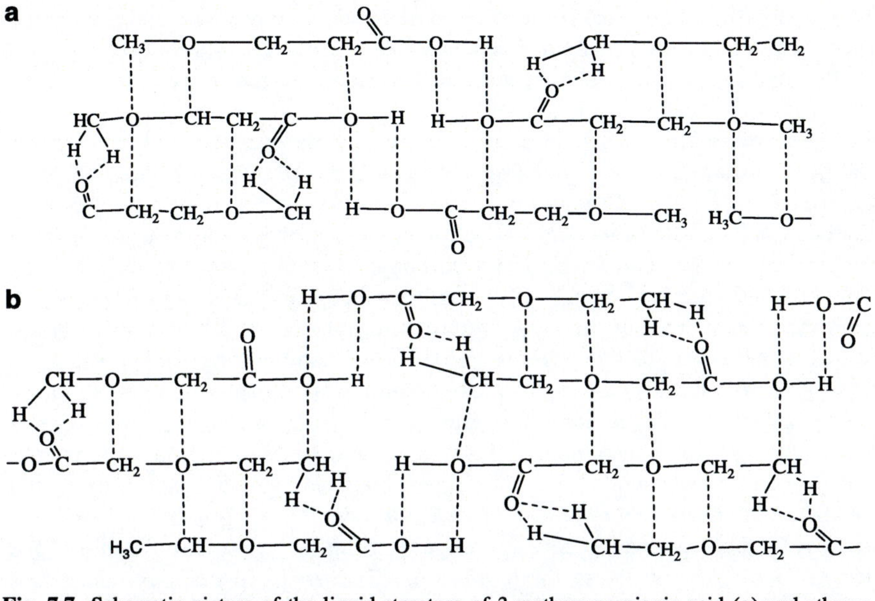

Fig. 7.7 Schematic picture of the liquid structure of 3-methoxypropionic acid (**a**) and ethoxyacetic acid (**b**)

Table 7.7 Energies (kJ mol^{-1}) of hydrogen bonds and specific interactions of peroxy monocarboxylic acid

Acids	Formula	$\Delta_{vap}H^0(T)$ [15]	T (K)	Bond energies DH – O··· H – O	D–O → CH_3	D = O··· H – CH
Peroxyacetic	$C_2H_4O_3$	44.2	288	9.05	5.63 × 4	0.60 × 6
Peroxypropionic	$C_3H_6O_3$	43.2	288	8.55	5.63 × 4	0.60 × 6
Peroxybutyric	$C_4H_8O_3$	45.5	288	9.70	5.63 × 4 6.78 × 2	0.60 × 6
Peroxyvaleric	$C_5H_{10}O_3$	57.2[a]	298	12.30	5.63 × 2 8.92 × 2	0.60 × 6
Peroxyhexanoic	$C_6H_{12}O_3$	69.1[a]	298	12.30	–	0.60 × 6
Peroxyheptanoic	$C_7H_{14}O_3$	80.8[a]	298	12.30	–	0.60 × 6
Peroxyoctanoic	$C_8H_{16}O_3$	92.7[a]	298	12.30	–	0.60 × 6
Peroxynonanoic	$C_9H_{18}O_3$	104.7[a]	298	12.30	–	0.60 × 6
Peroxydecanoic	$C_{10}H_{20}O_3$	117.1 ± 1.7	298	12.30	–	0.60 × 6
Peroxyundecanoic	$C_{11}H_{22}O_3$	124 ± 1.7	298	12.30	–	0.60 × 6
Peroxydodecanoic	$C_{12}H_{24}O_3$	131.4 ± 1.7	298	12.30	–	0.60 × 6

[a]Determined by interpolation

and peroxydecanoic acid (117.10 kJ mol^{-1}) is 1.7 kJ mol^{-1} at the accuracy of the vaporization enthalpy measurement ±1.7 kJ mol^{-1}, and finally, the difference for the dodecanoic acid (127.9 kJ mol^{-1}) and peroxydodecanoic acid (131.4 ± 1.7 kJ mol^{-1}) is of a negative value (−3.5). This permits a conclusion that at 12 bond vacancies in the liquid peroxydecanoic and peroxydodecanoic acids and at ten bond vacancies in the decanoic and dodecanoic acids, the overall values of the energies of the hydrogen bonds and specific interactions in the respective pairs of compounds are practically equal within experimental accuracy. This means that the interactions formed by the peroxy oxygen atoms essentially differ by their energy contribution to the enthalpy characteristic. At the same time, the low vaporization enthalpy of the peroxy acids is due to the decrease in the negative charge on the oxygen of the hydroxy group because of the electron density shift to the peroxide oxygen and further to the carbonyl oxygen. As a result, each oxygen atom is able to form hydrogen bonds of different stability. The electron density shift from the carbon atom of the methyl group along the chain to the carboxy group results in enhanced positive charges on the carbon and hydrogen atoms which are weakly compensated by the reverse dative bond. Consequently, the peroxy group oxygen possesses reduced donor properties in the course of formation of hydrogen bonds and specific interactions. Therefore, the difference in the vaporization enthalpy and consequently in the hydrogen bond energies depends also on the fragments involved in the interactions with the peroxy oxygen atom. The sharp increase in the vaporization enthalpy in going from the peroxybutyric acid to the peroxyvaleric acid is due to the number of CH_2 groups in the alkyl chain that primarily leads to the replacement of the fragment interacting with the peroxy oxygen atom forming the specific interaction of an enhanced stability, and on the contrary, they make a contribution with the energy of the formed interactions. These conclusions are confirmed by the dependence of the vaporization enthalpy of peroxy acids on the number of carbon atoms in the alkyl chain (Fig. 7.8).

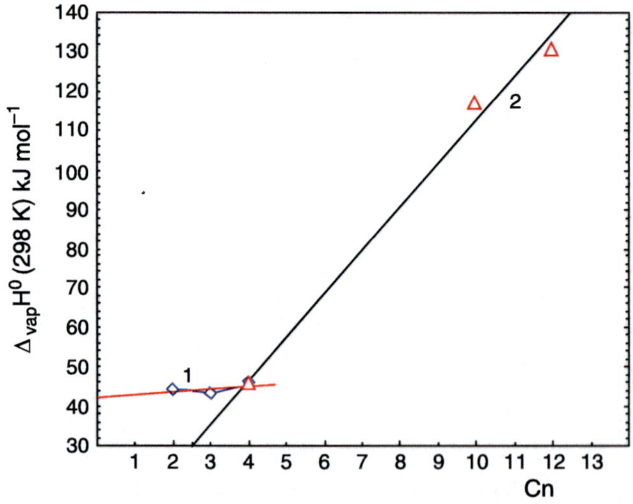

Fig. 7.8 Dependence of vaporization enthalpy on the number of carbon atoms in the alkyl chain of peroxy monocarboxylic acids (1) influence of dative bond and (2) increasing number of CH_2 groups

Note the gently sloping character of the plot for the first homologs of the peroxycarboxylic acids. It means that the reverse dative bond has virtually the same influence in the whole series peroxyacetic acid–peroxybutyric acid, and this is possible at the formation of the specific interactions with fragments containing equal number of carbon atoms. Then a steep increase in the vaporization enthalpy occurs from the peroxybutyric acid to peroxydecanoic acid, with the energy contribution to the vaporization enthalpy from a single methylene group equal 11.7 kJ mol^{-1}, twice larger than the analogous contribution to the vaporization enthalpy in ketones, alcohols, and carboxylic acids (Figs. 7.2 and 7.3). This suggests the participation in the interactions of a growing number of carbon atoms of the chain. This influence should weaken with the growing number of carbon atoms in the alkyl chain, and really the contribution to the enthalpy characteristic sharply decreases in the peroxydodecanoic acid (9.7 kJ mol^{-1}). The character of the dependence of the vaporization enthalpy of the peroxycarboxylic acids on the number of carbon atoms (Fig. 7.8) is analogous to similar dependences for various classes of organic compounds. Therefore, we were able to estimate by interpolation the vaporization enthalpy of unstudied compounds with the accuracy corresponding to the experimental measurements (Table 7.7).

The 12 bond vacancies in the molecules of the peroxyacarboxylic acids form specific interactions, resulting in a network structure of their liquid state, as illustrated by the schematic picture of the liquid peroxyacetic acid (Fig. 7.9). Thus, the structure of the liquid peroxyacetic acid is a network consisting of chains with two cycles generated by specific interactions and hydrogen bonds.

The chains are linked by weak hydrogen bonds. The interaction of the molecules of the nearest surrounding involves the formation of two stable hydrogen bonds by the terminal peroxy oxygen atoms.

The deficit of carbon atoms in the chain compels the oxygen atoms of the peroxy group that formed the hydrogen bonds to form additionally H-bonds of low stability $H-O\cdots H-CH_2$ with the positively charged hydrogen atom of the CH_3 group of

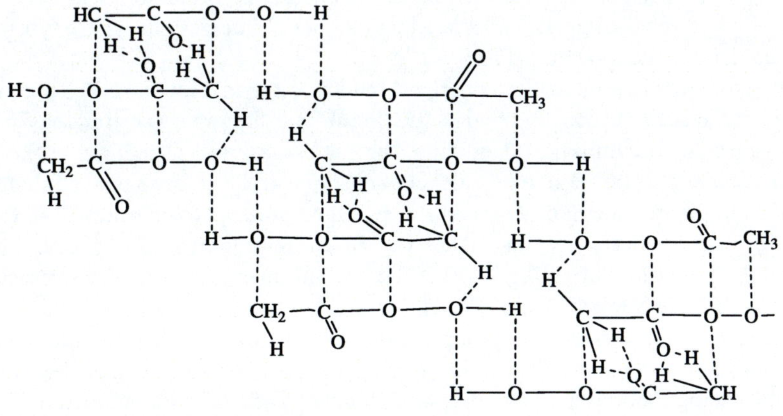

Fig. 7.9 Schematic picture of the liquid structure of peroxyacetic acid

contiguous chains. The second of these oxygen atoms forms with analogous terminal CH_3 group of the neighbor molecule or with the carbon atom of the carbonyl group a stable specific interaction H–O → CH_3– or H–O → C=O, respectively. Analogous specific interactions are formed by the oxygen of the peroxy group with the carbon atoms of the CH_3 and C=O groups of the adjacent molecules. The oxygen atom of the carbonyl group forms H-bonds of low stability with the remaining positively charged hydrogen atoms of the methyl group. Thus, in the liquid peroxyacetic acid, six H-bonds of low stability (0.60), two specific interactions H–O → CH_3, and two interactions H–O → C=O, whose energies should be taken equal to the energy of the specific interactions in the liquid dimethyl ether (5.63 kJ mol^{-1}), are formed. Consequently, the energy of the hydrogen bond should be estimated along (7.9)

$$DH-O\cdots H-O= \frac{\Delta_{vap}H^0(T)-2DH-O\rightarrow CH_3-2DH-O\rightarrow C=O-6D=O\cdots H-CH_2}{2}. \quad (7.9)$$

The results of the performed calculations (Table 7.7) show that the hydrogen bond energy in the liquid peroxyacetic acid is twice smaller than the hydrogen bond energy in the acetic acid (17.70 kJ mol^{-1}).

A significant contribution to the specificity of the intermolecular interactions in the peroxypropionic acid is provided by the CH_2 group whose hydrogen atoms take part in the formation of the H-bonds of reduced stability (Fig. 7.10). Thus, the involvement of the hydrogen atoms of the methylene group instead of those of the terminal methyl group in the formation of the H-bonds $=O\cdots H-CH$ with an energy of 0.60 kJ mol^{-1} cannot essentially lead to the increase in the vaporization enthalpy for the carbon atoms of the CH_2 and CH_3 groups that are involved in the specific interactions. The difference consisted only in the fact that the carbon atom of the carbonyl group interacts through the oxygen atom of the carbonyl.

Hence the structures of liquid peroxyacetic and peroxypropionic acids with the equal number of bond vacancies form equal number of interactions whose types are characterized by equivalent stability.

Therefore, the hydrogen bond energy should be estimated by (7.6). The obtained energy of the stable hydrogen bond (8.55 kJ mol^{-1}) is the same as estimated for the peroxyacetic acid within the accuracy of the used measurement method [14].

The peroxybutyric acid molecule contains two CH_2 groups: one of them is involved in the specific interaction with the oxygen of the peroxy group O → CH_2, and the second remains free and takes part in the formation of the specific interaction H–O → CH_3–CH_2 (Fig. 7.10). The second analogous specific interaction H–O → CH_2–CH_2 forms two methylene groups of the chain. In its turn, the increased number of the methylene groups in the peroxybutyric acid affects the redistribution of the electron density, increasing the latter on the oxygen atoms and thus increasing their donor properties and stabilizing the specific interactions.

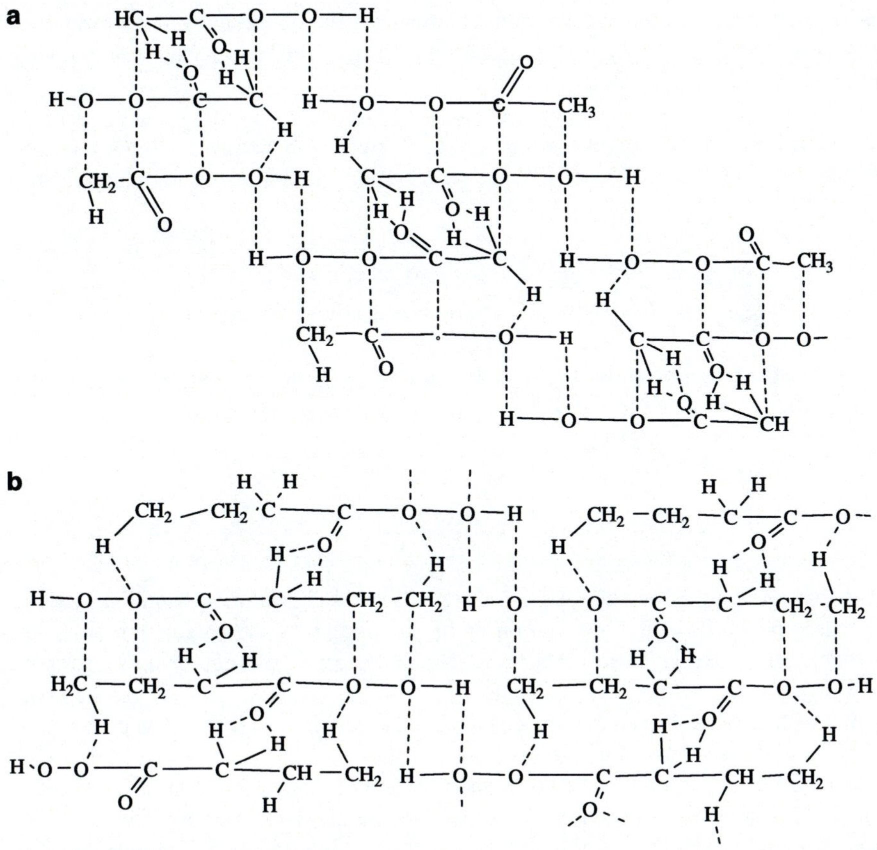

Fig. 7.10 Schematic picture of the liquid structure of peroxypropionic acid (**a**) and peroxybutyric acid (**b**)

In the structure of the liquid state of the peroxybutyric acid, the number of specific interactions remains the same, and the hydrogen bond energy should be estimated from (7.10)

$$DH-O\cdots H-O= \frac{\Delta_{vap}H^0(T)\text{po.a.}-2DH-O\rightarrow CH_3-CH_2-2DH-O\rightarrow CH_3-6D{=}O\cdots H-CH_2}{2}. \quad (7.10)$$

The calculated energy of the stable hydrogen bond in liquid peroxybutyric acid (9.70 kJ mol^{-1}) exceeds the energies of the specific interactions.

Taking into account that the effect of the reverse dative bond in the alkyl chain ends at the third carbon atom from the oxygen, the maximum value of the hydrogen bond energy is expected in the peroxyvaleric acid. Hence the subsequent growth of the vaporization enthalpy in the homologous series originates from the growing number of methylene groups.

The performed thermodynamic analysis of the peroxycarboxylic acids permits a conclusion on the regular stabilization of hydrogen bonds in the series due to the shift in the electron density and distribution of charges on the atoms of the molecule.

$$\text{Peroxypropionic acid } (8.55) > \text{Peroxyacetic acid } (9.05) >$$
$$\text{Peroxybutyric acid } (9.70) > \text{Peroxyvaleric acid } (12.30\,\text{kJ mol}^{-1}).$$

The low energies of the hydrogen bonds close to the energies of the specific interactions indicate that the peroxyacids do not dimerize in vapor.

7.7 Hydrogen Bonds in Unsaturated Monocarboxylic Acids

The unsaturated monocarboxylic acids are still poorly understood from the thermodynamic positions. The vaporization enthalpies found by the calculation from the vapor pressure are given in [7] for a moderate temperature range of investigations. For instance, the temperature of studying *cis*-2-butenoic acid (294 K) can be compared with the standard temperature of the butyric acid investigation. The presence or absence in the structure of the double bond –C=C– governs the vaporization enthalpy and hydrogen bond energy of the respective acids. The saturated and unsaturated carboxylic acids form the same type of specific intermolecular interactions H–O → CH_3–CH_2–C and H–O → CH_2=CH–C with the terminal group of the propenoic acid and the acids with the longer carbon–carbon chains. Considering the contribution to the vaporization enthalpy of this type of interaction (9.04 kJ mol^{-1}) (Chap. 6), the contribution of four H-bonds of low stability D=O → H–CH_2, and the contribution of the fourth CH group of the alkyl chain of the butenoic acid and the large number of methylene groups in further unsaturated acids, the hydrogen bond energy should be estimated by (7.11)

$$\mathrm{DH-O\cdots H-O} = \frac{\Delta_{\mathrm{vap}}H^0(T) - 2\mathrm{DH-O\rightarrow CH_3-CH=CH} - 4\mathrm{D=O\rightarrow H-CH_2} - n\cdot\mathrm{DCH}}{2}. \tag{7.11}$$

The results of the performed calculations compiled in Table 7.8 illustrate that the stability of the hydrogen bond in the *cis*-2-butenoic (*cis*-crotonic) acid (16.85 kJ mol^{-1}) is reduced by 0.65 kJ mol^{-1} compared to that in the liquid butyric acid. The vaporization enthalpy of the *trans*-2-butenoic acid (56.7 kJ mol^{-1}) experimentally measured at 368 K leads to the enhanced value of the hydrogen bond energy

Table 7.8 Energies (kJ mol^{-1}) of hydrogen bonds of the liquid unsaturated carboxylic acids

Acid	Formula	Structure	$\Delta_{vap}H^0(T)$ [7, 15]	T (K)	DH–O → $CH_3(CH_2)_n$	DH – O···H – O
$\sum D = O \cdots H - CH_2 = 0.60 \times 4\,kJ\,mol^{-1}$						
Acrylic (propenoic)	$C_3H_4O_2$	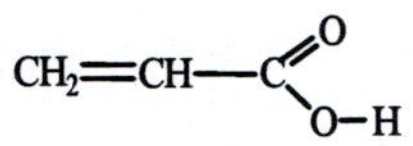	53.1 ± 4.2	298	$n = 2$ 9.05	16.30
Methacrylic (2-methylpropenoic)	$C_4H_6O_2$	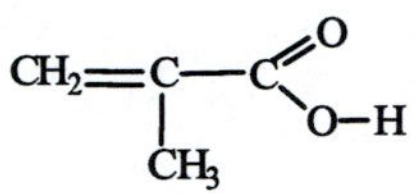	53.9 47.5 ± 0.4 51.6	336 298 313	9.05 $n = 2$ $DCH_3 = 0.8$	16.30
Butyric	$C_4H_8O_2$	CH_3-$(CH_2)_2$-COOH	58.8	298	8.94	17.50
cis-2-Butenoic	$C_4H_6O_2$	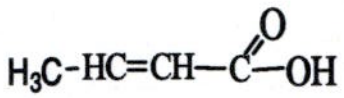	55.8	321	9.05 $n = 3$ DCH = 1.7	16.85
trans-2-Butenoic	$C_4H_6O_2$	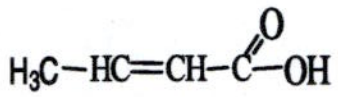	56.7	368	9.05 $n = 3$ DCH = 1.7	17.25
trans-2-methyl-2-butenoic	$C_5H_8O_2$	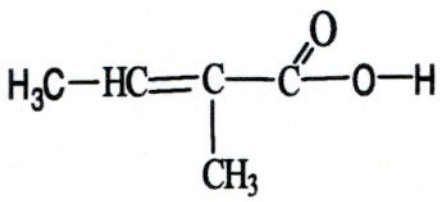	61.2	365	9.05 $n = 3$ DCH = 1.7 $DCH_3 = 4.5$	17.25

(continued)

Table 7.8 (continued)

Acid	Formula	Structure	$\Delta_{vap}H^0(T)$ [7, 15]	T (K)	DH–O → $CH_3(CH_2)_n$	DH – O···H – O
3-Methylbutenoic	$C_5H_8O_2$	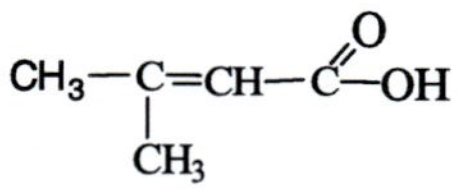	57.3	378	9.05 $n = 3$ $DCH_3 = 1.5$	16.85
2-Ethylpropenoic (2-ethylacrylic)	$C_5H_8O_2$	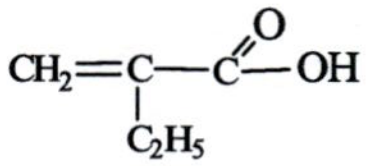	58.1 ± 0.4	298	9.05 $n = 2$ $DC_2H_5 = 5.0$	16.30

(17.25 kJ mol^{-1}) resulting from the *trans*-position of the methyl group. Evidently, taking into account the temperature dependence of the vaporization enthalpy of the *trans*-2-butenoic acid, this difference should increase. Still the obtained difference in the hydrogen bond energy in *cis*- and *trans*-2-butenoic acids illustrates the more stable hydrogen bond in the latter in agreement with the conceptions on the higher stability of the *trans*-isomers [17].

The reduced energy of the hydrogen bond of the acrylic acid shows that the influence of the reverse dative bond on the specific interactions ends at the butenoic acid, and consequently, the hydrogen bond energy in this acid is higher and is the maximum value for all the other unsaturated monocarboxylic acids with analogous alkene chains.

The lowest vaporization enthalpy of the methacrylic acid (Table 7.8) allows a conclusion that under the conditions of the performed experiment, the vapor contains both monomeric and dimeric molecules at the essential prevalence of the former. The vaporization enthalpy measured at 336 K is somewhat lower than should be at the standard temperature. The energy contribution of the isostructural methyl group (0.8 kJ mol^{-1}) obtained from the difference of the vaporization enthalpies of the methacrylic and acrylic acids is close in value to the analogous contribution of such group in the 2-methylpropinoic acid (0.6 kJ mol^{-1}). The energy contributions of the appropriate isostructural groups are estimated from the difference of vaporization enthalpies of *trans*-2-methyl-2-butenoic acid and *trans*-2-butenoic acid, 3-methylbutenoic acid and 2-butenoic acid, and 2-ethylpropenoic acid and propenoic acid. The energies of the hydrogen bonds are calculated by (7.11) and correspond to the temperature of the performed measurements of the enthalpy characteristics.

Note the sharply differing energy contributions to the enthalpy characteristic of the isostructural methyl (0.8 kJ mol^{-1}) and ethyl groups (5.0 kJ mol^{-1}) attached to the second carbon atom at the double bond in the derivatives of the acrylic acid. The latter energy contribution approaches the energy of the specific interaction formed by the ethyl group of the liquid diethyl ether (6.78 kJ mol^{-1}). This permits a conclusion that the presence of two carbon atoms in the ethyl group and in the double bond significantly decreases the shift in electron density along the chain to the oxygen atom of the carboxy group.

7.8 Energies of Hydrogen Bonds in Dicarboxylic Acids

Dicarboxylic acids with two terminal carboxy groups possess high enthalpies of sublimation and vaporization, and are capable of forming dimers and more complex associates in vapor. Similar to monocarboxylic acids, the dicarboxylic acids should have high enthalpy of melting and polymorphous transitions. The published vaporization enthalpies of dicarboxylic acids are cited without indication of the accuracy, are of incomparable values (Table 7.9), and are, therefore, unsuitable for the thermodynamic analysis.

Table 7.9 Vaporization enthalpy (kJ mol^{-1}) of saturated dicarboxylic acids

Acid	Formula	$\Delta_{vap}H^0(T)$ [15]	T (K)
Glutaric	$C_5H_8O_4$	98.1	443
Adipic	$C_6H_{10}O_4$	92.0	447
Heptanedioic	$C_7H_{12}O_4$	88.8	451
Octanedioic (suberic)	$C_8H_{14}O_4$	91.4	460
Nonanedioic (azelailic)	$C_9H_{16}O_4$	89.3	460
Sebacic	$C_{10}H_{18}O_4$	85.5	471

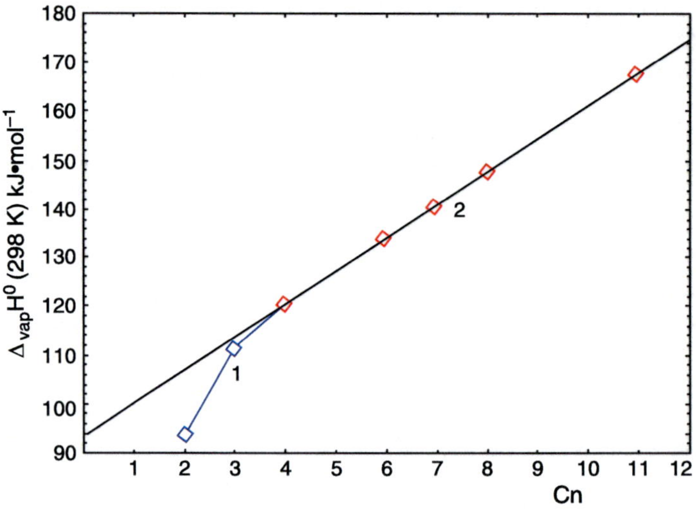

Fig. 7.11 Dependence of sublimation enthalpy on the number of carbon atoms in the molecules of saturated dicarboxylic acids (1) weakening influence of the reverse dative bond and (2) increasing number of CH_2 groups

It is significant that the dependence of the sublimation enthalpy in the homologous series of the saturated dicarboxylic acids on the number of carbon atoms in the molecule (Fig. 7.11) is analogous to the plots obtained for ketones, alcohols, and monocarboxylic acids. The sharp increase in the sublimation enthalpy at the appearance of the first methylene group in the malonic acid shows the essential shift in electron density from this group to the oxygen atoms of the carboxy groups, with the simultaneous dissociation of the dimers and the stabilization of the hydrogen bonds. Further increase in the number of CH_2 groups in the succinic acid results in lesser shift in electron density in the molecule, decrease in the growth of the hydrogen bond stabilization, increase in the degree of dimer dissociation, and the termination of the influence of the reverse dative bond. Consequently, the location of the vaporization enthalpy of succinic acid is on the linear dependence of the enthalpy characteristic caused by the subsequent energy contribution by the growing number of the CH_2 groups.

Thus, it should be assumed that the association in vapor should be terminated at the glutaric acid. This assumption is confirmed by the decrease in the energy contribution to the enthalpy characteristic of the second CH_2 group in the molecule of the succinic acid (10.4) when the contribution of this group to the sublimation enthalpy in every compound starting with 1,7-heptanedioic acid amounts to 6.75 kJ mol^{-1}. This value exceeds the energy contribution (6.15 kJ mol^{-1}) by the analogous group in the crystalline monocarboxylic acid.

The terminal carboxy groups of the dicarboxylic acids secure the formation of chains, where each molecule interacts with the terminal carboxy groups forming four hydrogen bonds. The involvement of the oxygen atoms of these groups and hydrogen atoms of the CH_2 groups in the formation of the specific intermolecular interactions with the molecules of contacting chains provides their cross-linking (Fig. 7.12). As a result, the crystalline and liquid dicarboxylic acids form a network structure. The two carbonyl oxygen atoms of the molecule form eight H-bonds of low stability with the molecules of the nearest environment of energy 1.00 kJ mol^{-1}.

The molecule of the succinic acid with a chain of four carbon atoms formed with the fragments CH_2–C– four equivalent specific interactions DH–O → CH_2–C– with unshared electron pairs of the oxygen atoms from the molecules of the nearest surrounding (Fig. 7.12) of energy equal to that in the solid diethyl ether. Consequently, the energy of the hydrogen bond should be obtained from (7.12)

$$\mathrm{DH-O\cdots H-O} = \frac{\Delta_{\mathrm{sub}}H^0(298) - 4\mathrm{DH-O\rightarrow CH_2-C} - 8\mathrm{D{=}O\cdots H-CH}}{4}. \tag{7.12}$$

Results of the performed calculation are given in Table 7.9.

Similar to the succinic acid, the presence of the CH_2 group in the molecule of malonic acid results in significant changes in the specificity of the intermolecular interactions in the liquid and solid states (Fig. 7.13).

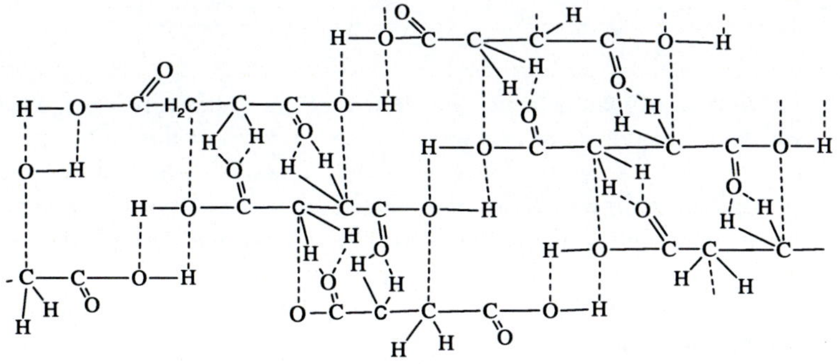

Fig. 7.12 Schematic picture of the liquid and solid structures of succinic acid

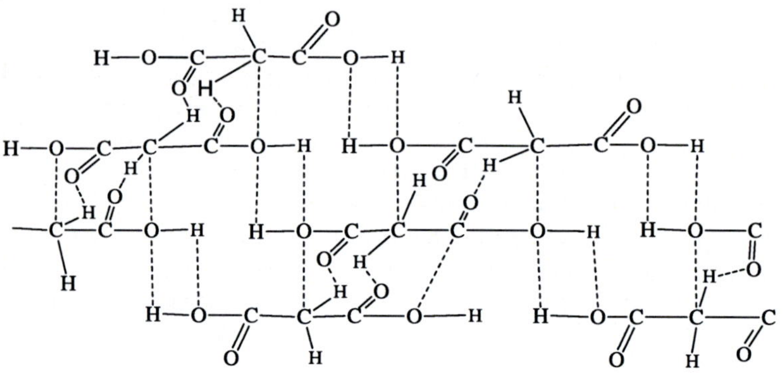

Fig. 7.13 Schematic picture of the liquid and solid structures of malonic acid

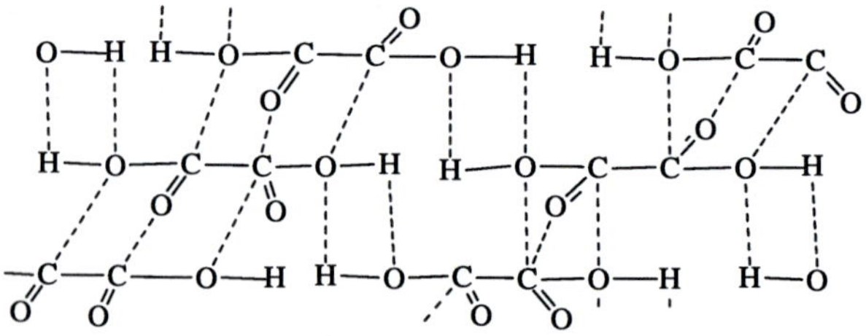

Fig. 7.14 Schematic picture of the liquid and solid structures of oxalic acid

In this molecule, a deficit exists of the bond vacancies with a positive charge. Therefore, the interacting molecules of the nearest surrounding form four hydrogen bonds, two specific interactions DH–O → CH_2–C–, two interactions DH–O → C–, and four hydrogen bonds of low stability $DC = O \cdots H - CH$. Hence the hydrogen bond energy is calculated by (7.13)

$$DH{-}O{\cdots}H{-}O{=}\frac{\Delta_{sub}H^0(298)-2DH{-}O{\rightarrow}CH_2{-}C{-}2DH{-}O{\rightarrow}C{-}4D{=}O{\cdots}H{-}CH}{4}. \quad (7.13)$$

In the molecule of oxalic acid, the deficit of the positively charged bond vacancies is even greater; therefore, one carbonyl oxygen of the molecule forms a specific interaction of low stability with the carbonyl carbon of the second molecule (Fig. 7.14). The remaining nonbonded second carboxy group is forced to take part in nonspecific interactions. Yet the crystal and liquid structures of the oxalic acid conserves the network structure characteristic of all dicarboxylic acids, with the terminal carboxy groups cross-linked by four specific intermolecular interactions H–O → C=.

The involvement of each carbon atom of the molecule in the formation of specific interactions governs the energy of the latter that should be taken equal to the interaction energy in the solid dimethyl ether (7.25 kJ mol^{-1}). The low electron density excess on the oxygen of the carbonyl group results in the low energy of the specific interaction formed by the CH_2 group; therefore, the energy of the interaction C=O → C=O should be estimated at 1.00 kJ mol^{-1}. The remaining free bond vacancies of the carbonyl oxygen atom take part in cooperative interactions whose energies are within the errors of the experimentally measured vaporization and sublimation enthalpies. The difference between the sublimation enthalpy and the overall energy of the specific intermolecular interactions equals the sum of the energy of four hydrogen bonds; therefore, for a single bond, the value is estimated from (7.14)

$$\mathrm{DH-O\cdots H-O} = \frac{\Delta_{\mathrm{sub}}H^0(298) - 4\mathrm{DH-O \to C=O} - 2\mathrm{DC=O \to C=O}}{4}. \tag{7.14}$$

The obtained value of the hydrogen bond energy in the solid oxalic acid (Table 7.10) is consistent with the values for other dicarboxylic acids, where a regular growth of the hydrogen bond energy is observed at the increase of the number of carbon atoms in the molecule till four. Maximum stabilization is attained in the succinic acid where the specific interaction is formed by two carbon atoms of the chain DH–O → CH_2–C–. Thus, further increase in the number of carbon atoms in the chain of the 1,5-pentanedioic acid results in the formation of two specific interactions DH–O → CH_2–C– and two DH–O → CH_2–CH_2–C– with a three-carbon fragment and, respectively, with enhanced, maximum possible stability due to the termination of the influence of the reverse dative bond on the shift in electron density from the third carbon atom of the chain. In the structure of the condensed state of the 1,5-pentanedioic acid and all subsequent compounds of the homologous series, eight H-bonds of low stability are conserved. The distinguishing feature of the 1,6-hexanedioic acid is the formation of four specific interactions DH–O → CH_2–CH_2–C–, which like those in 1,5-pentanedioic acid lead to the increase in sublimation enthalpy. The hydrogen bond energies of these compounds are estimated from the difference of the vaporization enthalpy and the overall energy of the specific interactions in 1,5-pentanedioic acid and 1,6-hexanedioic acid by (7.15) and (7.16), respectively.

$$\mathrm{DH-O\cdots H-O} = \frac{\Delta_{\mathrm{sub}}H^0(298) - 2\mathrm{DH-O \to CH_2-C} - 2\mathrm{DH-O \to CH_2-CH_2-C} - 8\mathrm{D{=}O\cdots H-CH}}{4}, \tag{7.15}$$

$$\mathrm{DH-O\cdots H-O} = \frac{\Delta_{\mathrm{sub}}H^0(298) - 4\mathrm{DH-O \to CH_2-CH_2-C} - 8\mathrm{D=O\cdots H-CH}}{4}. \tag{7.16}$$

The subsequent growth of the enthalpy characteristic in 1,7-heptanedioic acid and the dicarboxylic acids of this homologous series originates from the energy

Table 7.10 Energies (kJ mol^{-1}) of hydrogen bonds in the liquid saturated dicarboxylic acids

Acid	Formula	$\Delta_{sub}H^0(T)$ [16]	T (K)	D–O → CH_3	$D=O\cdots H-C$	$DH-O\cdots H-O$
Oxalic	$C_2H_2O_4$	93.7 ± 1.3	298	7.25 × 4	–	16.20
Malonic	$C_3H_4O_4$	111.4 ± 0.7	298	7.25 × 4 9.16 × 4	1.0 × 4	18.65
Succinic	$C_4H_6O_4$	120.3 ± 4.4	298	9.16	1.00 × 8	18.90
1,5-Pentanedioic (glutaric)	$C_5H_6O_4$	127.0[a]	298 298	11.96 × 2 9.15 × 2	1.0 × 8	18.90
1,6-Hexanedioic (adipic)	$C_6H_8O_4$	133.6 ± 1.3	298	11.96 × 4	1.0 × 8 $D(CH_2) = 2.00$	18.90
1,7-Heptanedioic (pimelic)	$C_7H_{12}O_4$	139.9 ± 1.0		11.96 × 4	1.0 × 8 $D(CH_2) = 8.25$	18.90
1,8-Octanedioic	$C_8H_{14}O_4$	147.8 ± 3.8	298	11.96 × 4	1.0 × 8 $D(CH_2) = 15.50$	19.10
1,9-Nonanedioic	$C_9H_{16}O_4$	153.8[a]	298	11.96 × 4	1.0 × 8 $D(CH_2) = 22.25$	18.90
1,10-Sebacic	$C_{10}H_{18}O_4$	160.7 ± 2.5	389	11.96 × 4	1.0 × 8 $D(CH_2) = 29.00$	18.90
1,11-Undecanedioic	$C_{11}H_{20}O_4$	167.4[a]	298	11.96 × 4	1.0 × 8 $D(CH_2) = 7.80$	18.90
Dodecanedioic	$C_{12}H_{22}O_4$	153.1 ± 2.4	438	11.96 × 4	$\sum D{=}O \rightarrow CH_3$	18.8
Methylmalonic	$C_4H_6O_4$	113.2 ± 2.2	298	7.25 × 2 9.16 × 2	1.0 × 4 $DCH_3 = 1.4$	18.65
Dimethylmalonic	$C_5H_8O_4$	111.7 ± 2.1	298	7.25 × 2 9.16 × 2	1.0 × 4 $DCH_3 = 1.4$	18.65
Ethylmalonic	$C_5H_8O_4$	112.8 ± 2.2	298	7.25 × 2 9.16 × 2	$DC_2H_5 = 1.4$ 1.0 × 4	18.65
Butylmalonic	$C_7H_{12}O_4$	124.6 ± 2.2	298	7.25 × 2 9.16 × 2	$DC_4H_9 = 13.4$ 1.0 × 4	18.65
cis-Butenedioic (maleic)	$C_4H_4O_4$	110 ± 2.5	368	9.16 × 4	1.0 × 8	16.3
trans-Butenedioic (fumaric)	$C_4H_4O_4$	123.6 ± 2.0	365	9.16 × 4	1.0 × 8	(21.2)

contribution to the enthalpy characteristic of the CH_2 groups. Consequently, the contribution of these groups is taken into account in the calculations of the hydrogen bond energies (7.17)

$$DH{-}O\cdots H{-}O= \frac{\Delta_{sub}H^0(298) - 4DH{-}O{\rightarrow}CH_2{-}CH_2{-}C - 8D{=}O\cdots H{-}CH - n\cdot DCH_2}{4}. \quad (7.17)$$

Inasmuch as two interacting molecules of the nearest surrounding contribute 6.75 kJ mol^{-1}, 3.35 kJ mol^{-1} corresponds to one CH_2 group of the dicarboxylic acids chain. Every hydrogen atom of this group interacts with a carbon atom of a similar group and forms a specific interaction DH_2–C–CH with an energy of

1.70 kJ mol^{-1}. This value exceeds the energy of the analogous interaction in various liquid organic substances by approximately 0.60–0.40 kJ mol^{-1}.

The obtained hydrogen bond energies of solid dicarboxylic acids are described by a regular series of stabilization

$$\begin{aligned}&\text{Oxalic acid}\,(16.20) < \text{Malonic acid}(16.20) > 1,4\text{ - Succinic acid}(18.90)\\&= 1,5\text{ - Pentanedioic acid}\,(18.90) = 1,6 - \text{Hexanedioic acid}\,(18.90)\\&= 1,11\text{ - Undecanedioic acid}\,(18.90\,\text{kJ mol}^{-1})\end{aligned}$$

similar to the stabilization series for solid monocarboxylic acids

$$\text{Acetic acid}\,(18.10) < \text{Propionic acid}\,(18.70) < \text{Butanoic acid}\,(18.60\,\text{kJ mol}^{-1}).$$

The trend in the stabilization of the hydrogen bonds in two first dicarboxylic acids is more pronounced due to, first, the involvement of all bond vacancies at increasing the carbon atom number from two to three in going to the malonic acid, and second, the shift in electron density to the oxygen atoms of the hydroxy groups and the stabilization of the specific interaction involving the fragment with two hydrogen atoms DH–O $\rightarrow$ CH_2–C=O. Dicarboxylic and monocarboxylic acids are alike in attaining the maximum energy of the hydrogen bond at two carbon atoms in the fragment forming the specific interaction. The equal energy of the hydrogen bond in the series malonic acid–1,11-undecanedioic acid and propionic acid–octadecanoic acid is a significant indication that the association in the vapor of the series of carboxylic acids under consideration does not affect the obtained energies of the hydrogen bonds.

The sublimation enthalpy of branched dimethylmalonic acid insignificantly differs from the enthalpy characteristic of the malonic acid. Therefore, taking into account the errors in the experimentally measured sublimation enthalpies of these compounds, these values are unfit for use in the thermodynamic analysis. The energy contributions of the isostructural methyl, ethyl, and butyl groups were obtained from the difference of the enthalpy characteristics of the appropriate compounds and the vaporization enthalpy of the malonic acid. The obtained energies of the contribution of the isostructural methyl and ethyl groups are comparable with the energies of interactions of these groups in the liquid ketones, alcohols, and carboxylic acids when these groups are located at the second carbon atom of the chain. The high energy of the specific interaction of the isostructural butyl group (13.2 kJ mol^{-1}) is about a half of the energy contributed by a similar fragment of dibutyl ether. The hydrogen bond energies of the solid derivatives of the malonic acid with the isostructural groups remain the same.

The information on the sublimation thermodynamics of unsaturated dicarboxylic acids is limited to the experimental data on fumaric and maleic acids that are reported with large errors. Some attention may be paid to the data in Table 7.10. The molecules of saturated and unsaturated carboxylic acids of normal structure are alike and have analogous condensed states. The same specific interactions and H-bonds of low stability are present. The difference consists in the participation in the specific interactions of the fragments with double bonds DH–O → CH–C– whose energy should be taken equal to the energy of the interaction in the liquid divinyl ether. The energy of these interactions in the liquid divinyl ether amounts to 6.57 kJ mol^{-1}, less by 0.20 kJ mol^{-1} than the energy of the analogous interaction in the liquid diethyl ether. Therefore, we may take the energy of the specific interaction in the solid divinyl ether and the butenedioic acid to be equal to the reduced value of the analogous interactions in solid ethyl ether, 8.90 kJ mol^{-1}. The results of the performed estimation of the hydrogen bond energy (Table 7.10) of the solid fumaric acid reasonably show the higher stability of the hydrogen bond in the *trans*-isomer compared to the *cis*-isomer.

7.9 Energies of Hydrogen Bonds and Specific Interactions in Cyclic Carboxylic Acids

7.9.1 Saturated Cyclic Carboxylic Acids and Benzoic Acids

Bond vacancies of molecules and the energy of the forming specific intermolecular interactions govern the structure of the liquid and solid states of the substances. The relative energies of the various types of specific interactions and hydrogen bonds provide a possibility of polymorphous transitions. The number and energy of ruptured bonds are related to the enthalpy and entropy of this process. The quasi-crystalline structure of the liquid state reflecting the removal of the structure of the liquid from that of the crystalline state reflects additionally the degree of the crystalline state deterioration. The most stable bonds and the considerable difference of their energy from the energy of the specific intermolecular interactions provide a possibility of existence in the vapor of dimers, trimers, and tetramers of the molecules [6]. Considering the structure of solid and liquid substances from various classes of organic compounds and analyzing the thermodynamic characteristics of the processes of the polymorphous transitions, melting, sublimation, vaporization, and dissociation of the molecular associates in vapor, we take as the basis the chemical nature of these processes [11–13]. An important criterion of the existence in the vapor of the associated molecules is the influence of temperature and vapor pressure. These factors have the opposite influence on the association in vapor [6]. Whereas the increased temperature favors dissociation, the growing pressure accompanying this process shifts the equilibrium to the side of association in dimers and more complex formations. The low sublimation and vaporization enthalpies of the carboxylic acids containing a cyclic fragment, namely, in the case of cyclopropanecarboxylic and cyclobutanecarboxylic acids, when the cyclopropane and cyclobutane have relatively high values of the enthalpy characteristic (Table 7.11), provide a possibility to state that no associated forms are present in the vapor of these acids.

Table 7.11 Energy contribution (kJ mol^{-1}) of CH_2 and CH groups in saturated and unsaturated cyclic hydrocarbons

Compounds	Formula	$\Delta_{vap}H^0(T)$ [7, 15]	T (K)	DCH$_2$; D CH
Liquid				
Cyclopropane	C_3H_6	19.9	254	6.6
Cyclobutane	C_4H_8	25.2	272	6.4
Cyclobutene	C_4H_6	24.6	260	6.1
Cyclopentane	C_5H_{10}	28.5 ± 0.1	298	5.7
Cyclohexane	C_6H_{12}	33.0 ± 0.1	298	5.5
Cycloheptane	C_7H_{14}	38.5 ± 0.2	298	5.5
Cyclooctane	C_8H_{16}	43.3 ± 0.2	298	5.43
Cyclooctene	C_8H_{14}	41.6	300	5.2
1,2-Butadiene	C_4H_6	23.9	298	5.8
1,3-Butadiene	C_4H_6	23.0	285	5.6
Cyclopentene	C_5H_6	28.9	299	5.8
1,3-Cyclopentadiene	C_5H_6	28.4	298	5.7
1,3-Cyclohexadiene	C_6H_8	32.6	322	5.45
1,4-Cyclohexadiene	C_6H_8	33.9	313	5.65
1,3,6-*cis*-Hexaditriene	C_6H_8	33.3	314	5.55
2,4-Hexadiene	C_6H_{10}	33.2	308	5.53
1-Hexyne	C_6H_{10}	33.4	280	5.57
Benzene	C_6H_6	33.8 ± 0.1	298	5.65
Solid [16]				
Cyclopropane	C_3H_6	27.6	298	9.2
Cyclobutane	C_4H_8	36.4	(145)	9.1
Cyclopentane	C_5H_{10}	42.6	(122)	8.5
Cyclohexane	C_6H_{12}	37.6	265	6.25
Benzene	C_6H_6	44.4	298	7.4

Thermodynamic calculation of the hydrogen bond energies in the carboxylic acids containing saturated and unsaturated hydrocarbon rings in order to apply the additivity principle requires a detailed analysis of the pattern of structural changes on replacing the hydrogen atoms in the CH_2 and CH groups by methyl, ethyl, carboxy, and hydroxy groups, and an estimation of their energy contribution to the enthalpy characteristic of the acid. The difficulty arising in performing these calculations originates from the necessity to use the sublimation and vaporization enthalpy of cyclopropane, cyclobutane, and benzene minus the energy contributed by their fragment minus the energy of the intermolecular interactions formed by the substituted hydrogen atoms.

It was mentioned before that the hydrogen atom located in the *trans*-position to the hydroxy group [18] formed a weak H-bond of energy 0.6 and 1.0 kJ mol^{-1} in the liquid and solid methanol, respectively. At the same time, the methyl or, e.g., ethyl group that substituted the hydrogen atom forms a specific interaction with the other energy. Therefore, it is necessary from the vaporization or sublimation enthalpy of the compound employed in the thermodynamic calculation to estimate the energy of the interactions formed by the hydrogen of the H–CH_2, H–HC, or H–C group of the respective compound (Chaps. 4 and 5). The existence of correlations between the weak specific interactions and H-bonds formed by the hydrogen atoms of the H–CH_2-, H–CH-, and H–C-groups, in particular, $DC = O \cdots H - C -$, is confirmed by the small difference between and hydrogen bond energies of saturated and unsaturated acids and alcohols. A similar conclusion is obtained by the analysis of the contribution to the vaporization enthalpy of the CH_2 groups of the initial members of the homologous series of cycloalkanes (Table 7.11). The contribution to the vaporization and sublimation enthalpy (kJ mol^{-1}) of the CH_2 and CH groups in the saturated and unsaturated cyclic compounds decreases in the series

$$\text{Cyclopropane}\,(6.6) > \text{Cyclobutane}\,(6.4) > \text{Cyclobutene}\,(6.1\,\text{kJ mol}^{-1}).$$

However, the vaporization enthalpy of these compounds was established at different nonstandard temperatures. The contribution of the CH_2 group in the series cyclopentane–cyclooctane where the vaporization enthalpy was measured under standard conditions was constant within the accuracy of the experimental measurements. Therefore, the contribution of each methylene group does not depend on their number in the molecule of cycloalkane. It is also fundamentally important that the contribution to the vaporization enthalpy of the CH_2 and CH groups in unsaturated cyclopentene, 1,2-butadiene, 1,3-butadiene, containing one and two double bonds –CH=CH–, and 1-hexyne with a triple bond –C≡C– remains constant within the accuracy of the measurement and equals 5.55 kJ mol^{-1}. A special importance has the fact of equal contribution to the vaporization enthalpy of the CH_2 group in the series of cyclic compounds and of the CH group of benzene (5.55 kJ mol^{-1}). Considering the scanty data on the contribution to the sublimation enthalpy of methylene groups (Table 6.10), it should be remarked that its maximum corresponds to cyclopropane (9.2 kJ mol^{-1}), exceeding the analogous contribution to the vaporization enthalpy by 2.6 kJ mol^{-1}. The contribution of the CH group to the sublimation enthalpy of solid benzene is 7.4 kJ mol^{-1}, exceeding the analogous contribution into the vaporization enthalpy by 1.75 kJ mol^{-1}. Evidently, the effect of the crystal field acting on the two hydrogen atoms in the CH_2 group of cyclopropane is higher than that on the CH group in benzene. The data of Table 7.10 show that the stabilizing effect of the crystal field in the cyclobutane, cyclopentane, and cyclohexane decreases with the growing number of CH_2 groups in the ring, and it should sharply diminish at the temperature of 298 K. Consequently, the charges on the carbon and hydrogen atoms in the cyclopropane essentially differ and exceed the charges in the cyclohexane. The outstanding behavior of the hydrogen atoms of the CH_2 groups of the cyclopropane is due to the electron density distribution in the ring ensuring the identical behavior of each hydrogen atom. It is remarkable that the contribution to the sublimation enthalpy of the CH_2 group is significantly less than the analogous contribution of the CH group of the benzene ring. The data on the sublimation enthalpies of the cyclic hydrocarbons (Table 7.11) cannot provide unambiguous choice of the energy contribution to the enthalpy characteristic from the CH_2 and CH groups. Detailed analysis of the vaporization enthalpy of the same compounds (Table 7.11) provides a possibility to conclude the following:

First, the contribution to the vaporization enthalpy of each CH_2 or CH group in the saturated and unsaturated cyclic hydrocarbons containing in the ring from four to eight carbon atoms practically does not depend on the number of these groups in the molecule.

Second, the contribution of CH groups in the unsaturated cyclic hydrocarbons is independent of the number of double bonds in the molecule.

Third, the contribution of CH_2 and CH groups in saturated and unsaturated hydrocarbons with double and triple bonds and in benzene is practically constant and can be taken to equal 5.55 kJ mol^{-1} (Table 7.11). Fourth, the most significant contribution to the enthalpy characteristic provides the hydrogen atom of CH_3 and CH_2 groups located in the *trans*-position; the other hydrogens provide essentially lesser contribution. These contributions have the key significance for they provide a possibility to establish the energy contributed by one hydrogen to the vaporization enthalpy and disregard the other two in the methylene group.

The contribution of the carbon atom to the enthalpy characteristic is equal to the difference between the energy contributed by the CH group (5.55 kJ mol^{-1}) and the hydrogen atom located in the *trans*-position (0.60, 1.0 kJ mol^{-1}), and amounts to 5.0 and 6.4 kJ mol^{-1} in the liquid and crystalline states, respectively. In sublimation of benzene, the contribution of the CH group is 7.40 kJ mol^{-1}, of the hydrogen atom is 1.0 kJ mol^{-1}, and of the carbon atom is 6.4 kJ mol^{-1}, respectively.

This energy value may correspond only to the specific intermolecular interaction involving the essentially unshared $2s^2$-electron pair of the carbon. Considering from this viewpoint the data on the contributions of the CH_3, CH_2, and CH groups, we should conclude that the carbon atoms in compounds with the double bond –C=C– (and obviously, with the triple bond –C≡C–, e.g., 1-hexyne) form specific interactions along the same mechanism as in the alkyl derivatives [19], via essentially unshared $2s^2$-electron pair of the carbon atom. In this connection, the specificity of the intermolecular interactions in alkanes becomes understandable, whose main point consists in the interaction of the essentially unshared $2s^2$-electron pair of the carbon atom transforming it into the pentacoordinate state. This is possible for the shift in electron density in the alkane chain results in inequality of the charges (although insignificant) on the odd and even carbon atoms in the alkyl chain or in the ring of the cyclic hydrocarbon. Therefore, the contacting molecules are always so located that the carbon atoms with different values of negative charge are present in the nearest environment, and then one of them acts as a donor and the other as an acceptor. The hydrogen atoms in these groups also have different charges, and thus they secure the maximum energy of the specific interaction and vaporization enthalpy. This mechanism should underlie the interaction of the corresponding groups in alkanes, alkenes, and obviously the molecules of cyclic hydrocarbons.

The sublimation enthalpies were measured with a sufficient precision at the standard conditions for two compounds (Table 7.12). The kinship of the chemical nature of the processes of vaporization, sublimation [12, 13], and specific intermolecular interactions in the liquid and solid compounds is an important argument supporting the identity of the nature of the interaction between the carbon atoms of

Table 7.12 Energies (kJ mol^{-1}) of hydrogen bonds and specific interactions in liquid and solid cyclic carboxylic acids

Acids	Formula	$\Delta_{vap}H^0(T)$ [7, 15]	T (K)	Bond energies		
				DH – O···H – O	DH–O → CH_2–CH	DH – O···H – C
Liquid						
Cyclopropanecarboxylic acid	$C_4H_6O_2$	58.9 ± 0.3	340	11.8	6.78	0.6 × 4
Cyclopropane	C_3H_6	19.9	312	–		
Benzoic acid	$C_7H_6O_2$	78.9	298	17.35	5.63	0.6 × 4
Benzene	C_6H_6	33.8	298			
Solid						
Cyclobutane-1,1-dicarboxylic	$C_6H_8O_4$	112.2 ± 07	–	11.4	8.83	1.0 × 6
Cyclobutane-1,2-dicarboxylic	$C_6H_8O_4$	120.0 ± 0.9	–	13.6	8.83	1.0 × 6
Cyclobutane	C_4H_8	25.2	272	–	–	–
Benzoic acid	$C_7H_6O_2$	89.0; 89.3	298	19.45	6.60 × 2	1.0 × 4
Benzene	C_6H_6	44.4	298			
1,3,5-Benzenetricarboxylic acid	$C_9H_6O_6$	159.4	573	12.60	6.60 × 6	1.0 × 12

the benzene ring and other classes of organic compounds. Taking into account the energy of the H-bond of low stability $D = O \cdots H - C$ (1.0) and interactions DH–C → H–C having the energy 1.00 and 0.60 kJ mol^{-1} in the crystalline and liquid states, respectively, we obtain the contribution to the enthalpy characteristic of the carbon atom of the benzene ring equal to 6.40 and 5.00 kJ mol^{-1}, respectively. The energy contribution of the same carbon atoms to the vaporization enthalpy amounts to 5.00 kJ mol^{-1}.

Therefore, in the thermodynamic calculations of the hydrogen bond energies in the carboxylic acids containing a cyclic fragment, the contributions should be accounted for of the carbon atom (5.0 and 6.4 kJ mol^{-1}) and of the hydrogen atom in the *trans*-position of the CH_3, CH_2, and CH groups (0.60 and 1.0 kJ mol^{-1}) in the liquid and solid states, respectively. Consequently, the carbon atom introduced into the structure of the benzoic acid with the carboxy group increases the enthalpy of vaporization and sublimation by 5.0 and 6.4 kJ mol^{-1}, respectively. Therefore, the number of substituted hydrogen atoms and their contribution to the enthalpy characteristic should be considered. Here, the estimation of the energy of specific intermolecular interactions of benzoic acid derivatives and of the cyclopropanecarboxylic and cyclobutanecarboxylic acids in the *trans*-position should be made clear.

The cyclopropanecarboxylic acid with a single hydroxy group forms two hydrogen bonds, two specific intermolecular interactions DH–O → CH_2–, and four hydrogen bonds of low stability $D = O \cdots H - CH$ (Fig. 7.15). However, the substituted hydrogen atom of the ring that has contributed to the vaporization enthalpy of cyclopropane 0.60 kJ mol^{-1} no more takes part in the interactions; therefore, this value should be subtracted from the vaporization enthalpy of the latter. The overall energy of all interactions is equal to the difference between the vaporization enthalpies of the cyclopropanecarboxylic acid and cyclopropane, which contains all the specificities of the interactions in its ring, save $DH_2 - C \rightarrow H - C = D = O \cdots H - C = 0.60\,\text{kJ mol}^{-1}$, and it is presented in (7.18a) that is transformed into (7.18b)

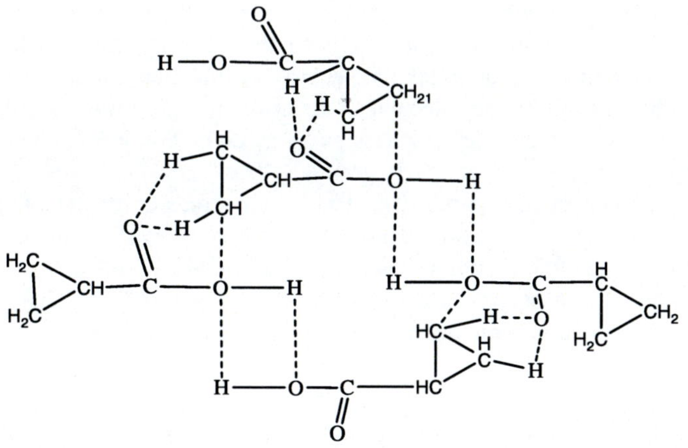

Fig. 7.15 Schematic picture of the liquid and solid structures of cyclopropanecarboxylic acid

$$2DH - O \cdots H - O + 2DH - O \rightarrow CH_2 + 4D = O \cdots H - CH$$
$$= \Delta_{vap}H^0(T)\text{cyc.pr.} - \Delta_{vap}H^0(T)\text{pr.} + DH_2 - C \rightarrow H - CH, \qquad (7.18a)$$

$$DH - O \cdots H - O = \frac{\Delta_{vap}H^0(T)\text{cyc.pr.} - \Delta_{vap}H^0(T)\text{pr.} + 2DH - O \rightarrow CH_2 + 3D = O \cdots H - CH}{2}. \qquad (7.18b)$$

The energy of the bond formed by the hydrogen from the *trans*-position of the CH_2 group and the carbonyl oxygen atom is 0.60 kJ mol^{-1}. We take the energy of the specific interaction DH–O → CH_2 as equal to the same type of interaction in the liquid diethyl ether (6.78 kJ mol^{-1}).

The result of the performed thermodynamic calculation (Table 7.12) gave the value of the hydrogen bond energy of 11.8 at 340 K, by 5.5 kJ mol^{-1} less than the energy of this bond in the propionic acid.

Thus, the cyclopropane fragment essentially affects the electron density distribution in the molecule of the cyclopropanecarboxylic acid and reduces the negative charge on the oxygen atom and positive charge on the hydrogen atom of the hydroxy group. At the same time, about 1.5-fold greater energy required for the cleavage of the specific interactions of the cyclopropane ring (19.3) at the vaporization of the cyclopropanecarboxylic compared to the energy of the stable hydrogen bond (11.8 kJ mol^{-1}) excludes the association of this acid in the vapor state.

The thermodynamic analysis of the solid cyclobutane-1,1-dicarboxylic acid and cyclobutane-1,2-dicarboxylic acid leads to interesting conclusions. The equal number of the bond vacancies in the molecules of these compounds and different location of the carboxy groups at one or two carbon atoms of the cyclobutane ring, respectively, govern the similarity of the structures of their liquid state. Each molecule forms four hydrogen bonds by two hydroxy groups and four specific intermolecular interactions by two methylene groups of the ring with the remaining free bond vacancies of the oxygen atoms of the hydroxy groups of two other contacting molecules. We again take the energy of this interaction as equal to the energy of the specific intermolecular interaction in the solid dimethyl ether. The oxygen atoms of the carboxy groups form eight H-bonds of low stability with the hydrogen atoms of the CH_2 groups (1.0 kJ mol^{-1}). Two such H-bonds compensate the energy of two substituted hydrogen atoms of the cyclobutane ring. The branched structure of the cyclobutane-1,1-dicarboxylic acid molecule suggests the formation of a network structure in the crystalline and liquid states (Fig. 7.16).

Cyclobutane-1,2-dicarboxylic acid forms analogous structures. The difference in these structures providing dissimilar hydrogen bond energies consists in the involvement of diverse methylene groups of the cyclobutane ring of these acids in the formation of specific interactions. The location of the carboxy groups at two carbon atoms of the ring in the molecule of the cyclobutane-1,2-dicarboxylic acid results in the identical shift in electron density from the carbon atoms of the CH_2 groups to the oxygen atoms of the hydroxy groups. Yet the location of both carboxy

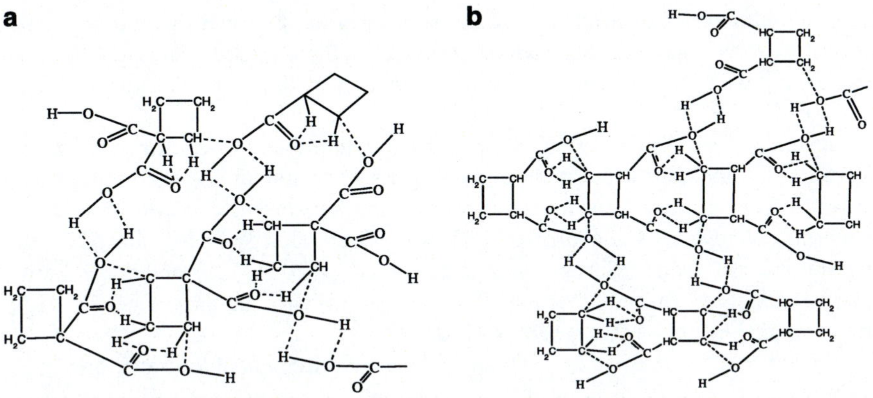

Fig. 7.16 Schematic picture of the liquid and solid structures of cyclobutane-1,1-dicarboxylic acid (**a**) and cyclobutane-1,2-dicarboxylic acid (**b**)

groups at the same carbon atom of the ring in the cyclobutane-1,1-dicarboxylic acid molecule leads to the shift in electron density to this carbon atom and further to the carboxy groups. Since this carbon atom cannot absorb as much electron density as two carbon atoms separately of the ring of the cyclobutane-1,2-dicarboxylic acid molecule, the oxygen atoms of the hydroxy groups of the latter molecule have enhanced negative charge and the respective hydrogens, the enhanced positive charge compared with the cyclobutane-1,1-dicarboxylic acid. Taking into consideration the obvious difference in the charges, we may conclude that in the crystal structure of the cyclobutane-1,2-dicarboxylic acid, the hydrogen bonds are more stable. The interactions of two substituted hydrogen atoms of the ring are reimbursed by the two hydrogen bonds of low stability. Taking into account the number and energy of the specific interactions and weak hydrogen bonds, we calculate the energy of the stable hydrogen bonds by (7.19)

$$\mathrm{DH-O\cdots H-O} = \frac{\Delta_{\mathrm{sub}}H^0(T)\mathrm{c.a.} - \Delta_{\mathrm{sub}}H^0(T)\mathrm{c.bu.} - 4\mathrm{DH-O} \rightarrow \mathrm{CH_2-C} - 6\mathrm{D{=}O\cdots H-CH}}{4}. \tag{7.19}$$

The hydrogen bond energies reported in Table 7.12 illustrate their stabilization in the cyclobutane-1,2-dicarboxylic acid. It follows from the results of the thermodynamic analysis that the lower stability of the hydrogen bonds in the liquid cyclobutane-1,1-dicarboxylic acid is due to the reduced negative charge on the oxygen atom and the reduced positive charge on the hydrogen atom of the hydroxy groups. It may be concluded that the search for the cause-and-effect relation during the thermodynamic analysis ensures the deeper understanding of the features of the specific interactions compared to the steric effect explaining everything but concealing the essence. A significant feature of the cyclobutane-1,1-dicarboxylic and cyclobutane-1,2-dicarboxylic acids consists in the fact that the energy of the cleavage of the specific interactions formed by the cyclobutane ring (25.2) is approximately two times greater than the energy of the stable hydrogen bonds (11.4 and 13.6 kJ mol^{-1}), thus excluding the association of these acids in the vapor.

Armitage and Gray [20] suggested that the liquid–vapor equilibrium of the benzoic acid includes dimerization. Analogous understanding was formulated in [20–26]. The results of the investigations of the benzoic acid sublimation using a thermal conductivity manometer [21], torsion effusion [22], effusion [27–30], transpiration [31], and calorimetry were discussed in [20, 31, 32], and the enthalpy characteristics of the sublimation were shown to vary in the range 88–92.5, and Cox [33] extended this range to 94 kJ mol^{-1}.

The existence of these reliable studies on the measurement of the vapor pressures of benzoic acid performed by different research teams using Knudsen method, torsion, effusion, calorimetric, and other methods supports the above assumption. The examination of the benzoic acid sublimation in the sublimation-calorimetric system incorporating Calvet microcalorimeter [34] afforded the value of the sublimation enthalpy in the range 88.9 $\pm$ 0.3–89.3 $\pm$ 0.3 kJ mol^{-1} in the good agreement with the results of earlier studies. These results confirmed the presence in the vapor of the benzoic acid of the monomer and dimer forms in the studied temperature range and evaluated the dissociation enthalpy of the dimers (68 $\pm$ 5 kJ mol^{-1}). Taking into account that in the benzoic acid dimer two hydrogen bonds are present (Fig. 7.17), the energy of each of these bonds equals 34 $\pm$ 5 kJ mol^{-1}.

The high enthalpy characteristic of the dimer dissociation and the high measurement error exceeding 15-fold than in the measurement of the sublimation enthalpy of the benzoic acid results from the low partial content of the dimers in the vapor phase. The enthalpy of the dimer dissociation significantly exceeding the experimental value of the overall sublimation enthalpy of monomers and dimers reflects the low degree of association in the vapor. The partial pressure of the dimeric form decreases with increasing temperature and overall pressure. Therefore, the semilog plot of the dimeric form's pressure should be curvilinear with the maximum value at low temperature. Consequently, the calculation of the sublimation enthalpy of the dimer from the semilog plot is impossible. Only these problems were encountered by the evaluation of the thermodynamic characteristics of sublimation [29] of monomer and dimer forms of the benzoic acid and the lack of these data in the literature. Activation of the dissociation process of the dimeric molecules results in a decreased slope of the semilog plot corresponding to the overall sublimation process that leads to the underestimated enthalpy characteristics. Therefore,

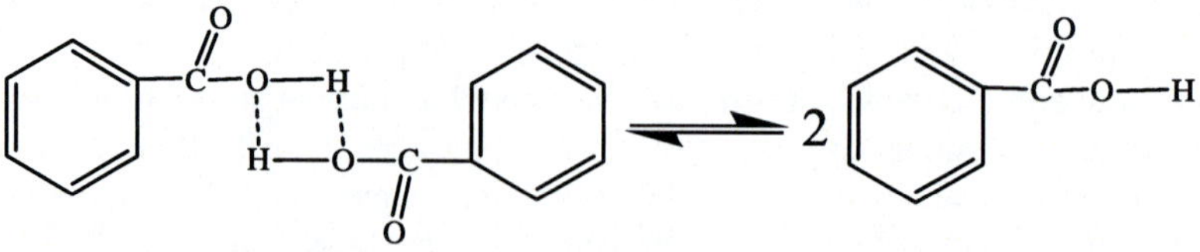

Fig. 7.17 Dissociation of dimeric molecule of benzoic acid

the sublimation enthalpy of benzoic acids published in the above studies (88–94 kJ mol^{-1}) is the lower limiting value.

The relations between the enthalpy characteristics of the vaporization at the association in vapor can be demonstrated by the example of alkylaluminums [35–37]. The trimethylaluminum dimer goes over into the vapor with the least enthalpy value for two specific interactions Al–CH_3 → Al remaining intact (Table 5.1). The going over to the vapor of monomers requires the cleavage of all existing specific interactions, and in this case, the enthalpy characteristic is high, and the overall vaporization process of dimers and monomers with a small fraction of monomers results in the intermediate enthalpy value. Nearly complete dimerization in vapor corresponds to the dissociation enthalpy equal 88.4 kJ mol^{-1}; consequently, the dissociation of one bond formed by the essentially unshared $2s^2$-electron pair of the pentacoordinate carbon of the methyl group requires 42.4 kJ mol^{-1}. The dissociation of triethylaluminum dimer requires less energy and at about equal partial fraction of monomeric and dimeric forms that results in other relations in the vaporization enthalpies.

Let us turn now to establishing the hydrogen bond energy in the liquid and solid benzoic acid. Its liquid structure is a network consisting of chains connected by two hydrogen bonds, two specific intermolecular interactions between the oxygen atom of the hydroxy group and the CH group of the benzene ring, and two weak hydrogen bonds between the carbonyl oxygen atom and the hydrogen atom of the CH group of the ring (Fig. 7.18). The nets are located in layers over each other with stacking position of the benzene rings.

The energy expenditure related to the benzene ring required for destruction of the liquid or crystalline state of benzene and benzoic acid at vaporization or sublimation has close endothermic effect related to the cleavage of the specific interactions formed by the CH groups. Therefore, the contribution to the enthalpy characteristic of the carboxy group is equal to the difference between the vaporization enthalpies of the benzoic acids and benzene minus the energy contribution of the substituted hydrogen atom (0.60 and 1.10 kJ mol^{-1} in the liquid and solid benzene, respectively).

The insignificant difference in the vaporization enthalpy of saturated and unsaturated compounds is a necessary and sufficient condition for taking the energy of the specific intermolecular interactions DH–O → CH_3, DH–O → CH_2, and DH–O → CH in the liquid and solid carboxylic acids equal to the energy of the specific intermolecular interactions in the corresponding ethers. At the same time, the

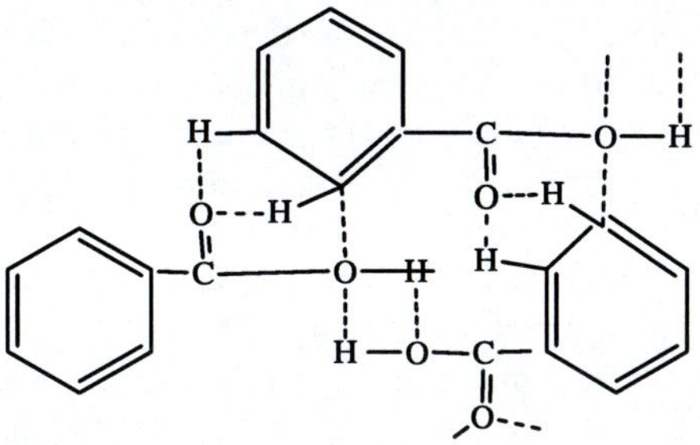

Fig. 7.18 Schematic picture of the liquid structure of benzoic acid

energies of the weak hydrogen bonds $=O\cdots H-C$ in the liquid and crystalline compounds (0.60 and 1.0 kJ mol^{-1}) outweigh the energy of the interactions of the substituted hydrogen atoms of the benzene rings. We take the energy of the specific interaction of the CH group of the benzene ring with the oxygen of the hydroxy group as equal to the energy of the interaction in the liquid or solid dimethyl ether (Table 7.12).

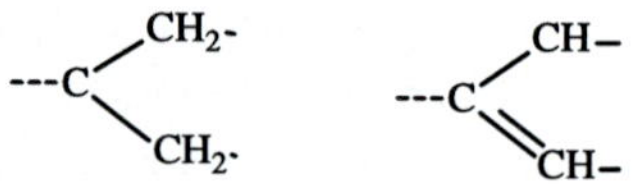

The hydrogen bond energy is estimated by the equation analogous to that used in the calculation of the hydrogen bond energy in the cyclopropanecarboxylic acid (7.20).

$$2DH-O\cdots H-O+2DH-O\rightarrow C-H=\Delta_{\rm sub}H^0(T)\text{bz.a.}-\Delta_{\rm sub}H^0(T)\text{bz.} \tag{7.20}$$

Hence in keeping to the method of the additive schemes of the chemical thermodynamics, the hydrogen bond energies at two different sublimation enthalpies equal 89.7 and 94 kJ mol^{-1}, and at vaporization enthalpy of liquid benzoic acid 78.9 kJ mol^{-1} amount to 15.10, 17.20, and 17.35 kJ mol^{-1}, respectively. It is perhaps the only case where the hydrogen bond energy in the solid state is equal or less than that in the liquid state. Consequently, the experimentally obtained enthalpy characteristics of the benzoic acid sublimation are essentially underestimated due to the errors in the procedure of processing the partial vapor pressures. It should also be noted that the melting enthalpy of benzene (10.8 kJ mol^{-1}) equals that of the benzoic acid (10.8 kJ mol^{-1}) that corresponds to identical stabilization of these compounds by the crystal field. This means that the stabilization of the benzoic acid occurs only by the action of the crystal field of benzene, and the carboxy group takes no part in the stabilization. It is, therefore, curious that disregarding the numerous data on the vaporization and sublimation enthalpies of benzene, still the reliable experimental results are lacking [7, 15, 16].

Taking into account the cited facts and the vaporization of dimeric molecules accompanied by the cleavage of all specific interactions of low stability and the weak hydrogen bonds, the stabilization of the hydrogen bond of the dimeric molecule (34 $\pm$ 5) at a decrease in the coordination number of the molecule is natural [6]. It is also sufficient for going into vapor of the benzoic acid dimer and its stable existence in the definite temperature range. At the same time, their partial content over the liquid benzoic acid is insignificant and weakly affects the vaporization enthalpy value, and consequently the established hydrogen bond energy is a real value.

The molecule of 1,3,5-benzenetricarboxylic acid forms in the crystalline and liquid states six hydrogen bonds. The remaining bond vacancies of the three oxygen atoms of the hydroxy groups form the specific intermolecular interactions with the molecules of the nearest surrounding whose number therefore doubles.

The carbonyl oxygen atoms form weak hydrogen bonds with the hydrogen atoms of the CH groups which outweigh the energy of the interactions of the substituted hydrogen atoms of the benzene rings. It permits a conclusion that the sum of all interactions may be established from the difference of the sublimation enthalpies of 1,3,5-benzenetricarboxylic acid and benzene. Employing the already taken energies of the specific interactions in the calculation of the hydrogen bond energy of the benzoic acid (7.20) and accounting for three substituted benzene hydrogen atoms, we estimated by (7.21) the hydrogen bond energy in the solid 1,3,5-benzenetricarboxylic acid (Table 7.11).

$$\begin{aligned} & 6\mathrm{DH-O\cdots H-O} + 6\mathrm{DH-O\cdots C-H-} + 9\mathrm{D{=}O\cdots H-C} \\ & = \Delta_{\mathrm{sub}}H^0(T)\mathrm{bz.a.} - \Delta_{\mathrm{sub}}H^0(T)\mathrm{bz.} \end{aligned} \quad (7.21)$$

The obtained value of the energy of this bond at 573 K (10.9 kJ mol^{-1}) should essentially increase at reducing it to the standard conditions due to the temperature dependance. Note that the respective value at the standard conditions should be affected both by the destabilizing effect of the growing number of the carboxy group and, on the contrary, by the stabilization due to the symmetrical location of the substituents in the benzene ring.

7.9.2 *Methylbenzoic Acids*

The results of the thermodynamic analysis of benzoic acid provide a possibility to begin the analysis of its methyl derivatives. The experimentally established sublimation enthalpies of these compounds provide important information on the hydrogen bonds and the specific intermolecular interactions. The substitution of a hydrogen in the benzene ring by a methyl group led to the essentially higher sublimation enthalpy of the monomethylbenzoic acid compared to the benzoic acid (Table 7.13). Consequently, the methyl group provides for the redistribution of the electron density, increasing the negative and positive charges on the carbon and hydrogen of the carboxy group, respectively. Therewith an important feature consists in the conservation of the structure of the solid benzoic acid (Fig. 7.19) where the specific intermolecular interaction H–O → C–H is replaced by the interaction H–O → CH_3–C formed by the essentially unshared $2s^2$(c)-electron pair of the carbon atom of the methyl group.

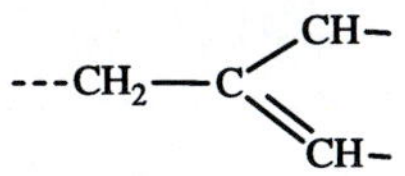

The significant growth of the sublimation enthalpy (by 15–18 kJ mol^{-1}) indicates the stabilization of the hydrogen bond.

Table 7.13 Energies (kJ mol^{-1}) of hydrogen bonds and specific interactions in solid methylbenzoic acids (298 K)

Methylbenzoic acid	Formula	$\Delta_{sub}H^0(T)$ [7, 15]	$DH-O\cdots H-O$
$\Delta_{sub}H^0(T)bz = 44.4$,	$DH-O \rightarrow CH = 7.25$,	$D=O \rightarrow CH_3 = 1.6$,	$D-O\cdots H-CH = 1.0$
2-Methylbenzoic	$C_8H_8O_2$	95.9 ± 0.1	17.00
3-Methylbenzoic	$C_8H_8O_2$	97.0 ± 0.3	17.50
4-Methylbenzoic (toluic)	$C_8H_8O_2$	98.3 ± 0.3	18.20
2,6-Dimethylbenzoic	$C_9H_{10}O_2$	99.1 ± 0.3	16.80
3,5-Dimethylbenzoic	$C_9H_{10}O_2$	102.2 ± 0.3	18.40
2,4-Dimethylbenzoic	$C_9H_{10}O_2$	103.5 ± 0.3	19.0
2,3-Dimethylbenzoic	$C_9H_{10}O_2$	104.6 ± 0.3	19.80
2,5-Dimethylbenzoic	$C_9H_{10}O_2$	105.0 ± 0.6	19.80
3.4-Dimethylbenzoic	$C_9H_{10}O_2$	106.4 ± 0.3	20.50
2,4,6-Trimethylbenzoic	$C_{10}H_{12}O_2$	103.6 ± 0.3	17.40
2,3,6-Trimethylbenzoic	$C_{10}H_{12}O_2$	104.4 ± 0.2	17.80
2,3,5-Trimethylbenzoic	$C_{10}H_{12}O_2$	106.7 ± 0.3	18.90
2,3,4-Trimethylbenzoic	$C_{10}H_{12}O_2$	109.3 ± 0.3	20.50
2,4,5-Trimethylbenzoic	$C_{10}H_{12}O_2$	109.8 ± 0.5	20.40
3, 4,5-Trimethylbenzoic	$C_{10}H_{12}O_2$	111.0 ± 0.5	21.10
2,3,5,6-Tetramethylbenzoic	$C_{11}H_{14}O_2$	106.1 ± 0.3	17.30
2,3,4,6-Tetramethylbenzoic	$C_{11}H_{14}O_2$	109.7 ± 0.5	19.10
2,3,4,5-Tetramethylbenzoic	$C_{11}H_{14}O_2$	115.9 ± 0.6	22.20
Pentamethylbenzoic	$C_{12}H_{16}O_2$	113.4 ± 0.3	19.60
$\Delta_{sub}H^0(T)bz = 44.4$,	$DH-O \rightarrow CH = 9.10$,	$D=O \rightarrow CH_3 = 1.6$,	$DH-O \rightarrow CH = 9.10$,
2-Ethylbenzoic	$C_9H_{10}O_2$	101.1 ± 0.4	18.20
3-Ethylbenzoic	$C_9H_{10}O_2$	99.2 ± 0.4	17.65
4-Ethylbenzoic	$C_9H_{10}O_2$	98.4 ± 0.2	17.30
3,5-Diethylbenzoic	$C_{11}H_{14}O_2$	104.1 ± 4.7	18.45
2-Isopropylbenzoic	$C_{10}H_{12}O_2$	101.0 ± 0.4	17.40
4-Isopropylbenzoic	$C_{10}H_{12}O_2$	101.1 ± 0.3	17.45
3-Isopropylbenzoic	$C_{10}H_{12}O_2$	104.1 ± 0.3	18.90

$$2\text{-Methylbenzoic acid } (95.9) < 3\text{-Methylbenzoic acid } (97.0) <$$
$$4\text{-Methylbenzoic acid } (98.3\,\text{kJ mol}^{-1}).$$

The growth of the sublimation enthalpy in the above series is caused by the relative position of the methyl and carboxy groups in the benzene ring; namely, the stabilization of the hydrogen bond increases with increasing distance between these groups and attains maximum in the 4-methylbenzoic acid.

The capability of the carboxy group to form two hydrogen bonds, and two specific intermolecular interactions by the oxygen atom of the hydroxy group and additionally by two electron pairs of the carbonyl oxygen atom to form four H-bonds of low stability with the molecules of the nearest surrounding makes it possible to reveal the cause of the unequal energy contribution to the sublimation enthalpy of each successive methyl group substituting the hydrogen atom of the benzene ring. The schematic picture of the structure of solid 2-methylbenzoic acid (Fig. 7.19) illustrates the formed interactions and underlies the calculation of the hydrogen bond energy by (7.22)

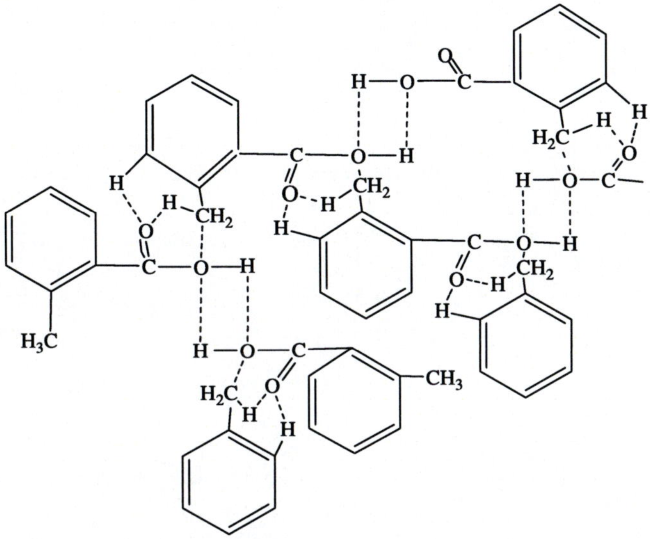

Fig. 7.19 Schematic picture of the liquid structure of 2-methylbenzoic acid

$$2DH-O\cdots H-O+2DH-O\rightarrow CH_2-C+4D=O\cdots$$
$$H-C=\Delta_{sub}H^0(T)\text{bz.a.}-\left(\Delta_{sub}H^0(T)\text{bz}-2DH-C\rightarrow H-C\right). \qquad (7.22)$$

The energy of the hydrogen bond of the low stability $DC=O\cdots H-C-$ outweighs the contribution to the sublimation enthalpy of the substituted hydrogen atom of the benzene ring; therefore, the calculation is performed by (7.23)

$$2DH-O\cdots H-O=\Delta_{sub}H^0(T)\text{bz.a.}-\left(\Delta_{sub}H^0(T)\text{bz}-2DH-C\rightarrow H-C\right)-$$
$$2DH-O\rightarrow CH_3-C-4D=O\cdots H-CH_n, \qquad (7.23)$$

where n is the number of hydrogen atoms.

The substitution of the hydrogen atom in the benzene ring by the methyl group increases the negative charge of the carbon atom of the ring and the positive charge of the carbon and hydrogen atoms of the methyl group, approaching the energy of the specific interaction formed by the three adjacent carbon atoms to the energy of the analogous interaction in solid or liquid dipropyl ether (11.60, 8.92 kJ mol^{-1}).

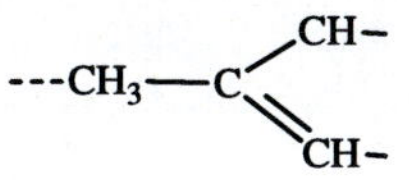

The results of the performed calculations (Table 7.13) of the hydrogen bond energies in the solid monomethylbenzoic acids illustrate the stabilization of the hydrogen bonds compared to the hydrogen bonds in the benzoic acid. The stability

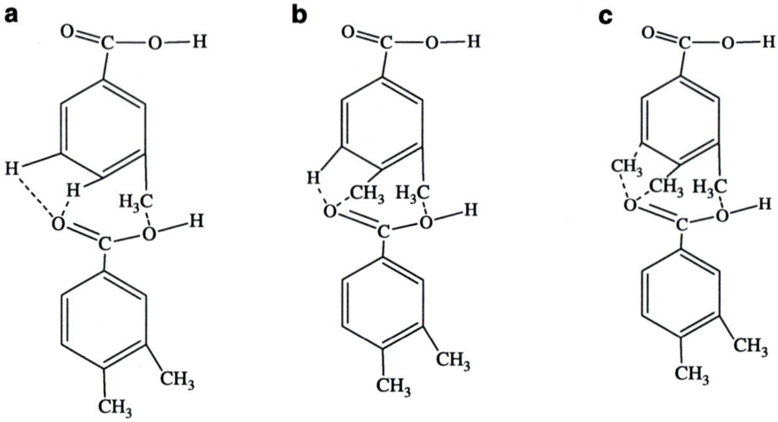

Fig. 7.20 Fragment of the solid structure of 3-dimethylbenzoic acid (**a**), 3,4-dimethylbenzoic acid (**b**), 3,4,5-trimethylbenzoic acid (**c**)

of the hydrogen bonds varies in the series of the methylbenzoic acids in the same sequence as the sublimation enthalpy. The stabilization of the hydrogen bonds in this series originates from the growing effect of the methyl group on the electron density shift with growing distance from the carboxy group.

The structure of solid dimethylbenzoic acids is similar to that of the methylbenzoic acids, and the difference consists in the growing number of weak specific intermolecular interactions $=O \rightarrow CH_3$ (Fig. 7.20) from two to four.

In the calculation of the hydrogen bond energy of the dimethylbenzoic acids, two stable specific intermolecular interactions $H\text{–}O \rightarrow CH_3$ and two weak interactions $=O \rightarrow CH_3$ between the molecules of the nearest surrounding should be taken into account (Fig. 7.20). However, the latter compensates the interactions of two substituted hydrogen atoms of the benzene ring. Therefore, the calculations follow (7.24), where the energy of the specific interactions formed by the carbonyl oxygen atom and the second methyl group is included.

$$DH\text{–}O\cdots H\text{–}O= \frac{\Delta_{sub}H^0(T)\text{mbz.a.} - \left(\Delta_{sub}H^0(T)\text{bz} - 3DH\text{–}C\rightarrow H\text{–}C\right) - 2DH\text{–}O\rightarrow CH_3 - 2D{=}O\cdots H\text{–}C - 2DH\text{–}C\rightarrow H\text{–}C}{2}. \quad (7.24)$$

The results of the thermodynamic calculations compiled in Table 7.13 show that the hydrogen bond energies in the dimethylbenzoic acids vary from 16.8 to 20.5 kJ mol^{-1} and exceed by 2.5 kJ mol^{-1} the most stable bond in the 4-methylbenzoic acid. Hence the increase in the number of methyl group to two in the dimethylbenzoic acids stabilizes the hydrogen bonds in agreement with the notion on the influence of the methyl groups on the increase in the negative and positive charges on the oxygen and hydrogen atoms of the hydroxy group, respectively.

The reduced energy of the hydrogen bond in the solid 2-methylbenzoic acid (17) compared to the most realistic energy of the hydrogen bond in the benzoic acid (17.2 kJ mol^{-1}) indicates that the location of the methyl group contiguously to the carboxy group somewhat hampers the shift in electron density to the hydroxy group. In this respect, 2,6-dimethylbenzoic acid is interesting for both its methyl groups are in the *ortho*-position to the carboxy group. The sublimation enthalpy of this compound exceeds the enthalpy characteristic of the 2-methylbenzoic acid only by 3.2 kJ mol^{-1}. This small difference in the enthalpy characteristics originates from the position of both methyl groups close to the carboxy group, and they equally take part in the redistribution of the electron density and decrease the negative charge of the oxygen and the positive charge of the hydrogen of the hydroxy group. The consequence of this fact is the formation of a hydrogen bond of reduced stability (16.80 kJ mol^{-1}). This shows that since the electron pair of the oxygen of the hydroxy group forms the interaction with the first methyl group, the second methyl group in keeping with the rule of the system striving to the energy minimum forms by its essentially unshared $2s^2$-electron pair of the carbon atom a specific interaction $=O \rightarrow CH_3$ with one of the electron pairs of the carbonyl oxygen, thus simultaneously decreasing the negative charge of the oxygen of the hydroxy group. Therewith a hydrogen bond of low stability $D = O \cdots H - C$ of energy 1.0 kJ mol^{-1} is replaced. This means that the difference between the sublimation enthalpies of 2-methylbenzoic acid and 2,6-dimethylbenzoic acid (3.2 kJ mol^{-1}) corresponds to the energy of two specific interactions $DC{=}O \rightarrow CH_3$ formed by the molecules of the nearest surrounding. Therefore, the energy of one specific interaction $D{=}O \rightarrow CH_3$ equals 1.6 kJ mol^{-1}. This figure is well consistent with the previously estimated energy of $DC{=}O \rightarrow CH_3 = 1.4$ kJ mol^{-1}. The good agreement between the estimation of the energy of this type of interaction before and now for the 2,6-dimethylbenzoic acid is very important, since

first, the energy value $D{=}O \rightarrow CH_3 = 1.4$ kJ mol^{-1} is an additional confirmation of the correct approach to the estimation of the energy of this specific interaction;

second, it substantiates the reliability of the obtained values of the hydrogen bond energies.

The difference between the sublimation enthalpies of the 3-methylbenzoic acid and 3.5-dimethylbenzoic acid is somewhat larger (5.2 kJ mol^{-1}). The first methyl group in the molecule of 3,5-methylbenzoic acid also occupies the most energy feasible bond vacancy of the oxygen atom of the hydroxy group and forms a stable specific interaction $DH{-}O \rightarrow CH_3$, leaving for the second methyl group the free bond vacancy of the electron pair of the carbonyl oxygen atom. As a result, it forms a weak specific interaction $D{=}O \rightarrow CH_3$ (1.6 kJ mol^{-1}). The remaining energy (5.2–3.2 kJ mol^{-1}) of 2.0 kJ mol^{-1} apparently is spent for the stabilization of the hydrogen bonds. Probably, the specific interaction also undergoes some stabilization of the secondary range. In the compound containing two methyl groups most remote from the carboxy group, the 3,4-dimethylbenzoic acid forms the same specific interactions, although the difference between the sublimation enthalpies of the 4-methylbenzoic acid and 3,4-dimethylbenzoic acid amounts to 8.1 kJ mol^{-1}.

However, most of this energy corresponds to the stabilization of the hydrogen bonds. On the contrary, the numerical values of the sublimation enthalpies of the dimethylbenzoic acids (Table 7.13) illustrate the small difference (1–2 kJ mol^{-1}) in the enthalpy characteristics of sublimation of 2,3-dimethylbenzoic acid and 2,4-dimethylbenzoic acid, and 2,4-dimethylbenzoic acid and 2,5-dimethylbenzoic acid caused by the analogous energy expenditure for the hydrogen bond stabilization.

Stabilization of the hydrogen bonds in the methylbenzoic acids and dimethylbenzoic acids is described by regular series:

$$\text{2 - Methylbenzoic acid } (17.0) < \text{3 - Methylbenzoic acid } (17.5) <$$
$$\text{4 - Methylbenzoic acid } (18.2\,\text{kJ mol}^{-1}),$$

$$\text{2,6 - Dimethylbenzoic acid } (16.8) < \text{3.5 - Dimethylbenzoic acid } (18.4) <$$
$$\text{2,4 - Dimethylbenzoic acid } (19.0) < \text{2,3 - Dimethylbenzoic acid } (19.8) =$$
$$\text{2,5 - Dimethylbenzoic acid } (19.8) < \text{3,4 - Dimethylbenzoic acid } (20.5\,\text{kJ mol}^{-1}).$$

Hence the common conception of the rigid benzene ring system and of the overlapping of the ionization regions of MO of s- and p-types presumably due to the high symmetry of C_6H_6 and preventing the mixing of the $1b_{1u}$(s)-orbital with the MO of the p-type [1] requires certain correction. This is connected with the influence of the methyl group in the methylbenzoic acids and its position in the ring on the shift in electron density in the molecule, which is so significant that the hydrogen bond energy grows from 17.0 in the 2-methylbenzoic acid to 18.2 in the 4-methylbenzoic acid. Even larger stabilization of the hydrogen bond occurs in the dimethylbenzoic acids that varies in a rather wide range (16.8–22.2 kJ mol^{-1}) depending on the location of the methyl groups in the benzene ring. Fundamentally important is the increase in the sublimation enthalpy of the 3,4-dimethylbenzoic acid compared with the 2,6-dimethylbenzoic acid by 7.3 kJ mol^{-1}, and all this energy is spent for the stabilization of the hydrogen bonds.

Now let us consider the trimethylbenzoic acids. In their solid structure, the molecules of the nearest surrounding form additionally two weak specific intermolecular interactions $=O \rightarrow CH_3$ involving the third methyl group and the carbonyl oxygen atom (Fig. 7.25). The energy of this type of intermolecular interaction also equals 1.6 kJ mol^{-1}. As a result, both electron pairs of the oxygen atoms are involved in the interaction, and four specific interactions of low stability arise whose energy should be taken into account in the thermodynamic calculations. The results of the performed thermodynamic calculations compiled in Table 7.13 and obtained using the equivalence of the energy contributions accounted for in (7.25) illustrate that in some cases, hydrogen bond stabilization occurs, although less significant than in dimethylbenzoic acids, save in the 3,4,5-trimethylbenzoic acid.

$$\text{DH} - \text{O} \cdots \text{H} - \text{O} = \Delta_{\text{sub}}H^0(T)\text{mbz.a.} - \left(\Delta_{\text{sub}}H^0(T)\text{bz} - 4\text{DH} - \text{C} \rightarrow \text{H} - \text{C}\right) -$$
$$2\text{DH} - \text{O} \rightarrow \text{CH}_3 - 4\text{D} = \text{O} \cdots \text{CH}_3. \qquad (7.25)$$

Further increase in the number of methyl groups in the methylbenzoic acid molecules to four and five causes the formation of interactions involving only the hydrogen atoms of the methyl groups having the energy of 1.6 kJ mol^{-1}. The equations for the calculation of hydrogen bond energy differ only in the number of existing interactions of carbon atoms of the benzene ring –C → H–CH_2 in the structures of the solid state, (7.26) and (7.27)

$$\text{DH}-\text{O}\cdots\text{H}-\text{O}= \frac{\Delta_{sub}H^0(T)\text{mbz.a.}-\left(\Delta_{sub}H^0(T)\text{bz}-5\text{DH}-\text{C}\to\text{H}-\text{C}\right)-2\text{DH}-\text{O}\to\text{CH}_3-4\text{D}{=}\text{O}\to\text{CH}_3-2\text{D}-\text{C}\cdots\text{H}-\text{CH}_2}{2}, \quad (7.26)$$

$$\text{DH}-\text{O}\cdots\text{H}-\text{O}= \frac{\Delta_{sub}H^0(T)\text{mbz.a.}-\left(\Delta_{sub}H^0(T)\text{bz}-6\text{DH}-\text{C}\to\text{H}-\text{C}\right)-2\text{DH}-\text{O}\to\text{CH}_3-4\text{D}{=}\text{O}\to\text{CH}_3-4\text{D}-\text{C}\to\text{H}-\text{CH}_2}{2}. \quad (7.27)$$

The established energies of the hydrogen bonds (Table 7.13) are described by the regular stabilization series depending on the methyl group's position in the benzene ring of the trimethylbenzoic acids and tetramethylbenzoic acids:

2,4,6 - Trimethylbenzoic acid (17.4) < 2,3,6 - Trimethylbenzoic acid (17.8) <
2,3,5 - Trimethylbenzoic acid (18.9) < 2,4,5 - Trimethylbenzoic acid (20.4) <
2,3,4 - Trimethylbenzoic acid (20.5) < 3,4,5 - Trimethylbenzoic acid (21.1 kJ mol^{-1}),

2,3,5,6 - Tetramethylbenzoic acid (17.3) < 2,3,4,6 - Tetramethylbenzoic acid (19.1) <
2,3,4,5 - Tetramethylbenzoic acid (22.2 kJ mol^{-1}).

The highest stability of the hydrogen bonds is attained in the following methyl derivatives of benzoic acid at the growing number of methyl groups:

4 - Methylbenzoic acid (18.2) < 3,4 - Dimethylbenzoic acid (20.5)
= 2,4,6 - Trimethylbenzoic acid (20.5)
< 2,3,4,5 - Tetramethylbenzoic acid (22.2 kJ mol^{-1}).

In the compounds with less stable hydrogen bonds at the growing number of methyl groups, the regular series is finished by the 2,4,5-trimethylbenoic acid

2,6 - Dimethylbenzoic acid (16.8) < 2 - Methylbenzoic acid (17.0) <
2,3,5,6 - Tetramethylbenzoic acid (17.3) <
2,4,5 - Trimethylbenzoic acid (17.4 kJ mol^{-1}).

In contrast, in the compounds with the highest stability of the hydrogen bond, the most stable hydrogen bond was found in 2,3,4,5-tetramethylbenzoic acid.

The results of the thermodynamic analysis of the methylbenzoic acids permit the following conclusions:

First, the growing number of methyl groups in the benzene ring leads to the stabilization of the hydrogen bonds.

Second, the carbonyl oxygen atom with its two electron pairs forms unstable specific intermolecular interactions $=O \rightarrow CH_3$ having the energy of 1.6 kJ mol^{-1}.

The replacement of the methyl group by ethyl in the methylbenzoic acids (Fig. 7.21) results in the reverse dependence of the sublimation enthalpy of the ethylbenzoic acids compared with the methylbenzoic acids.

$$4\text{-Methylbenzoic acid} < 3\text{-Ethylbenzoic acid} < 2\text{-Ethylbenzoic acid}.$$

This is due to the influence of the second carbon atom of the ethyl group on the shift in electron density depending on its position in the benzene ring (Table 7.13). Interestingly, in contrast to the ethylbenzoic acids, the sublimation enthalpies of the isopropylbenzoic acids vary depending on the substituent position in the benzene ring in the same order as the methylbenzoic acids.

The difference between the sublimation enthalpy of 3-ethylbenzoic acid and 3,5-ethylbenzoic acid of 4.9 kJ mol^{-1} originates from the presence in the crystal structure of the molecules of the nearest surrounding of two weak specific interactions $=O \rightarrow CH_3\text{–}CH_2$, and each of them can be estimated at 2.5 kJ mol^{-1}. The enhanced energy compared to that in the interaction of the methyl group $=O \rightarrow CH_3$ (1.6 kJ mol^{-1}) seems natural and is due to the presence of two carbon atoms in the ethyl group. Taking into account the existing specific interactions and hydrogen bonds in the crystalline ethylbenzoic acid and 3,5-ethylbenzoic acid, the thermodynamic calculations of the hydrogen bond energies should be performed along (7.28)

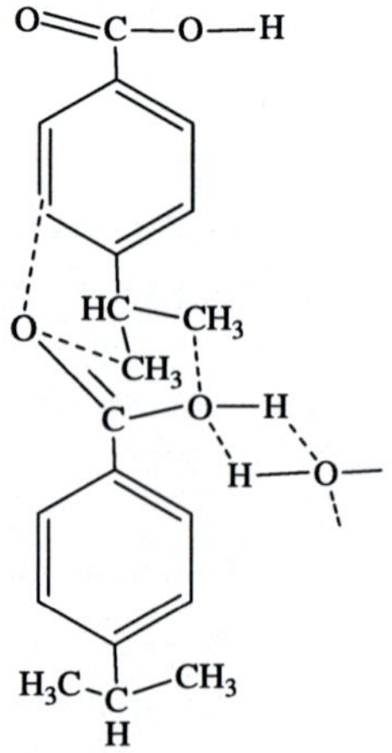

Fig. 7.21 Schematic picture of the solid structure of isopropylbenzoic acids

and (7.29), where all the existing specific intermolecular interactions are taken into consideration.

$$DH-O\cdots H-O=\frac{\Delta_{sub}H^0(T)\text{etbz.a.}-\Delta_{sub}H^0(T)\text{bz}-2DH-O\rightarrow CH_3-CH_2+2H-C}{2}\text{(ethylbenzoic acids)}, \quad (7.28)$$

$$DH-O\cdots H-O=$$

$$\frac{\Delta_{sub}H^0(T)\text{etbz.a.}-\Delta_{sub}H^0(T)\text{bz}-2DH-O\rightarrow CH_3-CH_2-2D{=}O\rightarrow CH_3-CH_2+3DCH_3}{2}\text{(3,5-diethylbenzoic acid)}. \quad (7.29)$$

The results of the performed calculations (Table 7.13) illustrate the small difference in the hydrogen bond energies in compounds with methyl and ethyl groups.

As shown in Fig. 7.21, in the isopropylbenzoic acids, the isostructural methyl groups form weak specific interactions of energy 1.6 kJ mol^{-1}. Therefore, the calculation of the hydrogen bonds energies should be performed by (7.30), accounting for all the existing specific intermolecular interactions

$$DH-O\cdots H-O=$$

$$\frac{\Delta_{sub}H^0(T)\text{ipbz.a.}-(\Delta_{sub}H^0(T)\text{bz}-2DH-C\rightarrow H-C)-2DH-O\rightarrow CH_3-CH_2-2D{=}O\rightarrow CH_3}{2}. \quad (7.30)$$

The results of the performed thermodynamic calculations (Table 7.12) provide a possibility to construct the regular series of changes in the energies of the hydrogen bonds:

$$\text{4-Ethylbenzoic acid (17.30)} < \text{3-Ethylbenzoic acid (17.65)} <$$
$$\text{2-Ethylbenzoic acid (18.20 kJ mol}^{-1}) < \text{3,5-Ethylbenzoic acid (18.40 kJ mol}^{-1}),$$

$$\text{2-Isopropylbenzoic acid (17.40)} \approx \text{4-Isopropylbenzoic acid (17.45)} <$$
$$\text{3-Isopropylbenzoic acid (18.40 kJ mol}^{-1}),$$

$$\text{3,5-Dimethylbenzoic (18.40)} = \text{3,5-Diethylbenzoic acid (18.40 kJ mol}^{-1}),$$

$$\text{2-Ethylbenzoic acid} > \text{2-Ethylbenzoic acid} > \text{2-Isopropylbenzoic acid;}$$

$$\text{3-Methylbenzoic acid (17.50)} < \text{3-Ethylbenzoic acid (17.65)} <$$
$$\text{3-Isopropylbenzoic acid (18.40 kJ mol}^{-1}),$$

$$\text{4-Ethylbenzoic acid (17.30)} < \text{4-Isopropylbenzoic acid (17.45)} <$$
$$\text{4-Methylbenzoic (18.20 kJ mol}^{-1}).$$

7.9.3 Hydroxybenzoic Acids

The feature distinguishing the hydroxybenzoic acids from methylbenzoic acids is the direction of the electron density shift in their molecules under the action of the oxygen atoms of the hydroxy and carboxy groups. Notwithstanding the rigid system of the benzene ring, the electron density shifts to the oxygen atoms. The higher energy of the hydrogen bonds in the carboxylic acids than in the alcohols suggests that this would remain in the hydroxybenzoic acids. The considerable difference in the sublimation enthalpies of hydroxybenzoic acids and dihydroxybenzoic acids related to the number and energy of the specific interactions (Table 7.14) is the consequence of the governing role of the position of the hydroxy group in the ring with respect to hydrogen bond stabilization.

Comparison of the sublimation enthalpy of these compounds shows that its variation corresponds to the following regular series:

$$2\text{-Hydroxybenzoic acid} < 4\text{-Hydroxybenzoic acid} < 3\text{-Hydroxybenzoic acid},$$

$$2,5\text{-Hydroxybenzoic acid} < 2,4\text{-Hydroxybenzoic acid} < 2,3\text{-Hydroxybenzoic acid},$$

$$2,6\text{-Hydroxybenzoic acid} < 3,5\text{-Hydroxybenzoic acid} < 3,4\text{-Hydroxybenzoic acid}.$$

Consequently, the position of the hydroxy group at the second carbon atom of the ring results in lower value of the sublimation enthalpy and, therefore, in the least stable hydrogen bonds. The location of the hydroxy group in position 3 of the ring corresponds to higher sublimation enthalpy. Hence the sublimation enthalpies are exclusively sensitive to the structure of the hydroxybenzoic acids and dihydroxybenzoic acids, and in an explicit form show the overall hydrogen bond energies. It is important to see here that the highest sublimation enthalpy values are obtained not at the largest distance between the hydroxy and carboxy groups in the hydroxybenzoic acid, but only when the hydroxy group is in position 3 of the benzene ring and in positions 3, 4 in the 3,4-dihydroxybenzoic acid. A similar dependence in the variation of the enthalpy characteristic was observed in the methyl derivatives

Table 7.14 Energies (kJ mol^{-1}) of hydrogen bonds in solid hydroxybenzoic acids (298 K)

Hydroxybenzoic acids	Formula	$\Delta_{sub}H^0(T)$ [7, 15]	$DH - O \cdots H - O$
2-Hydroxybenzoic	$C_7H_6O_3$	96.3 ± 0.5	6.4
4-Hydroxybenzoic	$C_7H_6O_3$	114.1 ± 0.7	10.8
3-Hydroxybenzoic	$C_7H_6O_3$	125.6 ± 0.7	13.5
2,5-Hydroxybenzoic	$C_7H_6O_4$	109 ± 3	4.7
2,6-Hydroxybenzoic	$C_7H_6O_4$	111 ± 7	5.0
2,3-Hydroxybenzoic	$C_7H_6O_4$	116.4 ± 4	6.0
2,4-Hydroxybenzoic	$C_7H_6O_4$	126 ± 4	7.5
3,5-Hydroxybenzoic	$C_7H_6O_4$	135 ± 9	9.0
3,4-Hydroxybenzoic	$C_7H_6O_4$	153 ± 9	12.0

of the benzoic acids. This corresponds to the condition of the electron density shift in the system ensuring maximum charges on the atoms of the hydroxy groups at their minimum reciprocal influence; therefore, the hydrogen bonds formed by the carboxy and hydroxy groups reach the maximum energy. Obviously, the small change in the energy of the specific interactions also exists. However, the accounting for all these special features in the thermodynamic calculations at the existence of unknown energies of two different types of hydrogen bonds and specific intermolecular interactions requires searching for more adequate additive approach.

Each hydroxy group of the hydroxybenzoic acid forms two hydrogen bonds, and the remaining free bond vacancies of the oxygen atoms form specific intermolecular interactions with the CH groups of the benzene rings of the molecules of the nearest environment. The additional hydroxy groups in the molecules of the hydroxybenzoic acid reduce the electron density not only on the oxygen atoms of the carboxy and hydroxy groups, but also on the carbon atoms of the benzene ring. Therefore, these atoms form weaker interactions compared to the CH groups in the methylbenzoic and benzoic acids. Therefore, the energy of the interaction H–O → CH in the solid hydroxybenzoic acids and the specific interaction in solid methyl ether are well consistent and should be estimated at 6.6 kJ mol^{-1}. The number of these interactions in the molecules of hydroxybenzoic acids and dihydroxybenzoic acids is two and three, respectively, and it doubles in the interacting molecules. The carbonyl oxygen forms a hydrogen bond of low stability with the hydrogen atom of the CH group (1.0 kJ mol^{-1}). They do not contribute any energy to the enthalpy characteristic, and this should be taken into consideration in the course of the thermodynamic analysis. The energy of the H-bonds of low stability outweighs the energy of the substituted hydrogen atoms; therefore, it should not be taken into account in thermodynamic calculations. In view of the bond vacancies of the molecules of the hydroxybenzoic acids, the structure of their solid state may be represented by the schematic picture shown in Fig. 7.22.

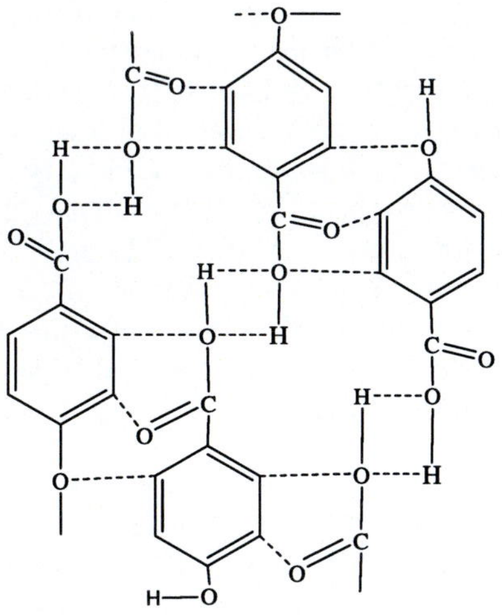

Fig. 7.22 Schematic picture of the solid structure of 4-dihydroxybenzoic acid

Thermodynamic calculations of the energies of two types of hydrogen bonds of dissimilar energy at the known energies of the specific interactions may provide the overall value and the mean energy of the hydrogen bonds formed by the hydroxy and carboxy groups. All the mentioned types of interactions are taken into account in (7.31)

$$\mathrm{DH{-}O{\cdots}H{-}O{=}}$$
$$\frac{\Delta_{\mathrm{sub}}H^{0}(T)\mathrm{hbz.a.}-(\Delta_{\mathrm{sub}}H^{0}(T)\mathrm{bz}-2\mathrm{DH{-}C{\to}H{-}C})-4\mathrm{DH{-}O{\to}CH}-4\mathrm{D{=}O{\cdots}H{-}C}}{4}. \tag{7.31}$$

The results of the thermodynamic calculations (Table 7.14) show a regular increase in the average energy of the hydrogen bonds in the series from 6.4 to 13.5 kJ mol^{-1}.

$$2\text{-Hydroxybenzoic acid }(6.4) < 4\text{-Hydroxybenzoic acid }(10.8) <$$
$$3\text{-Hydroxybenzoic acid }(13.5\,\mathrm{kJ\,mol^{-1}}).$$

The maximum energy of the hydrogen bond in the 3-hydroxybenzoic acid is due to the minimum reciprocal influence of the hydroxy and carboxy groups. It is just the reason why the mean hydrogen bond energy (13.5 kJ mol^{-1}) is close to the hydrogen bond energy in the solid benzoic acid (15.9 kJ mol^{-1}). Thus, the real energies of the hydrogen bonds formed by the hydroxy group of the solid 3-hydroxybenzoic and benzoic acids are not very different, although the former is apparently lower. At the same time, the essentially reduced average values of the hydrogen bonds in the solid 4-hydroxybenzoic acid and especially 2-hydroxybenzoic acid are the consequence of the reciprocal influence of the hydroxy and carboxy groups on the distribution of the electron density on the atoms of these groups, particularly at position 2 in the ring of the hydroxy group. Apparently, the most decreased energy of the hydrogen bond formed by the hydroxy group should be observed in the 2-hydroxybenzoic acid. The lowest average value of the hydrogen bond energy (6.4 kJ mol^{-1}) is possible on condition that the hydrogen bond formed by the carboxy group is also destabilized and possesses reduced energy. Taking as constant the hydrogen bond energy formed by the carboxy group in benzoic and 4-hydroxybenzoic acid, we obtain the energy of the hydrogen bond formed by the hydroxy group equal to 5.7 kJ mol^{-1}. This value cannot be regarded as real since it is lower than the mean energy of the hydrogen bond in the 2-hydroxybenzoic acid. Therefore, we should conclude that both dissimilar types of hydrogen bonds are destabilized by the reciprocal influence, and the hydrogen bond formed by the hydroxy group suffers a stronger destabilization.

The number of hydrogen bonds and specific interactions in the solid dihydroxybenzoic acids increases to six, but the number of the H-bonds of low stability $= \mathrm{O} \cdots \mathrm{H} - \mathrm{C}$ is retained (Fig. 7.23). Therefore, the mean value of energy of the two unlike types of hydrogen bonds should be calculated from the equation accounting for all formed interactions (7.32)

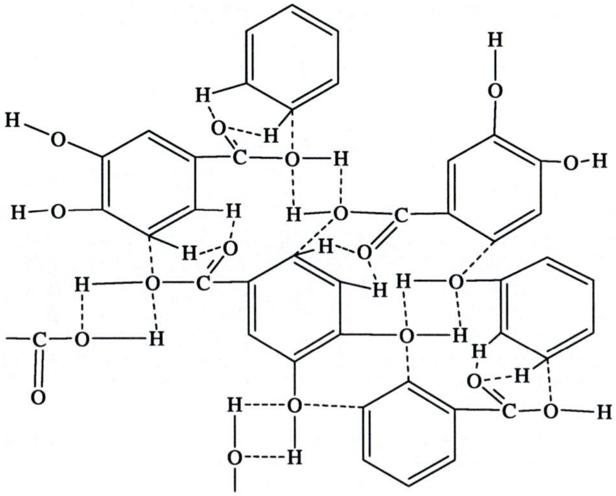

Fig. 7.23 Schematic picture of the solid structure of 3,4-dihydroxybenzoic acid

$$\mathrm{DH{-}O\cdots H{-}O} = \frac{\Delta_{\mathrm{sub}}H^0(T)\mathrm{hbz.a.} - (\Delta_{\mathrm{sub}}H^0(T)\mathrm{bz} - 2\mathrm{DH{-}C{\to}H{-}C}) - 6\mathrm{DH{-}O{\to}CH} - 4\mathrm{D{=}O\cdots H{-}C}}{6}. \quad (7.32)$$

The data compiled in Table 7.14 show that the maximum energies of the hydrogen bonds are observed in the dihydroxybenzoic acids at the hydroxy groups at the third and fourth carbon atom of the benzene ring (12.0 kJ mol^{-1}), and at the third and fifth atoms (9.0 kJ mol^{-1}). The data obtained lead to the following conclusions:

First, the carboxy and hydroxy groups produce a reciprocal destabilizing influence on the hydrogen bonds they form.

Second, the destabilizing effect increases with the growing number of the hydroxy groups and respectively of the hydrogen bonds.

7.9.4 Methoxybenzoic Acids

An important aspect favoring the understanding of the effect of the methoxy group oxygen atom in the benzoic acid derivatives on the hydrogen bond formed by the carboxy group is the establishment of the influence of the methoxy group on the energy of the hydrogen bonds in the solid methoxybenzoic acids. The published enthalpy characteristics of sublimation of the methoxybenzoic acids underlie the thermodynamic analysis aimed at the estimation of the hydrogen bond energy. The replacement of hydrogen in the hydroxy group of hydroxybenzoic acids by methyl, ethyl, or other alkyl group strongly changes the number of bond vacancies: Ten free

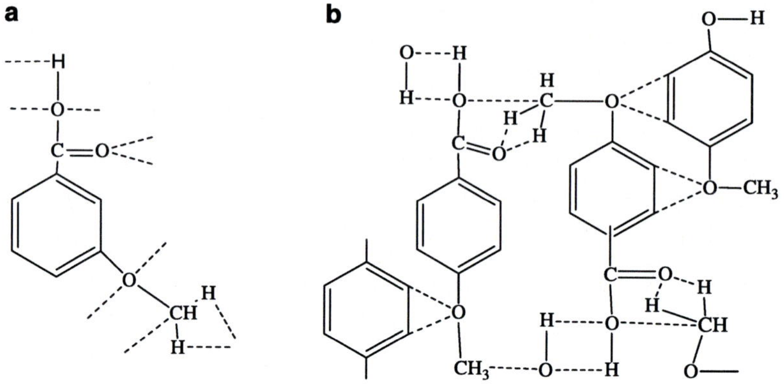

Fig. 7.24 Bonding vacancies (**a**) and schematic picture of the solid structure of 4-methoxybenzoic acid (**b**)

bond vacancies exist in the molecules of methoxybenzoic acids (Fig. 7.24). Specific interactions may be additionally formed by the carbon and hydrogen atoms of the CH groups of the benzene ring. The presence in the methoxy group of an ether oxygen atom with its high negative charge should affect the energy of its specific interaction formed with the carbon atom of the benzene ring in the methoxybenzoic acid molecule. The schematic structure of the solid phase of the monomethoxybenzoic acid (Fig. 7.24) illustrates the formation of two stable hydrogen bonds by the hydroxy groups $H - O \cdots H - O$, of four hydrogen bonds of low stability between the carbonyl group oxygen and hydrogen atoms of the methyl group (1.6 kJ mol^{-1}), of four interactions –O → CH of the oxygen atom of the methoxy group and the carbon of the benzene ring, and two specific interactions of the methyl group with the oxygen of the hydroxy group. The energy of the latter interaction equals the energy of the analogous interaction in the solid dimethyl ether (6.60 kJ mol^{-1}). Thus, the methoxybenzoic acid with a single methoxy group forms in the liquid and solid states a network structure cross-linked by the specific interactions and hydrogen bonds of different stability.

Note that in the thermodynamics of the hydrogen bond energies, the equal number of bond vacancies and interactions formed with the molecules of the nearest surrounding should be taken into account. It is also necessary to remember that in the benzene ring of the methoxybenzoic acids, a hydrogen substitution in the ring occurs; therefore, from the sublimation enthalpy of benzene, the energy should be subtracted equal to the contribution to the sublimation enthalpy by each substituted hydrogen atom $n{\cdot}D$H–C → H–C (1.00 kJ mol^{-1}). In the molecule of the dimethoxybenzoic acid with two methoxy groups, the number of specific interactions formed by the methyl groups with the oxygen atoms of their and carboxy groups increases to four. The free bond vacancies of the methoxy group oxygen of the contacting molecules form additionally two interactions –O → CH with the CH group of the benzene ring.

Hence taking into account the number of each existing specific interaction in solid and liquid mono- and dimethoxybenzoic acids, the hydrogen bond energies in monomethoxybenzoic acids should be calculated by (7.33):

$$\mathrm{DH{-}O\cdots H{-}O} = \frac{\Delta_{sub}H^0(T)\mathrm{mbz.a.} - \left(\Delta_{sub}H^0(T)\mathrm{bz} - 2\mathrm{DH{-}C{\rightarrow}H{-}C}\right) - 4\mathrm{D{-}O\cdots H{-}C} - 2\mathrm{D{-}O{\rightarrow}CH_3} - 4\mathrm{D{=}O\cdots H{-}CH_2}}{2}, \tag{7.33}$$

and those in the dimethoxybenzoic acids by (7.34):

$$\mathrm{DH{-}O\cdots H{-}O} = \frac{\Delta_{sub}H^0(T)\mathrm{mtbz.a.} - \left(\Delta_{sub}H^0(T)\mathrm{bz} - 3\mathrm{DH{-}C{\rightarrow}H{-}C}\right) - 6\mathrm{D{-}O\cdots H{-}CH} - 4\mathrm{DH{-}O{\rightarrow}CH_3} - 4\mathrm{D{=}O\cdots H{-}CH_2}}{2}. \tag{7.34}$$

Taking into account the increased number of the specific interactions >C–O → CH_3 and >C–O → CH in the solid 3,4,5-trimethoxybenzoic acid, its hydrogen bond energy is estimated by (7.35):

$$\mathrm{DH{-}O\cdots H{-}O} = \frac{\Delta_{sub}H^0(T)\mathrm{mtbz.a.} - \left(\Delta_{sub}H^0(T)\mathrm{bz} - 3\mathrm{DH{-}C{\rightarrow}H{-}C}\right) - 6\mathrm{D{-}O\cdots H{-}CH} - 4\mathrm{DH{-}O{\rightarrow}CH_3} - 4\mathrm{D{=}O\cdots H{-}CH_2}}{2}. \tag{7.35}$$

Among the existing specific interactions and hydrogen bonds, the energy values of the hydrogen bond $H - O \cdots H - O$ and of the specific interaction D–O → CH remain unknown. Therefore, we utilize the sublimation enthalpies of the methoxybenzenes for evaluating the energy of the specific interaction D–O → CH

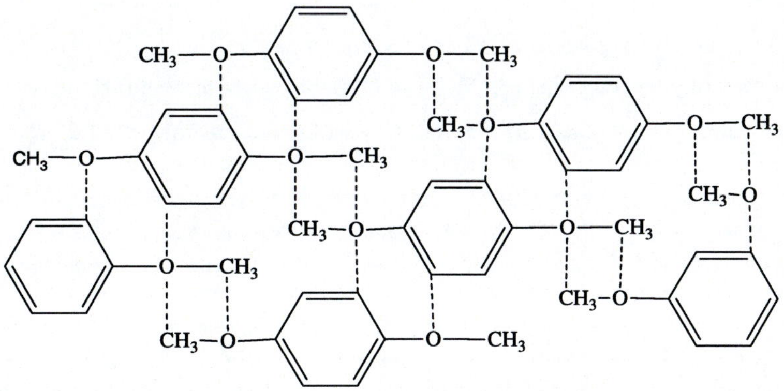

Fig. 7.25 Schematic picture of the solid structure of 4-methoxybenzoic acid

formed by the oxygen atom of the methoxy group and the CH group of benzene. The schematic picture of the solid 1,4-dimethoxybenzene given in Fig. 7.25 illustrates four interactions –O → CH_3 and four specific interactions D–O → CH formed by the remaining free bond vacancies of the methoxy group oxygen and the CH group of the benzene ring.

This permits to estimate the energy of this type of interaction from the difference of the sublimation enthalpy of 1,4-dimethoxybenzene and sum of the energies of n·D–O → CH_3 and the energy contribution to the enthalpy characteristic of the benzene ring:

$$\mathrm{D-O\rightarrow CH} = \frac{\Delta_{\mathrm{sub}}H^0(T)\mathrm{mtbz.a.} - (\Delta_{\mathrm{sub}}H^0(T)\mathrm{bz} - 2\mathrm{DH-C\rightarrow H-C}) - 4\mathrm{D-O\rightarrow CH_3}}{4}. \quad (7.36)$$

A similar equation (7.37) is written for the evaluation of the same specific interaction in the solid trimethoxybenzenes:

$$\mathrm{D-O\rightarrow CH} = \frac{\Delta_{\mathrm{sub}}H^0(T)\mathrm{mtbz.a.} - (\Delta_{\mathrm{sub}}H^0(T)\mathrm{bz} - 3\mathrm{DH-C\rightarrow H-C}) - 6\mathrm{D-O\rightarrow CH_3}}{6}. \quad (7.37)$$

The results of the performed calculations of the energy of the specific interaction D–O → CH (Table 7.15) were used in the calculation of the energy of the hydrogen bonds in the solid methoxybenzoic acids. The obtained energy values (Table 7.15) indicate a regular increase in the stability of the hydrogen bond in the methoxybenzoic acid depending on the position of the methoxy group in the benzene ring:

2-Methoxybenzoic acid (18.05) < 3-Methoxybenzoic acid (19.45) <
4-Methoxybenzoic acid (20.40 kJ mol^{-1}),

2,5-Dimethoxybenzoic acid (14.35) < 2,3-Dimethoxybenzoic acid (14.60) <
2,6-Dimethoxybenzoic acid (17.15) < 2,4-Dimethoxybenzoic acid (18.00) <
3,5-Dimethoxybenzoic acid (18.85) < 3,4-Dimethoxybenzoic acid (21.25 kJ mol^{-1}).

The estimated hydrogen bond energies confirm the existing notion on the stabilization of this bond in the series of monomethoxybenzoic acids and attaining the maximum value at the *para*-position of the methoxy group with respect to the carboxy group in the 4-methoxybenzoic acid (Table 7.15). The maximum values of the hydrogen bond energies in the dimethoxybenzoic acids are attained at the location of the two methoxy groups at 3 and 4 carbon atoms and with a little less stability in the 3,5-dimethoxybenzoic acid. However, the location at the same carbon atoms of the benzene ring of three methoxy groups in the molecule

Table 7.15 Energies (kJ mol^{-1}) of hydrogen bonds and specific interactions in the solid methoxybenzoic acids (298 K)

Methoxybenzoic acid	Formula	$\Delta_{sub}H^0(T)$ [16]	DH–O → CH_3	DC = O···H – CH_2 + DC = O···HC	DC–O → CH	DH – O···H – O
1,4-Dimethoxybenzene	$C_8H_{10}O_2$	84.1	6.60 × 4	–	3.30 × 4	–
1,2,3-Trimethoxybenzene	$C_9H_{12}O_3$	98.0 ± 0.3	6.60 × 6	–	2.80 × 6	–
1,3,5-Trimethoxybenzene	$C_9H_{12}O_3$	100.6 ± 1.4	6.60 × 6	–	3.40 × 6	–
2-Methoxybenzoic acid	$C_8H_8O_3$	104.7 ± 0.3	6.60 × 2	1.60 × 4	3.30 × 2	18.05
3-Methoxybenzoic acid	$C_8H_8O_3$	107.5 ± 0.4	6.60 × 2	1.60 × 4	3.30 × 2	19.45
4-Methoxybenzoic acid	$C_8H_8O_3$	109.8 ± 0.6	6.60 × 2	1.60 × 4	3.30 × 2	20.4
2,3-Dimethoxybenzoic acid	$C_9H_{10}O_4$	116.6 ± 0.3	6.60 × 4	1.60 × 4	3.30 × 4	14.60
2,4-Dimethoxybenzoic acid	$C_9H_{10}O_4$	123.4 ± 0,4	6.60 × 4	1.60 × 4	3.30 × 4	18.0
2,6-Dimethoxybenzoic acid	$C_9H_{10}O_4$	121.7 ± 0.4	6.60 × 4	1.60 × 4	3.30 × 4	17.15
2,5-Dimethoxybenzoic acid	$C_9H_{10}O_4$	116.1 ± 0.7	6.60 × 4	1.60 × 4	3.30 × 4	14.35
3,4-Dimethoxybenzoic acid	$C_9H_{10}O_4$	129.8 ± 0.8	6.60 × 4	1.60 × 4	3.30 × 4	21.25
3,5-Dimethoxybenzoic acid	$C_9H_{10}O_4$	127.1 ± 0.6	6.60 × 4	1.60 × 4	3.3 × 4	19.85
3,4,5-Trimethoxybenzoic acid	$C_{10}H_{12}O_5$	131.2 ± 0.8	6.60 × 6	1.60 × 4	3.4 × 6	15.4

of 3,4,5-trimethoxybenzoic acid results in the destabilization of the hydrogen bond to the least energy 15.40 kJ mol^{-1} among the methoxybenzoic acids under consideration.

It permits a conclusion that the energy of the hydrogen bond in the presence of three methoxy and one carboxy group is affected not only by the position of the methoxy groups in the ring; the presence of four substituents results in the appearance of the endocoordination effect between the oxygen atoms of the methoxy groups and in the decrease in the dissociation energy of the bond between the carbon and oxygen atoms >C–O, resulting in the destabilization of the hydrogen bond.

References

1. Nefedov VI, Vovna VI (1989) Electronic structure of organic and elementorganic compounds. Nauka, Moscow
2. Ohno K, Takano S, Mase K (1986) Penning ionization electron spectroscope molecules combining the C=O group aldehydes and carboxylic acids. J Phys Chem 90:2015–2019
3. Ohno K, Mase K (1986) Effect of through-space–through-bond interaction on electron distribution. J Phys Chem 90:2015–2019
4. Eyermann CI, Jally WL (1983) The use of core and valence of ionization potentials to quantity the bonding and antibonding character of molecular orbitals compounds of nitrogen and phosphorus. J Phys Chem 87:3080–3082
5. Tomoda S, Achida Y, Nomoto K (1983) Photoelectron spectroscopic study of simple hydrogen-bonded dimers. J Chem Phys 74:113–120
6. Baev AK (1987) Chemistry of gas-heterogenic systems of elementorganic compounds. Science and Technology, Minsk
7. Vasiliev IA, Petrov VM (1984) Thermodynamic properties of oxygen organic compounds. Chemistry, Leningrad
8. Zaichikov AM (2003) Mixing enthalpy of *N,N*-disubstittuted amides of formic and acetic acids. Russ J Phys Chem 77:617–622
9. Zaichikov AM (2005) Thermodynamic investigation of oxyethylated glycols and theirs mixings with tertiary amides of carboxylic acids. Russ J Gen Chem 75:1320–1328
10. Zaichirov AM (2008) Structural-thermodynamic parameters and intermolecular interactions in aqua solutions of amides formic acids. Russ J Struct Chem 49:299–308
11. Baev AK (1972) Complex formation ability halogenides of second – six groups periodical system. General and applied chemistry, vol 5. Vaisheshika Educating, Minsk, pp 35–51
12. Baev AK (1969) To problems of chemical nature of phase transformation general and applied chemistry, vol 1. Vaisheshika Educating, Minsk, pp 197–206
13. Baev AK (1969) Phase condition and complex formation ability of halogenide metals. General and applied chemistry, vol 1. Vaisheshika Educating, Minsk, pp 207–218
14. Pak C, Lee HM, Kim JC, Rim DW, Rim KS (2005) Theoretical investigation of normal to strong hydrogen bonds. Struct Chem 16:187–202
15. Chickos JS, Acree WE Jr (2003) Enthalpies of sublimation of organic and organometallic compounds, 1910–2001. J Phys Chem Rev Data 32:537–698
16. Chickos JS, Acree WE Jr (2002) Enthalpies of vaporization of organic and organometallic compounds, 1880–2002. J Phys Chem Rev Data 31:519–879
17. Mathieu J, Panico R (1972) Mecansnes reactionnels de chimie organique. Hermann, Paris
18. Ivanov EV, Puchovskie YuP, Abrosimov VK (2004) About formation of weak hydrogen bonds CH···O in liquid methanol. In: Abstract. IX international conference. The problem of solvation and complex formation in solutions, Plyos, Russia, p 345

19. Baev AK (1997) Thermodynamic analysis of structural transformation at systems Zn $(CH_3)_2$–$M(CH_3)_2$ (Te, S, Se). Russ J Inorg Chem 42:662–672
20. Armitage JW, Gray P (1962) Temperature coefficient vaporization heats and curvature lines of vapor pressure formic acid, acetates acid and another compounds formed dissociated vapor. Trans Faraday Soc 58:1746
21. de Krui CG, Oonk HA (1973) The determination of enthalpies of sublimation by mean of thermal conductivity manometer. J Chem Ing Tech 45:455
22. Stull DR, Westrum EF Jr, Sinke GC (1969) The chemical thermodynamic of organic compounds. Wiley, New York
23. de Kruif CG (1980) Enthalpies of sublimation and vapour pressures of 11 polycyclic hydrocarbons. J Chem Thermodyn 12:243
24. de Kruif CG, Kuipers T, van Miltenburg JC, Schaake RCF, Stevens G (1981) The vapour pressure of solid and liquid naphthalene. J Chem Thermodyn 13:1081
25. de Kruif CG, Schaake RCF, van Miltenburg JC, van der Klauw K, Blok JQ (1982) Thermodynamic properties of the normal alkanoic acids III. Enthalpies of vaporization and vapour pressures of 13 normal alkanoic acids. J Chem Thermodyn 14:791–798
26. Galis-van Ginkel CHD, Galis GHM, Timmermans CWM, de Kruif CG, Oonk HA (1978) Enthalpies of sublimation and dimerization in the vapour phase of formic, acetic, propanoic and butanoic acids. J Chem Thermodyn 10:1083
27. Malaspina L, Gigli R, Bardi G (1973) Microcalorimetric determination of the enthalpy of sublimation of benzoic acid and anthracene. J Chem Phys 59:387
28. Colomina LM, Laynez JL, Peres-Ossorio R, Turrion C (1972) Enthalpies of combustion and formation of six methyl esters of benzene carboxylic acids. J Chem Thermodyn Data 4:499–506
29. van Ginkel CHD, de Kruif CG, de Waal FEB (1975) The need for temperature control in effusion experiments (and application to heat of sublimation determination). J Phys E Sci Instrum 8:490
30. Colomina LM, Jimi P, Turrion C (1982) Thermodynamic properties of benzoic acid derivatives. X. Enthalpies of combustion and formation of 2,3-, 2,4-, 2,6-, 3,4- and 3,5-dimetoxybenzoic acid. Academic Press, London
31. Maraweiz E (1972) Vaporization enthalpy *n*-alkanes from C_{10} till C_{20}. J Chem Thermodyn Data 4:139–144
32. Cox JD (1964) The calorimetry of combustions and related reactions: organic compounds. Pure Appl Chem 8:143–156
33. Cox JD (1982) Notation for states and processes, significance of the word standard in chemical thermodynamics, and remarks on commonly tabulated forms of thermodynamic functions. Pure Appl Chem 54:1239–1250
34. Murata S, Sakiym M, Seki S (1982) Enthalpy of sublimation of benzoic acid and dimerization in the vapor phase in the temperature range from 320–370 K. J Chem Thermodyn Data 14:723–731
35. Baev AK, Vitalev SE (1989) Thermodynamic vaporization and association of triethylaluminium. Vesti Akad Navuk Belarusii 2:58–63
36. Vitalev SE, Baev AK (1993) Thermodynamic vaporization and association of trimethylaluminum. Russ J Izv Vischich Ucheb Zaved Chem Chem Technol 36:36–40
37. Baev AK (2001) Thermodynamic evaporation of alkyl compounds of elements second – six groups. Russ J Izv Vischich Ucheb Zaved Chem Chem Technol 44:3–13

Chapter 8
Specific Intermolecular Interactions of Hydrocarbons

8.1 Specific Interactions in Hydrocarbons with Tetrahedral Structure of Molecules

Thermodynamic analysis of the main classes of organic compounds is incomplete when the problem of bonding is not considered in the crystalline and liquid alkanes, alkenes, and alkynes, and the specificity of the intermolecular interactions in these homologous series is not revealed. The reason for this is the necessity in discussion of the failure of the sp^3-hybridization model of the carbon atom to demonstrate the most obvious contradictions of this model to the rules of the variation of the thermodynamic properties and the dissociation features of compounds. The hybridization concept advanced by Pauling resulted in the development of the structural chemistry of organic compounds. However, already in 1958 Talrose [1] reported a concept of cation $[CH_5]^+$ which later became the base of the carbocation chemistry. In 1961, Coylson [2] remarked that the hybridization could not be regarded as a "phenomenon" for the same reason why the resonance of various structures, e.g., of covalent and ionic structures, could not be considered a "phenomenon." Still Coylson stated that the hybridization concept provides a possibility to conserve the notion of a localized bond and coupling of the orbitals involved in the interactions of atoms. Fourteen years later, it was suggested in [3] that the hybridization concept may be used in compounds with the coordination number smaller or equal to four, ≥ 4. According to Nefedov and Vovna [4], the sp^3-hybridization is formally allowed in alkanes except for methane. Notwithstanding the progress in the development of the structural chemistry in the historical aspect, we substantiate in Chap. 2 the inconsistency of the sp^3-hybridization model of carbon atom for the development of the theory of specific intermolecular interactions and hydrogen bonds in organic and organoelemental compounds, in the condensed state and nonelectrolyte solutions where the carbon atom is present in the pentacoordinate state [5]. This point was further extended in [6–13]. The easy-for-comprehension notion of the sp^3-hybridization of the carbon atom made this model attractive and thus makes rejection of the model difficult notwithstanding its

A.K. Baev, *Specific Intermolecular Interactions of Organic Compounds*,
DOI 10.1007/978-3-642-21622-0_8, © Springer-Verlag Berlin Heidelberg 2012

contradiction to the rules of variation of the fundamental properties of organoelemental compounds of diverse groups and periods of the Periodic System. The shift in electron density from a hydrogen atom to a carbon atom in the organic compounds ensuring the positive and negative charges, and acceptor and donor properties of the hydrogen and carbon atoms, respectively, necessarily results in the formation in the nearest surrounding of the specific intermolecular interaction H–C → H–C, in particular, between the atoms of methane and benzene molecules. Just because of the presence of a positive charge on the hydrogen and a negative charge on the oxygen of the hydroxy group in methanol [14], the existence of hydrogen bonds of the type C–H···O–H possessing low stability was suggested. The formation of these bonds occurs by the same mechanism as that of the hydrogen bonds between the hydroxy groups of methanol. These interactions consequently affect the spectral characteristics [15]. Analysis of the experimentally observed difference in the spectral bands allowed a conclusion [16] that one such bond per molecule formed in the condensed state of methanol. The quantum-chemical calculations by the density functional theory indicated that the specific interaction C–H···O–H alongside the strong hydrogen bond H–O···H–O significantly affects the formation of methanol trimers in the vapor phase [17]. The comparison of the results of the quantum-chemical calculations of the structure of methanol monomer, the electron density on the atoms, and spectral characteristics confirmed the nonequivalence of the methyl group protons in the methanol molecule [18]. The characteristics corresponding to the proton located in the *trans*-position to the hydroxy group differ from the characteristics of the other two protons of the methyl group. The difference in the charges on the hydrogen atoms of methyl group is not an exclusive property of methanol. It is known for many organic compounds. However, a characteristic feature of methane is the presence of a single carbon atom possessing the negative charge, which forms the most stable specific interaction with a hydrogen atom of the molecule from the nearest surrounding in the liquid and solid states. Therefore, the hydrogen atoms with differing positive charges form between themselves interactions of low stability not exceeding in the energy the error in the measurement of the enthalpy characteristic. As a result, the methane molecules of the nearest surrounding form two specific interactions with the transfer of a part of the electron density from the essentially unshared $2s^2$-electron pair to the hydrogen atom H–C → H–C, providing for the carbon atom the pentacoordinate state (Fig. 8.1).

Therefore, the tetrahedral structures of the molecules of methane and 2,2-dimethylpropane in the condensed state would not differ in the types of specific interactions. The presence in the molecule of 2,2-dimethylpropane of four methyl groups results in the formation of eight specific interactions H–C → H–C where the carbon atom is also pentacoordinate.

Taking into account that the enthalpy of a phase transition is related to the number and energy of the ruptured specific interactions [19], we are able to estimate the energy of the specific interaction in question from the enthalpy characteristic of the vaporization by dividing the latter by the number of existing interactions, namely, $\Delta_{vap}H^0(T)/2 = D\text{C–H} \leftarrow \text{C–H}$ and $\Delta_{vap}H^0(T)/8 = D\text{C–H} \leftarrow \text{C–H}$ for methane and 2,2-dimethylpropane, respectively. The obtained values of the energy

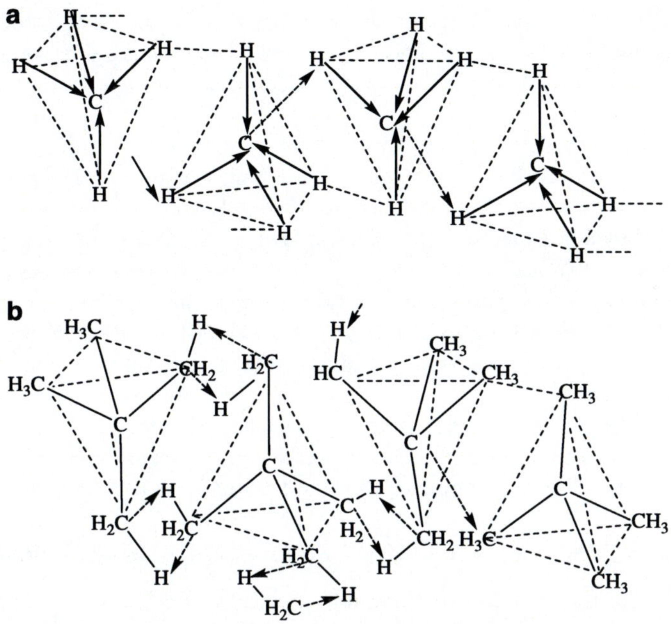

Fig. 8.1 Element of schematic structure of liquid and solid compounds with tetrahedral structure of molecule (**a**) methane, (**b**) 2,2-dimethylpropane

Table 8.1 Energies (kJ mol^{-1}) of specific interactions in compounds with tetrahedral structure of molecules

Compound	Formula	$\Delta_{sub}H^0(T)$ [20]		$\Delta_{vap}H^0(T)$ [21]		DH–C → H–C	
		Sublimation	*T* (K)	Vaporization	*T* (K)	Sublimation	Vaporization
Methane	CH_4	10.0	84	8.6	175	5.0	4.3
2,2-Dimethylpropane	C_5H_{12}	–	–	21.6	298	–	2.7

of the specific interaction for methane in solid and liquid states at the temperature of 84 and 175 K (Table 8.1) are essentially overestimated compared to the values at the standard conditions. At the same time, the low energy of the specific interaction (2.7 kJ mol^{-1}) in 2,2-dimethylpropane molecule is due to small positive and negative charges on the hydrogen and carbon atoms in this molecule.

8.2 Energies of Specific Interactions in Saturated Hydrocarbons

The analysis of the thermodynamic properties of the initial compounds in the homologous series of alkanes makes it possible to discuss the effect of the growing number of carbon atoms in the chain on the specificity of the formation of intermolecular interactions. First of all, note that the reliably established melting points of alkanes [22] fit to two plots of the dependence on the number of carbon atoms in the hydrocarbon chain $T_{\mathrm{m.p.}}(\mathrm{K}) = f(\mathrm{C}_n)$ (Chap. 1). The melting points of compounds with even and odd carbon atoms in the chain correspond to separate plots describing this dependence. The plot for compounds with even number of carbon atoms is situated at higher temperature, and a condition is valid that the difference between the melting points of compounds with odd and even number of carbon is always smaller than the difference between the melting points of compounds with even and odd number of carbon atoms. Inasmuch as the melting is a chemical process and it implies the partial rupture of the specific interactions and consequently with the distortion of the crystal structure [23, 24], then the value of the melting point corresponds to the stability of the specific interactions. Therefore, we are able to assume that the molecules with even and odd number of carbon atoms have different specificities inherent to each series in the shift in electron density. The influence of the even and odd carbon atoms appears also in the vaporization enthalpies [25] of organometal alkyls of tellurium, tin [26], and zinc [27, 28]:

$$\begin{gathered}\mathrm{Te(CH_3)_2} - (5.9) - \mathrm{Te(C_2H_5)_2} - (3.9) - \mathrm{Te}(n - \mathrm{C_3H_7})_2 - (7.9) \\ - \mathrm{Te}(n - \mathrm{C_4H_9})_2 - \left(6.1\,\mathrm{kJ\,mol^{-1}}\right) - \mathrm{Te}(n - \mathrm{C_5H_{11}})_2.\end{gathered}$$

In this connection, the dissociation energies of the C–C bonds [29] in the molecules of saturated hydrocarbons are illustrative (Fig. 8.2).

Disregarding the number of carbon atoms in the molecules of hexane and heptane, the rupture of the methyl groups requires a maximum energy. The energy of the cleavage of the second and fifth carbon atoms of the hexane chain containing even number of carbon atoms has an essentially lower value (342.5 kJ mol^{-1}) than the energy required for the cleavage of the second and sixth carbon atoms of heptane (371.4 kJ mol^{-1}). Yet the cleavage of the central carbon atoms of these compounds requires approximately the same energy. This character of the distribution of the energy of the C–C bond rupture is due to the distribution of the electron density in the molecules of these compounds, and ensures the high energy at the cleavage of the terminal methyl groups and different value of the energy required for the cleavage of the carbon atoms connected to the terminal methyl groups.

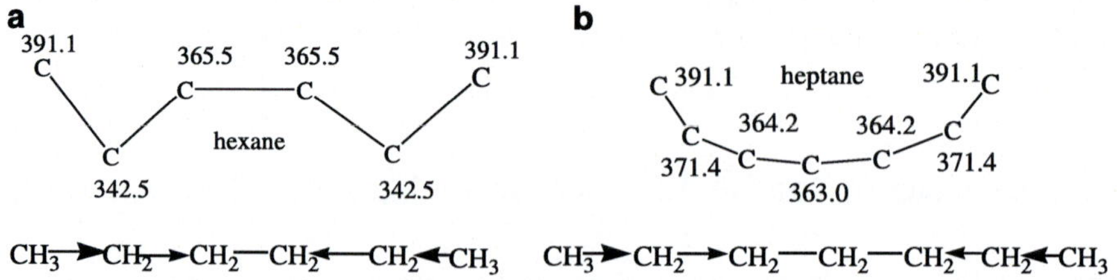

Fig. 8.2 Distribution of dissociation energy (kJ mol^{-1}) of the carbon–carbon bond in the molecule of hexane (**a**) and heptane (**b**)

These data also show that the small difference in the energy required for the cleavage of all subsequent central carbon atoms of the chain is due to the small difference in the negative charges on the even and odd carbon atoms.

The maximum energy of the methyl group's cleavage indicates that the electron density shifts from the central carbon atoms along the chain. The minimum energy required for the cleavage of the second carbon atom of hexane containing the even number of carbon atoms allows a conclusion that the maximum electron density is shifted from the carbon atom bound to it. At the same time, in the heptane molecule possessing odd number of carbon atoms, the central carbon atoms of the chain supply practically equal electron density to the second carbon atom from each end of the chain, and this atom shifts part of this density to the carbon atom of the methyl group.

At the same time, in the event of the odd number of carbon atoms in the chain, equal values of the energy required for the C–C bond rupture correspond to each of the three central carbon atoms, and consequently their negative charges are equal and related to the same stability of the intermolecular interactions they form. The maximum energy of the cleavage of the first C–C bond in hexane and heptane indicates that the atoms forming it have enhanced difference in the charges. In other words, in the liquid and solid states, the alkane molecules can mutually coordinate in such position that the carbon atoms with an enhanced negative charge would be near the carbon atom with the reduced charge. Hence the carbon atoms of the terminal methyl groups possessing the enhanced negative charge form the specific interaction along the donor–acceptor mechanism by transmission of part of the electron density from the essentially unshared $2s^2$-electron pair of the carbon atom with a large negative charge to the carbon atom with reduced negative charge on the similar unshared electron pair $DH_2C–H_3C \rightarrow CH_2–CH_3$ (Fig. 8.3).

In their turn, the central carbon atoms of the methylene groups in the chain make certain contribution into the enthalpy characteristic, taking part in the interactions with the hydrogen atoms of the molecules of the nearest surrounding in the liquid and solid states. The correlation of the energy parameters of molecules with the donor properties of the carbon atom and with the energies of the intermolecular interactions allows a conclusion that the specific interactions formed by the CH_2 groups contribute equal or similar energy to the enthalpy characteristics of these compounds. This rule is confirmed for the homologous series of all considered classes of organic compounds. In case of describing the enthalpy characteristics of a

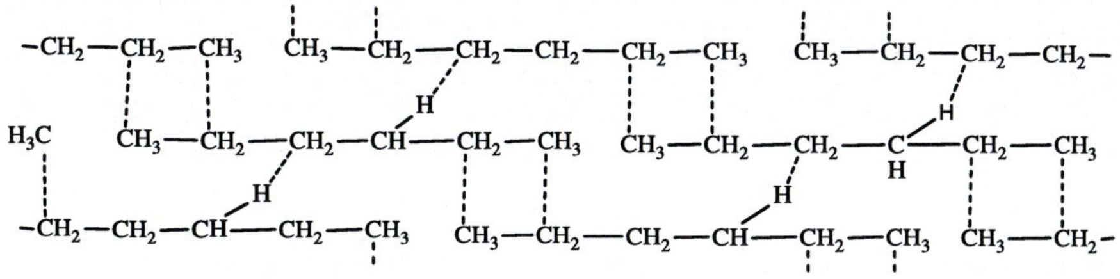

Fig. 8.3 Schematic picture of liquid structure of hexane

large number of members of a homologous series in dependence on the number of carbon atoms in the chain and providing the measurements were performed with sufficient accuracy for several series, a constant energy contribution from the additional methylene group was observed.

This rule is valid for alkanes (Fig. 8.4) where equal energy contribution of each methylene group is observed in the series hexane–dodecane, dodecane–octadecane, and octadecane–dodeicosane.

The interatomic distances in alkane molecules are equal to those shown in Fig. 8.5.

The character of the variation of the dissociation energy of the –C–C– bonds should remain in the molecules of butane and propane and should result in the different electron density and negative charge on the carbon atoms with enhanced value on the methyl groups.

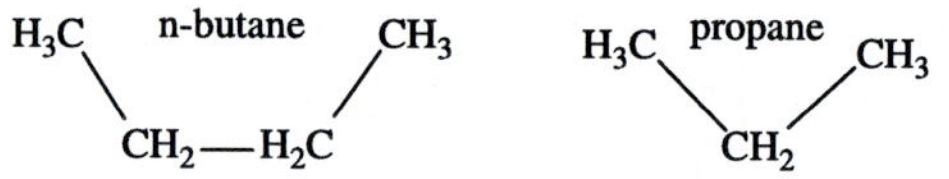

This allows a conclusion on the similarity of the interactions in the liquid butane and hexane. In the propane molecule having two methyl groups with an enhanced negative charge on the carbon atom and high donor properties, a single carbon atom

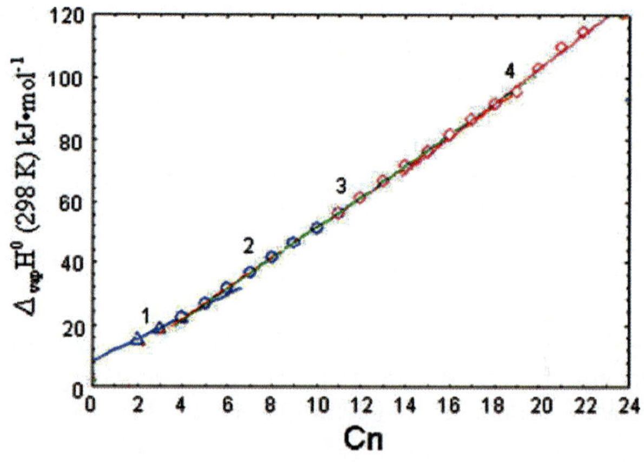

Fig. 8.4 Dependence of vaporization enthalpy on the number of carbon atoms in the chain of alkane (298 K)

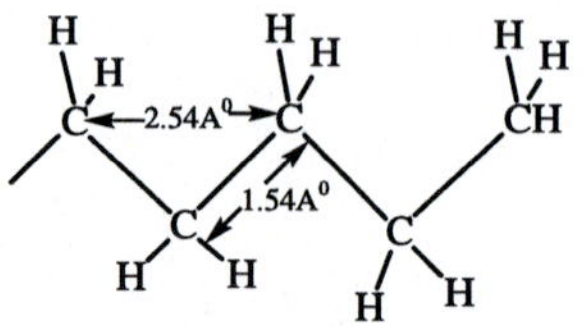

Fig. 8.5 Interatomic distances in the carbon chain of pentane

of the methylene group is present with a reduced negative charge which can play a role of an acceptor of the shifted electron density in the formation of the specific intermolecular interaction CH_2–CH_3 → CH_2–CH_3. The shift in electron density from a single carbon atom of the CH_2 group to two methyl groups resulted in their reduced negative charges compared to the carbon atoms of the methyl groups of *n*-butane where the electron density was shifted from two methylene groups. Therefore, the specific interaction of the same type DCH_2–CH_3 → CH_2–CH_3 should be more stable in the liquid *n*-butane. The absence of the second acceptor in the propane molecule causes the methyl groups of the interacting molecules of the nearest surrounding to form other less stable specific interactions. Evidently, this duality in the interactions of the two methyl groups of propane is reflected in the energies of the specific interactions and their contribution to the enthalpy characteristic. Analogous type of the specific interaction exists in liquid and solid methane and is formed by the ethane molecule (Fig. 8.6).

In the latter molecule, each carbon atom bears equal negative charges due to the shift in electron density from the hydrogen atoms, and consequently, the intermolecular interactions are caused by the acceptor properties of the hydrogen and donor properties of the carbon atoms. The latter transmits a part of the electron density from the essentially unshared $2s^2$-electron pair to the hydrogen atom, thus forming an additional coordination at the carbon atom (Fig. 8.6).

The presence of three hydrogen atoms in the methyl group of ethane suggests a formation with the molecule of the nearest surrounding of six specific interactions. Taking into consideration that the ethane molecule contains two identical methyl groups, each of them contributes to the vaporization enthalpy energy equal to 6.65 kJ mol^{-1}, and consequently the average value of the energy of the interactions formed by hydrogen atoms equals 1.30 and 1.70 kJ mol^{-1} in the liquid and solid states, respectively. However, this approach is of a formal character since we have shown before that in the cyclic saturated and unsaturated hydrocarbons, the energy contribution to the enthalpy characteristic is practically independent of the number of hydrogen atoms linked to the carbon atom (Sect. 7.9.1). It was shown in [18] and in Sect. 8.1 that the hydrogen atom in the methyl group of methanol located in the *trans*-position to the hydroxy group makes a larger contribution to the enthalpy characteristic than the other hydrogen atoms of the same methyl group. Basing on these facts and on the energy contributed by the methyl group of ethane (7.65 kJ mol^{-1}) essentially exceeding the energy of the specific interactions in the liquid methanol formed involving the oxygen atom, we are obliged to understand that the hydrogen atoms of the methyl group in ethane also are unequal. Therefore, interacting with the hydrogen atoms of the methyl groups in the molecules of the nearest surrounding, they form bonds of low stability whose energies are within the error of the experimental measurements of the enthalpy characteristics. The equal negative charges of the carbon atoms of two methyl groups, excessive number of hydrogen atoms with the insignificantly differing positive charge participating in the interaction, and a high energy contribution to the vaporization enthalpy from the methyl group indicate the formation through the pentacoordinate carbon atom of a sufficiently stable specific interaction DH–C → H–C.

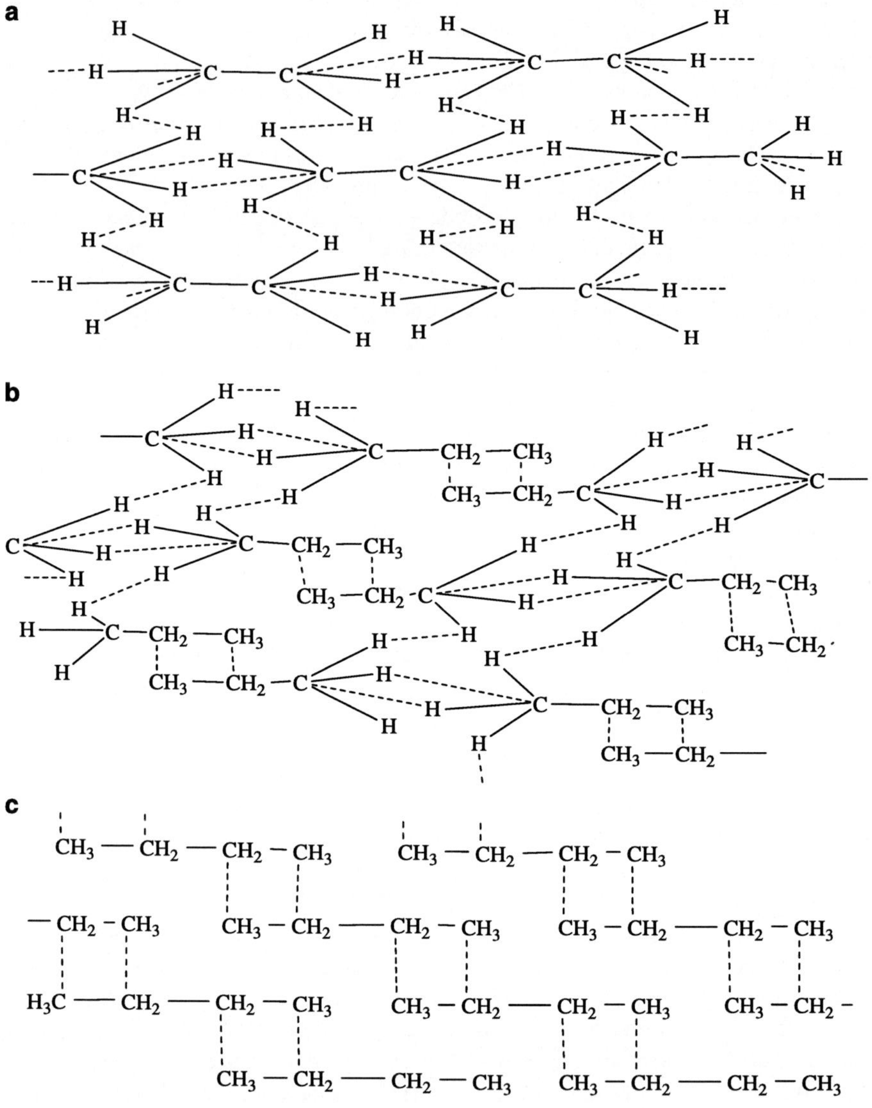

Fig. 8.6 Fragment of schematic picture of liquid structure (**a**) ethane, (**b**) propane, (**c**) *n*-butane

Its energy is estimated from the vaporization enthalpy by dividing by the number of formed stable interactions $\Delta H^0(T)/4$. The obtained value of the energy of this type of interaction (3.78 kJ mol^{-1}) has a higher stability than that existing in the 2,2-dimethylpropane.

Taking into account the temperature dependence of the vaporization enthalpy of methane (8.6 kJ mol^{-1} at 177 K) and the specific interaction energy of the liquid methane (4.3 kJ mol^{-1}), it is expectable that the energies of the specific interactions of liquid methane and ethane have similar values at standard conditions. The vaporization enthalpy of ethane measured at 288 K does not change notably the calculated value of this type of interaction. Its value which is less than the energy contribution of the CH_2 group of the saturated hydrocarbons (5.00 kJ mol^{-1}) indicates the validity of the assumptions and the performed calculations. Taking this energy value for this type of specific interaction formed by the methyl group of propane, we calculate the energy of the second type of specific interaction $DH_3C \rightarrow CH_2$ by (8.1) at 5.58 kJ mol^{-1}.

$$DH_3C \rightarrow CH_2 = \frac{\Delta_{evap}H^0(T) - 2DH_3C \rightarrow CH_2}{2}. \qquad (8.1)$$

The presence of two methyl and two methylene groups in the *n*-butane molecule indicates the formation of a single type of the specific interaction $DH_3C \rightarrow CH_2$ in the liquid and the crystalline state; therefore, the energy of the specific interaction should be calculated directly from the enthalpy characteristic $\Delta_{evap}H^0(T)/4$, whose value is given in Table 8.2. Symptomatically the obtained value of the energy of this type of specific interaction $DH_3C \rightarrow CH_2$ is equal to the energy of the same interaction existing in the liquid propane. This means that the sum of energies of two types of the specific interactions formed, on the one hand, in liquid *n*-butane $DH_3C \rightarrow CH_2$ and ethane DH–C → H–C, respectively, and, on the other hand, the sum of the same types of interactions present in the liquid propane have equal values. This fact additionally confirms the validity of the assumptions concerning the nature of the specific interactions.

The transition from the structure of the liquid ethane having specific interactions DH–C → H–C to the propane possessing two types of the interactions $DH_3C \rightarrow CH_2$ and DH–C → H–C and further to *n*-butane with the specific interactions $DH_2C \leftarrow CH_3$ does not lead to sharp changes in the standard vaporization enthalpies, and consequently, the dependence $\Delta_{vap}H^0(T) = f(C_n)$ is well described by a straight line (Fig. 8.4). The position of the enthalpy characteristic of methane considerably lower than this line (Fig. 8.7) is due to low charges on the hydrogen and carbon atoms, low stability of the specific interactions formed by the methane molecule,

Table 8.2 Energies (kJ mol^{-1}) of specific interactions in liquid and solid alkanes

Compound	$\Delta_{sub}H^0(T)$ [11]		$\Delta_{vap}H^0(T)$ [12]		DH–C → H–C		$DH_3C \rightarrow CH_2$	
	Sublimation	*T* (K)	Vaporization	*T* (K)	Sublimation	Vaporization	Sublimation	Vaporization
Ethane	20.5	90	15.3	288	5.12[a]	3.78	–	–
Propane	28.9	86	18.8	293	5.12[a]	3.78	9.33[a]	5.58
n-Butane	35.9	107	22.4	298	–	–	9.0[a]	5.60
					ΔCH_2 *n*			
n-Pentane	42.0	143	26.4	298	8.15 $n = 1$	5.05 $n = 1$	9.0	5.6
n-Hexane	50.8	178	31.5 ± 0.1	298	8.15 $n = 2$	5.05 $n = 2$	9.0	5.6

[a]Determined at low temperature

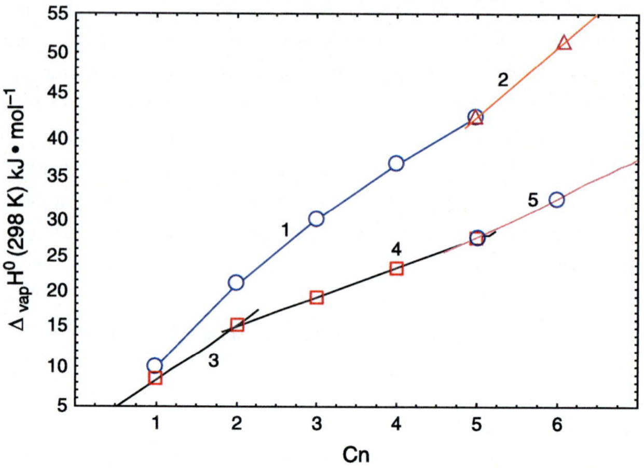

Fig. 8.7 Dependence of enthalpy characteristics of alkanes on the number of carbon atoms in the chain (1) sublimation at 84–175 K, (2) sublimation at 298 K, (3, 4, 5) vaporization, (1, 3, 4) evaluation of the influence of changing temperature and types of specific interactions, (2, 5) influence of the number of CH_2 groups

and a large number of the bond vacancies. As a result, the interaction between the hydrogen atoms in the molecules of the nearest surrounding is of low energy comparable with the error in the measurement of the thermodynamic properties.

The character of the dependence of sublimation enthalpy on the carbon atom number in the alkane chain in the first members of the homologous series under consideration (Fig. 8.7) illustrates the temperature dependence of the enthalpy characteristic and the change in the types of specific interactions and in their energy in the series ethane–*n*-butane. The description of this influence for the vaporization enthalpy by a straight line corresponds to the adequate contribution of energy from the interactions of the types DH–C → H–C and $DH_3C \rightarrow CH_2$, existing separately in ethane and *n*-butane, respectively, and in a combination in the liquid and crystalline propane. The subsequent increase in the enthalpy characteristic (lines 2 and 5, Fig. 8.7) illustrates the growing contribution of the CH_2 groups in alkane molecules.

The constant value of the energy of the specific intermolecular interaction $DH_2C \leftarrow CH_3$ in the first members of the alkanes series, propane and *n*-butane, is an important fact in combination with the above-cited data on the dissociation energy of the C–C bond in hexane and heptane, and corresponds to the total alkanes series. The validity of this conclusion is confirmed by the dependence of the vaporization enthalpy of alkanes presented in Fig. 8.4. The location of CH_3 group (s) enthalpies of ethane, propane, and *n*-butane, and of *n*-pentane–*n*-decane on two crossing straight lines is caused by the replacement of less stable specific interactions DC–H → C–H formed by the methyl groups of ethane and propane by more stable $DH_3C \rightarrow CH_2$ formed by all compounds of the alkane series with over four carbon atoms in the chain. In this connection, attention should be directed to the increasing number of methylene groups in the alkane molecules beginning with *n*-pentane. This means that the methylene group contributes to the enthalpy characteristic a considerably greater energy than the specific interactions DC–H → C–H formed by the methyl groups of ethane and propane. The energy contribution of the CH_2 group can be estimated from the difference in the vaporization enthalpy of two contiguous members of the homologous series of alkanes (Fig. 8.4). The description of the vaporization enthalpies of the series *n*-pentane–*n*-decane by a straight line (Fig. 8.2) is a consequence of the equal energy contribution to the enthalpy characteristic from each methylene group in the compounds of this series. Therefore, the value of this contribution can be estimated as an average for compounds *n*-pentane–*n*-decane and *n*-undecane–*n*-octadecane amounting to 4.80 and 5.05 kJ mol^{-1}, respectively; and for the sublimation process in the series *n*-pentane–*n*-decane, it is equal to 8.15 kJ mol^{-1}; for enthalpies measured at temperatures, it is 150–216 K; and for compounds *n*-dodecane–*n*-eicosane at the standard temperature, it is 8.67 kJ mol^{-1}. Therefore, the performed thermodynamic analysis and established energy parameters of alkane make it possible to estimate the valid values of the sublimation enthalpy of alkanes with a large number of carbon atoms in the chain using (8.2)

$$\Delta H^0(298)C_nH_{2n}+2 = \Delta_{vap}H^0(298)C_{15}H_{32} + \Delta CH_2(8.67)\cdot(n-15), \qquad (8.2)$$

and analogous equation for the vaporization process.

8.3 Energies of Specific Interactions in Saturated Cyclic Hydrocarbons

The first members of the cyclic saturated hydrocarbons, cyclopropane have planar molecules and cyclopentane cover forming planar structure of the liquid state, where the methylene group of the maximum negative charge on the carbon atom is located over the methylene group of another cycle with a carbon of reduced negative charge.

This position ensures a uniform distribution of the structural and energy parameters of molecules and is related to the formation of stable specific interactions (Fig. 8.8). The cyclohexane and the cyclic hydrocarbons with larger number of carbon atoms have a closed carbon–carbon chain like that in the open-chain hydrocarbons. However, the cyclization introduces considerable changes in the properties of these compounds. It should be mentioned in this connection that the cyclic structure lacks terminal methyl groups with enhanced negative charge on the carbon atom, generating the shift in electron density along the chain. This fact results in significant changes in the specificity of the interactions in the cyclic compounds, for the replacement of the terminal methyl groups by the CH_2 groups

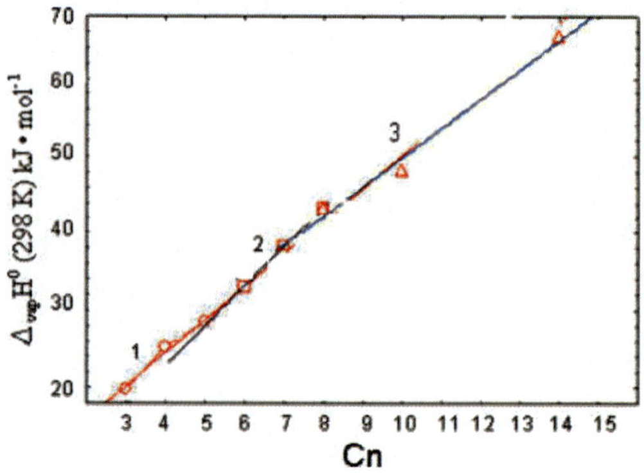

Fig. 8.8 Dependence of vaporization enthalpies of cyclic hydrocarbons on the number of carbon atoms in the ring (1) influence of the changing types of specific interactions, (2, 3) influence of the increased number of CH_2 groups

results in the replacement of two types of the specific interactions of propane with different energies $DH_3C \rightarrow CH_2 > DH\text{–}C \rightarrow H\text{–}C$ by stronger interactions in cyclopropane formed by the methylene groups. As a consequence, the vaporization enthalpy of cyclopropane (19.9 kJ mol^{-1}) is higher by 1.1 kJ mol^{-1}. At the same time, the specific interactions formed by the CH_2 groups are stabilized in cyclobutane and cyclopentane whose vaporization enthalpy exceeds that of butane and pentane, respectively, by 2.8 and 2.1 kJ mol^{-1}. A characteristic feature of cyclic hydrocarbons distinguishing them from alkanes is the clearly pronounced decrease in the rate of growth of the vaporization enthalpy, meaning the decrease in the energy contribution of the CH_2 groups at the growing number of carbon atoms in the ring, clearly seen in the dependences of the enthalpy characteristics on the number of carbon atoms in the ring (Figs. 8.4 and 8.8). The strongest decrease in the energy contribution to the enthalpy characteristic is observed in the first representatives of the cyclic hydrocarbons, in cyclobutane and cyclopentane, which alters by the equivalent energy contribution of the CH_2 groups in cyclohexane and cycloheptane. This shows that in the flat molecules of cyclopropane– cyclopentane, on the one hand and, on the other hand, in cyclohexane and cycloheptane with the structure of the chain similar to alkanes (Fig. 8.4), the alternation of the charges of carbon atoms along the ring is essentially different. For compounds in the series cyclooctane–cyclotetradecane, the energy contribution to the vaporization enthalpy is equal within the experimental error apparently because of the attainment of a uniform influence of the interspecies interactions of associated molecules.

The structure of cyclic saturated hydrocarbons in the liquid and crystalline states can be represented by a schematic picture (Fig. 8.9) taking into account the specificity of the interactions of the CH_2 groups of the molecules in the nearest surrounding and the formation of the interaction C–H → C–H along the donor–acceptor mechanism where the carbon atom is in the pentacoordinate state.

The changing energy contribution to the enthalpy characteristic (Table 8.3) indicates the destabilization of the interaction with the growing number of methylene groups in the ring of saturated hydrocarbons.

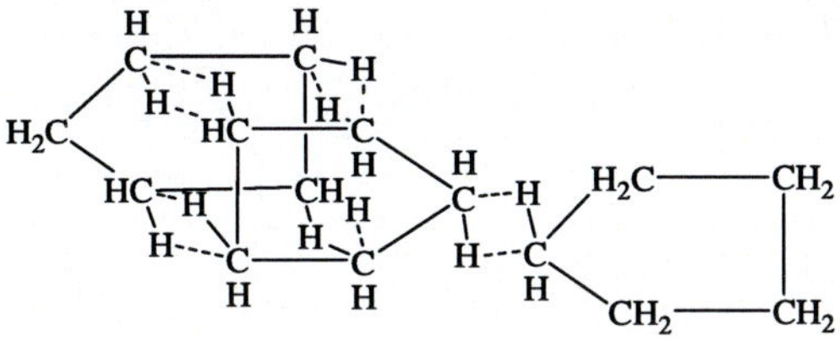

Fig. 8.9 Fragment of schematic picture of liquid structure of cyclopentane

Table 8.3 Energies (kJ mol^{-1}) of specific interactions in liquid and solid cyclic saturated hydrocarbons

Compounds	Formula	Sublimation [20]		DCH_2	Vaporization [21]		DCH_2
		$\Delta_{sub}H^0(T)$	T (K)		$\Delta_{vap}H^0(T)$	T (K)	
Cyclopropane	C_3H_6	27.4	298	9.1	19.9	312	6.6
Cyclobutane	C_4H_8	36.4	145	9.1	25.2	272	6.3
Cyclopentane	C_5H_{10}	42.1	122	8.04	28.5 ± 0.1	298	5.7
Cyclohexane	C_6H_{12}	–	–	–	33.0 ± 0.1	298	5.5
Cycloheptane	C_7H_{14}	–	–	–	37.2 ± 0.2	272	5.3
Cyclooctane	C_8H_{16}	–	–	–	43.3 ± 0.2	298	5.4
Cyclododecane	$C_{12}H_{24}$	100.2[a]	298	8.8	48.2	358	4.0
Cyclotetradecane	$C_{14}H_{28}$	117.6[a]	298	8.8	65.3 ± 0.2	298	4.1

[a]Established by interpolation

The energy of the specific interaction DC–H → C–H of the series of liquid saturated cyclic hydrocarbons is described by the equation

$$\Delta_{vap}H^0(T) = \mathrm{DH}_n\mathrm{C{\to}C{-}H} \cdot 2n, \qquad (8.3)$$

where n is the number of carbon atoms in the molecule. This expression can be used for the estimation of the enthalpy characteristics of yet unstudied cyclic saturated hydrocarbons.

8.4 Energies of Specific Interactions in Branched Saturated Hydrocarbons

As shown in the previous chapters, the effect of the side methyl and ethyl groups consists in their involvement in the distribution of the electron density and charges on atoms and functional groups taking part in the formation of specific interactions. The contribution of these groups to the vaporization enthalpy is practically not connected with the steric effect. Therefore, the revealing of the influence of the side groups in the liquid and crystalline saturated hydrocarbons is especially important for the development of the thermodynamic theory of specific interactions. As seen from Table 8.4, the characteristic feature inherent to various homologous series of saturated hydrocarbons is the increase in the enthalpy at going to a molecule with

Table 8.4 Energies (kJ mol^{-1}) of specific interactions in liquid cyclic saturated hydrocarbons with side methyl and ethyl groups (298 K)

Compound	Formula	$\Delta_{vap}H^0(T)$ [21]	$DCH_2CH_3 \rightarrow CH_2–CH_3CH_2$	$\sum DCH_3$	$DC–H \leftarrow C–H$
Cyclopropane	C_3H_6	19.9	–	6.63	–
1,1-Dimethylcyclopropane	C_4H_{10}	25.1 ± 0.8	–	5.2:2	1.3
n-Butane	C_4H_{10}	22.4	5.6	–	–
2-Methylbutane	C_5H_{12}	24.8	–	2.4	1.20
2,2-Dimethylbutane	C_6H_{14}	27.8 ± 0.1	–	5.2:2	1.3
2,2,3-Trimethylbutane	C_7H_{16}	32.0	–	9.6:3	1.6
2,2,3,3-Tetramethylbutane	C_8H_{18}	43.4 ± 0.2	–	21.0:4	2.6
Cyclobutane	C_4H_8	25.2 (272 K)		6.3	–
n-Pentane	C_5H_{12}	26.4	–	–	–
2-Methylpentane	C_6H_{14}	29.9 ± 0.1	–	3.5	1.75
3-Methylpentane	C_6H_{14}	30.3 ± 0.1	–	3.9	1.95
2,3-Dimethylpentane	C_7H_{16}	34.3 ± 0.1	–	7.9:2	2.2
2,4-Dimethylpentane	C_7H_{16}	32.7 ± 0.1	–	6.3:2	1.6
3,3-Dimethylpentane	C_7H_{16}	33.0 ± 0.1	–	6.6:2	1.65
3-Ethylpentane	C_7H_{16}	35.1 ± 0.1	4.35	8.7	–
Cyclopentane	C_5H_{10}	28.5	–	5.7	–
Methylcyclopentane	C_6H_{12}	31.6 ± 0.1	–	3.1	1.55
1,1-Dimethylcyclopentane	C_7H_{14}	33.8	–	5.3:2	1.20
1,2-Dimethylcyclopentane	C_7H_{14}	35.8	–	7.3:2	1.85
1,3-Dimethylcyclopentane	C_7H_{14}	34.5	–	6.0:2	1.50
Ethylcyclopentane	C_7H_{14}	36.5	4.0	8.0	–
Butylcyclopentane	C_7H_{14}	46.0	5.6 ($2DCH_2 \times 2 = 6.3$)	17.5	–
Cyclohexane	C_6H_{12}	33.6 ± 0.1	5.6	–	–
Methylcyclohexane	C_7H_{14}	35.1 ± 0.4	–	1.5	0.70 0.75
Ethylcyclohexane	C_9H_{18}	39.20.4	2.8	5.6	–
Propylcyclohexane	C_9H_{18}	45.2	5.6	11.6	11.2 + 0.4

an isostructural methyl group, from the linear to cyclic structure when two carbon atoms of the terminal methyl groups are replaced by a C–C bond, at replacement of a hydrogen atom of the ring by a methyl group, at growing number of methyl substituents in the ring, and at replacement of the latter by an ethyl group.

$$\text{Pentane} < \text{Methylpentane} < \text{Cyclopentane} < \text{Methylcyclopentane} < \text{Dimethylpentane} < \text{Ethylpentane}.$$

The butane derivatives show inadequate growth of the energy contribution to the vaporization enthalpy with the increasing number of methyl groups attached to the chain from one in the 2-methylbutane to three in the 2,2,3-trimethylbutane and further to four methyl groups in the 2,2,3,3-tetramethylbutane. This means that the position of the methyl groups at the second carbon atom and at the third carbon atom has a nonuniform effect on the stabilization of the formed specific interactions.

In the *n*-pentane derivatives, the prevailing effect on the stabilization of interactions causes methyl groups to attach to the second and the central carbon atoms and not to the second and fourth carbon atoms (cf. 2,3-dimethylpentane and 2,4-dimethylpentane). The compounds with the isostructural groups have vaporization enthalpy higher than the compounds of the normal structure; therefore, it is possible to estimate the energy contribution of these groups to the enthalpy characteristic from the difference between the vaporization enthalpies of compounds with the side group and that of the normal structure along (8.4)

$$\Delta H^0(298)\text{iso.g.} = \Delta_{\text{vap}}H^0(298)\text{iso.h.} - \Delta_{\text{vap}}H^0(298)\text{h.} \tag{8.4}$$

Branched compounds with a low energy contribution to the vaporization enthalpy indicate the low stability of two specific interactions formed by the isostructural groups H–C → H–C taking part in cross-linking the chains of molecules in liquid and crystalline states as shown in the schematic picture presented in Fig. 8.10.

Therefore, the energy of the specific interaction should be calculated by (8.5)

$$\text{DH}-\text{C} \rightarrow \text{H–C} = \frac{\Delta_{\text{vap}}H^0(298)\text{iso.h.} - \Delta_{\text{vap}}H^0(298)\text{h.}}{2} = \frac{\Delta_{\text{vap}}H^0(298)\text{iso.g.}}{2}. \tag{8.5}$$

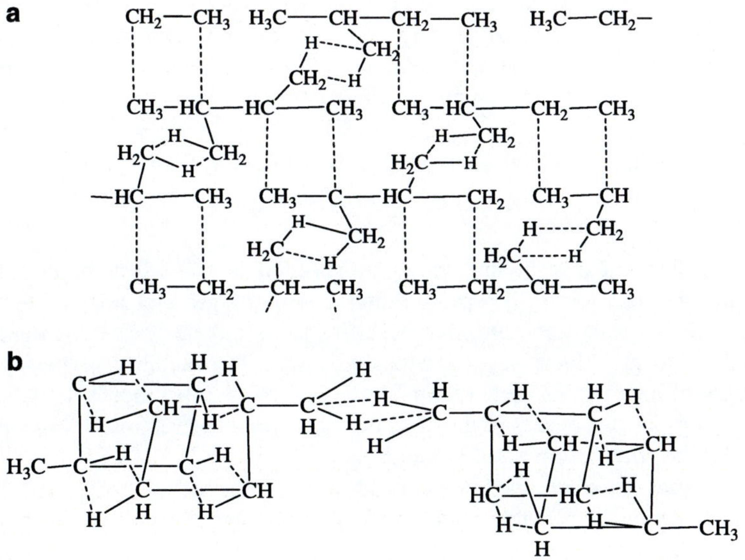

Fig. 8.10 Fragment of schematic picture of liquid and solid structures (**a**) 2-methylbutane and (**b**) methylcyclobutane

At two and more side groups in the molecule, their number is taken into account in the calculation

$$\mathrm{DH{-}C} \rightarrow \mathrm{H{-}C} = \frac{\Delta_{\mathrm{vap}}H^0(298)\mathrm{iso.h.} - \Delta_{\mathrm{vap}}H^0(298)\mathrm{h.} - \left(\Delta_{\mathrm{vap}}H^0(298)\mathrm{iso.g}/n\right)}{2}. \tag{8.6}$$

The results of the performed calculations (Table 8.4) illustrate the stabilization of the specific interactions in the liquid derivatives of *n*-butane with the growing number of side methyl groups.

$$\begin{aligned}&2\text{-Methylbutane}\ (1.2) < 2,2\text{-Dimethylbutane}\ (1.3)\\&< 2,2,3\text{- Trimethylbutane}\ (1.6) < 2,2,3,3\text{- Tetramethylbutane}\ (2.6\,\mathrm{kJ\,mol^{-1}}).\end{aligned}$$

The maximum double stabilization of the specific interactions is attained at the location of the fourth side methyl group on the last free position at the third carbon atom. Stabilization of the specific interaction of the liquid derivatives of *n*-pentane with one side methyl group is achieved by the transition of the methyl group from the second to the central carbon atom.

$$2\text{- Methylpentane}\ (1.75) < 3\text{- Methylpentane}\ (1.95\,\mathrm{kJ\,mol^{-1}}).$$

In the presence of two isostructural groups, their position at the second and the fourth carbon or both or at the third carbon results in low stability of the specific interactions; the most stable interactions are observed when these groups are situated at the second and third carbon atoms.

$$\begin{aligned}&2,4\text{- Dimethylpentane}\ (1.6) \approx 3,3\text{- Dimethylpentane}\ (1.65)\\&< 2,3\text{- Dimethylpentane}\ (2.2\,\mathrm{kJ\,mol^{-1}}).\end{aligned}$$

The replacement of a methyl group by an ethyl in the *n*-pentane derivatives causes the stabilization of the specific interaction, and at its position at the third carbon atom, it virtually acquires the properties close to the properties of analogous groups located at both sides of the central atom. The stability of the specific interaction formed by this ethyl group $DH_3C \rightarrow CH_2$ (4.35 kJ mol^{-1}) is enhanced, but it is lower than the energy of the same type of specific interactions (5.6 kJ mol^{-1}) formed by the ethyl fragment of *n*-pentane.

The energies of the specific interactions formed by the methyl groups attached to the cyclopentane ring depend on the position of the methyl groups in the ring

$$\begin{aligned}&1,1\text{- Dimethylcyclopentane}\ (1.20) < 1,3\text{- Dimethylcyclopentane}\ (1.50)\\&< \text{Methylcyclopentane}\ (1.55) < 1,2\text{- Dimethylcyclopentane}\ (1.85\,\mathrm{kJ\,mol^{-1}}).\end{aligned}$$

The regular series of stability of interactions formed by isostructural methyl groups in derivatives of *n*-pentane and cyclopentane illustrate the enhanced stability of the specific interactions of the liquid branched pentane derivatives.

The specific interaction $DCH_2–CH_3 \rightarrow CH_2–CH_3$ of reduced stability is formed by the ethyl group of ethylcyclopentane (4.0 kJ mol^{-1}) compared to an analogous type of interaction of similar group in 3-ethylpentane (4.35).

The stabilization of the specific interactions formed by various alkyl groups attached to the cyclohexane ring grows with the increasing number of carbon atoms in the substituent.

$$\text{Methylcyclohexane DH–C} \rightarrow \text{H–C} = 0.75 < \text{Ethylcyclohexane DCH}_2\text{–CH}_3 \rightarrow \text{CH}_2\text{–CH}_3 = 2.8 < \text{Propylcyclohexane DCH}_2\text{–CH}_3 \rightarrow \text{CH}_2\text{–CH}_3 = 5.6\,\text{kJ}\,\text{mol}^{-1}.$$

The butyl substituent of the butylcyclopentane possessing four carbon atoms shows an individual character in the formation of the specific interactions, and in the calculation of the energy contribution to the enthalpy characteristic, the contribution of each CH_2 group of the remaining $CH_2–CH_2$ fragment should be taken into account.

The results of the performed thermodynamic analysis (Table 8.4) show that the specific interaction formed by the isostructural methyl DH–C → H–C and ethyl $DCH_2–CH_3 \rightarrow CH_2–CH_3$ groups are less stable than those formed by the methylene groups in the rings and chains of the saturated hydrocarbons. The energies of the specific interactions formed by the side groups depend on their location in the chain or the ring.

8.5 Energies of Specific Interactions in Alkenes

The thermodynamic analysis of the published data on the vaporization enthalpy of unsaturated hydrocarbons should be systematized by various homologous series based on the dependence of the enthalpy characteristic on the number of carbon atoms in the hydrocarbon chain. A good description of vaporization enthalpy of compounds of the homologous series of 1-butene (Fig. 8.11) suggests that the obtained relation should be divided according to four series of compounds: ethylene–1-pentene, 1-pentene–1-decene, 1-decene–1-hexadecene, and 1-hexadecene–1-eicosene.

The character of the plot corresponding to the first series of unsaturated hydrocarbons regard the influence of the replacement of a certain type of specific interactions by the other, more stable ones. The second and further series correspond to the growth of the vaporization enthalpy owing to the energy contribution by increasing number of methylene groups, and this contribution within each series is practically constant.

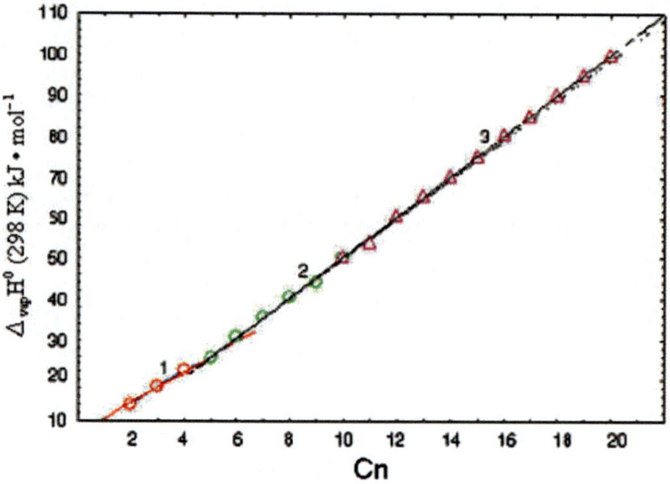

Fig. 8.11 Dependence of vaporization enthalpies on the number of carbon atoms in 1-alkenes (1) influence of the changing types of specific interactions, (2, 3) influence of the increased number of CH_2 groups

The numerous investigations compiled in [4] show that compounds with double bonds C=C have a rigid flat structure around this bond with no rotation around the double bond. Therefore, the ethylene molecule is flat. The ethylene molecule having two bond vacancies of carbon atoms and two hydrogen atoms forms four specific interactions H–C → H–C with the molecules of the nearest surrounding in the liquid and crystalline states. Ethylene like ethane forms a chain structure whose chains are linked by the specific interactions, ensuring the pentacoordinate state of the carbon atoms (Fig. 8.12). The chains are linked by weak interactions formed by the second carbon atom. The formation of four specific interactions of the same type provides a possibility to establish their energy directly from the vaporization enthalpy $\Delta H^0(T)/4$ (Table 8.5). The energies of the specific interactions DH–C → H–C tend to decrease the stability due to reduction in the charge on the carbon atom at diminished number of hydrogen atoms in the ethylene molecule caused by the replacement of the ordinary bond by the double bond.

$$\begin{aligned}\text{DC–H} &\leftarrow \text{C–H}\,(\text{ethane}, 3.78\,\text{kJ}\,\text{mol}^{-1}) < \text{DC–H}\\ &\leftarrow \text{C–H}\,(\text{ethylene}, 3.50\,\text{kJ}\,\text{mol}^{-1}).\end{aligned} \tag{8.7}$$

The carbon atom of the methyl group in propylene due to the shift in electron density from three hydrogen atoms has an enhanced negative charge compared to the CH_2 and CH groups bound by the double bond $-CH_2{=}CH-$. It is therefore presumable that two specific interactions $D{=}HC{-}CH_3 \rightarrow CH{-}CH_3$ formed by the methyl groups should be more stable than the interactions DH–C → H–C. Thus, the propylene molecule like that of propane forms two different types of specific

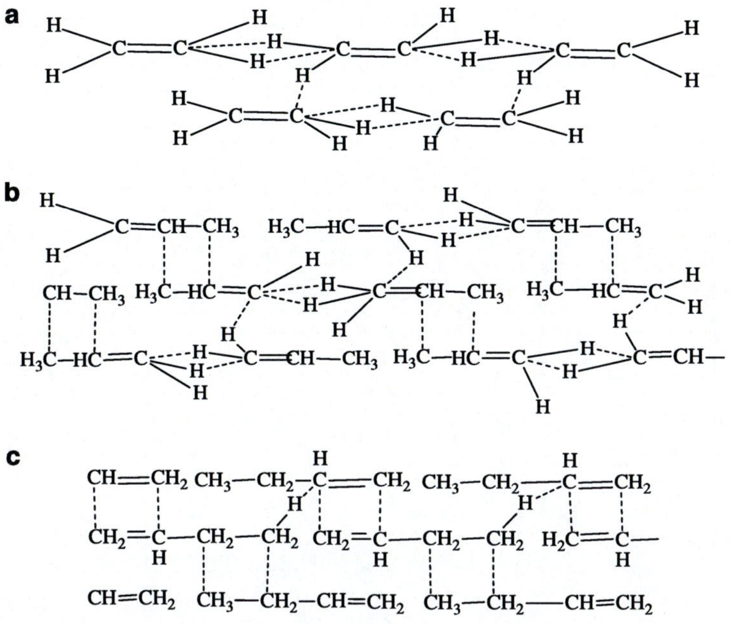

Fig. 8.12 Schematic picture of structure of liquid alkenes (**a**) ethylene, (**b**) propylene, (**c**) 1-butene

interactions, two DH–C → H–C and two D=HC–CH_3 → CH–CH_3. As a result, in the liquid state, chains are formed linked by weak interactions formed by the hydrogen atoms. Taking the energy contribution into the vaporization enthalpy of the specific interaction DH–C → H–C equal to that in propane, we obtain subtracting from the vaporization enthalpy of propylene the sum of energies of these two specific interactions the energy of the second type of specific interactions existing in the liquid propylene (8.8)

$$D = HC\text{–}CH_3 \rightarrow CH\text{–}CH_3 = \frac{\Delta_{vap}H^0(T) - 2DH\text{–}C \rightarrow H\text{–}C}{2}. \qquad (8.8)$$

The obtained value of this specific interaction (Table 8.5) equals 5.7 kJ mol^{-1} at the temperature below standard.

In the crystal and the liquid state, the 1-butene molecule forms two types of specific interactions (Fig. 8.12c), among which one is formed by the carbon atoms at the ordinary bonds DH_2C–CH_3 → CH_2–CH_3, and the other type of interactions involves the carbon atoms at the double bond DHC=CH_2 → CH=CH_2. Taking the energy of the former specific interaction in liquid butene equal to that in liquid butane, we calculate the energy of the specific interaction of the second type by (8.9) at 5.6 and 5.7 kJ mol^{-1} at 298 and 282 K, respectively.

Table 8.5 Energies (kJ mol^{-1}) of specific interactions of liquid alkenes

Compound	Formula	Vaporization [21] $\Delta_{vap}H^0(T)$	Vaporization [21] T (K)	$DH\text{–}C \leftarrow C\text{–}H$	$D{=}HC\text{–}HC_3 \rightarrow CH\text{–}CH_3$	$DCH_2\text{–}CH_3 \rightarrow CH_2\text{–}CH_3$	$\sum DCH_3$
Ethylene C_2H_4	$H_2C{=}CH_2$	14.0	267	3.5		–	–
Propylene C_3H_6	$H_2{>}CH{-}CH{=}CH_2$	18.5	285	3.5	5.7	–	–
1-Butene C_4H_8	$H_3C{-}CH_2{-}CH{=}CH_2$	22.8 22.4	282 298	–	5.7 5.6	5.7 5.6	–
1-Pentene C_5H_{10}	$H_3C{-}(CH_2)_2{-}CH{=}CH_2$	25.5 ± 0.1	298	–	5.6	5.6	3.1
2-Pentene C_5H_{10}	$H_3C{-}CH_2{-}CH{=}CH{-}CH_3$	26.8	298	–	5.6	5.6	3.8
1-Hexene C_6H_{12}	$H_3C{-}(CH_2)_3{-}CH{=}CH_2$	30.6	298	–	5.6	5.6	8.15
1-Heptene C_7H_{14}	$H_3C{-}(CH_2)_4{-}CH{=}CH_2$	35.6*	298	–	5.6	5.6	13.2

$$\mathrm{D} = \mathrm{HC{-}CH_3} \rightarrow \mathrm{CH{-}CH_3} = \frac{\Delta_{\mathrm{vap}}H^0(T) - 2\mathrm{DH_2C{-}CH_3} \rightarrow \mathrm{CH_2{-}CH_3}}{2}. \quad (8.9)$$

The energy 5.7 kJ mol^{-1} corresponds to the interaction D=HC–CH_3 → CH–CH_3 in the liquid propylene at 282 K. Consequently, first, the energies of two different types of specific interactions D=HC–CH_3 → CH–CH_3 and DCH_2–CH_3 → CH_2–CH_3 in liquid 1-butene have equal values due to the equal charges on the carbon atoms of the methyl, methylene, and methine groups. Second, the specific interaction D=HC–CH_3 → CH–CH_3 in liquid propylene and 1-butene has equal stability. Third, in the same way, the energy of the specific interaction DH–C ← C–H is also equal in liquid ethylene and propylene.

Hence we are able to consider the energy value to be equal in two types of interactions CH_3 → CH–CH_3 and DCH_2–CH_3 → CH_2–CH_3 for all unsaturated hydrocarbons from 1-butene series. This is confirmed by the character of the dependence $\Delta_{\mathrm{vap}}H^0(T) = f(C_n)$, whose break at the vaporization enthalpy of 1-pentene corresponds to the end of the variation in the types of specific interactions and their energies in the liquid compounds. Therefore, the vaporization enthalpy is equal further to the sum of energies of two types of the specific interactions and the sum of energy contributions of CH_2 groups to the enthalpy characteristic.

$$\Delta_{\mathrm{vap}}H^0(298) = 2\mathrm{D} = \mathrm{HC{-}CH_3} \rightarrow \mathrm{CH{-}CH_3} + 2\mathrm{DCH_2{-}CH_3} \rightarrow \mathrm{CH_2{-}CH_3} + \sum \mathrm{DCH_3}. \quad (8.10)$$

The results of the calculations are compiled in Table 8.5. The energy contribution of methylene group is calculated from the difference between enthalpy characteristics of compounds with different number of carbon atoms on the part of the dependence $\Delta_{\mathrm{vap}}H^0(T) = f(C_n)$ described by a straight line within the accuracy of measuring the vaporization enthalpy. The energy contribution of the CH_2 group of alkenes having from five to ten carbon atoms in the chain (5.05 kJ mol^{-1}) is the same as in saturated hydrocarbons (5.06 kJ mol^{-1}). Consequently, the ordinary –C–C– and double –C=C– bonds in the saturated and unsaturated hydrocarbons of 1-butene series play equal roles in the shift in electron density from the CH_2 groups, ensuring the leveling of charges on the carbon atoms of CH and CH_2 groups and the equal value of energies of the specific interactions D=HC–CH_3 → CH–CH_3 and DCH_2–CH_3 → CH_2–CH_3. Taking into account the contribution of the methylene group to the enthalpy characteristic and the formation by this group in the liquid state of four specific interactions DH–C → H–C, the energy of this type of interaction is estimated at 1.25 kJ mol^{-1}.

The obtained conclusions can be supplemented by the analysis of unsaturated hydrocarbons with two double bonds.

Like propane and propylene, the two methylene groups in the allene molecule form two different types of specific interactions DH–C → H–C and D=HC=H_2C → CH=CH_2 with different energy due to the absence in the molecule of an electron acceptor stronger than hydrogen atoms (Fig. 8.13). Analogous to propane and

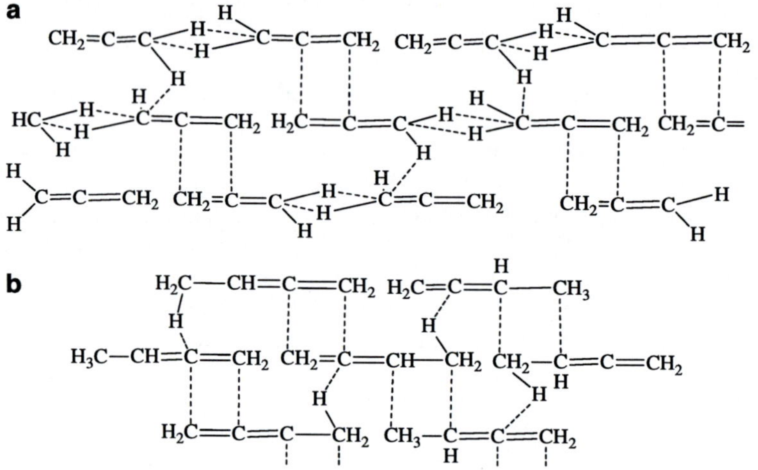

Fig. 8.13 Schematic picture of liquid structure (**a**) allene, (**b**) 1,2-butadiene

propylene, the specific interaction D=HC=H$_2$C → CH=CH$_2$ present in liquid butane and allene should have equal energies. At the same time, they should be equal in energy to that of the specific interactions formed in liquid diene compounds with the greater number of carbon atoms, provided the interactions are identical. This is a fundamental condition, as a butadiene molecule is capable of formation of three types of specific interactions depending on the location of the double bonds. Depending on the position of two double bonds in the molecules of hydrocarbons with a large number of carbon atoms, three types of specific interactions can form, e.g., in the liquid 1,3-hexadiene. Thus, the problem consists in the estimation of the energies of the specific interactions in the liquid 1,2-butadiene and 1,3-butadiene. The schematic picture of liquid allene and 1,2-butadiene shown in Fig. 8.13b illustrates the formation of chains with the strong specific interactions and the interactions of low stability linking the chains.

The allene and 1,2-butadiene molecules form two types of specific intermolecular interactions, where one is the common, DHC=H$_2$C → HC=CH$_2$. The energy of the second type of interaction D=CH–H$_3$C → CH–CH$_3$ in the liquid 1,2-butadiene is analogous to the interaction in the liquid 1-butene of the energy 5.6 kJ mol^{-1}. Therefore, subtracting from the vaporization enthalpy of 1,2-butadiene the sum of the energies of these specific interaction, we obtain the energy of the second type of interactions

$$\mathrm{D=C=CH_2 \rightarrow C=CH_2} = \frac{\Delta_{\mathrm{vap}}H^0(298) - 2\mathrm{D=CH{-}H_3C \rightarrow CH-CH_3}}{2}. \tag{8.11}$$

The single type of specific interactions in the liquid 1,3-butadiene makes it possible to estimate their energy from the vaporization enthalpy $\Delta_{vap}H^0(298)/4$. This type of specific interaction exists in the liquid allene, and consequently from the difference between the vaporization enthalpy of allene and the energy of the specific interactions DC=H_2C → C=CH_2, we estimate by (8.11) the energy of the second type of interaction DC–H ← C–H. The value obtained is overestimated, as the energy of the first type of interaction is not evaluated at the standard temperature. Therefore, the more reasonable energy is obtained using the energy of the interaction DHC=H_2C → CH=CH_2 in the liquid 1,3-butadiene.

The results of similar calculations of the energies of specific interactions (Table 8.6) provided a possibility to establish the energy contribution to the vaporization enthalpy of the growing number of CH_2 groups in pentadienes and hexadienes. This value is in good agreement with that established based on the dependence $\Delta_{vap}H^0(298) = f(C_n)$.

As seen from the plots in Fig. 8.14, the vaporization enthalpies of the hydrocarbons from the series of 1,3-butadiene with terminal double bonds fit well to a straight line; therefore, we used extrapolation for estimating the enthalpy characteristics of some unstudied compounds.

As already mentioned, the cyclic hydrocarbons are characterized by the alternation of the charges on the carbon atoms of the ring, ensuring the formation of stable specific interactions in their liquid and solid states and the layer structure of their crystals. This concept is confirmed by the fact that the vaporization enthalpies of cyclic unsaturated hydrocarbons fit to two straight lines presented in Fig. 8.14. The vaporization enthalpy of compounds with an even number of carbon atoms in the ring is described by plot (3) due to the uniform alternation of the positive and negative charges on their atoms. At the same time, in the event of odd carbon atoms in the ring, one among them has a reduced charge and forms a specific interaction with reduced stability. The results of the experimental measurements of the vaporization enthalpy of compounds in question of sufficient accuracy [21] sufficed for establishing this fact.

Interestingly, the mean values of the contributions of the CH_2 and CH groups (Tables 8.3 and 8.7) in the saturated and unsaturated hydrocarbons are practically constant. Therefore taking into account the data reported in Sects. 7.9.1 and 8.3, the energy contributions of the CH_2 and CH groups to the enthalpy characteristics of the cyclic hydrocarbons are described by the following expressions:

$$\text{Cyclopentane}\,(5.7) = \text{Cyclopentene}\,(5.7) = 1,3\text{-Cyclopentadiene}\,(5.7\,\text{kJ mol}^{-1}),$$

$$\begin{aligned}\text{Cyclohexane}\,(5.6) &= \text{Cyclohexene}\,(5.6) = \text{Benzene}\,(5.63)\\ &\approx 1,3\text{- Cyclohexadiene}\,(5.55\,\text{kJ mol}^{-1}),\end{aligned}$$

$$\text{Cycloheptane}\,(5.3) = \text{Cycloheptene}\,(5.24) < 1,3,5\text{- Cycloheptatriene}\,(5.53\,\text{kJ mol}^{-1}).$$

Table 8.6 Types and energies (kJ mol^{-1}) of specific interactions in diene hydrocarbon series

Compound	Formula	Vaporization [21]		DC–H ← C–H	DCH$_2$–H$_3$C → CH$_2$–CH$_3$	D=CH–H$_3$C → CH–CH$_3$	DC=H$_2$C → C=CH$_2$	∑DCH$_2$
		$\Delta_{vap}H^0(T)$	T (K)					
Allene	$H_2C{=}C{=}CH_2$	22.6	259	4.95 4.80[a]	–	–	6.35 6.5[a]	–
1,3-Butadiene	$CH_2{=}CH{-}CH{=}CH_2$	26.1	268	–	–	6.5	–	–
1,2-Butadiene	$H_3C{-}CH{=}C{=}CH_2$	23.9	298	–	5.6	–	6.35	–
1,2-Pentadiene	$H_3C{-}CH_2{-}CH{=}C{=}CH_2$	28.7	298	–	5.6	–	6.35	4.9
1,4-Pentadiene	$H_2C{=}CH{-}CH_2{-}HC{=}CH_2$	25.2	298	–	5.6	–	6.35	1.3
1,3-Hexadiene	$CH_3{-}CH_2{-}HC{=}CH{-}CH{=}CH_2$	32.0	309	–	5.6	6.5	–	7.8
1,4-Hexadiene	$CH_2{=}CH{-}CH_2{-}(CH_2)_2{-}CH{=}CH_2$	30.2	313	–	5.6	–	–	7.4
1,5-Hexadiene	$CH_2{=}CH{-}H_2C{-}CH_2{-}CH{=}CH_2$	30.5	288	–	5.6	–	–	7.7
1,6-Heptadiene	$CH_2{=}CH{-}CH_2{-}(CH_2)_2{-}CH{=}CH_2$	34.8[b]	298	–	5.6	–	–	12.4

1,7-Octadiene	$CH_2{=}CH{-}CH_2{-}(CH_2)_3{-}CH{=}CH_2$	39.7[b]	298	–	5.6	–	–	17.3
1,8-Nonadiene	$CH_2{=}CH{-}CH_2{-}(CH_2)_4{-}CH{=}CH_2$	44.5[b]	298	–	5.6	–	–	22.1

[a]Corresponding to 298 K
[b]Determined by extrapolation

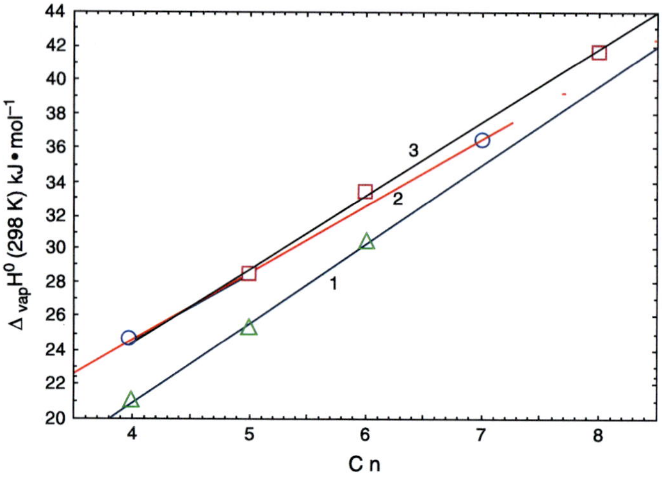

Fig. 8.14 Dependence of vaporization enthalpy on the number of carbon atoms in the chain (1) homologous series of 1,3-butadiene, (2) unsaturated cyclic hydrocarbons

Table 8.7 Energy contribution of methylene groups (kJ mol^{-1}) into vaporization enthalpy of unsaturated cyclic hydrocarbons

Compound	Formula	$\Delta_{vap}H^0(T)$ [21]	T (K)	$\sum DCH_2$
Cyclobutene		24.7	260	6.17
Cyclopentene	C_5H_8	28.4	300	5.7
Cyclohexene	C_6H_{10}	33.5 ± 0.5	298	5.6
Cycloheptene	C_7H_{12}	36.7 ± 0.1	298	5.24
Cyclooctene	C_8H_{14}	41.8	300	5.36
1,3-Cyclopentadiene	C_5H_6	28.4 ± 0.2	298	5.7
1,3-Cyclohexadiene	C_6H_8	33.8	298	5.55
1,3,5-Cycloheptatriene	C_5H_8	38.7 ± 0.2	298	5.53

Consequently, first, the energy contribution of the CH_2 groups to the vaporization enthalpy does not depend on the presence in the molecule of a double bond and of their number, and second, this contribution in cyclohexane derivatives with one and two double bonds and in benzene at the regular decrease in the number of CH_2 groups and increase in the number of CH groups remains constant. These conclusions indicate that this energy contribution originates from the carbon atom and one hydrogen atom attached to it. Therefore, we have a reason to suggest that the second hydrogen atom of the methylene group forms interactions of low stability, probably with a hydrogen atom of the similar group but with dissimilar charge; the energy of this interaction is within the error limit of the experimental methods used in the measurement of the thermodynamic properties. The reported data show that each CH_2 group of the cyclic hydrocarbon forms in the liquid and crystalline states the specific interaction H–C → C–H with the molecules of the nearest surrounding, involving the essentially unshared $2s^2$-electron pair of the pentacoordinate carbon atom and additionally involving the second hydrogen atom in the formation of an interaction of low stability, with the energy probably equal to the previously estimated value of 0.6 kJ mol^{-1}.

Thus irrespective of the presence in the molecule of a cyclic hydrocarbon of only CH_2- or both CH_2 and CH groups, their energy contributions to the enthalpy characteristic are the same. Consequently, each CH_2 and CH group in the interacting molecules of the nearest surrounding forms a stable specific interaction involving a single hydrogen atom. As a result, the molecules of the nearest surrounding form two specific interactions with equal energy value $DH_2C \rightarrow$ H–CH=DCH_2–H–C=DH–C → H–C. Taking the probable value of the contribution from the second hydrogen atom at 0.60 kJ mol^{-1}, we obtain the energy of the specific interaction DH–C → C–H, e.g., in cyclopentane derivatives equal to 2.55 kJ mol^{-1}.

The results of the performed thermodynamic analysis of unsaturated hydrocarbons with one and two double bonds in the series of ethylene and of butadiene permit the following conclusions:

- The energies of the specific interactions of the same type are constant in the saturated and unsaturated hydrocarbons with one and two double bonds because of a weak influence of the carbon atoms of the terminal fragment.
- The energy contribution to the vaporization enthalpy from the CH_2 group in the saturated (in the series *n*-pentane–*n*-decane) and unsaturated (in the series 1-pentene–1-decene) hydrocarbons equals 5.05 kJ mol^{-1}.
- The energy contribution to the vaporization enthalpy from the CH_2 group in compounds with two cumulated double bonds, 1,2-pentadiene–1,2-decadiene (4.8 kJ mol^{-1}), corresponds to the effect of these double bonds on the decreased shift in electron density in the molecule.

8.6 Energies of Specific Interactions in Alkynes

The vaporization enthalpies of acetylene hydrocarbons are described by the dependence $\Delta_{vap}H^0(298) = f(C_n)$ (Fig. 8.15) similar to the dependences of saturated and unsaturated hydrocarbons with one and two double bonds. The dependence describes, on the one hand, the enthalpy characteristics of acetylene, propyne, and 1-butyne with the changing types of specific interactions, and on the other hand, the energy contribution to the vaporization enthalpy from the CH_2 groups in the subsequent compounds of this homologous series. Evidently, the considerable difference in the mentioned dependences originates from the respective energies of the specific interactions formed in the series, on the one hand, of saturated and unsaturated hydrocarbons with one and two double bonds, and, on the other hand, in the acetylene series, and from the energy contributions of the methylene groups. It is possible to conclude that the relative strength of the specific interactions of the first members of the mentioned series grows in going from the ethylene series to

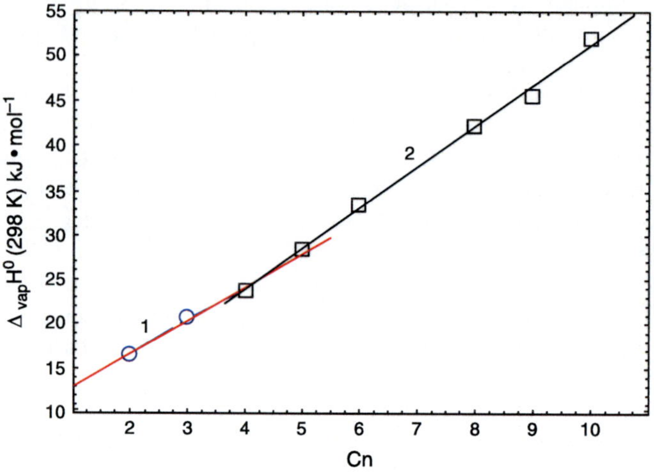

Fig. 8.15 Dependence of vaporization enthalpies on the number of carbon atoms in the 1-propyne series (1) influence of the changing types of specific interactions, (2) influence of the increased number of CH_2 groups

acetylene series, conserving small difference in the energy contributed by CH_2 groups. Applying this dependence, we estimated by interpolation and extrapolation the vaporization enthalpies of 1-heptyne, 1-undecyne, and 1-dodecyne. Considering the dependence of the enthalpy characteristic in the series 1-hexyne–1-dodecyne on the energy contribution of the methylene group, we estimated this contribution at 4.8 kJ mol^{-1}. Taking into account that in the liquid state this group forms four specific interactions DH–C → H–C, the energy of this interaction equals 1.2 kJ mol^{-1}. Thus, we can conclude that the energy of this type of interaction in liquid unsaturated hydrocarbons in the homologous series of 1-butene (1.25 kJ mol^{-1}) and 1-butyne has close values.

Similar to ethane and ethylene, the acetylene molecule forms in the liquid state four specific interactions of the same type with the molecules of the nearest surrounding whose energy is evaluated from the vaporization enthalpy $\Delta_{vap}H^0(T)/4$. The acetylene series compounds with the triple bond at the first carbon atom form two types of specific interactions, resulting in chains in the liquid state linked by weak interactions (Fig. 8.16). The propyne molecule having three carbon atoms forms one specific interaction, the same as that in acetylene, H–C → H–C≡, and also the second type of interaction, D≡(H)C → C≡.

The interaction of the first type should possess the same energy as in liquid acetylene. However, the value measured at 210 K is overestimated for the standard temperature. Therefore, it is possible to estimate the energy of these type of specific interactions after establishing the energy of the second type of interaction formed in propyne and homologs of propyne. The 1-butyne molecule also forms the specific interaction of the second type whose energy was estimated in the preceding paragraph at 5.6 kJ mol^{-1}. The results of performed calculations were used in the

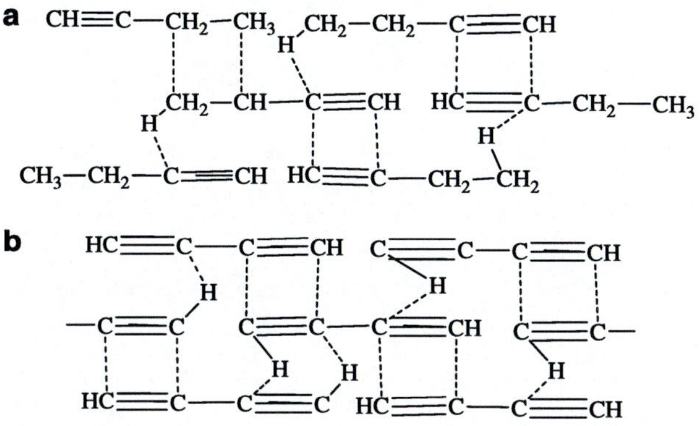

Fig. 8.16 Schematic picture of liquid structure (**a**) 1-butyne, (**b**) 1,3-butadiyne

estimation of the energy contributions from the growing number of methylene groups in the molecules of the propyne series (Table 8.8).

The position of the triple bond in 2-butyne, 2-pentyne in the internal part of the molecule, and symmetrically at the first and third carbon atoms in the molecule of 1,3-butadiyne provides a possibility to estimate the energy of the specific interaction D≡C–H$_3$C → C(≡)–CH$_3$ and D≡C ← CH directly from the vaporization enthalpy $\Delta_{vap}H^0(T)/4$, taking into account the formation of four specific interactions with the molecules of the nearest surrounding in the liquid state (Fig. 8.16). In 2-pentyne, 3-hexyne, and 2,4-hexadiyne the energy contribution to the vaporization enthalpy of the fragments ≡C–, –C≡C–, and ≡C–C≡ is found from the difference between the enthalpy characteristic and four specific interactions, respectively.

$$\Delta_{vap}H^0(T) - (2D{\equiv}C-, H_3C{\rightarrow}C(\equiv) - CH_3 + 2DCH_2\text{–}H_3C \rightarrow CH_2\text{–}CH_3),$$

$$\Delta_{vap}H^0(T)\text{–}4DCH_2\text{–}H_3C \rightarrow CH_2\text{–}CH_3 \quad \text{and}$$
$$\Delta_{vap}H^0(T) - 4D{\equiv}C-H_3C \rightarrow C(\equiv)\text{–}CH_3.$$

Note that the energy of the specific interaction DH–C → H–C in the liquid state of

$$\text{Propane}\ (3.78) < \text{Propylene}\ (4.8) = \text{Propyne}\ (4.8\ \text{kJ mol}^{-1})$$

is the smallest in propane having three hydrogen atoms on the terminal carbons. It leads to the conclusion that the presence in the molecule of a double or triple bond results in a stronger shift in electron density from two or even one hydrogen atom, respectively, than from three hydrogen atoms of the methyl group in propane molecule. Consequently, the hydrogen atoms of the CH_2 group in propene and of CH group in propyne possess enhanced positive charges, carbon atoms, and higher

Table 8.8 Types and energies (kJ mol^{-1}) of specific interactions in hydrocarbons with triple bonds

Compound	Formula	Vaporization [21]		DH–C → C–H	$DCH_2–H_3C → CH_2–CH_3$	D≡(H)C → C≡	$\sum DCH_2$
		$\Delta_{vap}H^0(T)$	T (K)				
Acethylene C_2H_2	$H-C\equiv C-H$	16.7	210[a]	4.18	–	–	–
1-Propyne C_3H_4	$H_3C-C\equiv CH$	22.1	275	4.8	–	6.25	–
1-Butyne C_4H_6	$CH_3-CH_2-C\equiv CH$	23.7	288	–	5.6	6.25	–
1-Pentyne C_5H_8	$CH_3-(CH_2)_2-C\equiv CH$	28.4	298	–	5.6	6.25	4.7
1-Hexyne C_6H_{10}	$CH_3-(CH_2)_3-C\equiv CH$	33.5	276	–	5.6	6.25	4.9 × 2
1-Heptyne C_7H_{12}	$CH_3-(CH_2)_4-C\equiv CH$	37.8[b]	298	–	5.6	6.25	13.8
1-Undecyne $C_{11}H_{20}$	$CH_3-(CH_2)_8-C\equiv CH$	57.0[b]	298	–	5.6	6.25	33.1
1-Dodecyne $C_{12}H_{22}$	$CH_3-(CH_2)_9-C\equiv CH$	61.7[b]	298	–	5.6	6.25	37.20

				$DH_3–C \rightarrow C\equiv$	$D\equiv C \leftarrow CH_3$	$D\equiv C–C\equiv$	
2-Butyne C4H6	$CH_3—C\equiv C—CH_3$	26.7	298	6.67	–	–	–
2-Pentyne C_5H_8	$CH_3—CH_2—C\equiv C—CH_3$	30.8	298	6.67[a]	6.7	–	D≡C–=6.3
3-Hexyne C_6H_{10}	$H_3C–CH_2–C\equiv C–CH_2–CH_3$	31.6 34.9[a]	275 298	5.6 6.67[a]	6.7[a]	–	D–C≡C–=6.3
1,3-Butadiyne C_4H_2	$CH\equiv C—C\equiv CH$	26.1	268	–	–	6.5	–
2,4-Hexadiyne C_6H_6	$H_3C—C\equiv C–C\equiv C—CH_3$	42.5	298	6.67	–	7.95 × 2	–

[a]Solid
[b]Determined by interpolation and extrapolation

negative charges; therefore, the acceptor and donor properties are correspondingly higher than that in the propane molecule.

Detailed analysis of thermodynamic properties and structure of saturated and unsaturated hydrocarbons with one and two double bonds and compounds with the triple bonds C≡C leads to the following conclusions:

- The energies of specific interactions of one type are practically constant within the total homologous series of saturated and unsaturated hydrocarbons, providing a possibility to use the additivity principle for a large number of compounds from various homologous series
- The correspondence between the structure and the energy results in obtaining valid thermodynamic properties by extrapolation and interpolation
- The energies of the specific intermolecular interactions in saturated and unsaturated hydrocarbons differ insignificantly due to the weak effect of the double bonds on the shift in electron density in the molecule.

8.7 Energies of Specific Interactions in Branched Unsaturated Hydrocarbons

The thermodynamic analysis carried out in Sect. 8.3 based on numerous experimental measurements of the vaporization enthalpy of branched saturated hydrocarbons and the conclusions therefrom can be supplemented with an analogous analysis of the enthalpy characteristics of unsaturated hydrocarbons possessing side alkyl groups. Note that the definite location of the side groups in the main alkyl chain affects the properties of compounds. Evidently, the branched molecule is an indivisible quantum-chemical complex with mutual influence of fragments on the distribution of the electron density that provides for the capability of functional fragments of interspecies interactions. In this connection, the applied approach considering the formation by the carbon atom of the isostructural methyl group of a specific interaction with the hydrogen atom of a similar group is preferable, since this approach is based on the nature of the interacting species and reveals the essence of the interacting carbon atom existing in the pentacoordinate state. This ability of the carbon atom is independent of the class of organic compound; its role is revealed in the energy of the formed specific interaction. The branched unsaturated hydrocarbons obey the general rule, and they also have higher vaporization enthalpy compared to that of the compounds with normal structure (Table 8.9). The energy contribution of the isostructural methyl group depends essentially on its position in the chain at the carbon atom, ensuring the enhanced difference in the negative charge on the carbon atom and positive charge on the hydrogen atom of the methyl group. The noted difference is revealed in the energy values of the specific interactions H–C → H–C formed by the analogous groups of the contacting molecules of the nearest surrounding. As shown before, the energy values of these interactions are estimated from the

Table 8.9 Energies (kJ mol^{-1}) of specific interactions and contribution of methyl and ethyl groups into vaporization enthalpy in branched unsaturated hydrocarbons at 298 K

Compound	Formula	$\Delta_{vap}H^0(T)$ [21]	$\sum DCH_3$	DHC → H–C	$DCH_2–H_3C \rightarrow CH_2–H_3C$
1-Butene	C_4H_8	20.1	–	–	–
2-Methyl-1-butene	C_4H_{10}	25.9 ± 0.1	5.8	2.9	–
3-Methyl-1-butene	C_4H_{10}	23.9	3.8	1.9	–
2-Butene	C_4H_8	21.3	–	–	–
2-Methyl-2-butene	C_4H_{10}	27.1	5.8	2.9	–
1,3-Dimethyl-1-butene	C_6H_{12}	32.5	12.4	3.1	–
2,3-Dimethyl-1-butene	C_6H_{12}	29.1	9.0	2.25	–
3,3-Dimethyl-1-butene	C_6H_{12}	26.8	6.1	1.52	–
2,3-Dimethyl-2-butene	C_6H_{12}	33.6	12.3	3.1	–
2-Ethyl-1-butene	C_4H_{12}	31.0	10.9	2.7	–
3-Methyl-1,2-butadiene	C_4H_8	27.3	3.4	1.7	–
2,3-Dimethyl-1,3-butadiene	C_6H_{10}	32.2 (288 K)	11.1	2.8	–
1-Pentene	C_5H_{10}	25.5	–	–	–
1-Methyl-1-pentene	C_6H_{12}	30.5	5.0	1.30	–
2-Methyl-1-pentene	C_6H_{12}	30.5	5.0	1.30	–
3-Methyl-1-pentene	C_6H_{12}	28.6	3.1	1.55	–
4-Methyl-1-pentene	C_6H_{12}	28.6 ± 0.2	3.1	1.55	–
2-Methyl-2-pentene	C_6H_{12}	31.6	4.8	2.4	–
3-Methyl-2-pentene	C_6H_{12}	31.3	4.5	2.25	–
4-Methyl-2-pentene	C_6H_{12}	29.5	2.7	1.35	–
3,3-Dimethyl-1,5-hexadiene	C_8H_{14}	35.2 (303 K)	4.7	1.2	–
1-Ethyl-1,5-hexadiene	C_8H_{14}	35.2 (308 K)	4.7	2.35	–
Cyclopentene	C_5H_8	24.4			–
1-Methylcyclopentene	C_6H_{10}	32.6 ± 0.2	7.8	3.9	–
3-Methylcyclopentene	C_6H_{10}	31.0 ± 0.2	6.4	3.2	–
4-Methylcyclopentene	C_6H_{10}	33.2	8.8	4.4	–

(continued)

Table 8.9 (continued)

Compound	Formula	$\Delta_{vap}H^0(T)$ [21]	$\sum DCH_3$	DHC → H–C	$DCH_2–H_3C \rightarrow CH_2–H_3C$
1,3-Dimethylcyclopentene	C_7H_{12}	35.0	10.6	2.65	–
1-Ethylcyclopentene	C_7H_{12}	38.5	14.1	–	7.05
3-Ethylcyclopentene	C_7H_{12}	36.3	11.9 300 K	–	5.95
4-Ethylcyclopentene	C_7H_{12}	36.5	12.1 300 K	–	6.05
1-Methyl-1-cyclohexene	C_7H_{12}	37.5 ± 0.2	4.0	2.0	–
3-Methyl-1-butyne	C_5H_8	25.8	2.1	1.05	–

difference between the vaporization enthalpies of the compound with the side alkyl group and the compound of normal structure

$$\text{DH–C} \rightarrow \text{H–C} = \frac{\Delta_{\text{vap}}H^0(298)\text{iso} - \Delta_{\text{vap}}H^0(298)}{2n}, \qquad (8.12)$$

where "n" is the number of side alkyl groups. This equation is also valid for cyclic compounds.

The results of performed calculations of the overall values of the energy contribution (Table 8.9) lead to a conclusion on the stronger binding of the isostructural methyl and ethyl groups compared to these groups in the branched saturated hydrocarbons (Table 8.5).

The energies of the specific interactions formed by the methyl groups attached to the chain of butyne, butane, and butadiene are described by the stabilization series.

$$\begin{aligned}&\text{3- Methyl - 1- butyne}\,(1.05) < \text{3- Methyl-1- butene}\,(1.9)\\ &< \text{2- Methyl-1- butene}\,(2.9) = \text{2- Methyl - 2 - butene}\,(2.9\,\text{kJ mol}^{-1}),\end{aligned}$$

$$\begin{aligned}&\text{3, 3- Dimethyl - 1- butene}\,(1.50) < \text{2, 3- Dimethyl - 1- butene}\,(2.25)\\ &< \text{2, 3- Dimethyl - 2- butene}\,(2.70)\\ &< \text{1, 3- Dimethyl - 1- butene}\,(3.1\,\text{kJ mol}^{-1}),\end{aligned}$$

$$\text{2, 3- Dimethyl - 2- butene}\,(2.70) < \text{2- Methyl - 2- butene}\,(2.9\,\text{kJ mol}^{-1}),$$

$$\begin{aligned}&\text{3- Methyl - 1, 2- butadiene}\,(1.7)\\ &< \text{2, 3- Dimethyl - 1, 3- butadiene}\,(2.8\,\text{kJ mol}^{-1}),\end{aligned}$$

$$\text{1, 2- Butadiene}\,(1.7) < \text{2, 3- Dimethyl - 1, 3- butadiene}\,(2.8\,\text{kJ mol}^{-1}).$$

The minimum energy corresponds to the liquid 3-methyl-1-butyne (1.05), and maximum energy is attained in liquid 1,3-dimethyl-1-butene (3.1) and 2,3-dimethyl-1,3-butadiene (2.8 kJ mol^{-1}), approaching to the interaction energy in methane at 175 K (4.3 kJ mol^{-1}). In the butadiene derivatives, a lower stability of the specific interactions is observed than in the corresponding derivatives of 1-butene. At the same time, 3,3-dimethyl-1,5-hexadiene (1.2 kJ mol^{-1}) formed the interaction whose energy exceeds only by 0.15 kJ mol^{-1} the lowest interaction energy in 3-methyl-1-butyne (1.05 kJ mol^{-1}), although in these molecules enhanced number of atoms donating electrons is present. Consequently, the stability of the specific interactions formed by isostructural methyl groups is determined by the shift in electron density in the chain depending on the location in the chain of the side group and on the place of the double bonds.

The branched derivatives of 1-pentene illustrate the insignificant stabilization with remoteness of the methyl group from the double bond.

$$\text{1-Methyl-1- pentene}\,(1.3) = \text{2-Methyl-1pentene}\,(1.30) < \text{3- Methyl-1- pentene}\,(1.55) = \text{4-Methyl-1- pentene}\,(1.55\,\text{kJ mol}^{-1}),$$

$$\text{4-Methyl - 2- pentene}\,(1.35) < \text{3- Methyl-2- pentene}\,(2.2) < \text{2-Methyl-2- pentene}\,(3.05\,\text{kJ mol}^{-1}).$$

However, this rule is reversed in the branched derivatives of 2-pentene, where the specific interaction of enhanced stability is observed in liquid 2-methyl-2-pentene formed by the methyl group attached to the carbon atom of the double bond. Thus, the stability of the interactions under consideration is governed by the length of the carbon chain and the location of the double bond and the isostructural methyl group. Branched hydrocarbons with a side ethyl group form specific interactions of lower stability than the majority of branched derivatives of 1-butene and 2-butene with a side methyl group, and than 2-methyl-2-pentene (3.05 kJ mol^{-1})

$$\text{1- Ethyl - 1,5- hexadiene}\,(2.35) < \text{2- Ethyl - 1- butene}\,(2.7\,\text{kJ mol}^{-1}).$$

Consequently, the isostructural ethyl group in compounds with unsaturated carbon–carbon chain behaves similarly to ethane that forms the interaction DH–C → H–C of the energy 3.82 kJ mol^{-1}, and the presence of the –C–C– bond results in the reduced acceptor and donor properties.

This hypothesis is confirmed by the energies of the specific interactions formed by the ethyl groups in liquid branched unsaturated and saturated cyclic hydrocarbons (Table 8.5) caused by successive uniform alternation of charges on the odd and even carbon atoms of the ring.

$$\text{3-Ethylcyclopentene}\,(5.95) \approx \text{4-Ethylcyclopentene}\,(6.05) < \text{1-Ethylcyclopentene}\,(7.05\,\text{kJ mol}^{-1}).$$

The maximum energy value of the specific interaction $DCH_2–H_3C \rightarrow CH_2–CH_3$ is attained at the ethyl group location by the first carbon atom at the double bond. In the substituted unsaturated cyclic hydrocarbons, the specific interactions formed by the methyl substituent are more stable.

$$\text{1,3- Dimethylcyclopentene}\,(2.65) < \text{3-Methylcyclopentene}\,(3.2) < \text{1- Methylcyclopentene}\,(3.9) < \text{4 -Methylcyclopentene}\,(4.4\,\text{kJ mol}^{-1}).$$

It is presumable that the difference in the regular trend in the stabilization of the specific interactions formed by the isostructural methyl and ethyl groups consists in their influence on the shift in electron density in the molecules depending on the location of the substituent in the ring. On the contrary, the influence of the cyclic structure of saturated and unsaturated hydrocarbons effects the change in the interaction type of the methyl group, as has been observed for the ethyl group, resulting in the approach of the interaction energy to that existing in liquid hydrocarbons (Tables 8.6 and 8.8).

References

1. Talrose VL, Lubimova AL (1952) Secondary process in the ion source of the mass spectrograph. Doclady Chem Technol Doclady Chem USSR 86:99–104
2. Coylson AC (1961) Valence. Oxford University Press, Oxford, p 428
3. Schustorovich EM (1975) Chemical bonds at co-coordinated compounds. Knowledge, Moscow
4. Nefedov VI, Vovna VI (1989) Electronic structure of organic and elementorganic compounds. Nauka, Moscow
5. Baev AK (1996) Five-coordinated carbon atom at intermolecular interactions. Russ J Coord Chem 22:399–402
6. Baev AK (2001) Pentacoordinated carbon atom at intermolecular interaction in solutions of alkyl compounds of non-transition elements. In: Book of abstract of VIth international symposium on carbonion chemistry, Marburg, Germany, p 75
7. Baev AK (2001) Thermodynamics and specific intermolecular interactions in solutions of non-electrolytes with pentacoordinated carbon atom. In: Book of abstract of 27th international conference on solution chemistry, Vaals, Netherlands, p 36
8. Baev AK (2001) Specific interactions in solvates of solutions of non-electrolytes with isostructural molecules components and pentacoordinated carbon atom. In: Book of abstract of VIIIth international conference. The problems of solution and complex formation in solution, Ivanovo, Russia, p 246
9. Baev AK (2002) Thermodynamic solutions of non-electrolytes with pentacoordinated carbon atom, at specific interactions. In: Book of abstract of XIVth international conference on chemical thermodynamics, St. Peterburg, Russia, p 132
10. Baev AK, Korolkov DV (2003) Phenomenon of dativy bond in elementorganic compounds. In: Book of abstract the XVth FECHEM conference on organometallic chemistry, University of Zurich, Switzerland, p 350
11. Baev AK (2003) Novel approaches to the structure of functional solvents and elementorganic compounds with saturated hydrocarbon ligands and their solvates in binary systems. In: Book of abstract of 27th international conference on solution chemistry, Debrecen, Hungary, p 79
12. Baev AK (2004) Insolvency of sp^3-gibrydixation model of carbon atom and chemistry of non-electrolytes solutions. In: Book of abstract of IXth international conference. The problems of solvation and complex formation in solution, Ivanovo, Russia, pp 62–64
13. Baev AK (2009) Phenomenological significance of no identical sp^3-hubridization model and nature of interaction in chemistry of liquid non-electrolytes and chemical thermodynamics. In: Book of abstract of the Xth international conference on physical and coordination chemistry of porphyrines and them analogous, Ivanovo, Russia, p 86
14. Legen AC, Roberns BP, Wakkwork AL (1990) Equilibrium conformation of four- and five-membered cycle molecules in the gas phase determination and classification. Chem Phys Lett 191:97

15. Hobza P, Havlas Z (2000) Blue-shifting hydrogen bonds. Chem Rev 100:4254–4264
16. Krupin SV, Puhovskoi VA (2004) Investigation of new configuration hydrogen bonding dimer of methanol by method of molecular orbital and functional density. In: Book of abstract. IX international conference. The problem of solvation and complex formation in solutions, Plyos, Russia, p 136
17. Mandado M, Graria AM, Musquera RA (2002) Approximate transferability in alkanols. J Mol Struct 584(1–3):221–234
18. Ivanov EV, Puchovskie YuP, Abrosimov VK (2004) About formation of weak hydrogen bonds CH···O in liquid methanol. In: Book of abstract. IX intenational conference. The problems of solvation and complex formation in solution, Plyos, Russia, p 345
19. Baev AK (1972) Complex formation ability halogenides of second – six groups periodical system. General and applied chemistry, vol 5. Vaisheshika Educating, Minsk, pp 35–51
20. Chickos JS, Acree WE Jr (2003) Enthalpies of sublimation of organic and organometallic compounds, 1910–2001. J Phys Chem Rev Data 32:537–698
21. Chickos JS, Acree WE Jr (2002) Enthalpies of vaporization of organic and organometallic compounds, 1880–2002. J Phys Chem Rev Data 31:519–879
22. Papulov YuG, Smuliak VM (1981) Physical properties and chemical structures. Kalinin State University, Kalinin
23. Baev AK (1969) Problems of chemical nature of phase transformation. General and applied chemistry, vol 1. Vaisheshika Educating, Minsk, pp 197–206
24. Baev AK (1969) Phase condition and complex formation ability of halogenide metals. General and applied chemistry, vol 1. Vaisheshika Educating, Minsk, pp 207–218
25. Podkovyrov AI, Baev AK, Kozyrkin BI (1990) Thermodynamics of vaporization of alkylchalcogenides. In: Book of abstract. VI conference of USSR on thermodynamic of organic compounds, Minsk, Belarus, p 73
26. Kozyrkin BI, Larionova LM, Socolovscei AE, Baev AK (1990) Thermo stability and volatility of tin alkyl compounds. Russ J Appl Chem 50:1798
27. Socolovskei AE, Baev AK (1984) Thermodynamic study of vaporization process of dimethyl- and diethylzink. Russ J Gen Chem 54(1):103
28. Socolovskei AE, Baev AK (1984) Thermodynamic study of vaporization process of butylzink organic compounds. Russ J Phys Chem 58:2692–2694
29. Magaril PZ (1970) Mechanism and kinetics homogeny thermal of transformation hydrocarbon. Chemistry, Moscow, p 223
30. Vasiliev IA, Petrov VM (1984) Thermodynamic properties of oxygen organic compounds. Chemistry, Leningrad
31. Lebedev YuA, Miroschnichenko EA (1981) Thermochemistry of the vaporization of organic compounds. The vaporization and sublimation enthalpy and saturated pressure. Nauka, Moscow

Index

A.K. Baev, *Specific Intermolecular Interactions of Organic Compounds*,
DOI 10.1007/978-3-642-21622-0, © Springer-Verlag Berlin Heidelberg 2012

H

T

V